VOLUME SEVENTY THREE

ADVANCES IN BOTANICAL RESEARCH

Plant Nematode Interactions: A View on Compatible Interrelationships

ADVANCES IN BOTANICAL RESEARCH

VOLUME SEVENTY THREE

ADVANCES IN BOTANICAL RESEARCH

Plant Nematode Interactions: A View on Compatible Interrelationships

Volume Editors

CAROLINA ESCOBAR and

CARMEN FENOLL

Biotechnology and Plant Molecular Biology Group,
Department of Environmental Sciences,
Faculty of Environmental Sciences and Biochemistry,
Universidad de Castilla-La Mancha, Toledo, Spain

ELSEVIER

AMSTERDAM • BOSTON • HEIDELBERG • LONDON
NEW YORK • OXFORD • PARIS • SAN DIEGO
SAN FRANCISCO • SINGAPORE • SYDNEY • TOKYO

Academic Press is an imprint of Elsevier

Academic Press is an imprint of Elsevier
125, London Wall, EC2Y 5AS, UK
The Boulevard, Langford Lane, Kidlington, Oxford OX5 1GB, UK
225 Wyman Street, Waltham, MA 02451, USA
525 B Street, Suite 1800, San Diego, CA 92101-4495, USA

First edition 2015

Notices

Knowledge and best practice in this field are constantly changing. As new research and experience broaden our understanding, changes in research methods, professional practices, or medical treatment may become necessary.

Practitioners and researchers must always rely on their own experience and knowledge in evaluating and using any information, methods, compounds, or experiments described herein. In using such information or methods they should be mindful of their own safety and the safety of others, including parties for whom they have a professional responsibility.

To the fullest extent of the law, neither the Publisher nor the authors, contributors, or editors, assume any liability for any injury and/or damage to persons or property as a matter of products liability, negligence or otherwise, or from any use or operation of any methods, products, instructions, or ideas contained in the material herein.

ISBN: 978-0-12-417161-9
ISSN: 0065-2296

CONTENTS

CONTRIBUTORS

Pierre Abad
INRA-CNRS-Université de Nice Sophia-Antipolis, Institut Sophia Agrobiotech, UMR 1355-7254, Sophia-Antipolis, France

Janice de Almeida Engler
Institut National de la Recherche Agronomique, UMR 1355 ISA/Centre National de la Recherche Scientifique, UMR 7254 ISA/Université de Nice-Sophia Antipolis, UMR ISA, Sophia-Antipolis, France

Marta Barcala
Laboratory of Plant Physiology, Department of Environmental Sciences, Universidad de Castilla-La Mancha, Toledo, Spain

Thomas J. Baum
Department of Plant Pathology and Microbiology, Iowa State University, Ames, IA, USA

David M. Bird
Department of Plant Pathology, NC State University, Raleigh, NC, USA

Holger Bohlmann
Department of Crop Sciences, Division of Plant Protection, University of Natural Resources and Life Sciences, Tulln, Austria

Patricia Buki
Department of Entomology, Nematology and Chemistry Units, Agricultural Research Organization (ARO), the Volcani Center, Bet Dagan, Israel

Javier Cabrera
Laboratory of Plant Physiology, Department of Environmental Sciences, Universidad de Castilla-La Mancha, Toledo, Spain

Fernando E. Díaz-Manzano
Laboratory of Plant Physiology, Department of Environmental Sciences, Universidad de Castilla-La Mancha, Toledo, Spain

Gilbert Engler
Institut National de la Recherche Agronomique, UMR 1355 ISA/Centre National de la Recherche Scientifique, UMR 7254 ISA/Université de Nice-Sophia Antipolis, UMR ISA, Sophia-Antipolis, France

Carolina Escobar
Laboratory of Plant Physiology, Department of Environmental Sciences, Universidad de Castilla-La Mancha, Toledo, Spain

Bruno Favery
INRA-CNRS-Université de Nice Sophia-Antipolis, Institut Sophia Agrobiotech, UMR 1355-7254, Sophia-Antipolis, France

Carmen Fenoll
Laboratory of Plant Physiology, Department of Environmental Sciences, Universidad de Castilla-La Mancha, Toledo, Spain

Diana Fernandez
IRD–Institut de Recherche pour le Développement, UMR 186 IRD-UM2-Cirad Résistance des Plantes aux Bioagresseurs, Montpellier, France

John Fosu-Nyarko
Nemgenix Pty Ltd, WA State Agricultural Biotechnology Centre, Murdoch University, Perth, WA, Australia

Michael Gardner
Division of Plant Sciences and Bond Life Sciences Center, University of Missouri, Columbia, MO, USA

Maíra Grossi de Sá
IRD–Institut de Recherche pour le Développement, UMR 186 IRD-UM2-Cirad Résistance des Plantes aux Bioagresseurs, Montpellier, France

Maria Fatima Grossi de Sa
Laboratório de Interação Molecular Planta-Praga, Embrapa Recursos Genéticos e Biotecnologia, PqEB, Brasília, Distrito Federal, Brazil

Florian M.W. Grundler
INRES – Molecular Phytomedicine, Rheinische Friedrich-Wilhelms-University of Bonn, Bonn, Germany

Tarek Hewezi
Department of Plant Sciences, University of Tennessee, Knoxville, TN, USA

Iberkleid Ionit
Department of Entomology, Nematology and Chemistry Units, Agricultural Research Organization (ARO), the Volcani Center, Bet Dagan, Israel; Department of Plant Pathology and Microbiology, Texas A&M University, College Station, TX, USA

John T. Jones
Cell and Molecular Sciences Group, The James Hutton Institute, Invergowrie, Dundee, UK

Michael G.K. Jones
School of Veterinary and Life Sciences, WA State Agricultural Biotechnology Centre, Murdoch University, Perth, WA, Australia

Mike Kolomiets
Department of Plant Pathology and Microbiology, The Robert H. Smith Faculty of Agriculture, Food and Environment, The Hebrew University of Jerusalem, Jerusalem, Israel

Tina Kyndt
Department Molecular Biotechnology, Ghent University (UGent), Gent, Belgium

Sophie Mantelin
Cell and Molecular Sciences Group, The James Hutton Institute, Invergowrie, Dundee, UK

Melissa G. Mitchum
Division of Plant Sciences and Bond Life Sciences Center, University of Missouri, Columbia, MO, USA

Sigal Brown Miyara
Department of Entomology, Nematology and Chemistry Units, Agricultural Research Organization (ARO), the Volcani Center, Bet Dagan, Israel

Chinh-Nghia Nguyen
INRA-CNRS-Université de Nice Sophia-Antipolis, Institut Sophia Agrobiotech, UMR 1355-7254, Sophia-Antipolis, France

Vũ Phong Nguyễn
IRD–Institut de Recherche pour le Développement, UMR 186 IRD-UM2-Cirad Résistance des Plantes aux Bioagresseurs, Montpellier, France; Nông Lâm University, Hồ Chí Minh City, Việt Nam

Charles H. Opperman
Department of Plant Pathology, NC State University, Raleigh, NC, USA

Anne-Sophie Petitot
IRD–Institut de Recherche pour le Développement, UMR 186 IRD-UM2-Cirad Résistance des Plantes aux Bioagresseurs, Montpellier, France

Michaël Quentin
INRA-CNRS-Université de Nice Sophia-Antipolis, Institut Sophia Agrobiotech, UMR 1355-7254, Sophia-Antipolis, France

Natalia Rodiuc
Laboratório de Interação Molecular Planta-Praga, Embrapa Recursos Genéticos e Biotecnologia, PqEB, Brasília, Distrito Federal, Brazil

Shahid Siddique
INRES – Molecular Phytomedicine, Rheinische Friedrich-Wilhelms-University of Bonn, Bonn, Germany

Peter Thorpe
Cell and Molecular Sciences Group, The James Hutton Institute, Invergowrie, Dundee, UK

Nhat My Truong
INRA-CNRS-Université de Nice Sophia-Antipolis, Institut Sophia Agrobiotech, UMR 1355-7254, Sophia-Antipolis, France

Anju Verma
Division of Plant Sciences and Bond Life Sciences Center, University of Missouri, Columbia, MO, USA

Paulo Vieira
NemaLab/ICAAM – Instituto de Ciências Agrárias e Ambientais Mediterrânicas, Universidade de Évora, Núcleo da Mitra, Évora, Portugal

Krzysztof Wieczorek
Division of Plant Protection, Department of Crop Sciences, University of Natural Resources and Life Sciences, Konrad-Lorenz Straße, Tulln, Austria

Valerie M. Williamson
Department of Plant Pathology, University of California-Davis, Davis, CA, USA

PREFACE

The complexity of biological systems has a striking example in the interaction of plants with sedentary endoparasitic nematodes. The statement that *the sum of the parts is not equal to the whole* underlying biological research turns fully meaningful when looking at these perfectly designed nematode-feeding cells inside plant roots, the giant cells and the syncytia. These cells still constitute an open and intriguing world that researchers are continuously revisiting: a micro-world where nematodes negotiate with the plant cells in a continuous cross talk and debate that are full of surprises. The capacity of nematodes to induce the generation of new cell functionalities and their subtle interloping with molecular transduction pathways already established in other plant programs to develop non-previously planned new cells are captivating processes that makes us aware of the flexibility of plant cells. We would like this book to be a guide for the personal discovery of one of the most complex molecular dialogues between the animal and plant kingdoms: a guidance to discover yet the limitations of cell biology, biochemistry, genomics and genetic techniques for the full study of the underlying mechanisms and processes leading this fascinating plant-pathogen interplay. Finally, we would like this book to be also an inspiration for your own enthrallment on the open fields of research in continuous dynamism facing compatible plant–nematode interactions.

This book provides an updated view of some of the processes subjacent to compatible plant–nematode interactions, particularly focussed in sedentary endoparasitic nematodes. This knowledge constitutes the starting point in the search for new control strategies for these agricultural pests. Among the most damaging groups of plant parasitic nematodes for agriculture are the root knot nematodes (RKNs, *Meloidogyne* spp.) and the cyst nematodes (CNs; *Heterodera* spp. and *Globodera* spp.), receiving their names from the characteristic structures formed within the roots after their infection (the galls or knots and the cysts, respectively). These sedentary parasites induce the formation of their feeding sites by directing the transdifferentiation of normal plant root cells into nematode feeding cells, named giant cells for RKNs, and syncytia for CNs. They show a sophisticated parasitism behaviour, as they develop an intimate relationship with their hosts, inducing highly specialized "pseudo-organs" to provide them with a continuous source of food. Plant damage caused by these parasites is mostly due to the reduced availability of water and nutrients caused by nematode feeding and disruption of root anatomy. Cyst nematodes also produce wounding, particularly during migration, that predisposes the plant to other soil pathogens attack, such as pathogenic bacteria or fungi, sometimes carried by the nematode itself. Therefore, endoparasitic nematodes pose a major threat to many crops worldwide.

The volume starts with two introductory chapters that provide an overview of the parasitism mode of RKNs and CNs, the two major groups of sedentary endoparasitic nematodes, presenting general features of their feeding sites. Dramatic rearrangements in the cell wall and cytoskeleton, alteration of cell cycle mechanisms such endoreduplication and/or mitosis, readjustment of enzymes involved in carbohydrate synthesis and degradation and alteration of lipid metabolism are among the main processes modified during the plant–nematode interaction that are revised in Chapters 3 to 6. These processes are especially relevant to allow the feeding cells act as sinks to provide nutrients for nematode life cycle completion. Several evidences show that nematodes induce their feeding sites by interfering with developmental mechanisms present in the plant. The molecular mechanisms used by microorganisms to interfere with plant processes are surprisingly subtle but can effectively modify predefined plant developmental patterns. Hence, nematodes may partially 'hijack' fundamental mechanisms of plant development, such as those described for molecular transducers common to lateral root (LR) and gall formation in RKNs (revised in Chapter 7).

Most of the research revised in these chapters was conducted in a few plant species, particularly *Arabidopsis thaliana*, all of them dicotiledoneus. However, sedentary endoparasitic nematodes are also major pests of important monocotiledoneus crops, thus the study of the compatible interactions with species such as maize or rice is an emerging field that is revised in Chapter 8. The above chapters review how tools such as transcriptomic analyses combined with genetics and molecular cell biology revealed dramatic and specific changes in gene expression in syncytia and giant cells. Nevertheless, to date, only a few players of the complex regulatory networks taking place during feeding cells development have emerged, and the understanding of the subtle interaction with their hosts and their necessary cross talk is still fragmentary. Yet, integrative analysis of proteomic and transcriptomics together with genetics and molecular and cell biology tools are facilitating its comprehension. Host endogenous small RNAs and RNA-silencing pathways appear also to play key roles in regulating gene expression changes and reprogramming cells in infected roots into specific syncytial or giant-cell types. Through massive sequencing approaches, different types of transcriptional and post-transcriptional small RNA-mediated gene silencing in the feeding sites are starting to be deciphered. Furthermore, two artificial RNA interference systems, host-induced gene silencing and virus-induced gene silencing are shown as powerful tools for functional genomics of plant–nematode interactions, and all discussed in Chapter 9.

The mechanisms that nematodes use to invade roots and to induce and maintain their feeding sites are unclear, but some seem to involve nematode effector molecules secreted from their oesophageal glands, which play various roles during the parasitism. Comparative genomics analysis tools from the sequenced plant-parasitic nematode genomes helped to identify gene candidates for playing a role in the compatible and incompatible parasitic interaction. Additionally, it confirmed information obtained from EST sequencing and increased our understanding regarding the extent of Horizontal Gene Transfer as a source of novelty during the evolution of parasitism (reviewed in Chapter 10). The advances in the study of nematode effectors such as the *CLE*-like and the *CEP* peptides and their functions during the interaction are reviewed in detail in Chapters 11, 12 and 13. The induction of feeding sites by nematode secretions constitutes one of the most remarkable examples of cell morphogenesis directed by pathogen effectors in plants. Besides, one of the effective strategies of these nematodes to alter the plant responses for their benefit is a battery of molecular effectors that potentially suppress or decrease plant defenses during migration and establishment.

Nowadays, nematode control strategies in agriculture include chemicals (nematicides and fumigants), biological control with nematode antagonists, physical methods such as solarisation and fallowing, cultural methods as crop rotation, as well as the use of resistant plants. The use of chemicals is gradually vanishing due to their environmental risks and toxicity, as effective pesticides against these complex eukaryotes will mostly be potentially harmful to other organisms. Biological control has resulted in a low efficiency unless applied in combination with other techniques. Crop rotation with non-host species or resistant cultivars has provided locally good results, and the use of resistant cultivars combined with classical inbreed programs is probably the most extended economically viable and less contaminant control system. However, the increasing isolation of virulent nematode populations to those resistant plants seriously question their durability, and scientists have now the responsibility to lay a creative ground to developing new tools for nematode control. In this context, nematode control strategies based on biotechnology are called to play an increasingly important role on sustainable crop production (discussed in Chapter 14).

There are still many challenges in understanding the complexity of these interactions. For instance, most research concentrates in examining feeding sites at a few selected times after infection, but feeding site cell status is continuously changing as it differentiates, possibly controlled by nematode nutritional needs. Therefore, inferred conclusions from the analysis of nematode-induced feeding sites at a few selected infection points should be taken cautiously. Holistic functional approaches on the nematode and the plant sides, combined with genetics and cell biology techniques, are generating vast lists of differentially expressed genes as candidates to play roles in the compatible interactions of sedentary endoparasitic nematodes with their hosts. The understanding of the mechanisms subjacent to root penetration and feeding site differentiation and maintenance, as well as a deeper knowledge of the nematodes modes of parasitism, is providing a variety of tools for the design of new control methods of these devastating agricultural pests. The further development of meaningful mega data analyses will contribute to integrated systemic views of the different events in this complex process.

Carolina Escobar
Carmen Fenoll

CHAPTER ONE

Overview of Root-Knot Nematodes and Giant Cells

Carolina Escobar[1,a], Marta Barcala[a], Javier Cabrera, Carmen Fenoll
Laboratory of Plant Physiology, Department of Environmental Sciences, Universidad de Castilla-La Mancha, Toledo, Spain
[1]Corresponding author: E-mail: carolina.escobar@uclm.es

Contents

Abstract

Root-knot nematodes (RKNs) are ubiquitous parasites with an amazing capacity to interact with a very large variety of plant species. They are sedentary endoparasitic nematodes that depend on the induction of a permanent feeding site in living roots to complete their life cycle. RKNs interfere with the genetic programmes of their hosts to transform root vascular cells into giant cells (GCs) through the injection of nematode effectors from their oesophageal glands. Dramatic rearrangements in GCs cytoskeleton, alteration of cell cycle mechanisms, such as mitosis and endoreduplication, readjustment of enzymes involved in carbohydrate synthesis and degradation are among those processes modified in GCs. GCs act as sinks to provide nutrients for life cycle completion from J2 larvae to adult females. The female produces an egg offspring protected by a gelatinous matrix and the free-living stage, J2, hatch from these eggs, completing the nematode life cycle. The model species *Arabidopsis thaliana* allowed easy in vivo observations of the interaction by video-enhanced contrast light microscopy on infected roots, and the wide range of existing genetic and molecular tools of this plant model has extended its use. Holistic approaches to tackle gene expression combined with cell biology techniques, as isolation of GCs by laser capture microdissection, allowed GC-specific transcriptomic analysis, generating vast lists of differentially expressed

[a]Both authors have contributed equally to this work.

Advances in Botanical Research, Volume 73
ISSN 0065-2296
http://dx.doi.org/10.1016/bs.abr.2015.01.001

genes. However, the design of consistent functional hypothesis about these genes and their products will require the development of in silico analysis tools for comparisons among the transcriptomes of plant–nematode compatible interactions. The understanding of the processes subjacent to GC differentiation and maintenance, as well as a deeper knowledge of RKN mode of parasitism, will provide tools for new control methods of these devastating agricultural pests.

1. INTRODUCTION TO PLANT PARASITIC NEMATODES

Nematodes are pluricellular organisms, classified within the large phylum Nematoda that encompasses unsegmented roundworms. Nematodes are widespread in almost all ecosystems and habitats throughout the planet, including different soils, marine and fresh waters. These ubiquitous organisms have proved an amazing adaptability to diverse and extreme environments from deserts to the arctic pole. They also show varied lifestyles (with representatives from free-living to parasitic species) and food resources (plants, bacteria, animals and fungi) (Perry & Moens, 2011). There are nematodes detrimental to agriculture, parasites of animal and humans, but also beneficial species, such as the entomopathogenic nematodes used in crop protection as insect control agents (Lacey & Georgis, 2012; Ravichandra, 2008), as well as free-living nematodes involved in soil nutrient turnover.

So far, more than 25.000 spp. have been included in the phylum (Zhang, 2013) but this number is constantly increasing as new species are discovered or redescribed (Elling, 2013). Classic taxonomy proposed two classes, based on morphological and anatomical characters (Chromadorea and Adenophorea), which diverged over 550 million years ago. Recently, a more comprehensive phylogenetic classification based mainly on molecular analysis of small subunit of ribosomal DNA (ssUrDNA) was proposed: Chromadorea and Enoplea (De Ley & Blaxter, 2002; De Ley & Blaxter, 2004; van Megen et al., 2009) (Table 1). Nematode species included within the Chromadorea class in the suborder Tylenchina (Table 1) have an especial relevance due to their great economic impact on agriculture. Plant parasitic nematodes affect frequently the root system, where they produce extensive damage such as galling and necrosis. As an indirect consequence of infection, aboveground plant parts are altered, showing a reduced growth, leaf chlorosis, poor yield and wilting. Crop losses, are sometimes underestimated because plant symptoms after the infection are unspecific and can be erroneously identified as resulting from nutritional deficiencies or abiotic stress.

Table 1 Phylogenetic Classification of Plant Parasitic Nematodes According to De Ley and Blaxter (2002)

Class	Subclass	Order	Suborder	Superfamily	Family	Subfamily	Genus
Chromadorea	Chromadoria	Rhabditida	Tylenchina	Tylenchoidea	Hoplolaimidae	Hoplolaiminae	Rotylenchus Hoplolaimus
						Rotylenchulinae	Rotylenchulus
					Heteroderidae	Heteroderinae	Cactodera Globodera Heterodera Punctodera
						Meloidogyninae	Meloidogyne
					Pratylenchidae	Pratylenchinae	Pratylenchus
						Nacobbinae	Nacobbus
				Criconematoidea	Tylenchulidae	Tylenchulinae	Tylenchulus
Enoplea	Dorylaimia	Dorylaimida	Dorylaimina	Dorylaimoidea	Longidoridae	Longidorinae	Longidorus Paralongidorus
						Xiphineminae	Xiphinema
	Enoplia	Enoplida					

Nematodes also represent an important economic issue in some leisure business such as golf courses, turfs along the world and in ornamental crops (Crow, 2005, 2007; Crow & Luc, 2014; Rahman Khan, Khan, & Mohide, 2005).

Most plant parasitic nematodes suffer four moults throughout their development from the juvenile stage (stages 1–4, J1–J4) until reaching the adult stage. Transition from J1 to J2 usually takes place within the egg, and after this first moult the egg hatches releasing the J2, which represents for the majority of the species the infective stage (Perry & Moens, 2011). J2 larvae are mostly microscopic (from 250 μm to 12 mm in length) and live in soils without feeding until they find a suitable host. Then, J2 invade and feed on living plants through a protrusible oral stylet that they use to puncture cells and to feed from them. Throughout their developmental stages, nematodes usually maintain a vermiform, worm-like shape. However, in several nematode species, such as *Meloidogyne* spp., *Heterodera* spp., *Rotylenchus* spp. and *Tylenchulus* spp., adult females adopt a swollen, pear-like or kidney-like shape (Decraemer & Hunt, 2013).

Plant parasitic nematodes are classified according to their lifestyle and feeding habits. Those that penetrate the host root to feed from different inner cell types are classified as endoparasites, whereas the nematodes that feed externally by inserting their mouth stylets into root cells from the root surface are called ectoparasites. They are further subclassified into sedentary, when they have a sessile stage, or migratory (Decraemer & Hunt, 2013). Examples of genera included in all these categories are found among the major agriculturally relevant nematode species. For instance, the sedentary ectoparasite *Tylenchulus semipenetrans* (citrus nematode) is responsible for losses in citrus and olive trees and, to a lesser extent, grapevines. The lance and the needle nematodes (*Hoplolaimus* spp. and *Longidorus* spp. respectively) are migratory ectoparasites that cause considerable losses in turf grasses and lawns, corn crops and grape vineyards. Migratory ectoparasitic nematodes are particularly relevant, as some act as virus vectors (e.g. *Xiphinema* spp., a grapevine pathogen). Among the endoparasitic nematodes, there are migratory species (e.g. *Pratylenchus* spp., a major problem in fruit trees) and sedentary ones, which constitute a most relevant group in agriculture.

Sedentary endoparasitic nematodes show the most sophisticated parasitism behaviour; they develop an intimate relationship within their hosts, inducing highly specialized 'pseudo-organs' to provide them with a continuous source of food. This group is represented by the root-knot nematode (RKN; *Meloidogyne* spp.) and the cyst nematodes (e.g. *Heterodera* and *Globodera* spp.), receiving their names from the characteristic structures formed in the roots after their infection: the galls or knots and the syncytia. Recently, phylogeny methods based on ssUrDNA (van Megen et al., 2009) support the idea that the similar parasitism behaviour of root-knot and cyst nematodes has been acquired by convergent evolution between both groups rather than the existence of a common ancestor (Castagnone-Sereno, Danchin, Perfus-Barbeoch, & Abad, 2013; Castagnone-Sereno, Skantar, & Robertson, 2011; Perry & Moens, 2011).

Plant damage caused by plant parasitic nematodes is mostly due to the reduced availability of water and nutrients because of nematode feeding and disturbance of root anatomy. Nematode-produced wounding also predispose the plant to other soil pathogens attack, what is sometimes favoured by pathogenic bacteria or fungi carried by the nematode itself (Back, Haydock, & Jenkinson, 2002; Jones & Goto, 2011; Stanton & Stirling, 1997). For example, wilt fungus *Fusarium oxysporum* can interact with RKNs in complex diseases, affecting tomato, cabbage or watermelon (Bergeson, Van Gundy, & Thomason, 1970; Fassuliotis & Rau, 1969; Jenkins & Coursen, 1957; Sumner & Johnson, 1972) and *Ralstonia solanacearum* bacteria can increase tomato wilt when RKNs are present (Valdez, 1978). For cyst nematodes, complex diseases are found mainly in potato and soybean crops (Back et al., 2002).

RKN species are polyphagous and can feed on almost all vascular plants tested (Jones & Goto, 2011), whereas cyst nematodes often show a more specific host preference and usually can parasitize a single plant species (e.g. *Globodera* spp. only infect potato). This Chapter will focus on the general biology of the RKN, while cyst nematodes will be reviewed in Chapter 2.

2. GENERAL ASPECTS OF ROOT-KNOT NEMATODES (RKNs)

Meloidogyne is a genus formed by more than 90 species (Jones et al., 2013), some of them including several races (Eisenback & Triantaphyllou, 1991; Ravichandra, 2008). Only a few species are referred as major agricultural pests, as they were considered the most abundant and damaging: *Meloidogyne incognita, Meloidogyne javanica, Meloidogyne arenaria* from Mediterranean and tropical areas, and the temperate species *Meloidogyne hapla.* Additionally, species previously considered minor agricultural pests as *Meloidogyne enterolobii*, *Meloidogyne paranaensis* or *Meloidogyne exigua* (from tropical and subtropical regions), and *Meloidogyne fallax*, *Meloidogyne minor* or *Meloidogyne chitwoodi* (from temperate regions) are emergent parasites that receive increasing attention (Elling, 2013; Moens, Perry, & Starr, 2009) as they are raising as important agriculture threats. Some of them, such as *M. chitwoodi*, *M. enterolobii* or *M. fallax*, have been included in the 2013 quarantine pest list from the European and Mediterranean Plant Protection Organization.

As previously indicated, RKNs are extremely polyphagous parasites. *Meloidogyne* spp. such as *M. incognita*, *M. javanica*, *M. hapla*, *M. arenaria*, *M. enterolobii*, *M. fallax* or *M. chitwoodi* show a broad host range, being able to parasitize vegetable crops, fruit trees and ornamental plants, whereas other species show a more restricted host range, as *M. minor* (grasses, potato and tomato) or *Meloidogyne hispanica* (peach, sugar beet, tomato). In accordance to this, species with narrower host ranges show more restricted geographical localizations, but as their host range widens, they show a global distribution (Triantaphyllou, 1985).

Control strategies in agriculture cover the use of chemicals (nematicides and fumigants), biological control with nematode antagonists, physical methods, such as solarization and fallowing, cultural methods as crop rotation, as well as the use of resistant plants. The use of chemicals is gradually vanishing due to their toxicity and environmental contamination potential. The frequently used methyl bromide, a broad spectrum and economically viable pesticide, has been banned in the European Union since 2010 (Kearn, Ludlow, Dillon, O'Connor, & Holden-Dye, 2014) and other countries are reducing its use. Organophosphate- and carbamate-based nematicides are also restricted. Those belonging to the fluoroalkenyl thioether group are effective against RKN showing a lower impact on the environment as compared to organophosphate- and carbamate-based nematicides and new nematicides derived from biologically active compounds such as those found in garlic are being developed (Kearn et al., 2014). However, effective chemical pesticides against these complex eukaryotes will mostly be potentially harmful for other organisms.

Biological control has resulted in a low effective strategy unless applied in combination with other techniques (Viaene, Coyne, & Kerry, 2006). The use of nematode antagonists that can be predators, parasites or pathogens such as the fungi *Verticillium* spp. and *Fusarium* spp., or the bacteria *Pasteuria penetrans*, is at its initial days. Despite being an ecofriendly strategy, few commercial products containing viable organism for biological control are available (Stanton & Stirling, 1997; Timper, 2011).

Crop rotation with nonhost species or resistant cultivars has provided good results for RKN control. Despite few poor or nonhost plant species are available, cover crops as marigolds (*Tagetes* spp.) or perennial grasses (such as bahiagrass (*Paspalum notatum*) and bermudagrass (*Cynodon dactylon* L. *Pers.*)) have been effective to control populations of *M. arenaria*, *M. hapla*, *M. incognita* and *M. javanica* (Hooks, Wang, Ploeg, & McSorley, 2010; Netcher & Taylor, 1979). With regard to resistant cultivars, several genes from tomato (*Mi* genes; Ammiraju, Veremis, Huang, Roberts, & Kaloshian, 2003; Rossi et al., 1998; Veremis, van. Heusden, & Roberts, 1999; Yaghoobi, Kaloshian, Wen, & Williamson, 1995), prunus (*Ma* and *RMia* genes; Claverie et al., 2004; Lu, Sossey-Alaoui, Reighard, Baird, & Abbott, 1999), carrot (*Mj* genes; Ali et al., 2014) and pepper (*Me* genes; Djian-Caporalino et al., 2007) have been described to confer resistance to many *Meloidogyne* spp. However so far only the tomato *Mi-1* gene has been cloned and successfully transferred to commercial cultivars (Devran & Sögüt, 2010). *Mi-1* confers resistance to three *Meloidogyne* spp. (*M. javanica, M. incognita* and *M. arenaria*), but this resistance is easily overcome when soil temperature increases (reviewed by Williamson (1998)). In addition, the isolation of virulent *Meloidogyne* spp. populations in tomato cultivars carrying the *Mi-1* gene questioned the durability of the *Mi*-resistance (Jacquet et al., 2005) and prompted the suggestion of a relationship between resistance breakdown and *Mi* gene dosage (Jacquet et al., 2005). Moreover, the durability of the *Me* gene seems to be influenced not only by allelic dosage but also by the genetic background, since other genes or quantitative trait loci may be contributing to resistance (Djian-Caporalino et al., 2014).

All these strategies should be combined in an integrated pest management (IPM) plan for effective control of RKN population in the field. A detailed evaluation of the cropping systems and accurate diagnosis of RKN species must be performed for an IPM successful implementation. Differences regarding host preferences that exhibit races of a determined species (e.g. for *M. incognita* all 4 races described can infect tomato cv. Rutgers, whereas only races 3 and 4 can parasite cotton cv. Deltapine (Hartman & Sasser, 1985; Mahdy, 2002)) should be considered. Therefore, designing an IPM is very laborious and overall it needs to be locally designed. Consequently, there is still a clear need to deeply understand the molecular basis of the RKN—plant interaction, including the development and maintenance of the specific feeding structures induced in the plant host, galls and giant cells (GCs). This knowledge together with that of the nematode biology could establish an emerging creative ground to develop new tools for RKN control.

3. THE MORPHOLOGY AND REPRODUCTION OF RKNs

RKNs display a conserved basic body plan throughout their life stages, with morphological features used for species identification. Briefly, J2 outermost body structure consists of a body wall encompassing three layers: the cuticle, the hypodermis (also known as epidermis) and the somatic muscles. The cuticle is a flexible, semipermeable exoskeleton with a noncellular, multilayer structure that is newly synthesized and secreted by the epidermis in each moult. Cuticle layers (cortical, medial and basal layer) can vary in thickness throughout the nematode life stages or can even be absent as is the case of the medial layer in adult females (Decraemer & Hunt, 2013; Eisenback, 1985). The cuticle is a collagenous matrix covered by an outer coat (epicuticle) mainly made of glycoproteins and other surface-associated proteins. This coat is probably involved in host immunity response (Decraemer & Hunt, 2013; Eisenback, 1985). The cuticle allows solute diffusion and water and gas exchange with the medium to compensate the lack of either respiratory or circulatory system. In females, cuticular morphological features of the perineum (the region surrounding the vulva and anus) are used for the perineal pattern analysis, i.e. a characteristic pattern of ridges and annulations stablish differences among RKN species. Beyond the musculature, digestive, reproductive and nervous systems are found within the RKN body. The digestive system consists of a mouth with a retractile stylet (Figure 1(A)—(C)) connected to an oesophagus (or pharynx) which ends in an intestine and a rectum. Within the oesophagus there is a median bulb or metacorpus containing a metacorporal valve (Figure 1(A)—(C)) responsible for the suction force necessary for nutrient uptake and for pumping out gland secretions coming from the dorsal and subventral glands. These glands play a main role during parasitism, including invasion, establishment and feeding site development. During the preparasitic stage, the predominant glands are the two subventral glands, involved in releasing cell wall-degrading enzymes such as cellulases or pectinases (Davis, Hussey, & Baum, 2004; Jaubert, Laffaire, Abad, & Rosso, 2002). However, during the parasitic stage, once the nematode establishes, the dorsal gland become more active. Morphological changes of these glands reflect their predominance during each stage, and thus the subventral glands reach their maximum size before invasion and begin to shrink as a nematode settles. On the contrary, the dorsal gland maximum size is described for the adult female stage (Hussey & Mims, 1990). The oesophageal gland secretions (dorsal and subventral) are released in spherical granules that vary in size, composition and morphology not only depending on nematode developmental stage, but also depending on nematode species (Hussey & Davis, 2004).

The intestine serves as storage organ where many lipid granules can be easily observed under light microscopy (Figure 1(A) and (B)). The digestive system ends in the rectum, with an anus at the end in females whereas in males is joined to the reproductive system to form the cloaca. In females, rectal glands opening to the anus are responsible for the secretion of the gelatinous matrix where eggs are embedded as they are deposited. In addition, the adult female body is almost filled by the gonads, a pair of tubular organs that converge in a vagina that opens to the outside by a vulva. In preparasitic J2, the reproductive system consists on a genital primordium that will develop into either ovaries or testis as soon as the J2 starts to feed.

In J2, the nervous system mainly controls the somatic musculature and sensory perception through the chemoreceptor organs (amphids and sensilla at the head, and phasmids at the posterior end). A distinctive feature of the nervous system is the nerve ring, that encircles the oesophagus behind the medium bulb (Eisenback, 1985), and is the coordinating centre for the nervous system.

Meloidogyne spp. usually reproduce by mitotic parthenogenesis (e.g. *M. incognita*, *M. javanica* or *M. arenaria*) although some species, as *M. hapla* (race A) or *M. chitwoodi*, multiplies by facultative meiotic parthenogenesis (Berg, Fester, & Taylor, 2008; Eisenback & Triantaphyllou, 1991). The female to male ratio is variable, though in general few males are produced and only under suboptimal conditions (e.g. insufficient nutrients, crowding or low temperature Davide & Triantaphyllou, 1967; Decker, 1989; Snyder, Opperman, & Bird, 2006). This decision is taken during the J2 parasitic stage (Triantaphyllou, 1973), but so far signals that promote this change have not been unravelled. Contrary to the adult female, the males are motile and vermiform, range from 1.100 to 2.000 μm in length and leave the host root right after the final moult (Eisenback & Triantaphyllou, 1991). Males can grow up to four times that of the J2 length (Figure 1(I)). They also display a distinctive visible feature of the reproductive system, the spicules, hook-like structures (Stanton & Stirling, 1997) to duct sperm during mating.

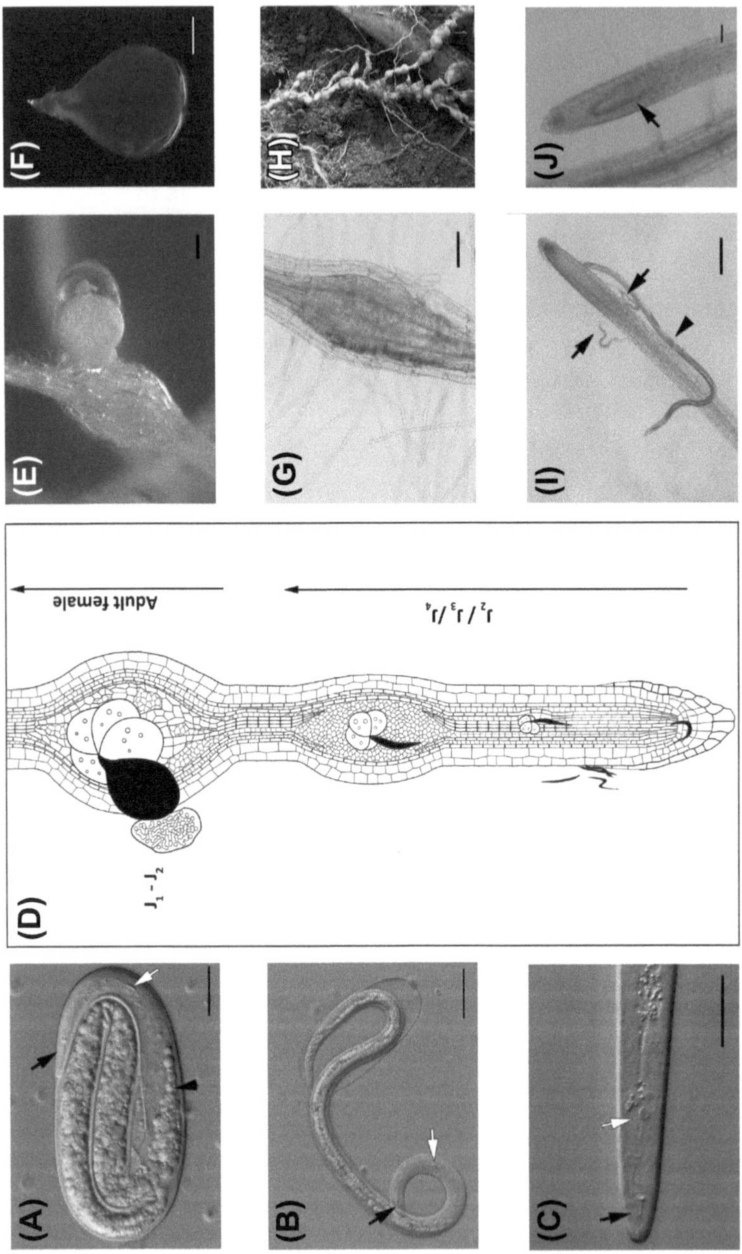
(A)
(B)
(C)
(D)
J_1 - J_2
J_2 / J_3 /J_4
Adult female
(E)
(F)
(G)
(H)
(I)
(J)

Figure 1 ***Root-knot nematode life cycle.*** A schematic diagram with pictures illustrating some of the key stages during the interaction. (A) Photograph of a developing *Meloidogyne javanica* J2 inside the egg. (B) Recently hatched *M. javanica* J2. (C) Close-up of an *M. javanica* J2 anterior body. For A, B and C stylet is indicated by a black arrow, median bulb by a white arrow and lipid globules by a black arrowhead. (D) Schematic diagram of an Root-knot nematode (RKN) life cycle as a time course of the progression of the infection represented in the same root. RKNs are black-coloured for easy location. Starting at the bottom of the diagram, a J2 penetrates the root at the elongation zone, migrates towards the tip and it turns 180° to enter the vascular cylinder, where it induces several giant cells (GCs). By 3 days post infection (dpi), an incipient gall has formed around the nematode including the GCs. The nematode gradually grows and develops into a female while GCs and galls enlarge, and eventually the pear-shaped mature female lays an egg mass that protrudes from the root surface. (E) Mature gall of *Arabidopsis thaliana* with adult female posterior region exposed outside the root and laying eggs within the gelatinous matrix. (F) Enlarged adult female of *M. javanica* showing the typical pear-like shape. (G) Incipient gall in *A. thaliana* plant at 3 dpi. (H) Overview of tomato roots infected with *M. javanica* showing profuse galling. (I) Two infective juveniles (J2, black arrow) an adult male (black arrowhead) show *M. javanica* motile stages nearby an *A. thaliana* root tip. (J) Initial stages of an *M. javanica* J2 migration in an Arabidopsis root, turning at the root tip to enter the vascular cylinder. Scale bars in A, B, C represent 20 μm, 0.2 mm in E, F and I, and 0.1 mm G and J. (See colour plate)

4. THE LIFE CYCLE OF RKNs

An RKN life cycle can be completed within 20–40 days, but its length is influenced by environmental conditions such as the temperature, to a lesser extent, soil moisture, and by the host species (Ravichandra, 2008; Rohini, Ekanayaka, & Di Vito, 1986). A typical RKN life cycle (Figure 1(D)) begins with the hatched J2s (Figure 1(B)), that are attracted towards the host roots (Figure 1(I)) after sensing chemical gradients of root diffusates (Teillet et al., 2013) with their chemosensory sensilla, the amphids (Perry & Moens, 2011). So far, only CO_2 has been identified as a prime long distance attractant for plant parasitic nematodes, including *M. incognita* (Robinson, 2002). Additional attractants are amino acids, sugars and other metabolites (Bird, 1959; Perry, 2001; Prot, 1980; Robinson, 2002). When a suitable root tip of a host is located in the soil, nematodes penetrate preferably behind the elongation zone, and migrate intercellularly down to the root tip. The reason why RKN move downwards towards the meristem in order to enter the vascular cylinder, is probably that the Casparian strip at the endodermis represents an insuperable barrier to their stylet. In fact, nematodes that do not orientate correctly within this region of the root are unable to induce a feeding site and eventually leave the root (von Mende, 1997; Wyss, Grundler, & Münch, 1992). The precise signals J2 might sense to orientate themselves once inside the root and to move towards the root tip are unknown. So far only CO_2 has been proved to be an attractant (Robinson, 2002), but it also has been suggested that pH gradients and even electric fields could guide them (Bird, 1959; von Mende, 1997). Molecular determinants in the plant cell surfaces (linked to cell walls or mobile in the apoplast) may as well be perceived by the migrating J2 to identify their pathway towards the root tip. Penetration and migration are accomplished by a combination of chemical and mechanical tools. Nematodes secrete a mixture of cell wall-degrading enzymes and use their stylet and head to push and separate the softened mid lamella that cements the root cells together (Perry & Moens, 2011). The secretion of cell wall-degrading enzymes to the apoplast for host invasion is a feature common to other plant endoparasitic nematodes during migration and also to pathogenic bacteria and fungi (Perry & Moens, 2011). RKNs secrete cellulases (endoglucanases), endoxylanases, pectatelyases and polygalacturonases produced by their subventral glands (Davis, Haegeman, & Kikuchi, 2011; Perry & Moens, 2011; Wieczorek et al., 2014). Phylogenetic studies suggest that plant parasitic nematodes acquired this capacity by ancient horizontal gene transfer from bacteria (Perry & Moens, 2011). It has also been suggested that acquisition of prokaryotic genes from the glycosyl hydrolase family 5 by sedentary endoparasitic nematodes could have occurred by pass on by a relative ancestor rather than by new horizontal gene transfer (Kyndt, Haegeman, & Gheysen, 2008; Rybarczyk-Mydłowska et al., 2012). Once the J2 reach the root tip, they rotate 180° (Figure 1(D) and (J)) to enter the vascular cylinder and move upwards until near the differentiation zone where they select several vascular cells to induce the formation of a feeding site (Bird, Opperman, & Williamson, 2009; Perry & Moens, 2011).

Upon feeding site development, the J2 becomes sedentary. The selected cells (usually five to eight cells) start to undergo dramatic morphological and metabolic changes, to become nutrient sinks. The most obvious morphological characteristic is their enlargement, and due to this feature, they were named by Treub (1886) GCs (Figures 1(D) and 2(A)). Additionally, cortex cells surrounding the GCs divide and become hypertrophied and the pericycle cells proliferate (Figure 2(A); Berg et al., 2008). The xylem in the vicinity is grossly disrupted and GCs are encaged by a newly developed intricate xylem network (Bartlem, Jones, & Hammes, 2014; Christie, 1936). Around GCs also protophloem is formed and proliferates dramatically (Absmanner, Stadler, & Hammes, 2013). Thus, the result is the formation of a unique pseudo-organ called gall containing the GCs (Figure 1(D), (G) and (H)). However, some *Meloidogyne* spp. have been described to cause small or no galling (e.g. *Meloidogyne artiellia, M. chitwoodi, M. fallax, M. minor* or *M. paranaensis*; Elling, 2013; Vovlas et al., 2005) in particular hosts, showing a limited hyperplasia and proliferation of surrounding tissues. *Meloidogyne kikuyensis* develops a different gall that resembles the nodules induced by rhizobium. This gall is located on one side of the root, and the GCs are encaged within the so-called feeding socket (Eisenback, Dodge, & Odge, 2012).

A detailed record of the first stages of parasitism was reported by using video-enhanced contrast light microscopy (Wyss et al., 1992). The ability of *Meloidogyne* spp. to parasite *Arabidopsis thaliana* (Sijmons, Grundler, Von Mende, Burrows, & Wyss, 1991) allowed a real progress in the understanding of the host–pathogen interaction. This was not only because Arabidopsis have thin, translucent roots that permit a direct observation of initials stages of parasitism inside the plant. In the last 10 years, scientists made important contributions to the knowledge of the molecular basis of the

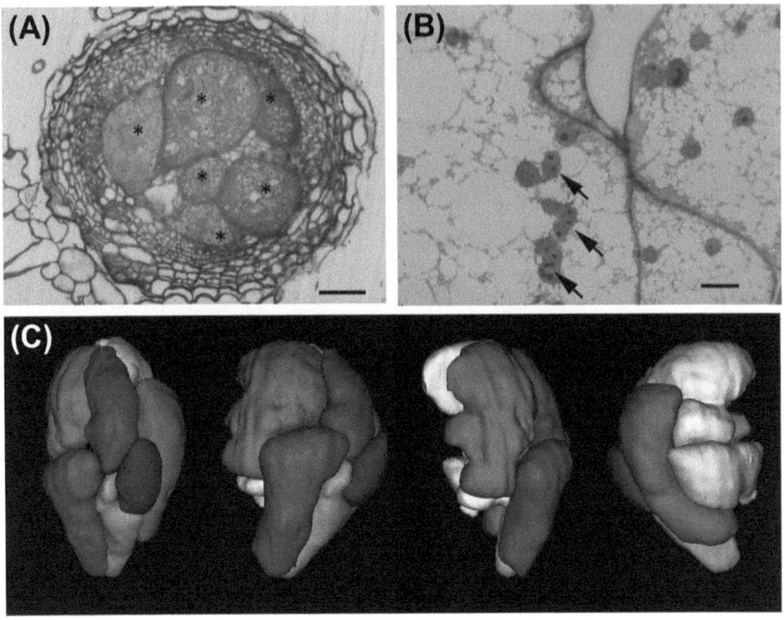

Figure 2 ***Morphology of giant cells induced by root-knot nematodes.*** Semi-thin sections of *Meloidogyne javanica* induced galls stained with toluidine. (A) Cross-section of a 3 days post infection (dpi) *Arabidopsis thaliana* gall showing giant cells (GCs) (*). Scale bar represents 50 μm. (B) *Nicotiana tabacum* longitudinal cross-section of a 7 dpi gall showing a partial view of three multinucleate GCs with ameboid nuclei (purple stain) and prominent nucleoli (black arrows) scale bar represents 20 μm. (C) 360° rotation views of a 3D reconstruction image of a fully developed GC system from an Arabidopsis gall 7 dpi with *M. javanica*. Note the irregular shape and ragged surface of the eight GCs that comprise this individual feeding site. (See colour plate)

plant–endoparasitic nematodes interaction by using other advantages provided by this simple model plant such as having a small genome, being easy to transform and with multiple genetic, functional, transcriptomic, proteomics tools already developed (reviewed in Gheysen and Fenoll (2011)). Very recently, a novel technique enabling nondestructive, long-term observations of live nematodes in planta based on the nematode fluorescent label with the lipid analogue PKH26, allowed to observe their behaviour, development, and morphology for the full duration of the parasite's life cycle by confocal microscopy (Dinh, Brown, & Elling, 2014).

Inside the gall, the nematode becomes sessile by atrophy of their somatic musculature, except for the head, and will alternate periods of feeding from the different GCs, having three consecutive moults (to J3, J4 and adult female). Only J2 will feed and after the last moult, the adult females resume feeding (Lewis & Perez, 2004). Neither J3 nor J4 have a functional stylet and hence they do not feed (Manzanilla-Lopez & Bridge, 2004).

Under favourable conditions and sufficient nutrients, J4 suffers the final moult to the female adult stage. By this time, the adult females have adopted the typical pear-like shape (Figure 1(D)–(F)), have enlarged over 500 times the J2 volume (Shepperson & Jordan, 1974), and begin to deposit hundreds of eggs containing the larvae in a gelatinous matrix of glycoproteinaceous nature that they secrete (Figure 1(D) and (E); Sharon & Spiegel, 1993). This matrix is a barrier to water loss (Wallace, 1968) and provides a protection to developing larvae from external pathogenic agents like bacteria and fungi. In *Meloidogyne* spp., the egg mass is exposed outside the root, due to the enlargement of the female, whose posterior body portion can protrude outside the gall (Figure 1(D) and (E)), making eggs more accessible to the rhizosphere microorganisms. Antimicrobial activity has been described for the gelatinous matrix (Orion, Kritzman, Meyer, Erbe, & Chitwood, 2001; Sharon & Spiegel, 1993), but if this matrix represents anything else apart from a physical barrier still must be elucidated. Anyhow, the egg mass does not represent a resistant form as in the case of the cyst nematodes. J2 usually hatches from the egg after its complete development to start a new life cycle (Figure 1(B)). Transition from J1 to J2 occurs inside the egg within the egg mass (Figure 1(A)).

5. GIANT CELLS (GCs): FROM VASCULAR CELLS TO NOURISHING CELLS

RKNs were described as plant pathogens from late 1880s (reviewed in Berg et al. (2008)). Initial research described their morphology and it is not until mid-1900s when the first studies focused on the nematode-induced plant morphological changes (Christie, 1936; Ravichandra, 2008). More detailed morphological features of the feeding cells induced in the plant hosts were already described in the 1960s by light and electron microscopy analysis (Bird, 1961; Huang & Maggenti, 1969). Nowadays, it still results a challenge to elucidate those cell processes involved in the dramatic morphological and physiological changes induced in the initial root cells transformed into a specialized structure for the nematode feeding, the GCs. During this process, the first evidence of a developing GC inside the root vascular cylinder is the appearance of binucleate cells near the nematode head (de Almeida Engler & Favery, 2011). Subsequently, new mitotic cycles with uncoupled cytokinesis will lead to the multinucleate status of the GCs (Figure 2(A) and (B)). According to the former histological description, mitosis promotion was confirmed by the specific expression of genes encoding mitotic cyclins and the corresponding cyclin-dependent kinases (CDKs) involved in transitions through S-G2-M phases. Some of them are *AtCYCB1;1*, *AtCYCA1;2*, *AtCDKB1;1*, *AtCDKA;1* and several D-type cyclin-coding genes, (de Almeida Engler et al., 1999; Barcala et al., 2010; Niebel et al., 1996). A clear increase in DNA content has been also confirmed (de Almeida Engler & Gheysen, 2013), probably due to repeated endoreduplication cycles, although other unconventional ways of DNA amplification (e.g. defective mitoses or nuclear fusion) are also suggested. These processes might help GC expansion. Yet, one of the characteristic features of GCs is those repeated cycles of mitosis, as they do not take place within syncytia. However, endoreduplication occurs in both feeding structures, as it does mitosis in the adjacent cell layers (see Chapter 4 for details on cell cycle). Finally, the nuclei of GCs are large, with irregular lobed shape and with large conspicuous nucleoli (Figure 2(B), Berg et al., 2008; Christie, 1936). As previously indicated, some *Meloidogyne* spp. are capable of inducing feeding sites with little galling. These feeding sites have been studied in detail and revealed fewer but larger nuclei (Vovlas et al., 2005), what could support the idea that increased DNA content might be necessary for GC expansion.

So far GC precursor cells have been described as vascular cells. From histological observations, parenchymatic cells within the stele that surround the nematode head are generally accepted as their initial cells after being triggered by oesophageal gland secretions (Berg et al., 2008). However, the precise cell type chosen by the nematode and GC ontogeny is still unclear. Accordingly, global transcriptomic similarities were encountered between early developing GCs dpi and suspension cells differentiating into xylem elements (Barcala et al., 2010). More recently, Cabrera, Diaz-Manzano, et al. (2014) described the crucial role of a transcription factor from the Lateral Organ Boundaries Domain family, LBD16, during GC development and confirmed the importance of the pericycle, a root meristematic tissue, during gall ontogeny similarly to that of lateral root formation. LBD16 is a molecular transducer integrated in a signalling cascade mediated by auxins for lateral root and gall formation. These findings strongly suggest that nematodes might alter pre-existing developmental pathways in the precursor cells of GCs, probably interfering with transduction cascades modulated by hormones, such as auxins or cytokinins (see further discussion in Chapter 7).

Kostoff and Kendall (1930) already suggested a putative role of nematode secretions during GC development. These secretions, that contain nematode effectors, are currently the focus of numerous studies and are addressed in several chapters in this volume. Transcriptomic and proteomic studies of isolated nematode glands confirmed the presence of putative secreted proteins and peptides with a possible function during invasion but also during the feeding site formation (reviewed in Quentin, Abad, & Favery (2013); Rosso & Grenier (2011)). The availability of whole genome sequence for *M. incognita* and *M. hapla* (Abad et al., 2008; Opperman et al., 2008, further details in Chapter 10) allowed in silico searches of putative effectors, but no clear picture of how the effectors are synthesized and secreted is available as yet (Berg et al., 2008; Mitchum et al., 2013; Quentin et al., 2013). Some effectors have been localized inside the feeding cells (Mj-NULG1a (Lin et al., 2013), Mi-EFF1 (Jaouannet et al., 2012; Zhang, Davies, & Elling, 2015)) whereas others locate in the apoplast (Vieira et al., 2011). In addition, comparisons among different nematode groups have revealed that some effectors are common to phytonematodes and others are lifestyle specific (Tucker & Yang, 2013), supporting the convergent parasitism style theory. Nematode subventral glands are more active during the preparasitic stage, and the effectors secreted in this stage are to assist during migration. In contrast, once the nematode becomes sedentary the dorsal gland is more active; and it has a main role in feeding site development and maintenance (Quentin et al., 2013). These effectors target host cellular processes such as cell cycle, transport or hormone signalling pathways, by mimicking or interfering with host regulators (Mitchum et al., 2013; Tucker & Yang, 2013). So far, few effectors have been functionally assessed by plant-mediated RNAi assays to attenuate nematode parasitism (reviewed in Dinh et al. (2014); Elling & Jones (2014) reviewed in chapters "Function of Root Knot Nematode Effectors and Their Targets in Plant Parasitism" and "Application of Biotechnology for Nematode Control in Crop Plants"). Identification of the plant targets of nematode effectors is also crucial to understand the plant regulatory networks that nematodes perturb for feeding site development and maintenance. Thus, effectors exhibit an enormous potential to develop biotechnological based strategies for the nematode control (see Chapter "Function of Root Knot Nematode Effectors and Their Targets in Plant Parasitism" in this book).

The GC becomes a typical highly metabolically active cell with a dense cytoplasm containing abundant organelles (endoplasmic reticulum (ER), ribosomes, mitochondria or Golgi bodies) (Berg et al., 2008; Christie, 1936). The large central vacuole is also fragmented into smaller ones

(Figure 2(A)) and chloroplast-like structures with starch accumulation are observed (Ji et al., 2013). Chaperones that may assist protein folding in cells with high metabolic activity, such as small heat shock proteins, are induced (Barcala et al., 2008). GCs constitute a sink of nutrients for the developing nematode and therefore, the metabolism of carbohydrates and amino acids is highly activated in these cells (Baldacci-Cresp et al., 2012; Gautam & Poddar, 2014; Machado et al., 2012). Sensitive metabolomics techniques recently confirmed that galls induced by *M. incognita* in *Medicago truncatula* present an elevated content of starch, sucrose, glucose, malate, fumarate and diverse amino acids (Phe, Tyr, Val, Glu, Asp) (Baldacci-Cresp et al., 2012). Similar results were found for roots of coffee and bitter gourd infected with *M. exigua* and *M. incognita*, respectively (Gautam & Poddar, 2014; Machado et al., 2012). Accordingly, the regulation of sugar, amino acid, water and Ca^{2+} membrane transporters is altered in the nematode feeding sites (NFS), as many transporters were differentially expressed in galls (Barcala et al., 2010; Hammes et al., 2005; Marella et al., 2013). Some genes coding amino acid transporters as *AtCAT6*, upregulated upon *M. incognita* infection, showed no evident functional role (Hammes, Nielsen, Honaas, Taylor, & Schachtman, 2006). In contrast, loss of function of other genes such as *AtAAP3* and *AtAAP6* impaired *M. incognita* infection (Marella et al., 2013). Sucrose is assumed to be the main source of carbohydrates for the nematode, and Arabidopsis mutants *sus1/sus4*, *cinv1* and *cinv1/cinv2* for the two main enzymes that cleave the sucrose, invertases (INVs) and sucrose synthases (SUSs), showed an increased in gall formation by *M. javanica* (Cabello et al., 2014). In agreement with this, *AtCINV2*, *AtSUS1*, and *AtSUS4* are upregulated in GCs and/or galls (Barcala et al., 2010, further information can be found in Chapter 5).

As gall development progresses, the GCs keep on enlarging. Their volume increase by 100 fold from 3 dpi to 40 dpi in *Arabidopsis* (Cabrera et al., 2015). This increase in volume probably corresponds to the stage when the adult female requires the highest nutrient supply for growth and egg production. Cells after 3D reconstruction appeared with almost no sphericity in accordance to the presence of abundant protuberances, crevices and lobules that provided an irregular shape (Figure 2(C); Cabrera et al., 2015). The irregular shape of GCs showed in Figure 2(C), as well as their enormous volume augment, should be accompanied of profound changes on the cytoskeleton organization. RKN induce long-term changes in the organization of the cytoskeleton during GCs expansion, i.e. microtubule and actin cytoskeleton disruption and rearrangements occur. A large number of unusual, randomly oriented actin bundles and cables were also observed (de Almeida Engler et al., 2004; Caillaud, Lecomte, et al., 2008; de Almeida Engler & Favery, 2011). Accordingly, upregulation of tubulin- and actin-coding genes was described by de Almeida Engler (2004). Other essential proteins involved in cytoskeleton dynamics such as formins may act as nucleating proteins stimulating the de novo polymerization of actin filaments controlling the assembly of actin cables. Those actin cables are probably required to guide the vesicle trafficking needed for an increasing demand of plasma membrane and cell wall biogenesis (Caillaud, Abad, & Favery, 2008). Other associated proteins such as actin depolymerizing factor (ADF; *At*ADF2) or microtubule-associated proteins (MAPs; *At*MAP65-3) are crucial for GC development and their corresponding genes were differentially regulated within GC (Caillaud, Lecomte, et al., 2008; Clement et al., 2009). Likewise, data from genome-wide transcriptomic analysis of galls and/or GCs showed upregulation of some formin-coding genes through gall growth and GC enlargement (de Almeida Engler et al., 2004; Barcala et al., 2010; Favery et al., 2004; Jammes et al., 2005), supporting the importance of the cytoskeleton during GC formation. Cytoskeleton reorganization is also particularly relevant during the special mitotic cycles occurring in GCs. In this respect, the presence of abnormal spindles and phragmoplasts in developing GCs was reported (Caillaud, Abad, et al., 2008; Banora et al., 2011). Misaligned phragmoplasts may also interfere with cell wall formation, thus resulting in multinucleate cells and wall stub formation (de Almeida Engler et al., 2004; de Almeida Engler & Favery, 2011; Jones & Payne, 1978). The presence of these wall stubs together with repeated mitotic cycles, are essential to distinguish GC differentiation from that of syncytia, also multinucleated but formed by fusion of adjacent cells, as suggested initially by Bird, 1961. Cytoskeleton dynamics is a complex and crucial process that requires sensitive cell biology techniques combined to molecular biology for a direct observation inside the galls girth. In the future, stress should be put to unveil these early processes of cytoskeleton reshuffling in GCs.

A network of microtubules positioned along to the plasma membrane lining the cell wall ingrowths (CWI) formed in GCs has been described (Berg et al., 2008). The formation of these CWIs was already described by Jones & Northcote, 1972, and are believed to increase and facilitate solute transport from adjacent xylem and phloem cells (Sobczak, Fudali, & Wieczorek, 2011). CWIs are concentrated in areas facing different tissues and not only vascular elements, but are particularly abundant opposite to xylem (Berg et al., 2008; Jones & Gunning, 1976; Jones & Northcote, 1972), what is in accordance to the upregulation of genes encoding membrane transporters as mentioned before. The CWIs develop from previous small patches of cell wall thickenings that increase their length and thickness. However, this process is not uniform across all the GCs cell wall, resulting in an irregularly thickened cell wall. Associated to the CWI progress, stacks of Golgi and ER are found, typical of transfer cell (TC) development, where Golgi vesicles release cell wall components (Berg et al., 2008) External signals for initiating cell wall deposition and CWI development are still unknown. Syncytia and GCs differ in their ontogeny and global transcriptional signatures, but both develop CWIs to facilitate high rates of apoplastic/symplastic molecular exchange, showing TC characteristics. The presence of CWI can allow GCs to compensate the considerable decrease in the surface/volume ratio as these cells expand (Cabrera et al., 2015). Similarities between transcriptional changes observed during the early stages of nematode feeding cells (NFC) formation and those described during differentiation of TCs suggest that auxin and ethylene might be putative signals triggering TC-like morphology of NFCs (Cabrera, Barcala, Fenoll, & Escobar, 2014; Rodiuc, Vieira, Banora, & de Almeida Engler, 2014). Although still scarce, there are some other data linking TC regulatory signals to NFSs. For instance, *ZmMRP-1* (Gomez, Royo, Guo, Thompson, & Hueros, 2002) encoding a primary sensor of the putative signals for TCs differentiation, is activated in galls as compared to the rest of the root (Barrero et al., 2009).

Another plant cell structure specialized in intercellular transport of molecules, the plasmodesmata (PD), are relevant for cyst nematode feeding site formation, as PDs represent the starting point for cell wall dissolution and therefore cell fusion (Grundler, Sobczak, & Golinowski, 1998; Hoth & Schneidereit, 2005; Hoth, Stadler, Sauer, & Hammes, 2008; Jones & Payne, 1978). With regard to RKNs, recent research conducted by Hofmann, Youssef-Banora, de Almeida-Engler, and Grundler (2010) reported symplastic connection between GCs and phloem, whose functionality may vary depending on GC developmental stage or even on host species (Grundler & Hofmann, 2011). In contrast, previous studies accepted that GCs were symplastically isolated (Jones & Dropkin, 1976) despite the existence of PD (Hoth et al., 2008). Further research is needed to clarify this important topic regarding GCs PD connections.

Sedentary plant parasitic nematodes produce within the feeding cell cytoplasm one of the most striking and so far poorly characterized key structures for nutrient withdrawal from the GCs and successful parasitism: the feeding tube (FT). FTs were first identified unequivocally by

(Rahman Razak & Evans, 1976), although they had been highlighted in the initial studies of Nemec (1911, 1932) as proteinaceous threads. In electron microscopy analysis, FTs are cylinders with an electron translucent lumen, connected to the stylet orifice in one side and blind at the distal end. They are described in different groups of sedentary endoparasites (Berg et al., 2008; Hussey & Mims, 1990; Rebois, 1980; Rumpenhorst, 1984; Sobczak, Golinowski, & Grundler, 1999); and even in the migratory ectoparasite *Trichodorus similis* (Wyss, Jank-Ladwig, & Lehmann, 1979). However, their structure may be genus specific (reviewed in Berg et al. (2008)). *Meloidogyne incognita* FTs showed a crystalline structure (Hussey & Mims, 1990; Nemec, 1932) suggesting a proteinaceous nature, whereas FTs from *Rotylenchulus reinformis* and *Heterodera schachtii* do not show such a regular structure (Rebois, 1980; Sobczak et al., 1999). In addition, electron energy loss spectroscopy analysis provided further support of the FT putative protein composition, revealing a high content of nitrogen and sulphur (reviewed in Berg et al. (2008)). It is accepted that FTs are formed by a rapid reaction of nematode secretions with unknown components of feeding site cytoplasm. However there are controversial hypothesis regarding whether FT composition is solely from nematode or plant material or a combination of both (Berg et al., 2008). FTs are formed each time the nematode pierces the cell wall for feeding and is then discarded; so, several FTs are encountered within a particular GC (Jones & Goto, 2011). During FT formation, the cell membrane seems to remain nearly intact, although it is unclear whether a small choke is open up, and only a small callose deposition has been reported at the cell wall disruption point, similarly to the feeding plug described for cyst nematodes. Functional FTs are intimately related to the endomembrane system, especially to ER, and this association has been described for FT from other genera, suggesting a relevant role for either FTs formation or nematode feeding (Berg et al., 2008). Whenever the stylet is retracted, the FT is abandoned, and no more endomembrane system can be observed in its proximity, thus suggesting that the nematode might provide a signal for active FT to recruit ER and other endomembrane complexes for their formation. FTs are thought to serve as molecular sieves during nutrient withdraw to avoid blockage of the stylet by large particles (proteins or even organelles) as the nematode pumps away the cell content. They may also serve to discriminate specific cell components, fine-tuning the composition of the nematode diet. To date, there is no indication that nematodes use FTs to inject secretions in the feeding cells. Several studies have been conducted to elucidate the maximum size of solutes that can be uptaken through cyst nematodes derived-FT pores by using fluorescent molecules (green fluorescent protein (GFP) (Goverse et al., 1998; Urwin, Moller, Lilley, McPherson, & Atkinson, 1997), monomeric red fluorescent protein (mRFP) (Valentine et al., 2007) or dextrans (Böckenhoff & Grundler, 1994)), but still contentious data have not allowed proposing a clear explanation. This issue has been recently approached by Eves-van den Akker et al. (2014). They remarked the structural differences between RKN and cyst nematodes FTs, as uniform discrete pores were formed in GCs and heteroporous in syncytia. To date, the specific composition, mechanisms of formation and action and detailed functions of FTs await for further insight.

6. HOLISTIC APPROACHES TO TACKLE GCs SPECIFIC GENE EXPRESSION

During gall and GC ontogeny a profound reprogrammation of gene expression takes place, as encountered in transcriptomic analysis such as microarray (Barcala et al., 2010; Jammes et al., 2005; Portillo et al., 2009) and massive sequencing (Ji et al., 2013; Cabrera et al., unpublished). Precise single cell isolation techniques as microaspiration or laser capture microdissection combined to global transcriptomic analysis constituted a step forward to the understanding of the specific transcriptomic signatures of GCs (Barcala et al., 2010; Fosu-Nyarko, Jones, & Wang, 2009; Portillo et al., 2013; Ramsay, Wang, & Jones, 2004; Wang, Potter, & Jones, 2003; Ji et al., 2013). It allowed bypassing the complexity of the gall transcriptome that included all the different tissues present in this pseudo-organ, and to stablish differences between whole gall and GC-specific transcriptomes. Recently, RNA-sequencing approaches for miRNA differential expression analysis increased the complexity of this scenario (Hewezi, Howe, Maier, & Baum, 2008; Kyndt et al., 2012; Cabrera et al. unpublished), as miRNAs have come up as key signal molecules, controlling and regulating many cellular processes at transcriptional, post-transcriptional and translational level (Yang, Xue, & An, 2007).

Those holistic approaches to gene expression generated vast lists of differentially expressed genes available in public databases and publications and valuable information of general tendencies for gene expression in the NFS. However, classifying detailed information of the regulation of particular genes or gene groups through cross-comparisons among complex data sets, or obtaining customized gene selections through sequential comparative and filtering is not an easy task. This had limited the design of consistent functional hypothesis about genes and gene products of GCs based on holistic gene expression data. One of the first data-mining spreadsheet tool, specifically designed for comparisons among transcriptomes of plant–nematode compatible interactions is NEMATIC (NEMatode–Arabidopsis Transcriptomic Interaction Compendium; Cabrera, Bustos, Favery, Fenoll, & Escobar, 2014 http://www.uclm.es/grupo/gbbmp/english/nematic.asp). It combines available transcriptomic data for the interaction between Arabidopsis and plant endoparasitic nematodes with data from different transcriptomic analyses regarding hormone and cell cycle regulation, development, different plant tissues, cell types and various biotic stresses, facilitating efficient in silico studies on plant–nematode biology. However, there is an increasing need to develop additional user friendly in silico analysis tools that may include other plant species and biological processes.

7. CONCLUSIONS

RKNs depend on a specifically developed cell type from their initial root vascular cells to complete its life cycle. Those GCs are induced and probably maintained by nematode secretions delivered through their stylets. Many questions regarding GC ontogeny and functioning remain unanswered. To date, only a few players of the complex regulatory networks taking place during GCs development have emerged, and the understanding of how these organisms can interact with their hosts in such a subtle manner is fragmentary. Yet, integrative analysis of proteomics and transcriptomics together with genetics and molecular and cell biology tools are facilitating its comprehension. However, the complexity of an evolving interaction makes its analysis a challenge, i.e. feeding site cell status is continuously changing as it differentiates, controlled by nematode nutritional needs. Therefore, comparisons and inferred conclusions from the analysis of galls/GCs at selected infection points should be taken cautiously.

Furthermore, valuable data were also obtained from the study of nematode putative effectors and their molecular interactions to their host targets, as well as the downstream responses, pointing out common and specific regulatory pathways manipulated by RKN and/or cyst nematodes.

ACKNOWLEDGEMENTS

This work was supported by the Spanish Government (AGL2010-17388 and AGL2013-48787-R to C. Escobar, and CSD2007-057 and PCIN-2013-053 to C. Fenoll). J.Cabrera was supported by fellowships from the Ministry of Economy and Competitiveness, Spain.

REFERENCES

Abad, P., Gouzy, J., Aury, J.-M., Castagnone-Sereno, P., Danchin, E. G. J., Deleury, E., et al. (2008). Genome sequence of the metazoan plant-parasitic nematode *Meloidogyne incognita*. *Nature Biotechnology, 26*, 909–915.

Absmanner, B., Stadler, R., & Hammes, U. Z. (2013). Phloem development in nematode-induced feeding sites: the implications of auxin and cytokinin. *Frontiers in Plant Science, 4*, 241.

Ali, A., Matthews, W. C., Cavagnaro, P. F., Iorizzo, M., Roberts, P. A., & Simon, P. W. (2014). Inheritance and mapping of *Mj-2*, a new source of root-knot nematode (*Meloidogyne javanica*) resistance in carrot. *Journal of Heredity, 105*, 288–291.

de Almeida Engler, J., De Vleesschauwer, V., Burssens, S., Celenza, J. L., Jr., Inze, D., Van Montagu, M., et al. (1999). Molecular markers and cell cycle inhibitors show the importance of cell cycle progression in nematode-induced galls and syncytia. *The Plant Cell, 11*, 793–808.

de Almeida Engler, J., & Favery, B. (2011). The plant cytoskeleton remodelling in nematode induced feeding sites. In J. Jones, G. Gheysen, & C. Fenoll (Eds.), *Genomics and molecular genetics of plant-nematode interactions* (pp. 369–393). Springer Netherlands.

de Almeida Engler, J., & Gheysen, G. (2013). Nematode-induced endoreduplication in plant host cells: why and how? *Molecular Plant-Microbe Interactions, 26*, 17–24.

de Almeida Engler, J., Van Poucke, K., Karimi, M., De Groodt, R., Gheysen, G., Engler, G., et al. (2004). Dynamic cytoskeleton rearrangements in giant cells and syncytia of nematode-infected roots. *Plant Journal, 38*, 12–26.

Ammiraju, J. S., Veremis, J. C., Huang, X., Roberts, P. A., & Kaloshian, I. (2003). The heat-stable root-knot nematode resistance gene *Mi-9* from *Lycopersicon peruvianum* is localized on the short arm of chromosome 6. *Theoretical and Applied Genetics, 106*, 478–484.

Back, M. A., Haydock, P. P. J., & Jenkinson, P. (2002). Disease complexes involving plant parasitic nematodes and soilborne pathogens. *Plant Pathology, 51*, 683–697.

Baldacci-Cresp, F., Chang, C., Maucourt, M. l. M., Deborde, C., Hopkins, J., Lecomte, P., et al. (2012). (Homo)glutathione deficiency impairs root-knot nematode development in *Medicago truncatula*. *PLoS Pathogens, 8*, e1002471.

Banora, M. Y., Rodiuc, N., Baldacci-Cresp, F., Smertenko, A., Bleve-Zacheo, T., Mellilo, M. T., et al. (2011). Feeding cells induced by phytoparasitic nematodes require gamma-tubulin ring complex for microtubule reorganization. *PLoS Pathogens, 7*, e1002343.

Barcala, M., García, A., Cabrera, J., Casson, S., Lindsey, K., Favery, B., et al. (2010). Early transcriptomic events in microdissected Arabidopsis nematode-induced giant cells. *Plant Journal, 61*, 698–712.

Barcala, M., Garcia, A., Cubas, P., Almoguera, C., Jordano, J., Fenoll, C., et al. (2008). Distinct heat-shock element arrangements that mediate the heat shock, but not the late-embryogenesis induction of small heat-shock proteins, correlate with promoter activation in root-knot nematode feeding cells. *Plant Molecular Biology, 66*, 151–164.

Barrero, C., Royo, J., Grijota-Martinez, C., Faye, C., Paul, W., Sanz, S., et al. (2009). The promoter of ZmMRP-1, a maize transfer cell-specific transcriptional activator, is induced at solute exchange surfaces and responds to transport demands. *Planta, 229*, 235–247.

Bartlem, D. G., Jones, M. G. K., & Hammes, U. Z. (2014). Vascularization and nutrient delivery at root-knot nematode feeding sites in host roots. *Journal of Experimental Botany, 65*, 1789–1798.

Bergeson, G. B., Van Gundy, S. D., & Thomason, I. J. (1970). Effect of *Meloidogyne javanica* on rhizosphere microflora and Fusarium wilt of tomato. *Phytopathology, 69*, 1245–1249.

Berg, R. H., Fester, T., & Taylor, C. G. (2008). Development of the root-knot nematode feeding cell. In R. H. Berg, & C. G. Taylor (Eds.), *Cell biology of plant nematode parasitism* (pp. 115–152). Berlin: Springer Berlin Heidelberg.

Bird, A. F. (1959). The attractiveness of roots to the plant parasitic nematodes *Meloidogyne javanica* and *M. hapla*. *Nematologica, 4*(4), 322–335.

Bird, A. F. (1961). The ultrastructure and histochemistry of a nematode-induced giant cell. *Journal of Biophysical and Biochemical Cytology, 11*, 701–715.

Bird, D., Opperman, C., & Williamson, V. (2009). Plant infection by root-knot nematode. In R. H. Berg, & C. G. Taylor (Eds.), *Cell biology of plant nematode parasitism* (pp. 1–13). Springer Berlin Heidelberg.

Böckenhoff, A., & Grundler, F. M. W. (1994). Studies on the nutrient uptake by the beet cyst nematode *Heterodera schachtii* by in situ microinjection of fluorescent probes into the feeding structures in *Arabidopsis thaliana*. *Parasitology, 109*, 249–255.

Cabello, S., Lorenz, C., Crespo, S., Cabrera, J., Ludwig, R., Escobar, C., et al. (2014). Altered sucrose synthase and invertase expression affects the local and systemic sugar metabolism of nematode-infected *Arabidopsis thaliana* plants. *Journal of Experimental Botany, 65*, 201–212.

Cabrera, J., Barcala, M., Fenoll, C., & Escobar, C. (2014). Transcriptomic signatures of transfer cells in early developing nematode feeding cells of Arabidopsis focused on auxin and ethylene signaling. *Frontiers in Plant Science, 5*, 107.

Cabrera, J., Bustos, R., Favery, B., Fenoll, C., & Escobar, C. (2014). NEMATIC: a simple and versatile tool for the in silico analysis of plant-nematode interactions. *Molecular Plant Pathology, 15*, 627–636.

Cabrera, J., Díaz-Manzano, F. E., Barcala, M., de Almeida Engler, J., Engler, G., Fenoll, C., et al. (2015). Phenotyping nematode feeding sites: three dimensional reconstruction and volumetric measurements of giant cells induced by root-knot nematodes in Arabidopsis. *New Phytologist*. http://dx.doi.org/10.1111/nph.13249.

Cabrera, J., Diaz-Manzano, F. E., Sanchez, M., Rosso, M.-N., Melillo, T., Goh, T., et al. (2014). A role for lateral organ boundaries-domain 16 during the interaction Arabidopsis-*Meloidogyne* spp. provides a molecular link between lateral root and root-knot nematode feeding site development. *New Phytologist, 203*, 632–645.

Caillaud, M. C., Abad, P., & Favery, B. (2008). Cytoskeleton reorganization, a key process in root-knot nematode-induced giant cell ontogenesis. *Plant Signaling & Behavior, 3*, 816–818.

Caillaud, M. C., Lecomte, P., Jammes, F., Quentin, M. M., Pagnotta, S., Andrio, E., et al. (2008). MAP65–3 microtubule-associated protein is essential for nematode-induced giant cell ontogenesis in Arabidopsis. *The Plant Cell, 20*, 423–437.

Castagnone-Sereno, P., Danchin, E. G., Perfus-Barbeoch, L., & Abad, P. (2013). Diversity and evolution of root-knot nematodes, genus *Meloidogyne*: new insights from the genomic era. *Annual Review of Phytopathology, 51*, 203–220.

Castagnone-Sereno, P., Skantar, A., & Robertson, L. (2011). Molecular tools for diagnostics. In J. Jones, G. Gheysen, & C. Fenoll (Eds.), *Genomics and molecular genetics of plant-nematode interactions* (pp. 443–464). Springer Netherlands.

Christie, J. R. (1936). The development of root-knot nematode galls. *Phytopathology, 26*, 1–22.

Claverie, M., Bosselut, N., Lecouls, A. C., Voisin, R., Lafargue, B., Poizat, C., et al. (2004). Location of independent root-knot nematode resistance genes in plum and peach. *Theoretical and Applied Genetics, 108*, 765–773.

Clement, M., Ketelaar, T., Rodiuc, N., Banora, M. Y., Smertenko, A., Engler, G., et al. (2009). Actin-depolymerizing factor2-mediated actin dynamics are essential for root-knot nematode infection of Arabidopsis. *The Plant cell, 21*, 2963–2979.
Crow, W. T. (2005). How bad are nematode problems on Florida's golf courses? *Florida Turf Digest, 22*, 10–12.
Crow, W. T. (2007). Understanding and managing parasitic nematodes on turfgrasses. In *Handbook of turfgrass management & physiology* (pp. 351–374). Boca, Raton: CRC Press.
Crow, W. T., & Luc, J. E. (2014). Field efficacy of furfural as a nematicide on turf. *Journal of Nematology, 46*, 8–11.
Davide, R. G., & Triantaphyllou, A. C. (1967). Influence of the environment on development and sex differentiation of root-knot nematodes. *Nematologica, 13*, 111–117.
Davis, E. L., Haegeman, A., & Kikuchi, T. (2011). Degradation of the plant cell wall by nematodes. In J. Jones, G. Gheysen, & C. Fenoll (Eds.), *Genomics and molecular genetics of plant-nematode interactions* (pp. 255–272). Springer Netherlands.
Davis, E. L., Hussey, R. S., & Baum, T. J. (2004). Getting to the roots of parasitism by nematodes. *Trends in Parasitology, 20*, 134–141.
De Ley, P., & Blaxter, M. (2002). Systematic position & phylogeny. In D. L. Lee (Ed.), *The biology of nematodes* (pp. 1–30). London: Taylor & Francis.
De Ley, P., & Blaxter, M. (2004). A new system for Nematoda: combining morphological characters with molecular trees, and translating clades into ranks and taxa. In R. Cook, & D. J. Hunt (Eds.), *Nematology monographs and perspectives* (pp. 633–653). Leiden: E.J. Brill.
Decker, H. (1989). *Plant nematodes and their control (phytonematology)*. Brill.
Decraemer, W., & Hunt, D. J. (2013). Structure and classification. In R. N. Perry, & M. Moens (Eds.), *Plant Nematology* (pp. 3–39). Wallingford, UK: CABI publ.
Devran, Z., & Sögüt, M. A. (2010). Occurrence of virulent root-knot nematode populations on tomatoes bearing the *Mi gene* in protected vegetable-growing areas of Turkey. *Phytoparasitica, 38*, 245–251.
Dinh, P. T., Brown, C. R., & Elling, A. A. (2014). RNA Interference of effector gene *Mc16D10L* confers resistance against *Meloidogyne chitwoodi* in arabidopsis and potato. *Phytopathology, 104*, 1098–1106.
Djian-Caporalino, C., Fazari, A., Arguel, M. J., Vernie, T., VandeCasteele, C., Faure, I., et al. (2007). Root-knot nematode (*Meloidogyne* spp.) *Me* resistance genes in pepper (*Capsicum annuum* L.) are clustered on the P9 chromosome. *Theoretical and Applied Genetics, 114*, 473–486.
Djian-Caporalino, C., Palloix, A., Fazari, A., Marteu, N., Barbary, A., Abad, P., et al. (2014). Pyramiding, alternating or mixing: comparative performances of deployment strategies of nematode resistance genes to promote plant resistance efficiency and durability. *BMC Plant Biology, 14*, 53.
Eisenback, J. D. (1985). Detailed morphology and anatomy of second-stage juveniles, males, and females of the genus *Meloidogyne* (root-knot nematode). In J. N. Sasser, & C. C. Carter (Eds.), *An advanced treatise on* Meloidogyne, *Vol. 1* (pp. 47–78). Raleigh: North Carolina State University Graphics.
Eisenback, J. D. D., Dodge, D. J., & Odge, D. J. D. (2012). Description of a unique, complex feeding socket caused by the putative primitive root-knot nematode, *Meloidogyne kikuyensis*. *Journal of Nematology, 44*, 148–152.
Eisenback, J. D., & Triantaphyllou, H. H. (1991). Root-knot nematodes: *Meloidogyne* species and races. In W. R. Nickle (Ed.), *Manual of agricultural nematology* (pp. 191–274). New York: Marcel Dekker.
Elling, A. A. (2013). Major emerging problems with minor *Meloidogyne* species. *Phytopathology, 103*, 1092–1102.
Elling, A. A., & Jones, J. T. (2014). Functional characterization of nematode effectors in plants. *Methods in Molecular Biology, 1127*, 113–124.

Eves-van den Akker, S., Lilley, C. J., Ault, J. R., Ashcroft, A. E., Jones, J. T., & Urwin, P. E. (2014). The feeding tube of cyst nematodes: characterisation of protein exclusion. *PLoS One, 9*, e87289.

Fassuliotis, G., & Rau, G. J. (1969). The relationship of *Meloidogyne incognita* acrita to the incidence of cabbage yellows. *Journal of Nematology, 1*, 219–222.

Favery, B., Chelysheva, L. A., Lebris, M., Jammes, F., Marmagne, A., de Almeida-Engler, J., et al. (2004). Arabidopsis formin *AtFH6* is a plasma membrane–associated protein upregulated in giant cells induced by parasitic nematodes. *The Plant Cell, 16*, 2529–2540.

Fosu-Nyarko, J., Jones, M. G. K., & Wang, Z. (2009). Functional characterization of transcripts expressed in early-stage *Meloidogyne javanica*-induced giant cells isolated by laser microdissection. *Molecular Plant Pathology, 10*, 237–248.

Gautam, S. K., & Poddar, A. N. (2014). Study on protein and sugar content in *Meloidogyne incognita* infested roots of bitter gourd. *International Journal of Current Microbiology and Applied Sciences, 3*, 470–478.

Gheysen, G., & Fenoll, C. (2011). Arabidopsis as a tool for the study of plant-nematode interactions. In J. Jones, G. Gheysen, & C. Fenoll (Eds.), *Genomics and molecular genetics of plant-nematode interactions* (pp. 139–156). Springer Netherlands.

Gomez, E., Royo, J., Guo, Y., Thompson, R., & Hueros, G. (2002). Establishment of cereal endosperm expression domains: identification and properties of a maize transfer cell-specific transcription factor, ZmMRP-1. *Plant Cell, 14*, 599–610.

Goverse, A., Biesheuvel, J., Wijers, G. J., Gommers, F. J., Bakker, J., Schots, A., et al. (1998). In planta monitoring of the activity of two constitutive promoters, CaMV 35S and TR2′, in developing feeding cells induced by *Globodera rostochiensis* using green fluorescent protein in combination with confocal laser scanning microscopy. *Physiological and Molecular Plant Pathology, 52*, 275–284.

Grundler, F. W., & Hofmann, J. (2011). Water and nutrient transport in nematode feeding sites. In J. Jones, G. Gheysen, & C. Fenoll (Eds.), *Genomics and molecular genetics of plant-nematode interactions* (pp. 423–439). Springer Netherlands.

Grundler, F. M. W., Sobczak, M., & Golinowski, W. (1998). Formation of wall openings in root cells of *Arabidopsis thaliana* following infection by the plant-parasitic nematode *Heterodera schachtii*. *European Journal of Plant Pathology, 104*, 545–551.

Hammes, U. Z., Nielsen, E., Honaas, L. A., Taylor, C. G., & Schachtman, D. P. (2006). AtCAT6, a sink-tissue-localized transporter for essential amino acids in Arabidopsis. *Plant Journal, 48*, 414–426.

Hammes, U. Z., Schachtman, D. P., Berg, R. H., Nielsen, E., Koch, W., McIntyre, L. M., et al. (2005). Nematode-induced changes of transporter gene expression in Arabidopsis roots. *Molecular Plant-Microbe Interactions, 18*, 1247–1257.

Hartman, K. M., & Sasser, J. N. (1985). Identification of *Meloidogyne* species on the basis of differential host test and perineal pattern morphology. In K. R. Barker, C. C. Carter, & J. N. Sasser (Eds.), *An advanced treatise on* Meloidogyne, *Vol. 2* (pp. 69–78). Raleigh: North Carolina State University Graphics.

Hewezi, T., Howe, P., Maier, T. R., & Baum, T. J. (2008). Arabidopsis small RNAs and their targets during cyst nematode parasitism. *Molecular Plant-Microbe Interactions, 21*, 1622–1634.

Hofmann, J., Youssef-Banora, M., de Almeida-Engler, J., & Grundler, F. M. W. (2010). The role of callose deposition along plasmodesmata in nematode feeding sites. *Molecular Plant-Microbe Interactions, 23*, 549–557.

Hooks, C. R. R., Wang, K.-H., Ploeg, A., & McSorley, R. (2010). Using marigold (*Tagetes* spp.) as a cover crop to protect crops from plant-parasitic nematodes. *Applied Soil Ecology, 46*, 307–320.

Hoth, S., & Schneidereit, A. (2005). Nematode infection triggers the de novo formation of unloading phloem that allows macromolecular trafficking of green fluorescent protein into syncytia. *Plant Physiology, 138*, 383–392.

Hoth, S., Stadler, R., Sauer, N., & Hammes, U. Z. (2008). Differential vascularization of nematode-induced feeding sites. *Proceedings of the National Academy of Sciences of the United States of America, 105*, 12617–12622.

Huang, C. S., & Maggenti, A. R. (1969). Wall modifications in developing giant cells of *Vicia faba* and *Cucumis sativus* induced by root knot nematode, *Meloidogyne javanica. Phytopathology, 59*, 931–937.

Hussey, R. S., & Davis, E. L. (2004). Nematode esophageal glands and plant parasitism. In Z. X. Chen, S. Y. A. Chen, & D. W. Dickson (Eds.), *Nematology advances and perspectives, volume I. Nematode morphology, physiology and ecology* (pp. 258–293). Wallingford, UK: CABI publ.

Hussey, R. S., & Mims, C. W. (1990). Ultrastructure of esophageal glands and their secretory granules in the root-knot nematode *Meloidogyne incognita. Protoplasma, 156*, 9–18.

Jacquet, M., Bongiovanni, M., Martinez, M., Verschave, P., Wajnberg, E., & Castagnone-Sereno, P. (2005). Variation in resistance to the root-knot nematode *Meloidogyne incognita* in tomato genotypes bearing the *Mi* gene. *Plant Pathology, 54*, 93–99.

Jammes, F., Lecomte, P., de Almeida-Engler, J., Bitton, F., Martin-Magniette, M. L., Renou, J. P., et al. (2005). Genome-wide expression profiling of the host response to root-knot nematode infection in Arabidopsis. *Plant Journal, 44*, 447–458.

Jaouannet, M., Perfus-Barbeoch, L., Deleury, E., Magliano, M., Engler, G., Vieira, P., et al. (2012). A root-knot nematode-secreted protein is injected into giant cells and targeted to the nuclei. *New Phytologist, 194*, 924–931.

Jaubert, S., Laffaire, J. B., Abad, P., & Rosso, M. N. (2002). A polygalacturonase of animal origin isolated from the root-knot nematode *Meloidogyne incognita. FEBS Letters, 522*, 109–112.

Jenkins, W. R., & Coursen, B. W. (1957). The effect of root-knot nematodes, *Meloidogyne incognita* acrita and *M. hapla*, on Fusarium wilt of tomato. *Plant Disease Report*, 182–186.

Ji, H., Gheysen, G., Denil, S., Lindsey, K., Topping, J. F., Nahar, K., et al. (2013). Transcriptional analysis through RNA sequencing of giant cells induced by *Meloidogyne graminicola* in rice roots. *Journal of Experimental Botany, 64*, 3885–3898.

Jones, M. G. K., & Dropkin, V. H. (1976). Scanning electron microscopy of nematode-induced giant transfer cells. *Cytobios, 15*, 149–161.

Jones, M. G. K., & Goto, D. B. (2011). Root-knot nematodes and giant cells. In J. Jones, G. Gheysen, & C. Fenoll (Eds.), *Genomics and molecular genetics of plant-nematode interactions* (pp. 83–100). Springer Netherlands.

Jones, M. G. K., & Gunning, B. E. S. (1976). Transfer cells and nematode induced giant cells in Helianthemum. *Protoplasma, 87*, 273–279.

Jones, J. T., Haegeman, A., Danchin, E. G. J., Gaur, H. S., Helder, J., Jones, M. G. K., et al. (2013). Top 10 plant-parasitic nematodes in molecular plant pathology. *Molecular Plant Pathology, 14*, 946–961.

Jones, M. G. K., & Northcote, D. H. (1972). Multinucleate transfer cells induced in coleus roots by the root-knot nematode, *Meloidogyne arenaria. Protoplasma, 75*, 381–395.

Jones, M. G., & Payne, H. L. (1978). Early stages of nematode-induced giant-cell formation in roots of *Impatiens balsamina. Journal of Nematology, 10*, 70–84.

Kearn, J., Ludlow, E., Dillon, J., O'Connor, V., & Holden-Dye, L. (2014). Fluensulfone is a nematicide with a mode of action distinct from anticholinesterases and macrocyclic lactones. *Pesticide Biochemistry and Physiology, 109*, 44–57.

Kostoff, D., & Kendall, J. (1930). Cytology of nematode galls on Nicotiana roots. *Zentralbl Bakteriol Parasitenk, 81*, 86–91.

Kyndt, T., Denil, S., Haegeman, A., Trooskens, G., Bauters, L., Van Criekinge, W., et al. (2012). Transcriptional reprogramming by root knot and migratory nematode infection in rice. *New Phytologist, 196*, 887–900.

Kyndt, T., Haegeman, A., & Gheysen, G. (2008). Evolution of *GHF5* endoglucanase gene structure in plant-parasitic nematodes: no evidence for an early domain shuffling event. *BMC Evolutionary Biology, 8*, 305.

Lacey, L. A., & Georgis, R. (2012). Entomopathogenic nematodes for control of insect pests above and below ground with comments on commercial production. *Journal of Nematology, 44*.

Lewis, E. E., & Perez, E. E. (2004). Aging and developmental behavior. In G. Randy, & A. L. Bilgrami (Eds.), *Nematode behaviour* (pp. 151–176).

Lin, B., Zhuo, K., Wu, P., Cui, R., Zhang, L-h., & Liao, J. (2013). A novel effector protein, MJ-NULG1a, targeted to giant cell nuclei plays a role in *Meloidogyne javanica* parasitism. *Molecular Plant-Microbe Interactions, 26*, 55–66.

Lu, Z. X., Sossey-Alaoui, K., Reighard, G. L., Baird, W. V., & Abbott, A. G. (1999). Development and characterization of a codominant marker linked to root-knot nematode resistance, and its application to peach rootstock breeding. *Theoretical and Applied Genetics, 99*, 115–122.

Machado, A., Campos, V., da Silva, W. R., Campos, V., Zeri, A., & Oliveira, D. (2012). Metabolic profiling in the roots of coffee plants exposed to the coffee root-knot nematode, *Meloidogyne exigua*. *European Journal of Plant Pathology, 134*, 431–441.

Mahdy, M. (2002). Biological control of plant parasitic nematodes with antagonistic bacteria on different host plants. In *Hohen Landwirtschaftlichen Fakultät*. Bonn: Rheinischen Friedrich-Wilhelms-Universität.

Manzanilla-Lopez, R. H. E. K., & Bridge, J. (2004). In Z. X. Chen, S. Y. Chen, & D. W. Dickson (Eds.), *Plant diseases caused by nematodes* (pp. 636–716). Wallingford, UK: CABI Publishing.

Marella, H. H., Nielsen, E., Schachtman, D. P., Taylor, C. G., Danforth, D., Science, P., et al. (2013). The amino acid permeases AAP3 and AAP6 are involved in root-knot nematode parasitism of Arabidopsis. *Molecular Plant-Microbe Interactions, 26*, 44–54.

van Megen, H., van den Elsen, S., Holterman, M., Karssen, G., Mooyman, P., Bongers, T., et al. (2009). A phylogenetic tree of nematodes based on about 1200 full-length small subunit ribosomal DNA sequences. *Nematology, 11*, 927–950.

von Mende, N. (1997). Invasion and migration behaviour of sedentary nematodes. In C. Fenoll, F. M. W. Grundler, & S. A. Ohl (Eds.), *Cellular and molecular aspects of plant-nematode interactions* (pp. 51–64). Springer Netherlands.

Mitchum, M. G., Hussey, R. S., Baum, T. J., Wang, X., Elling, A. A., Wubben, M., et al. (2013). Nematode effector proteins: an emerging paradigm of parasitism. *New Phytologist, 199*, 879–894.

Moens, M., Perry, R., & Starr, J. (2009). Meloidogyne species—a diverse group of novel and important plant parasites. In R. N. Perry, M. Moens, & J. L. Starr (Eds.), *root-knot nematodes* (p. 483). Wallingford, UK: CABI publ.

Nemec, B. (1911). Über die Nematodenkrankheiten der Zuckerrübe. *Zeitschrift für P Anzenkrankheiten, 21*, 1–10.

Nemec, B. (1932). Über die Galen von *Heterodera schachtii* auf der Zuckerrübe. *Studies from the Plant Physiological Laboratory of Charles University Prague, 4*, 1–14.

Netcher, C., & Taylor, D. P. (1979). Physiological variation with the genus Meloidogyne and its implications on integrated control. In *Root-knot nematodes (*Meloidogyne *species), systematics, biology and control* (pp. 269–294). London: Acad. Press. London.

Niebel, A., De Almeida Engler, J., Hemerly, A., Ferreira, P., Inzé, D., Van Montagu, M., et al. (1996). Induction of *cdc2a* and *cyc1At* expression in *Arabidopsis thaliana* during early phases of nematode-induced feeding cell formation. *Plant Journal, 10*, 1037–1043.

Opperman, C. H., Bird, D. M., Williamson, V. M., Rokhsar, D. S., Burke, M., Cohn, J., et al. (2008). Sequence and genetic map of *Meloidogyne hapla*: a compact nematode genome for plant parasitism. *Proceedings of the National Academy of Sciences of the United States of America, 105*, 14802–14807.

Orion, D., Kritzman, G., Meyer, S. l. F., Erbe, E. F., & Chitwood, D. J. (2001). A role of the gelatinous matrix in the resistance of root-knot nematode (*Meloidogyne* spp.) eggs to microorganisms. *Journal of Nematology, 33*, 203–207.

Perry, R., & Moens, M. (2011). Introduction to plant-parasitic nematodes; modes of parasitism. In J. Jones, G. Gheysen, & C. Fenoll (Eds.), *Genomics and molecular genetics of plant-nematode interactions* (pp. 3–20). Springer Netherlands.

Perry, R. N. (2001). An evaluation of types of attractants enabling plant-parasitic nematodes to locate plant roots. *Russian Journal of Nematology, 13*, 83–88.

Portillo, M., Cabrera, J., Lindsey, K., Topping, J., Andres, M. F., Emiliozzi, M., et al. (2013). Distinct and conserved transcriptomic changes during nematode-induced giant cell development in tomato compared with Arabidopsis: a functional role for gene repression. *New Phytologist, 197*, 1276–1290.

Portillo, M., Lindsey, K., Casson, S., García-Casado, G., Solano, R., Fenoll, C., et al. (2009). Isolation of RNA from laser-capture-microdissected giant cells at early differentiation stages suitable for differential transcriptome analysis. *Molecular Plant Pathology, 10*, 523–535.

Prot, J. C. (1980). Migration of plant-parasitic nematodes towards roots. *Revue de Nématologie, 3*, 305–318.

Quentin, M., Abad, P., & Favery, B. (2013). Plant parasitic nematode effectors target host defence and nuclear functions to establish feeding cells. *Frontiers in Plant Science, 4*.

Rahman Khan, M., Khan, S. M., & Mohide, F. (2005). Root-knot nematode problem of some winter ornamental plants and its biomanagement. *Journal of Nematology, 37*, 198–206.

Rahman Razak, A., & Evans, A. A. F. (1976). An intracellular tube associated with feeding by *Rotylenchulus reniformis* on cowpea root. *Nematologica, 22*, 182–189.

Ramsay, K., Wang, Z., & Jones, M. G. K. (2004). Using laser capture microdissection to study gene expression in early stages of giant cells induced by root-knot nematodes. *Molecular Plant Pathology, 5*, 587–592.

Ravichandra, N. G. (2008). *Plant nematology*. New Delhi, India: I. K. International Pvt Ltd.

Rebois, R. V. (1980). Ultrastructure of a feeding peg and tube associated with *Rotylenchulus reniformis* in cotton. *Nematologica, 26*, 396–405.

Robinson, A. (2002). Host finding by plant-parasitic nematodes. In E. E. Lewis, J. Campbell, & M. Sukhdeo (Eds.), *The behavioral ecology of parasites* (pp. 89–110). Wallingford, UK: CABI publ.

Rodiuc, N., Vieira, P., Banora, M. Y., & de Almeida Engler, J. (2014). On the track of transfer cell formation by specialized plant-parasitic nematodes. *Frontiers in Plant Science, 5*, 160.

Rohini, K., Ekanayaka, H. M., & Di Vito, M. (1986). Life cycle and multiplication of *Meloidogyne incognita* on tomato and eggplant seedlings. *Tropical Agriculturist, 142*.

Rossi, M., Goggin, F. L., Milligan, S. B., Kaloshian, I., Ullman, D. E., & Williamson, V. M. (1998). The nematode resistance gene *Mi* of tomato confers resistance against the potato aphid. *Proceedings of the National Academy of Sciences of the United States of America, 95*, 9750–9754.

Rosso, M. N., & Grenier, E. (2011). Other nematode effectors and evolutionary constraints. In J. Jones, G. Gheysen, & C. Fenoll (Eds.), *Genomics and molecular genetics of plant-nematode interactions* (pp. 287–307). Springer Netherlands.

Rumpenhorst, H. J. (1984). Intracellular feeding tubes associated with sedentary plant parasitic nematodes. *Nematologica, 30*, 77–85.

Rybarczyk-Mydłowska, K., Maboreke, H. R., van Megen, H., van den Elsen, S., Mooyman, P., Smant, G., et al. (2012). Rather than by direct acquisition via lateral gene transfer, GHF5 cellulases were passed on from early Pratylenchidae to root-knot and cyst nematodes. *BMC Evolutionary Biology, 12*, 221.

Sharon, E., & Spiegel, Y. (1993). Glycoprotein characterization of the gelatinous matrix in the root–knot nematode *Meloidogyne javanica*. *Journal of Nematology, 25*, 585–589.

Shepperson, J. R., & Jordan, W. C. (1974). Observations on in vitro survival and development of Meloidogyne. *Proceedings of the Helminthological Society of Washington, 41*, 254.

Sijmons, P. C., Grundler, F. M. W., Von Mende, N., Burrows, P. R., & Wyss, U. (1991). *Arabidopsis thaliana* as a new model host for plant-parasitic nematodes. *Plant Journal, 1*, 245–254.

Snyder, D. W., Opperman, C. H., & Bird, D. M. (2006). A method for generating *Meloidogyne incognita* males. *Journal of Nematology, 38*, 192–194.

Sobczak, M., Fudali, S., & Wieczorek, K. (2011). Cell wall modifications induced by nematodes. In J. Jones, G. Gheysen, & C. Fenoll (Eds.), *Genomics and molecular genetics of plant-nematode interactions* (pp. 395–422). Springer Netherlands.

Sobczak, M., Golinowski, W., & Grundler, F. (1999). Ultrastructure of feeding plugs and feeding tubes formed by *Heterodera schachtii*. *Nematology*, 363–374.

Stanton, J. M., & Stirling, G. (1997). Nematodes as plant parasites. In J. F. Browm, & H. J. Ogle (Eds.), *Plant pathogens and plant diseases* (pp. 127–141). Armidale, Australia: University of New England.

Sumner, D. R., & Johnson, A. W. (1972). The effect of nematodes and crop sequence Fusarium wilt of watermelon. *Phytopathology, 62*, 791.

Teillet, A., Dybal, K., Kerry, B. R., Miller, A. J., Curtis, R. H. C., & Hedden, P. (2013). Transcriptional changes of the root-knot nematode *Meloidogyne incognita* in response to *Arabidopsis thaliana* root signals. *PLoS One, 8*, e61259.

Timper, P. (2011). Utilization of biological control for managing plant-parasitic nematodes. In K. Davies, & Y. Spiegel (Eds.), *Progress in biological control, Biological control of plant-parasitic nematodes: Building Coherence between microbial ecology and molecular mechanisms* (pp. 259–289). Springer Netherlands.

Treub, M. (1886). Quelques mots sure les effets du parasitisme de l'Heterodera javanica dans les racines de la canne a sucre. *Annales du Jardin botanique de Buitenzorg, 6*, 93–96.

Triantaphyllou, A. C. (1973). Environmental sex differentiation of nematodes in relation to pest management. *Annual Review of Phytopathology, 11*, 441–462.

Triantaphyllou, A. C. (1985). Cytogenetics,cytotaxonomy and phylogeny of root-knot nematodes. In J. N. Sasser, & C. C. Carter (Eds.), *An advanced teatrise on meloidogyne. Vol. I, biology and control* (pp. 113–126). Raleigh, USA: North Carolina State University Graphics.

Tucker, M. L., & Yang, R. (2013). A gene encoding a peptide with similarity to the plant IDA signaling peptide (AtIDA) is expressed most abundantly in the root-knot nematode (*Meloidogyne incognita*) soon after root infection. *Experimental Parasitology, 134*, 165–170.

Urwin, P. E., Moller, S. G., Lilley, C. J., McPherson, M. J., & Atkinson, H. J. (1997). Continual green-fluorescent protein monitoring of cauliflower mosaic virus 35S promoter activity in nematode-induced feeding cells in *Arabidopsis thaliana*. *Molecular Plant-Microbe Interactions, 10*, 394–400.

Valdez, R. B. (1978). Nematodes attacking tomato and their control. In *First international symposium on tropical tomato* (pp. 136–152). AVRDC publ.

Valentine, T. A., Randall, E., Wypijewski, K., Chapman, S., Jones, J., & Oparka, K. J. (2007). Delivery of macromolecules to plant parasitic nematodes using a tobacco rattle virus vector. *Plant Biotechnology Journal, 5*, 827–834.

Veremis, J. C., van Heusden, A. W., & Roberts, P. A. (1999). Mapping a novel heat-stable resistance to Meloidogyne in *Lycopersicon peruvianum*. *Theoretical and Applied Genetics, 98*, 274–280.

Viaene, N., Coyne, D. L., & Kerry, B. R. (2006). Biological and cultural management. In R. Perry, & M. Moens (Eds.), *Plant nematology*. Wallingford, UK: CABI.

Vieira, P., Danchin, E. G. J., Neveu, C., Crozat, C., Jaubert, S., Hussey, R. S., et al. (2011). The plant apoplasm is an important recipient compartment for nematode secreted proteins. *Journal of Experimental Botany, 62*, 1241–1253.

Vovlas, N., Rapoport, H. F., Díaz, R. M. J., Castillo, P., Nematologia, B., Nazionale, C., et al. (2005). Differences in feeding sites induced by root-knot nematodes, *Meloidogyne* spp., in Chickpea. *Phytopathology, 95*, 368–375.

Wallace, H. R. (1968). The influence of soil moisture on survival and hatch of Meloidogyne javanica. *Nematologica, 14*, 231–242.

Wang, Z., Potter, R. H., & Jones, M. G. K. (2003). Differential display analysis of gene expression in the cytoplasm of giant cells induced in tomato roots by *Meloidogyne javanica. Molecular Plant Pathology, 4*, 361–371.

Wieczorek, K., Elashry, A., Quentin, M., Grundler, F. M. W., Favery, B., Seifert, G. J., et al. (2014). A distinct role of pectate lyases in the formation of feeding structures induced by cyst and root-knot nematodes. *Molecular Plant-Microbe Interactions, 27*, 901–912.

Williamson, V. M. (1998). Root-Knot nematode resistance genes in tomato and their potential for future use. *Annual Review of Phytopathology, 36*, 277–293.

Wyss, U., Grundler, F. M. W., & Münch, A. (1992). The parasitic behavior of second-stage juveniles in *Meloidogyne incognita* in roots of *Arabidopsis thaliana. Nematologica, 38*, 98–111.

Wyss, U., Jank-Ladwig, R., & Lehmann, H. (1979). On the formation and ultrastructure of feeding tubes produced by Trichodorid nematodes. *Nematologica, 25*, 385–390.

Yaghoobi, J., Kaloshian, I., Wen, Y., & Williamson, V. M. (1995). Mapping a new nematode resistance locus in *Lycopersicon peruvianum. Theoretical and Applied Genetics, 91*, 457–464.

Yang, T., Xue, L., & An, L. (2007). Functional diversity of miRNA in plants. *Plant Science, 172*, 423–432.

Zhang, Z. Q. (2013). Animal biodiversity: an update of classification and diversity in 2013. *Zootaxa, 3703*, 5–11.

Zhang, L., Davies, L. J., & Elling, A. A. (2015). A *Meloidogyne incognita* effector is imported into the nucleus and exhibits transcriptional activation activity in planta. *Molecular Plant Pathology, 16*(1), 48–60.

CHAPTER TWO

Introductory Chapter on the Basic Biology of Cyst Nematodes

Holger Bohlmann
Department of Crop Sciences, Division of Plant Protection, University of Natural Resources and Life Sciences, Tulln, Austria
E-mail: holger.bohlmann@boku.ac.at

Contents

Abstract

Cyst nematodes are a group of sedentary, biotrophic plant pathogenic nematodes. Their life cycle starts with the hatching of juveniles, often induced by metabolites exuded from the roots of their host plants. They invade the roots with the help of the stylet and cell wall degrading enzymes produced in the subventral gland cells and move intracellularly to the central cylinder where they induce a feeding site with effectors produced mainly in the dorsal gland cell. Starting from the initial syncytial cell, several hundred root cells are incorporated into a syncytium by local cell wall dissolutions. This syncytium is the only source of nutrients for the cyst nematodes which they take up through their stylet and a feeding tube produced in the syncytium at the tip of the stylet. Males become mobile again after the fourth moult and leave the roots to mate with females. The females stay attached to their feeding site during their whole life and produce hundreds of eggs after mating. The majority of eggs will be contained in the female body. When the female dies, its body will harden and become the cyst which protects the eggs. Cysts can survive in the soil for many years until the new generation of juveniles will hatch again under favourable conditions.

Advances in Botanical Research, Volume 73
ISSN 0065-2296
http://dx.doi.org/10.1016/bs.abr.2014.12.001

1. INTRODUCTION

Nematodes are widely distributed on earth and occur in almost all ecosystems. They can be free living, feeding on bacteria (such as the model organism *Caenorhabditis elegans*) or fungi or live as parasites of animals and plants. Animal parasites include, for instance, entomopathogenic nematodes which are used in plant protection against insect pests (Dillman & Sternberg, 2012). They can also cause important diseases of animals and humans. *Ascaris lumbricoides* may be found in more than 1 billion people (Dold & Holland, 2011). The guinea worm (*Dracunculus medinensis*), whose females can become as long as 80 cm, was traditionally removed by pulling it out from the wound and wind it up on a wooden stick (Muller, 1971). Some think that this is what is shown on the Rod of Asclepius, the medical symbol. Decraemer and Hunt (2006) have reported that 4100 species are regarded as plant parasitic nematodes which occur mainly on the roots of their host plants. Since plant parasitic nematodes live usually belowground and may not always induce obvious symptoms on the aboveground plant parts, it is clear that many more species are still to be discovered. According to their life style they can be divided into migratory and sedentary parasites. Molecular phylogenetic studies have revealed that the parasitic life style in these groups has evolved independently several times (Holterman et al., 2009). Cyst nematodes and root knot nematodes are the main groups of the sedentary parasites.

Nematodes belong to the phylum nematoda with the cyst nematodes in the order *Tylenchida*. Cyst nematodes are found in the subfamily *Heterodeninae* which was formerly placed in the family *Heteroderidae* (Evans & Rowe, 1998) with 6 genera and a total of 99 species. The largest genus *Heterodera* had 67 species and *Globodera* 12. Now, with increasing use of molecular markers in systematics, the cyst nematodes (subfamily *Heterodeninae*) have been placed in the family *Hoplolaimidae* with currently 8 genera with 115 species in the subfamily *Heterodeninae*: *Heterodera*, *Globodera*, *Cactodera*, *Punctodera*, *Dolichodera*, *Betulodera*, *Paradolichodera* and *Vittatidera* (Turner & Subbotin, 2013). The number of cyst nematode species will certainly increase in the future. The far largest genus is still *Heterodera* with now 82 species, which, together with the genus *Globodera* (12 species) contains many species of global agronomic importance. Accordingly, most of what we know about cyst nematodes comes from research on *Heterodera schachtii*, the sugar beet cyst nematode, and *Heterodera glycines*, the soybean cyst nematode and from the potato cyst nematodes, *Globodera rostochiensis* (called the "golden nematode" because the females have a yellow or golden colour) and *Globodera pallida* (called the "pale potato cyst nematode" because the females are cream coloured). Other cyst nematodes on important crop plants (reviewed by Nicol et al., 2011) include *Heterodera oryzicola, Heterodera elachista, Heterodera oryzae* and *Heterodera sacchari* on rice and *Heterodera zeae, Heterodera avenae* and *Punctodera chalcoensis* on maize. Cereal cyst nematodes are a global problem in wheat-producing countries. The cereal cyst nematode complex includes several closely related species, especially *H. avenae*, but also *Heterodera filipjevi* and *Heterodera latipons* (Nicol, Elekcioglu, Bolat, & Rivoal, 2007).

2. MORPHOLOGY

A detailed description of cyst nematode morphology and ultrastructure is given by Zunke and Eisenback (1998). In brief, juvenile cyst nematodes (Figure 1) are vermiform and measure between 330 and 700 μm while the males are approximately twice as large, between 450 and 1700 μm. Mature *Globodera* females are nearly round while females of *Heterodera* species have a lemon-shaped body. They vary in length from 300 to 990 μm and in width from 200 to 810 μm. Juveniles have a dome-shaped head region and a tapering tale. The body is covered with an elastic cuticle which is secreted by the hypodermis and may be coated with proteins, carbohydrates and lipids which could be important in the suppression or evasion of host defences (Curtis, 2007). Robertson et al. (2000) cloned a gene from *Globodera rostochiensis* which encoded a peroxidase. The protein, although lacking a signal peptide, was detected on the surface of juveniles and might be involved in protection against plant defence responses such as the production of reactive oxygen induced through the damage caused by the migrating juvenile. During infection of *Arabidopsis* roots by *H. glycines* juveniles the root cells produced hydrogen peroxide which could be detected histochemically (Waetzig, Sobczak, & Grundler, 1999).

During moulting, the old cuticle is removed and a new cuticle is formed. Nematodes lack a skeleton and the cuticle is therefore important to maintain the shape, together with the hydroskeleton formed by the inner pseudocoelom which is lined by longitudinal muscle cells. Movement is accomplished by alternating contraction of the ventral and dorsal muscle cells. Other

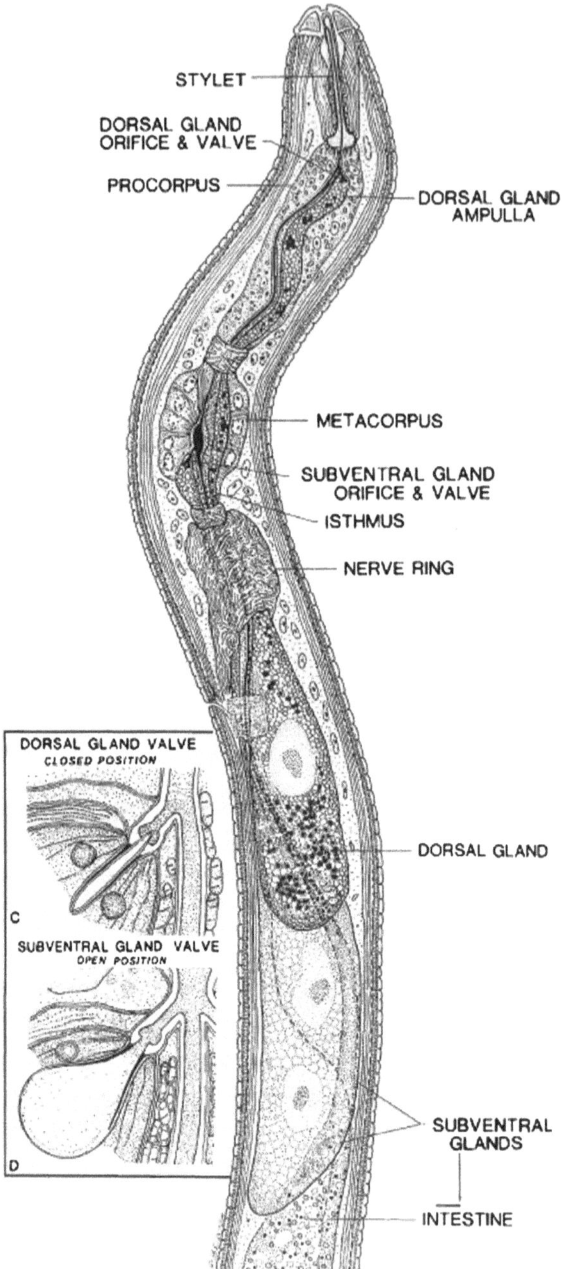

Figure 1 Longitudinal view (LV) through the anterior region. Insert: LV showing a closed valve or end apparatus within a dorsal gland ampulla and the open valve or end apparatus within one of a pair of subventral gland ampullae (Endo, 1984). *Copyright 1984 by the Helminthological Society of Washington, used with permission.*

specialized muscle cells exist at the mouth to move the stylet, at the oesophagus, and along the digestive tract and the reproductive system.

Nematodes have a simple central nervous system with a major anterior, circumpharyngeal nerve ring in the head region and dorsal and ventral nerve cords which are connected by commissures. It controls mainly movement and some sensory functions, such as host finding and penetration by the infective juveniles and locating the females for mating in case of the males. Females stay attached to their feeding site and have probably very limited sensory perceptions.

The head region contains the main chemoreceptor sense organs, the amphids. They are cup shaped with a cavity formed by sheath cells which contains the dendrites of the amphidial neurons. In case of *C. elegans* the function of all 12 amphidial neurons is known in detail. They are specialized for the reception of different stimuli (Bargmann, 2006). In case of cyst nematodes or other plant pathogenic nematodes we are far away from such a detailed knowledge but we can assume that the amphids are also involved in chemoreception of different semiochemicals which might be sex pheromones or substances exuded from plant roots. Amphids also produce secretions which might contain effectors involved in the suppression of host plant defence reactions. It was also found that amphid secretions are involved in producing the feeding plug which seals the plant cell wall where the nematode inserts its stylet (Endo, 1978; Sobczak, Golinowski, & Grundler, 1999). Secretions produced in amphids might also contain avirulence proteins (see below) since it was found that one protein produced in amphids was only found in an avirulent line but not in virulent lines of the root knot nematode *Meloidogyne incognita* (Semblat, Rosso, Hussey, Abad, & Castagnone-Sereno, 2001). In addition to the amphids the head region contains also several other sensilla that have been described in detail at the ultrastructural level (Endo, 1980). At the tale region there are two phasmids (Baldwin, 1985) which seem to be degraded in male cyst nematodes (Carta & Baldwin, 1990), indicating that they cannot be involved in female finding or mating. All these sensory structures are not functionally defined in cyst nematodes.

The digestive system of cyst nematodes is highly adapted to the plant pathogenic life style of the cyst nematodes and consists of a hollow tube extending from the mouth and includes the stylet, oesophagus (pharynx), intestine, rectum and anus. Nutrients are only taken up from the syncytium, the feeding site which is induced in the host roots, through the hollow stylet. The stylet is also used to inject secretions from the gland cells into the plant and to destroy plant cell walls during migration in the plant root. It can be moved forward by protractor muscles. Since there are no retractor muscles, the stylet is moved back simply by the elasticity of the oesophagus and the body pressure. The oesophagus consists of procorpus, metacorpus, and the muscular basal bulb at the posterior end. The metacorpus pumps secretions from the gland cells into the plant but also nutrients from the plant into the intestine.

Cyst nematodes have one dorsal oesophageal (pharyngeal) gland cell and two subventral gland cells. The gland cells extend into a storage ampulla with a valve that controls the opening of these glands, with the dorsal gland opening into the procorpus near the base of the stylet and the subventral glands opening into the metacorpus. These three glands almost fill the body width of juvenile nematodes. The gland cells contain numerous secretory granules and are involved in the production of various effector proteins that are important for the repression of plant defence reactions and for the induction of syncytia (Hewezi & Baum, 2013). The subventral glands are well developed in pre-parasitic juveniles and contain numerous secretory granules. These granules gradually disappear and the subventral glands decrease in size while the juvenile is migrating through the roots. At the same time, the dorsal oesophageal gland increases in size (Tytgat et al., 2002), indicating that it is producing effectors that are involved in the induction of syncytia. These differences in size have also been observed by Wyss (1992). He also noticed that at the end of the second-stage juvenile (J2) the dorsal gland was considerably larger in juveniles developing into females than in those developing into males which supports the view that the secretions from this gland are an important source for effectors that are involved in the induction and maintenance of syncytia which are much larger if associated with female nematodes than those associated with male nematodes (see below). A method has recently been published that allows the isolation of dorsal and subventral gland cells (Maier, Hewezi, Peng, & Baum, 2013). It will now be possible to analyse the function of these glands at the transcriptomic, proteomic and metabolomic level in detail and to get a complete catalogue of all effectors produced in these glands (Hewezi & Baum, 2013).

Cyst nematodes show sexual dimorphism. The males are vermiform, similar to the juveniles, but bigger. They have a pair of copulatory spicules that can be unsheathed from the cloaca with the help of protractor muscles and moved back with retractor muscles. During copulation the males move their spicules into the vulva of the female to transfer the sperm which is produced in a single testis. The reproductive system of female nematodes consists of two genital branches with ovary, followed by an oviduct, spermatheca for holding the sperm delivered by the male, and uterus. These branches connect to a vagina and finally a vulva. Cyst nematodes may lay their eggs within a gelatinous matrix or may retain them in their body, depending on the species. In case of *H. schachtii*, for instance, almost all eggs are retained within the female body while *H. glycines* females will lay a large number of eggs within the gelatinous matrix. When all eggs are developed, the female cuticle hardens, supported by the activity of polyphenol oxidase and the body is turned into the cyst which contains the eggs and retains the female shape. Cysts are usually easily visible by the naked eye.

3. HATCHING

The life cycle of cyst nematodes (Figure 2) starts with hatching of the J2 which are the dormant stage of the cyst nematodes and are protected by the cyst, the dead body of the female. The J2 not only have to leave the eggs

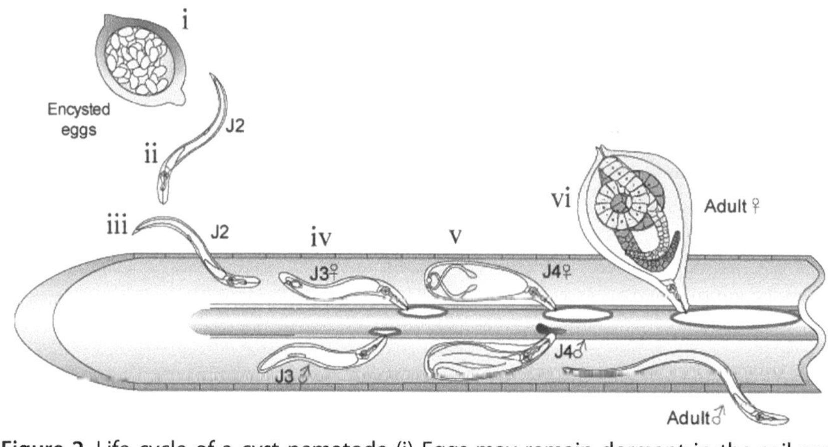

Figure 2 Life cycle of a cyst nematode (i) Eggs may remain dormant in the soil protected within the tanned cyst for many years. (ii) Under favourable conditions the J2 hatches and migrates toward a host root. (iii) The J2 penetrates the root and migrates intracellularly through the cortex toward the vascular cylinder where it initiates formation of a feeding site. Sex is determined toward the end of the J2 stage (iv and v). A multinucleate feeding site (syncytium) is established by cell wall dissolution. The female enlarges while the motile, vermiform adult male develops within the J4 cuticle. The male does not feed after the J3 stage and its syncytium begins to degrade. (vi) The male leaves the root and fertilizes the adult female, which grows to rupture the root surface. Eggs develop within the female body wall, which tans to form the cyst. *Reproduced from Lilley, Atkinson, and Urwin (2005), with permission.*

but also the cyst in those cases where the eggs are contained in a cyst. This is sometimes called emergence but often the whole event is called hatching. The dry cyst does not contain free water and can therefore survive freezing (Wharton & Ramløv, 1995). Within the cyst the nematode embryo develops to the first-stage juvenile (J1) which moults within the egg to the J2. While the cyst is a protective container for the eggs that were produced by the female, the egg protects the juveniles until they hatch. The J2 nematodes are thus protected by a double system. The eggshell has three layers: an inner lipid layer, a middle chitin layer and an outer vitelline layer (Burgwyn, Nagel, Ryerse, & Bolla, 2003). The lipid layer is semipermeable and allows water, small ions and gases to pass through. The chitin layer consists of proteins with a chitin microfibril core that provides strength to the egg but is also flexible to some extent to allow movement of the juveniles inside the egg. The outer vitelline layer is important during fertilization of the egg.

Hatching can be stimulated by root exudates (also called leachates or diffusates) from host plants. The dependence on certain host factors for hatching varies between cyst nematode species and is to some extent related to the host range. While species such as *H. schachtii* with a rather wide host range and *H. avenae* hatch to a large extent in water, *H. glycines* is partly dependent on root exudates and other species such as the potato cyst nematodes *G. rostochiensis* and *G. pallida* are almost entirely dependent on signals from their host plants (Perry, 2002) but there will still be a small proportion of J2 that hatch spontaneously in the absence of a host crop in the spring (Devine, Dunne, O'Gara, & Jones, 1999). The level of spontaneous hatch in *G. pallida* is lower than in *G. rostochiensis* (Turner & Evans, 1998). Ready hatching in water of *H. schachtii* and other species has been explained with a lower osmotic pressure of the fluid within their eggs as compared to *G. rostochiensis*. Juveniles of *H. schachtii* were found to move in 0.3 M sugar solutions (Perry, Clarke, & Hennessy, 1980). The juveniles within the egg survive in the trehalose containing perivitelline fluid in a partly dehydrated condition (Womersley & Smith, 1981). Trehalose-mediated anhydrobiosis is not specific to nematodes but has also been found in various other animals, for instance, tardigrades and rotifers. The trehalose concentration can be as high as 0.34 M trehalose in case of *G. rostochiensis* and even 0.5 M in case of *Heterodera goettingiana*. Since a high trehalose concentration inside the eggs prevents movement of the juvenile nematodes, a reduction of the trehalose concentration is a prerequisite for hatching. Hatching factors induce a Ca^{2+}-dependent change in the permeability of the inner lipid layer of the egg, leading to an influx of water and an efflux of trehalose, thus reducing the trehalose concentration. The J2 also take up water and become metabolically active and the lowered trehalose concentration within the egg allows their movement. Recent transcriptome analysis of *G. pallida* has shown that 526 genes were upregulated but only 6 downregulated at the transition from encysted eggs (containing dormant J2) and hatched J2 nematodes. The upregulated genes included 11 which coded for poly-A polymerases, enzymes that add the poly-A tail to pre-mRNAs. This reflects the general increase in gene transcription when the J2 emerge from dormancy (Cotton et al., 2014). After becoming mobile, the J2 use their stylet to break the eggshell. Juveniles of *G. rostochiensis* cut a slit in the eggshell through which they hatch from the egg. Although the pharyngeal gland cells become packed with granules before the J2 leave the egg, there is little evidence that enzymes from these glands are involved in supporting the stylet cuts in opening the egg. One exception is the expression of a chitinase gene which is induced in *G. pallida* hatched J2 as compared to dormant J2 which lead the authors to speculate that the encoded chitinase might be involved in degradation of the middle chitin layer of the eggshell (Cotton et al., 2014).

The soybean cyst nematode is one species which is partly dependent on hatching factors, chemical signals from the host plant that initiate hatching of the juveniles (Tsutsumi & Samurai, 1966). A hatching factor that stimulates hatch of *H. glycines* has been isolated from kidney bean roots and identified as glycinoeclepin A (Figure 3), a terpenoid (Masamune, Anetai, Takasugi, & Katsui, 1982). Roots were collected from a 1 ha field and resulted in just 50 μg of the substance, indicating that it should be active at very low concentrations. Indeed, glycinoeclepin A stimulates hatching at concentrations as low as 10^{-11} to 10^{-12} g/ml. Hatching factors could be used as agrochemicals to control nematodes, for instance, by applying glycinoeclepin A to fields when grown with a non-host crop. This should induce the hatching of *H. glycines* cysts, however, the juveniles would be unable to find a suitable host. It might also be possible to induce hatching in the spring before sowing soybean or in autumn after the crop has been harvested. The hatched juveniles would again be unable to find a suitable host root and would die after their nutrient resources are exhausted. Such a treatment of the fields with a hatching factor should lead to a significant reduction of the number of cysts in the soil. In line with such considerations, chemical synthesis of glycinoeclepin A has first been achieved by Murai, Tanimoto, Sakamoto, and Masamune (1986). Others have reported different synthetic pathways

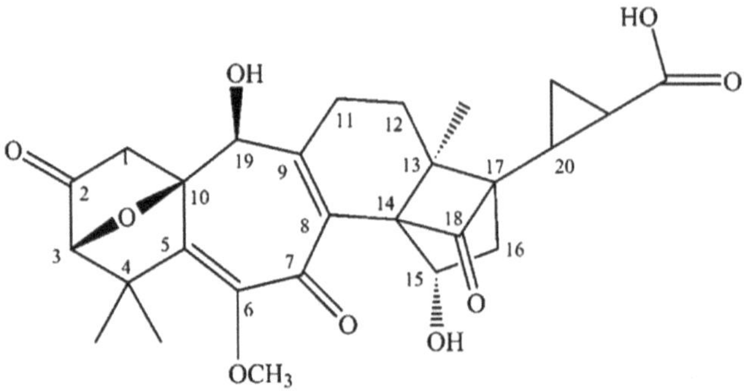

Solanoeclepin A

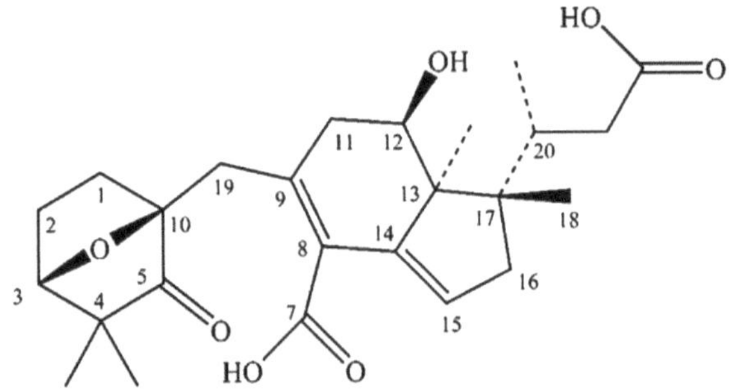

Glycinoeclepin A

Figure 3 Structure of solanoeclepin A and glycinoeclepin A. *Reproduced from Schenk et al. (1999), with permission.*

(Shiina, Tomata, Miyashita, & Tanino, 2010) but the costs at the moment would not allow the use of glycinoeclepin A as an agrochemical. Another possibility might be the use of active glycinoeclepin A analogues and their synthesis has already been reported but not their use as agrochemicals (Giroux & Corey, 2008; Kraus, Johnston, Kongsjahju, & Tylka, 1994).

A cyst nematode species which is largely dependent on hatching factors to stimulate hatching from the cysts is *G. rostochiensis*. Therefore, hatching factors have been intensively studied in this species. A hatching factor similar to glycinoeclepin A for *H. glycines* has been isolated from potato roots and named solanoeclepin A (Schenk et al., 1999) (Figure 3). Chemical synthesis of this compound has also been reported recently (Tanino et al., 2011). Two other potato-produced hatching factors are the glycoalkaloids α-solanine and α-chaconine. Potato root exudates contain still other hatching factors but these have not been fully characterized (Devine, Byrne, Maher, & Jones, 1996). Artificial hatching inducers have also been identified and include metavanadate and picrolonic acid (Clarke & Shepherd, 1966). The latter induced hatching of *G. rostochiensis* but not that of *G. pallida* and *Globodera tabacum* (Greet, 1974). A list of artificial hatching factors can be found in Sharma and Sharma (1998).

If there are hatching inducers, there should also be hatching inhibitors. Such metabolites which can antagonize the hatch of potato cyst nematodes induced by hatching inducers have been found in root exudates of non-host species, such as asparagus (Takasugi, Yachida, Anetai, Masamune, & Kagasawa, 1975) and white mustard (Forrest & Farrer, 1983). Hatching inhibitors were also found in exudates of young potato plants and partly resolved by gel permeation chromatography on Sephadex G-10 (Byrne, Twomey, Maher, Devine, & Jones, 1998). Hatching inhibitors and hatching inducers from potato roots thus act together to fine-tune the hatching of potato cyst nematodes only when the potato plant has reached a certain age. Also hatching inhibitors could be envisioned as environmentally safe agrochemicals against cyst nematodes, for instance, asparagusic acid. Asparagusic acid is not only a hatching inhibitor for *G. rostochiensis* and *H. glycines*, even in the presence of hatching inducers, but is also nematicidal to the J2 of *G. rostochiensis* and other nematodes (Takasugi et al., 1975). However, it seems that there have no attempt being made to use this substance, which is not toxic to humans, as an agrochemical (Mitchell & Waring, 2014). A number of different chemicals have also been identified as hatching inhibitors (Table 8.3 in Sharma & Sharma, 1998).

A third class of hatching factors has been identified in root exudates of potatoes (Byrne et al., 1998). These are called hatching factor stimulants. They have no hatching-inducing activity by themselves but can enhance the activity of hatching inducers.

While these different compounds are perceived by the nematodes as hatching factors, it is not known which primary function they have for the plants that produce them. Especially hatching inducers must have an important function for the plant. In case of hatching inhibitors, it could be possible that these are resistance factors of the plant, but at current there is no proof for such a function. However, one could envision that crop plants such as soybean or potato which strongly suffer from cyst nematode infections might be engineered to exude hatching inhibitors from their roots. Especially in case of the potato cyst nematodes which have a very narrow host range, the use of trap crops or antagonistic crops is being investigated. Trap crops induce the hatching of the nematodes but are destroyed before the nematodes are able to complete their life cycle thus reducing the nematode inoculum for the potato crop. Various plant species have been tested for their use as trap crops against potato cyst nematodes. Scholte (2000) found that *Solanum sisymbriifolium*, *Solanum quitoense*, *Solanum macrocarpon*, *Solanum viarum* and *Solanum mauritianum* combined complete resistance to *G. pallida* with a strong hatching effect on potato cyst nematodes. However, of these only *S. sisymbriifolium* grew well under the climatic conditions in Holland. A strong hatching effect was also found for *Solanum nigrum* but this species was not as resistant as the aforementioned *Solanum* species. Another problem with *S. nigrum* is that it can be a host for *Phytophthora infestans* which is another dangerous disease of potatoes (Lebecka, 2008). The host range to other pathogens must also be considered for other potential trap crops.

A good trap crop is a non-host which induces the hatching of the nematodes such as *S. sisymbriifolium* which is used against potato cyst nematodes in Europe (Dias, Conceição, Abrantes, & Cunha, 2012). However, a widespread use of *S. sisymbriifolium* could also lead to increased problems with *P. infestans* because it is also a host for this pathogen (Flier, 2003). *Solanum sisymbriifolium* produces hatching factors but these are different from those produced by potato roots. Solanoeclepin A was not detected in *S. sisymbriifolium* root exudates. *Globodera pallida* was not only attracted to but also invaded the roots of *S. sisymbriifolium*, however, the nematodes were unable to complete their life cycle (Sasaki-Crawley et al., 2010). The ideal trap crop would be one that is also a crop that could be harvested. In line with this, Franco, Main, and Oros (1999) have tested various lines of barley, quinoa, oca and others for their effect on the hatching of potato cyst nematodes. They found several lines that induced hatching at a level similar to that of potato while some antagonistic lines showed a permanent inhibitory effect. Such crops could be used in rotation with potato (Franco & Main, 2008).

4. HOST FINDING AND PENETRATION

The J2 are the mobile phase which can move in the soil to find the roots of a suitable host plant. At this stage the nematodes do not feed and they are thus solely dependent on their food reserves and must therefore locate a suitable host root and induce a feeding site before these food reserves have been exhausted. For the potato cyst nematodes *G. pallida* and *G. rostochiensis* this timeframe has been determined at 6–11 days under optimal conditions

(Robinson, Atkinson, & Perry, 1987). They are guided in their behaviour by stimuli originating from the host plants. It is generally assumed that they perceive these signals with their amphids. Different physical and chemical gradients (CO_2, temperature, pH, redox potential) might be used for host finding by the J2. According to Perry (2005) nematodes respond to three types of attractants. Long-distance attractants enable them to find a host plant for CO_2 which is produced by plants seems to be an important stimulus although it is not clear if CO_2 itself is perceived or rather the acidification by CO_2 (Wang, Bruening, & Williamson, 2009). However, CO_2 is also produced by decaying plant tissues and therefore other attractants are needed in addition. Having located the root area of a host plant, short-distance attractants come into play, which then direct the J2 to a certain root. Here CO_2 might again play a role but organic chemicals secreted by plant roots are probably more important. It is also possible that root volatiles might be perceived by the juvenile nematodes (Farnier et al., 2012). Finally, local attractants guide the J2 to a site on the root where it starts invading the root.

Once the J2 nematodes have reached a root they start to invade it, often behind the root tip at the elongation zone using their stylet but also cell wall degrading enzymes which are produced in the two subventral gland cells. These glands are highly active at the J2 stage but become inactive once the nematodes starts to induce its feeding site (Tytgat et al., 2002). The enzymes produced by these glands include pectate lyases and cellulases but also expansins (reviewed by Davis, Haegeman, & Kikuchi, 2011). The importance of enzymatic degradation of the plant cell walls during the migratory phase has been demonstrated by experiments which downregulated the expression of these genes with RNAi, resulting in lower infection efficiency (Chen, Rehman, Smant, & Jones, 2005; Vanholme et al., 2007). Supported by the secretions from the subventral gland cells, the J2 use their stylet to pierce the cell walls to produce a hole that allows them to enter the cell. They thus move intracellularly through the outer layers of the root which is in difference to the root knot nematodes which move intercellularly. This behaviour of cyst nematodes inside the root has been nicely documented by Wyss and colleagues (Wyss & Zunke, 1986).

5. INDUCTION OF A FEEDING SITE

Once the J2 reach the central cylinder they select a cambial or procambial cell which will become the initial syncytial cell (ISC). The J2 can probe several cells until it finds one that is suitable. It carefully inserts the stylet into a cell and waits for the cell response. If it encounters a strong defence reaction of the plant cell, such as a deposition of callose on the stylet (Golinowski, Sobczak, Kurek, & Grymaszewska, 1997; Sobczak et al., 1999) or if the protoplast collapses (Wyss, 1992) it will retract its stylet and probe another cell. This could be a neighbouring cell or the J2 could open the cell that showed a strong response with thrusts of the stylet and move through this cell to the next suitable candidate cell (Wyss, 1992). Once the J2 has found a suitable cell which will become the ISC the stylet is left inserted for approximately 7 h. During this preparation phase no pumping of the nematode metacorpal bulb was observed, indicating that no uptake from the plant cell and no injection of secretions occurred (Wyss, 1992). Subsequently, the stylet is withdrawn and reinserted to inject secretions into the cytoplasm. The nematode then starts to take up food from the feeding site. This occurs in phases which usually last for several hours and can be divided into three stages (Wyss, 1992). During stage II the stylet is withdrawn and reinserted. In stage II the nematode injects secretions into the plant cell which leads to the formation of feeding tubes. The ultrastructure of feeding tubes has been studied using electron microscopy for several nematode species. In case of *H. schachtii* it was found that the tubes have a size of about 1 × 4 μm (Sobczak et al., 1999). They are hollow tubes with an electron dense wall and an electron translucent lumen and are newly produced from secretions of the nematode during each feeding cycle. To what extent plant-derived proteins or other molecules might be involved in the formation of feeding tubes is unknown. It is thought that the feeding tubes are composed of proteins because they are osmiophilic and osmium rather labels proteins than polysaccharides (Berg, Fester, & Taylor, 2008). In stage I the nematode finally withdraws nutrients from the plant making use of the before produced feeding tubes. Afterwards the feeding tubes are disconnected from the stylet and remain in the cytoplasm for some time. The feeding tubes of cyst nematodes act as a kind of molecular sieve, allowing only molecules up to approximately 30 kDa to pass through (Böckenhoff & Grundler, 1994; Eves-van den Akker et al., 2014). Müller, Rehbock, and Wyss (1981) have calculated that an adult *H. schachtii* female can withdraw four times the volume of its syncytium per day, indicating that the nematodes might take up solutes containing all kinds of metabolites and small proteins.

It is generally believed that the sedentary parasitic nematodes produce effectors in their gland cells which will be introduced into the ISC to manipulate developmental processes within the plant and also to suppress resistant reactions of the plant (Hewezi & Baum, 2013). As biotrophic pathogens, it is important for the nematode survival to avoid a hypersensitive response of the plant to keep the ISC and the cells that are subsequently incorporated into the syncytium alive. The J2 can only induce a syncytium once; if they fail the nematode has to die. After inducing a syncytium the nematode becomes sedentary and will feed from the syncytium which will be its only source of nutrients.

If the nematode has successfully induced an ISC, it will then develop into a syncytium (Figure 4(A)) by integrating up to a few hundred surrounding cells. This is accomplished by cell wall modifying proteins such as expansins and cell wall degrading enzymes such as cellulases and pectinases produced by the plant itself, which lead to partial cell wall dissolutions between the syncytial elements (Figure 4(B)). Cell wall degrading enzymes are now no longer produced by the nematode and play thus no role in this process. Accordingly, the activity of the subventral gland cells which have produced cell wall degrading enzymes during infection of the root is reduced while the dorsal gland cell enlarges to produce various effectors.

The development of the syncytium is accompanied by drastic changes in cell morphology. Nuclei enlarge by endoreduplication which leads to polyploidy (de Almeida Engler & Gheysen, 2013). The large central vacuole degrades, leaving only a large number of small vacuoles, while the cytoplasm

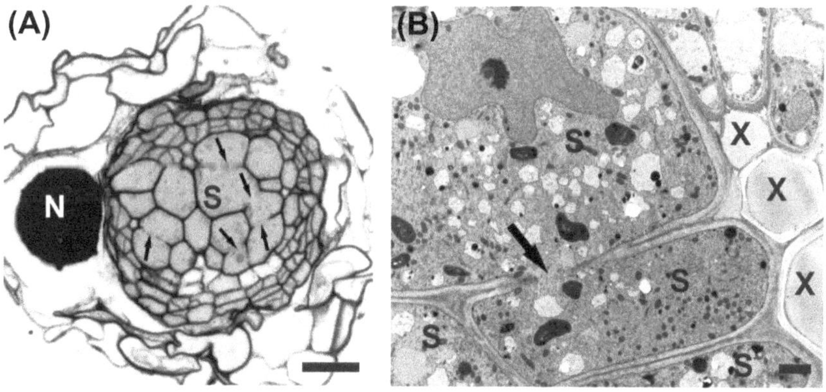

Figure 4 Structural features of cell walls of syncytia induced by *H. schachtii* in *Arabidopsis* roots at 5 days postinfection. (A) Anatomy of root containing syncytium (light microscopy). (B) Ultrastructure of root containing syncytium (transmission electron microscopy). Arrow indicates cell wall opening. N, nematode; S, syncytium; X, xylem vessel. Bars = 20 μm (A) and 1 μm (B). *From Bohlmann and Sobczak (2014).*

expands and becomes filled with numerous ribosomes, mitochondria, plastids and structures of the endoplasmic reticulum. Along with these visible changes of the syncytial elements also the metabolism changes. The syncytium has a high metabolic activity to provide the nutrients that are constantly withdrawn by the nematode. These changes in the metabolism are confirmed by several transcriptome studies of syncytia (Ithal et al., 2007; Szakasits et al., 2009).

Having established a syncytium, the nematodes can feed and develop further with two more moults to J3 and J4 stages and a final moult to the adult male and female nematodes. Sex determination in cyst nematodes is under environmental control and depends on nutrient supply by the host plant. Females have a higher nutrient demand than males because they grow much bigger and have to produce hundreds of eggs. The syncytia induced by male cyst nematodes are therefore smaller than those induced by female nematodes. In case of *H. schachtii* the volume of a syncytium which is associated with a male nematode has been calculated to be approximately 0.002 mm^3 while the volume of a syncytium associated with a female nematode was 0.026 mm^3, more than 10 times as large (Kerstan, 1969). An adult female of *H. schachtii* can withdraw four times the volume of the syncytium per day and it has been calculated that the total amount of food consumed by a female is 29 times that of a male (Müller et al., 1981). Therefore, more females can develop under favourable conditions while under unfavourable conditions more males develop (Grundler, Betka, & Wyss, 1991; Müller, 1985).

6. REPRODUCTION AND LIFE CYCLE

The females stay connected to their syncytium while the males become mobile again after a short J4 phase. They leave the roots to find a female for fertilization. They are guided by sex pheromones which are produced by the females. Several attempts have been made to identify such sex pheromones for cyst nematodes but with limited success (Aumann, Dietsche, Rutencrantz, & Ladehoff, 1998). Only for *H. glycines* vanillic acid was identified as a substance that has sex pheromone activity (Jaffe, Huettel, Demilo, Hayes, & Rebois, 1989). In the meantime chemical analysis has become more sensitive and it might now be possible to clearly identify sex pheromones for cyst nematodes.

The females grow bigger and their body enlarges due to the hundreds of eggs that are produced. Within the eggs the nematode develops to the J1 and finally the J2 stage. A cyst can contain eggs with every developmental stage from one-celled embryos to J1 and fully developed J2, depending on the age of the cyst (Tylka et al., 1993). The female eventually dies and her cuticle hardens and becomes the cyst that protects the eggs. Under favourable conditions the new generation of juveniles will hatch again to start a new cycle. Some of the eggs can be secreted in a gelatinous matrix which is especially found in *H. glycines*. A female of this species can produce 600 eggs and up to 200 can be deposited outside the cyst (Niblack, Lambert, & Tylka, 2006). Eggs outside the cyst have little protection and the nematodes in these eggs could most likely not survive for long and will hatch during the growing season in which they were produced.

The duration of the life cycle depends to a large degree on the soil temperature. In tropical regions or during the summer time in temperate regions a life cycle may be completed within 3 weeks. In case of *H. glycines* the optimal temperature for hatching is 24 °C (Slack & Hamblen, 1961). Alston and Schmitt (1988) found that *H. glycines* needed 4 weeks to complete its life cycle in the field from early June to early July and late in the season from September to November while in July and August only 3 weeks were needed. The number of generations per year depends on the duration of the life cycle but of course also on the availability of suitable host plants. In temperate regions it will mostly not be possible to complete more than one or two generations per year. In tropical regions with high soil temperatures and host plants grown throughout the year many more generations can develop during 1 year. For *H. oryzicola*, a duration of the life cycle of 30 days and 12 generations per year has been reported (Jayaprakash & Rao, 1982).

7. HOST RANGE

The host range of cyst nematodes differs between the species. One species with a large host range is *H. schachtii*. Steele (1965) tested 535 species from 283 genera in 49 plant families. He found that from these 218 species within 95 genera in 23 plant families were host plants. The largest numbers of host plants were found in the families *Brassicaceae* and *Chenopodiaceae*. Approximately 80% of the species within these families were scored as hosts. Later, the model plant *Arabidopsis thaliana* was also found to be a host plant for *H. schachtii* and this interaction has been developed into a model system for cyst nematode research (Sijmons, Grundler, von Mende, Burrows, &

Wyss, 1991). Roots of non-host plants can also be invaded by juveniles of *H. schachtii* but they usually do not develop in these plants. However, in rare cases also on these non-host plants single females have been reported (Steele, 1971). The large host range of the beet cyst nematode, together with the ability of the cysts to survive in soil for many years, makes control of this species through crop rotation difficult.

8. SURVIVAL

The cysts can survive in soil for many years until the J2 hatch under favourable conditions. *Heterodera glycines* eggs could, for instance, survive in soil for 11 years (Inagaki & Tsutsumi, 1971) but it has also been reported that potato cyst nematodes can survive in the cyst for as long as 20 years (Jones, Tylka, & Perry, 1998) which is one reason that cyst nematodes are such dangerous parasites and difficult to eradicate. However, the number of surviving nematodes under field conditions decreases each year if no host plants are available, due to spontaneous hatching and nematophagous fungi or other enemies. For *G. rostochiensis* it was found that viability of the eggs declined by 50% during the first year and by 40% during the second year. Approximately 80% of the loss was attributed to spontaneous hatching (Devine et al., 1999).

The soil is inhabited by myriads of animals and microorganisms. Cyst nematodes in the soil may be prey for various other predatory nematodes (Khan & Kim, 2007), insects, mites and tardigrades. They can also be parasitized themselves by fungi, attacking eggs, females and cysts (Kerry, 1988). Most spectacular are those fungi which have developed sophisticated hyphal structures to even capture nematodes moving through the soil (Nordbring-Hertz, Jansson, & Tunlid, 2006, pp. 1–11). All these antagonists are interesting for a biologist but are only of limited value for the control of cyst nematodes and other plant pathogenic nematodes in agriculture. However, the study of nematophagous fungi might reveal nematicidal proteins that could be used in transgenic approaches for plant resistance against nematodes (Yang et al., 2005, 2011).

9. PLANT RESISTANCE AGAINST CYST NEMATODES

Plants resistance is often governed by specific resistance genes (Dangl & Jones, 2001) that interact with avirulence genes of the pathogen, leading to a gene-for-gene relationship (Flor, 1971). Such a gene-for-gene relationship has also been found for resistance against plant pathogenic cyst nematodes which introduce effectors into plants with the help of their stylet. The gene-for-gene concept implies that pathogens develop races that differ in avirulence genes for a certain plant resistance gene. Such a relationship has been confirmed for *G. rostochiensis* and potato carrying the H1 resistance gene (Janssen, Bakker, & Gommers, 1991). Pathogenic races have been identified for those cyst nematode species which infect crops of agronomic importance such as *G. pallida, G. rostochiensis* and *H. glycines* (Cook & Rivoal, 1998). These pathotypes can only be distinguished with the help of a test panel of different species or cultivars (host differentials). Although not formally proven, these interactions will most likely also follow the gene-for-gene relationship. The emergence of new pathogenic races which are able to overcome the resistance of established cultivars is a constant threat to agriculture (Castagnone-Sereno, 2002).

The first plant resistance gene that was cloned is $Hs1^{Pro-1}$ from sugar beet against *H. schachtii* (Cai et al., 1997). It encodes an atypical resistance protein with a leucine-rich repeat (LRR) protein which is probably anchored in the cell membrane with a transmembrane domain. A second extracellular resistance protein involved in resistance against cyst nematodes was recently identified as Cf-2 (Lozano-Torres et al., 2012) which had before been identified as involved in resistance against the fungal pathogen *Cladosporium fulvum* (Krüger et al., 2002). Cf-2 guards the apoplastic Rcr3 protease according to the guard hypothesis (Dangl & Jones, 2001) against the effector Gr-VAP1 which is produced in the subventral glands. Secretions of the subventral glands are especially important during host invasion. One might therefore expect that Cf-2-mediated resistance leads to an early response against invading nematode juveniles but this was not the case. A hypersensitive response in cells surrounding the nematodes and in the syncytium was only observed starting at 7 days after infection (Lozano-Torres et al., 2012). The reason for this discrepancy is currently unknown but it could indicate that secretions from the subventral glands are also introduced into syncytia.

A few other resistance genes against cyst nematodes (reviewed by Kandoth & Mitchum, 2013) have been cloned from tomato (Ernst et al., 2002) and potato (Paal et al., 2004; van der Vossen et al., 2000) and these all belong to the widespread nucleotide binding (NB)-LRR class which code for intracellular proteins, indicating that these might act against effectors produced in the dorsal gland cell. More resistance genes from different crop plants or their wild relatives have been identified and mapped but not yet cloned (Tomczak, Koropacka, Smant, Goverse, & Bakker, 2008).

Resistance genes from soybean against *H. glycines* at the *Rhg1* and *Rhg4* loci have been chased by different research groups for many years. Now, after many unsuccessful attempts, it was found that resistance at these loci might depend on genes that do not belong to one of the canonical groups of resistance genes. Resistance at the *Rhg4* locus involves a serine hydroxymethyltransferase (Liu et al., 2012). Even more surprising, resistance at the *Rhg1* locus involves copy number variation of a stretch of three different genes in a 31-kilobase segment, one of them coding for an amino acid transporter (Cook et al., 2012). Susceptible varieties contain one copy of the 31-kilobase segment per haploid genome while resistant varieties can have 10 copies. These results emphasize the delicate balance between cyst nematodes and their host plants to keep the syncytium functional. Many plant genes are differentially regulated in syncytia for its development and maintenance. Downregulation of genes that are important for the function of syncytia can therefore result in reduced susceptibility (Ali et al., 2013; Siddique et al., 2009). It seems that the resistance at the *Rhg1* and *Rhg4* loci might also lead to resistance by disturbing the function of syncytia.

The response of resistant plants against cyst nematodes typically involves a late response of syncytia, leading to a slow and delayed hypersensitive response (Golinowski & Magnusson, 1991; Soliman, Sobczak, & Golinowski, 2005; Wyss, Stender, & Lehmann, 1984). Syncytia thus are functional for some time, which allows the development of males but not of females which have a much higher nutrient demand (see above). Such plants therefore lead to a reduction of the nematode population because very few females can develop to start the next infection cycle (Müller, 1998). A different response has been described for *Solanum canasense* against *G. pallida*. Only few juveniles are able to invade the roots and induce syncytia which are malfunctional (Castelli et al., 2006). This resistance seems to operate at different levels but the molecular mechanism behind it is unknown but could perhaps be mimicked in transgenic approaches for nematode resistance.

Induced resistance has been reported against *G. pallida* and *H. schachtii* using resistance-inducing bacteria (Hasky-Günther, Hoffmann-Hergarten, & Sikora, 1998; Reitz et al., 2000). Resistance can also be induced by chemicals. Beta-amino-butyric acid (BABA) was used to induce resistance against cereal cyst nematodes in wheat (Oka & Cohen, 2001). While resistance against other pathogens can be induced by jasmonic acid or salicylic acid, their role in plant interactions with cyst nematodes is not clear yet. Phytoalexins were found to be induced in roots of resistant but not susceptible soybean cultivars after infection with *H. glycines* (Huang & Barker, 1991). However, their role in resistance against cyst nematodes has to be further investigated.

REFERENCES

Ali, M. A., Plattner, S., Radakovic, Z., Wieczorek, K., Elashry, A., Grundler, F. M., et al. (2013). An *Arabidopsis* ATPase gene involved in nematode-induced syncytium development and abiotic stress responses. *The Plant Journal, 74*(5), 852–866. http://dx.doi.org/10.1111/tpj.12170.

de Almeida Engler, J., & Gheysen, G. (2013). Nematode-induced endoreduplication in plant host cells: why and how? *Molecular Plant–Microbe Interactions, 26*(1), 17–24. http://dx.doi.org/10.1094/mpmi-05-12-0128-cr.

Alston, D. G., & Schmitt, D. P. (1988). Development of *Heterodera glycines* life stages as influenced by temperature. *Journal of Nematology, 20*(3), 366–372.

Aumann, J., Dietsche, E., Rutencrantz, S., & Ladehoff, H. (1998). Physico-chemical properties of the female sex pheromone of *Heterodera schachtii* (Nematoda: *Heteroderidae*). *International Journal of Parasitology, 28*(11), 1691–1694. pii:S0020751998001416.

Baldwin, J. G. (1985). Fine structure of the phasmid of second-stage juveniles of *Heterodera schachtii* (Tylenchida: Nematoda). *Canadian Journal of Zoology, 63*(3), 534–542.

Bargmann, C. I. (2006). Chemosensation in *C. elegans*. *WormBook*, 1–29. http://dx.doi.org/10.1895/wormbook.1.123.1.

Berg, R. H., Fester, T., & Taylor, C. G. (2008). Development of the root-knot nematode feeding cell. In R. H. Berg, & C. G. Taylor (Eds.), *Cell biology of plant nematode parasitism*. Berlin: Springer.

Böckenhoff, A., & Grundler, F. M. W. (1994). Studies on the nutrient uptake by the beet cyst nematode *Heterodera schachtii* by in situ microinjection of fluorescent probes into the feeding structures in *Arabidopsis thaliana*. *Parasitology, 109*, 249–254.

Bohlmann, H., & Sobczak, M. (2014). The plant cell wall in the feeding sites of cyst nematodes. *Frontiers in Plant Science, 5*, 89. http://dx.doi.org/10.3389/fpls.2014.00089.

Burgwyn, B., Nagel, B., Ryerse, J., & Bolla, R. I. (2003). *Heterodera glycines*: eggshell ultrastructure and histochemical localization of chitinous components. *Experimental Parasitology, 104*(1–2), 47–53. pii:S0014489403001188

Byrne, J. M., Twomey, U., Maher, N. J., Devine, K. J., & Jones, P. W. (1998). Detection of hatching inhibitors and hatching factor stimulants *Globodera rostochiensis* in potato root leachate. *Annals of Applied Biology, 132*, 463–472.

Cai, D., Kleine, M., Kifle, S., Harloff, H. J., Sandal, N. N., Marcker, K. A., et al. (1997). Positional cloning of a gene for nematode resistance in sugar beet. *Science, 275*(5301), 832–834.

Carta, L. K., & Baldwin, J. G. (1990). Phylogenetic implications of phasmid absence in males of three genera in heteroderinae. *Journal of Nematology, 22*(3), 386–394.

Castagnone-Sereno, P. (2002). Genetic variability of nematodes: a threat to the durability of plant resistance genes? *Euphytica, 124*, 193–199.

Castelli, L., Bryan, G., Blok, V. G., Ramsay, G., Sobczak, M., Gillespie, T., et al. (2006). Investigations of *Globodera pallida* invasion and syncytia formation within roots of the susceptible potato cultivar Désirée and resistant species *Solanum canasense*. *Nematology, 8*, 103–110.

Chen, Q., Rehman, S., Smant, G., & Jones, J. T. (2005). Functional analysis of pathogenicity proteins of the potato cyst nematode *Globodera rostochiensis* using RNAi. *Molecular Plant–Microbe Interactions, 18*(7), 621–625. http://dx.doi.org/10.1094/Mpmi-18-0621.

Clarke, A. J., & Shepherd, A. M. (1966). Picrolonic acid as a hatching agent for the potato cyst nematode, *Heterodera rostochiensis* woll. *Nature, 211*(5048), 546.

Cook, D. E., Lee, T. G., Guo, X., Melito, S., Wang, K., Bayless, A. M., et al. (2012). Copy number variation of multiple genes at *Rhg1* mediates nematode resistance in soybean. *Science, 338*(6111), 1206–1209. http://dx.doi.org/10.1126/science.1228746.

Cook, R., & Rivoal, R. (1998). Genetics of resistance and parasitism. In S. B. Sharma (Ed.), *The cyst nematodes* (pp. 322–352). London: Chapman & Hall.

Cotton, J. A., Lilley, C. J., Jones, L. M., Kikuchi, T., Reid, A. J., Thorpe, P., et al. (2014). The genome and life-stage specific transcriptomes of *Globodera pallida* elucidate key aspects of plant parasitism by a cyst nematode. *Genome Biology, 15*(3), R43. http://dx.doi.org/10.1186/gb-2014-15-3-r43.

Curtis, R. H. (2007). Plant parasitic nematode proteins and the host parasite interaction. *Briefings in Functional Genomics and Proteomics, 6*(1), 50–58. http://dx.doi.org/10.1093/bfgp/elm006.

Dangl, J. L., & Jones, J. D. (2001). Plant pathogens and integrated defence responses to infection. *Nature, 411*(6839), 826–833. http://dx.doi.org/10.1038/35081161.

Davis, E. L., Haegeman, A., & Kikuchi, T. (2011). Degradation of the plant cell wall by nematodes. In J. Jones, G. Gheysen, & C. Fenoll (Eds.), *Genomics and molecular genetics of plant-nematode interactions* (Vol. 557). Springer Netherlands.

Decraemer, W., & Hunt, D. J. (2006). Structure and classification. In R. N. Perry, & M. Moens (Eds.), *Plant nematology* (pp. 4–32). Oxfordshire: CABI1.

Devine, K. J., Byrne, J., Maher, N., & Jones, P. W. (1996). Resolution of natural hatching factors for golden potato cyst nematode, *Globodera rostochiensis*. *Annals of Applied Biology, 129*(2), 323–334. http://dx.doi.org/10.1111/j.1744-7348.1996.tb05755.x.

Devine, K. J., Dunne, C., O'Gara, F., & Jones, P. W. (1999). The influence of in-egg mortality and spontaneous hatching on the decline of *Globodera rostochiensis* during crop rotation in the absence of the host potato crop in the field. *Nematology, 1*, 637–645.

Dias, M. C., Conceição, I. L., Abrantes, I., & Cunha, M. J. (2012). *Solanum sisymbriifolium* – a new approach for the management of plant-parasitic nematodes. *European Journal of Plant Pathology, 133*(1), 171–179.

Dillman, A. R., & Sternberg, P. W. (2012). Entomopathogenic nematodes. *Current Biology, 22*(11), R430–R431. http://dx.doi.org/10.1016/j.cub.2012.03.047.

Dold, C., & Holland, C. V. (2011). Ascaris and ascariasis. *Microbes and Infection, 13*(7), 632–637. http://dx.doi.org/10.1016/j.micinf.2010.09.012.

Endo, B. Y. (1978). Feeding plug formation in soybean roots infected with the soybean cyst nematode. *Phytopathology, 68*, 1022–1031.

Endo, B. Y. (1980). Ultrastructure of the anterior neurosensory organs of the larvae of the soybean cyst nematode, *Heterodera glycines*. *Journal of Ultrastructural Research, 72*(3), 349–366.

Endo, B. Y. (1984). Ultrastructure of the esophagus of larvae of the soybean cyst nematode, *Heterodera glycines*. *Proceedings of the Helminthological Society of Washington, 51*(1), 1–24.

Ernst, K., Kumar, A., Kriseleit, D., Kloos, D. U., Phillips, M. S., & Ganal, M. W. (2002). The broad-spectrum potato cyst nematode resistance gene (*Hero*) from tomato is the only member of a large gene family of NBS-LRR genes with an unusual amino acid repeat in the LRR region. *The Plant Journal, 31*(2), 127–136.

Evans, K., & Rowe, J. A. (1998). Distribution and economic importance. In S. B. Sharma (Ed.), *The cyst nematodes* (pp. 1–30). Dordrecht: Kluwer Academic Publishers.

Eves-van den Akker, S., Lilley, C. J., Ault, J. R., Ashcroft, A. E., Jones, J. T., & Urwin, P. E. (2014). The feeding tube of cyst nematodes: characterisation of protein exclusion. *PLoS One, 9*(1), e87289. http://dx.doi.org/10.1371/journal.pone.0087289.

Farnier, K., Bengtsson, M., Becher, P. G., Witzell, J., Witzgall, P., & Manduric, S. (2012). Novel bioassay demonstrates attraction of the white potato cyst nematode *Globodera pallida* (Stone) to non-volatile and volatile host plant cues. *Journal of Chemical Ecology, 38*(6), 795–801. http://dx.doi.org/10.1007/s10886-012-0105-y.

Flier, W. G. (2003). Epidemiological importance of *Solanum sisymbriifolium, S. nigrum* and *S. dulcamara* as alternative hosts for *Phytophthora infestans*. *Plant Pathology, 52*, 595–603.

Flor, H. H. (1971). Current status of the gene-for-gene concept. *Annual Review of Phytopathology, 9*, 275–296.

Forrest, J. M. S., & Farrer, L. A. (1983). The response of eggs of the white potato cyst nematode *Globodera pallida* to diffusate from potato and mustard roots. *Annals of Applied Biology, 103*, 283–289.

Franco, J., & Main, G. (2008). Management of nematodes of andean tuber and grain crops. In A. Ciancio, & K. G. Mukerji (Eds.), *Integrated management and biocontrol of vegetable and grain crops nematodes* (pp. 99–117).

Franco, J., Main, G., & Oros, R. (1999). Trap crops as a component for the integrated management of *Globodera* spp. (potato cyst nematodes) in Bolivia. *Nematropica, 29*, 51–60.

Giroux, S., & Corey, E. J. (2008). Enantioselective synthesis of a simple benzenoid analogue of glycinoeclepin A. *Organic Letters, 10*(24), 5617–5619. http://dx.doi.org/10.1021/ol8024633.

Golinowski, W., & Magnusson, C. (1991). Tissue response induced by *Heterodera schachtii* (Nematoda) in susceptible and resistant white mustard cultivars. *Canadian Journal of Botany, 69*, 53–62.

Golinowski, W., Sobczak, M., Kurek, W., & Grymaszewska, G. (1997). The structure of syncytia. In C. Fenoll, F. M. W. Grundler, & S. Ohl (Eds.), *Cellular and molecular aspects of plant-nematode interactions* (pp. 80–97). Dordrecht: Kluwer Academic Publisher.

Greet, D. N. (1974). The response of five round cyst nematodes (*Heteroderidae*) to five artificial hatching agents. *Nematologica, 20*, 363–364.

Grundler, F., Betka, M., & Wyss, U. (1991). Influence of changes in the nurse cell system (Syncytium) on sex determination and development of the cyst nematode *Heterodera schachtii* – total amounts of proteins and amino-acids. *Phytopathology, 81*(1), 70–74.

Hasky-Günther, K., Hoffmann-Hergarten, S., & Sikora, R. A. (1998). Resistance against the potato cyst nematode *Globodera pallida* systemically induced by the rhizobacteria *Agrobacterium radiobacter* (G12) and *Bacillus sphaericus* (B43). *Fundamental and Applied Nematology, 21*, 511–517.

Hewezi, T., & Baum, T. J. (2013). Manipulation of plant cells by cyst and root-knot nematode effectors. *Molecular Plant–Microbe Interactions, 26*(1), 9–16. http://dx.doi.org/10.1094/MPMI-05-12-0106-FI.

Holterman, M., Karssen, G., van den Elsen, S., van Megen, H., Bakker, J., & Helder, J. (2009). Small subunit rDNA-based phylogeny of the Tylenchida sheds light on relationships among some high-impact plant-parasitic nematodes and the evolution of plant feeding. *Phytopathology, 99*(3), 227–235. http://dx.doi.org/10.1094/phyto-99-3-0227.

Huang, J. S., & Barker, K. R. (1991). Glyceollin I in soybean-cyst nematode interactions: spatial and temporal distribution in roots of resistant and susceptible soybeans. *Plant Physiology, 96*, 1302–1307.

Inagaki, H., & Tsutsumi, M. (1971). Survival of the soybean cyst nematode, *Heterodera glycines* Ichinohe (Tylenchida: *Heteroderidae*) under certain storing conditions. *Applied Entomology and Zoology, 6*, 156–162.

Ithal, N., Recknor, J., Nettleton, D., Maier, T., Baum, T. J., & Mitchum, M. G. (2007). Developmental transcript profiling of cyst nematode feeding cells in soybean roots. *Molecular Plant–Microbe Interactions, 20*(5), 510–525. http://dx.doi.org/10.1094/MPMI-20-5-0510.

Jaffe, H., Huettel, R. N., Demilo, A. B., Hayes, D. K., & Rebois, R. V. (1989). Isolation and identification of a compound from soybean cyst nematode, *Heterodera glycines*, with sex pheromone activity. *Journal of Chemical Ecology, 15*, 2031–2043.

Janssen, R., Bakker, J., & Gommers, F. J. (1991). Mendelian proof for a gene-for-gene relationship between virulence of *Globodera rostochiensis* and the H resistance gene in *Solanum tuberosum* ssp. *andigena* CPC 1673. *Revue de Nematologie, 14*, 213–219.

Jayaprakash, A., & Rao, Y. S. (1982). Life history and behaviour of the cyst nematode, *Heterodera oryzicola* Rao and Jayaprakash, 1978 in Rice (*Oryza sativa* L.). *Proceedings of the Indian Academie of Sciences (Animal Sciences), 91*(3), 283–295.

Jones, P. W., Tylka, G. C., & Perry, R. N. (1998). Hatching. In R. N. Perry, & D. J. Wright (Eds.), *The physiology and biochemistry of free living and plant parasitic nematodes* (pp. 181–212). Oxon: CAB International.

Kandoth, P. K., & Mitchum, M. G. (2013). War of the worms: how plants fight underground attacks. *Current Opinion in Plant Biology, 16*(4), 457–463. http://dx.doi.org/10.1016/j.pbi.2013.07.001.

Kerry, B. (1988). Fungal parasites of cyst nematodes. *Agriculture, Ecosystems and Environment, 24*, 293–305.

Kerstan, U. (1969). Die Beeinflussung des Geschlechterverhaltnisses in der Gattung Heterodera II. Minimallebensraum – Selektive Absterberate der Geschlechter – Geschlechterverhältnis (*Heterodera schachtii*). *Nematologica, 15*, 210–228 (in German).

Khan, Z., & Kim, Y. H. (2007). A review on the role of predatory soil nematodes in the biological control of plant parasitic nematodes. *Applied Soil Ecology, 35*, 370–379.

Kraus, G. A., Johnston, B., Kongsjahju, A., & Tylka, G. L. (1994). Synthesis and evaluation of compounds that affect soybean cyst nematode egg hatch. *Journal of Agricultural and Food Chemistry, 42*, 1839–1840.

Krüger, J., Thomas, C. M., Golstein, C., Dixon, M. S., Smoker, M., Tang, S., et al. (2002). A tomato cysteine protease required for Cf-2-dependent disease resistance and suppression of autonecrosis. *Science, 296*(5568), 744–747. http://dx.doi.org/10.1126/science.1069288.

Lebecka, R. (2008). Host–pathogen interaction between *Phytophthora infestans* and *Solanum nigrum*, *S. villosum*, and *S. scabrum*. *European Journal of Plant Pathology, 120*, 233–240.

Lilley, C. J., Atkinson, H. J., & Urwin, P. E. (2005). Molecular aspects of cyst nematodes. *Molecular Plant Pathology, 6*(6), 577–588. http://dx.doi.org/10.1111/j.1364-3703.2005.00306.x. MPP306 [pii].

Liu, S., Kandoth, P. K., Warren, S. D., Yeckel, G., Heinz, R., Alden, J., et al. (2012). A soybean cyst nematode resistance gene points to a new mechanism of plant resistance to pathogens. *Nature, 492*(7428), 256–260. http://dx.doi.org/10.1038/nature11651 ([pii]).

Lozano-Torres, J. L., Wilbers, R. H. P., Gawronski, P., Boshoven, J. C., Finkers-Tomczak, A., Cordewener, J. H. G., et al. (2012). Dual disease resistance mediated by the immune receptor Cf-2 in tomato requires a common virulence target of a fungus and a nematode. *Proceedings of the National Academy of Sciences of the United States of America, 109*(25), 10119–10124. http://dx.doi.org/10.1073/pnas.1202867109.

Maier, T. R., Hewezi, T., Peng, J., & Baum, T. J. (2013). Isolation of whole esophageal gland cells from plant-parasitic nematodes for transcriptome analyses and effector identification. *Molecular Plant–Microbe Interactions, 26*(1), 31–35. http://dx.doi.org/10.1094/mpmi-05-12-0121-fi.

Masamune, T., Anetai, M., Takasugi, M., & Katsui, N. (1982). Isolation of a natural hatching stimulus, glycinoeclepin A for the soybean cyst nematode. *Nature, 297*, 495–496.

Mitchell, S. C., & Waring, R. H. (2014). Asparagusic acid. *Phytochemistry, 97*, 5–10. http://dx.doi.org/10.1016/j.phytochem.2013.09.014.

Muller, R. (1971). Dracunculus and dracunculiasis. *Advances in Parasitology, 9*, 73–151.

Müller, J. (1985). Der Einfluß der Wirtspflanze auf die Geschlechtsdeterminierung bei *Heterodera schachtii*. *Mitteilungen aus der Biologischen Bundesanstalt für Land und Forstwirtschaft Berlin Dahlem, 226*, 46–63 (in German).
Müller, J. (1998). New pathotypes of the beet cyst nematode (*Heterodera schachtii*) differentiated on alien genes for resistance in beet (*Beta vulgaris*). *Fundamental of Applied Nematology, 21*, 519–526.
Müller, J., Rehbock, K., & Wyss, U. (1981). Growth of *Heterodera schachtii* with remarks on amounts of food consumed. *Revue de Nematologie, 4*, 227–234.
Murai, A., Tanimoto, N., Sakamoto, N., & Masamune, T. (1986). Total synthesis of glycinoeclepin A. *Journal of the American Chemical Society, 110*(6), 1985–1986.
Niblack, T. L., Lambert, K. N., & Tylka, G. L. (2006). A model plant pathogen from the kingdom Animalia: *Heterodera glycines*, the soybean cyst nematode. *Annual Review of Phytopathology, 44*, 283–303. http://dx.doi.org/10.1146/annurev.phyto.43.040204.140218.
Nicol, J. M., Elekcioglu, I. H., Bolat, N., & Rivoal, R. (2007). The global importance of the cereal cyst nematode (*Heterodera* spp.) on wheat and international approaches to its control. *Communications in Agricultural and Applied Biological Sciences, 72*(3), 677–686.
Nicol, J. M., Turrner, S. J., Coyne, D. L., den Nijs, L., Hockland, S., & Zahne Maafi, Z. (2011). Current nematode threats to world agriculture. In J. Jones, G. Gheysen, & C. Fenoll (Eds.), *Genomics and molecular genetics of plant–nematode interactions* (pp. 21–43). Dordrecht, The Netherlands: Springer.
Nordbring-Hertz, B., Jansson, H. B., & Tunlid, A. (2006). *Nematophagous fungi. Encyclopedia of life sciences*. Chichester: John Wiley & Sons.
Oka, Y., & Cohen, Y. (2001). Induced resistance to cyst and root-knot nematodes in cereals by DL-β-amino-*n*-butyric acid. *European Journal of Plant Pathology, 107*, 219–227.
Paal, J., Henselewski, H., Muth, J., Meksem, K., Menendez, C. M., Salamini, F., et al. (2004). Molecular cloning of the potato *Gro1-4* gene conferring resistance to pathotype Ro1 of the root cyst nematode *Globodera rostochiensis*, based on a candidate gene approach. *The Plant Journal, 38*(2), 285–297. http://dx.doi.org/10.1111/j.1365-313X.2004.02047.x.
Perry, R. N. (2002). Hatching. In D. L. Lee (Ed.), *The biology of nematodes* (pp. 147–169). London: Taylor and Francis.
Perry, R. N. (2005). An evaluation of types of attractants enabling plant-parasitic nematodes to locate plant roots. *Russian Journal of Nematology, 13*, 83–88.
Perry, R. N., Clarke, J., & Hennessy, J. (1980). The influence of osmotic pressure on the hatching of *Heterodera schachtii*. *Revue de Nematologie, 3*, 3–9.
Reitz, M., Rudolph, K., Schroder, I., Hoffmann-Hergarten, S., Hallmann, J., & Sikora, R. A. (2000). Lipopolysaccharides of *Rhizobium etli* strain G12 act in potato roots as an inducing agent of systemic resistance to infection by the cyst nematode *Globodera pallida*. *Applied and Environmental Microbiology, 66*(8), 3515–3518.
Robertson, L., Robertson, W. M., Sobczak, M., Helder, J., Tetaud, E., Ariyanayagam, M. R., et al. (2000). Cloning, expression and functional characterisation of a peroxiredoxin from the potato cyst nematode *Globodera rostochiensis*. *Molecular and Biochemical Parasitology, 111*(1), 41–49.
Robinson, M. P., Atkinson, H. J., & Perry, R. N. (1987). The influence of soil moisture and storage time on the motility, infectivity and lipid utilization of second stage juveniles of the potato cyst nematodes *Globodera rostochiensis* and *G. pallida*. *Revue de Nematologie, 10*, 343–348.
Sasaki-Crawley, A., Curtis, R., Birkett, M., Powers, S., Papadopoulos, A., Pickett, J., et al. (2010). Signalling and behaviour of potato cyst nematode in the rhizosphere of the trap crop, *Solanum sisymbriifolium*. *Aspects of Applied Biology, 103*, 45–51.

Schenk, H., Driessen, R. A. J., de Gelder, R., Goubitz, K., Nieboer, H., Brüggemann-Rotgans, I. E. M., et al. (1999). Elucidation of the structure of solanoeclepin A, a natural hatching factor of potato and tomato cyst nematodes, by single-crystal X-ray diffraction. *Croatica Chemica Acta, 72*(2–3), 593–606.
Scholte, K. (2000). Screening of non-tuber bearing Solanaceae for resistance to and induction of juvenile hatch of potato cyst nematodes and their potential for trap cropping. *Annals of Applied Biology, 136*(3), 239–246.
Semblat, J. P., Rosso, M. N., Hussey, R. S., Abad, P., & Castagnone-Sereno, P. (2001). Molecular cloning of a cDNA encoding an amphid-secreted putative avirulence protein from the root-knot nematode *Meloidogyne incognita. Molecular Plant–Microbe Interactions, 14*(1), 72–79.
Sharma, S. B., & Sharma, R. (1998). Hatch and emergence. In S. B. Sharma (Ed.), *The cyst nematodes* (pp. 191–216). Dordrecht: Kluwer Academic Publishers.
Shiina, Y., Tomata, Y., Miyashita, M., & Tanino, K. (2010). Total synthesis of glycinoeclepin A: generation of a Novel Bridgehead Anion species. *Chemistry Letters, 39*(8), 835–837.
Siddique, S., Endres, S., Atkins, J. M., Szakasits, D., Wieczorek, K., Hofmann, J., et al. (2009). Myo-inositol oxygenase genes are involved in the development of syncytia induced by *Heterodera schachtii* in *Arabidopsis* roots. *New Phytologist, 184*(2), 457–472. http://dx.doi.org/10.1111/j.1469-8137.2009.02981.x.
Sijmons, P., Grundler, F. M. W., von Mende, N., Burrows, P., & Wyss, U. (1991). *Arabidopsis thaliana* as a new model host for plant-parasitic nematodes. *The Plant Journal, 1*, 245–254.
Slack, D. A., & Hamblen, M. L. (1961). The effect of various factors on larval emergence from cysts of *Heterodera glycines. Phytopathology, 51*, 350–355.
Sobczak, M., Golinowski, W. A., & Grundler, F. M. W. (1999). Ultrastructure of feeding plugs and feeding tubes formed by *Heterodera schachtii. Nematology, 1*, 363–374.
Soliman, A. H., Sobczak, M., & Golinowski, W. (2005). Defence responses of white mustard, *Sinapis alba*, to infection with the cyst nematode *Heterodera schachtii. Nematology,* 7, 881–889.
Steele, A. E. (1965). The host range of the sugar beet nematode, *Heterodera schachtii* Schmidt. *Journal of the American Society of Sugar Beet Technologists, 13*(7), 573–603.
Steele, A. E. (1971). Invasion of non-host plant roots by Larvae of the sugarbeet nematode, *Heterodera schachtii. Journal of the American Society of Sugar Beet Technologists, 16*, 457–460.
Szakasits, D., Heinen, P., Wieczorek, K., Hofmann, J., Wagner, F., Kreil, D. P., et al. (2009). The transcriptome of syncytia induced by the cyst nematode *Heterodera schachtii* in *Arabidopsis* roots. *The Plant Journal, 57*(5), 771–784. http://dx.doi.org/10.1111/j.1365-313X.2008.03727.x [pii]:TPJ3727.
Takasugi, M., Yachida, Y., Anetai, M., Masamune, T., & Kagasawa, K. (1975). Identification of asparagusic acid as a nematicide occuring naturally in the roots of asparagus. *Chemistry Letters, 1975*(1), 43–44.
Tanino, K., Takahashi, M., Tomata, Y., Tokura, H., Uehara, T., Narabu, T., et al. (2011). Total synthesis of solanoeclepin A. *Nature Chemistry, 3*(6), 484–488. http://dx.doi.org/10.1038/nchem.1044.
Tomczak, A., Koropacka, K., Smant, G., Goverse, A., & Bakker, E. (2008). Resistant plant responses. In R. H. Berg, & C. G. Taylor (Eds.), *Cell biology of plant nematode parasitism* (pp. 83–113). Berlin: Springer.
Tsutsumi, M., & Samurai, K. (1966). Influence of root diffusates of several host and non-host plants on the hatching of the soybean cyst nematode (*Heterodera glycines* ICHINOHE 1952). *Japanese Journal of Applied Entomology and Zoology, 10*, 129–137.
Turner, S. J., & Evans, K. (1998). The origins, global distribution and biology of potato cyst nematodes (*Globodera rostochiensis* (Woll.) and *G. pallida* (Stone)). In R. J. Marks, & B. B. Brodie (Eds.), *Potato cyst nematodes, biology, distribution and control* (pp. 7–26). Wallingford: CAB International.

Turner, S. J., & Subbotin, S. (2013). Cyst nematodes. In R. N. Perry, & M. Moens (Eds.), *Plant nematology* (pp. 109–143). Oxfordshire: CAB International.
Tylka, G. L., Niblack, T. L., Walk, T. C., Harkins, K. R., Barnett, L., & Baker, N. K. (1993). Flow cytometric analysis and sorting of *Heterodera glycines* eggs. *Journal of Nematology, 25*(4), 596–602.
Tytgat, T., De Meutter, J., Vanholme, B., Claeys, M., Verreijdt, L., Gheysen, G., et al. (2002). Development and pharyngeal gland activities of *Heterodera schachtii* infecting *Arabidopsis thaliana* roots. *Nematology, 4*, 899–908.
Vanholme, B., Van Thuyne, W., Vanhouteghem, K., De Meutter, J., Cannoot, B., & Gheysen, G. (2007). Molecular characterization and functional importance of pectate lyase secreted by the cyst nematode *Heterodera schachtii*. *Molecular Plant Pathology, 8*(3), 267–278. http://dx.doi.org/10.1111/J.1364-3703.2007.00392.X.
van der Vossen, E. A., van der Voort, J. N., Kanyuka, K., Bendahmane, A., Sandbrink, H., Baulcombe, D. C., et al. (2000). Homologues of a single resistance-gene cluster in potato confer resistance to distinct pathogens: a virus and a nematode. *The Plant Journal, 23*(5), 567–576.
Waetzig, G. H., Sobczak, M., & Grundler, F. M. W. (1999). Localization of hydrogen peroxide during the defence response of *Arabidopsis thaliana* against the plant-parasitic nematode *Heterodera glycines*. *Nematology, 1*, 681–686.
Wang, C., Bruening, G., & Williamson, V. M. (2009). Determination of preferred pH for root-knot nematode aggregation using pluronic F-127 gel. *Journal of Chemical Ecology, 35*(10), 1242–1251. http://dx.doi.org/10.1007/s10886-009-9703-8.
Wharton, D. A., & Ramløv, H. (1995). Differential scanning calorimetry studies on the cysts of the potato-cyst nematode *Globodera rostochiensis* during freezing and melting. *Journal of Experimental Biology, 198*(12), 2551–2555.
Womersley, C., & Smith, L. (1981). Anhydrobiosis in nematodes I. The role of glycerol, myo-inositol and trehalose during desiccation. *Comparative Biochemistry and Physiology Part B: Comparative Biochemistry, 70*, 579–586.
Wyss, U. (1992). Observations on the feeding behaviour of *Heterodera schachtii* throughout development, including events during moulting. *Fundamental and Applied Nematology, 15*, 75–89.
Wyss, U., Stender, C., & Lehmann, H. (1984). Ultrastructure of feeding sites of the cyst nematode *Heterodera schachtii* Schmidt in roots of susceptible and resistant *Raphanus sativus* L. var. Oleiformis pers cultivars. *Physiological Plant Pathology, 25*(1), 21–37.
Wyss, U., & Zunke, U. (1986). Observations on the behaviour of second stage juveniles of *Heterodera schachtii* inside host roots. *Revue de Nematologie, 9*, 153–165.
Yang, J., Huang, X., Tian, B., Wang, M., Niu, Q., & Zhang, K. (2005). Isolation and characterization of a serine protease from the nematophagous fungus, *Lecanicillium psalliotae*, displaying nematicidal activity. *Biotechnology Letters, 27*(15), 1123–1128. http://dx.doi.org/10.1007/s10529-005-8461-0.
Yang, J., Wang, L., Ji, X., Feng, Y., Li, X., Zou, C., et al. (2011). Genomic and proteomic analyses of the fungus *Arthrobotrys oligospora* provide insights into nematode-trap formation. *PLoS Pathogens, 7*(9), e1002179. http://dx.doi.org/10.1371/journal.ppat.1002179.
Zunke, U., & Eisenback, J. D. (1998). Morphology and ultrastructure. In S. B. Sharma (Ed.), *The cyst nematodes* (pp. 31–56). Dordrecht: Kluwer Academic Publishers.

CHAPTER THREE

Cell Wall Alterations in Nematode-Infected Roots

Krzysztof Wieczorek
Division of Plant Protection, Department of Crop Sciences, University of Natural Resources and Life Sciences, Konrad-Lorenz Straße, Tulln, Austria
E-mail: krzysztof.wieczorek@boku.ac.at

Contents

Abstract

Sedentary plant-parasitic nematodes induce sophisticated feeding sites in roots of many crop plants causing tremendous yield losses in agriculture worldwide. The most damaging genera are cyst and root-knot nematodes forming syncytia and giant cells embedded in root galls, respectively. In addition to many morphological changes, elaborated cell wall alterations play an important role during the formation of both feeding sites and thus are essential for survival and reproduction of these parasites. Plant-parasitic nematodes were the first animals shown to produce cell wall modifying and degrading proteins and enzymes. Already during the migration through the root tissue, infective juveniles secrete e.g. cellulases and expansins, which, in addition to mechanical cell wall disruption performed with the stylet, considerably

Advances in Botanical Research, Volume 73
ISSN 0065-2296
http://dx.doi.org/10.1016/bs.abr.2014.12.002

facilitate the infection process. At the begining of sedentary stage secretions from the dorsal gland trigger the formation of a feeding site in the central cylinder. These widely unknown substances change the host gene expression affecting, in addition to many others, also cell wall-related processes leading to various architectural modifications in host cells. Several cellulolytic and pectolytic enzymes as well as expansins are specifically expressed and play an important function during the formation of the feeding sites and development of the nematodes. This chapter summarizes data on cell wall modifications in plant tissues infected with cyst and root-knot nematodes as well as on involvement of different proteins, of both nematode and plant origin, during migration and sedentary phase of nematode parasitism.

1. INTRODUCTION

The plant cell wall is a thick and rigid but at the same time complex and dynamic structure that surrounds the protoplast. It creates an outer cover providing mechanical strength preventing bursting of the plant cell caused by the intracellular osmotic pressure. Plant cell wall controls cell and plant growth contributing to plant morphology, regulates the transport of diffusates through the apoplast and plays an important role in perception and signal transduction during plant development and defence responses. Its extensive intercalated network containing cellulose microfibrils, hemicelluloses, pectin and structural, enzymatic and catalytic proteins undergoes regulated architectural changes during normal plant growth and development (Carpita & Gibeaut, 1993). Each plant cell is covered by its own cell wall while pectic middle lamella connects cell walls of neighbouring cells. Growing cells synthetize a primary cell wall that consists mainly of cellulose, hemicelluloses and pectins. After the cessation of growth, the secondary cell wall is deposited on the inner surface of the primary wall conferring mechanical strength in specialized cell types such as xylem elements, sclerenchyma and fibre cells. The structure of the plant cell wall has to be flexible and able to react to different environmental and developmental conditions as well as various biotic and abiotic stresses. For that reason, plants possess a vast number of cell wall-related genes. In genome of *Arabidopsis*, more than 700 glycosyl transferase and hydrolase genes and more than 200 genes encoding enzymes related to pectin degradation can be found. Several hundred other genes are involved in synthesis, modification and degradation of the cell wall (Somerville et al., 2004). These numbers clearly indicate the complexity and specificity of the cell wall biology and show how sophisticated and flexible this organelle in fact is.

The structure and chemical composition of plant cell wall is a substantial barrier to plant-parasitic nematodes. However, beside of the protrusible stylet that provides a tool for physical breaching of the cell wall they developed sophisticated mechanisms to overcome this structural obstacle as well as other plant defence responses to reach their goals: successfully penetrate the root, migrate within the host, become sedentary attached to the specialized feeding site and produce offspring. On the one hand, plant-parasitic nematodes actively secrete gland proteins that digest host cell walls thereby facilitating the penetration and migration within the plant tissue. On the other hand, during the sedentary stage of the parasitism the nematodes inject secretions from dorsal gland cell containing different mainly unknown effector proteins into the plant cells to trigger the formation of a feeding organ in the plant root, from which nutrients are withdrawn. These effectors help the nematode to force the plant to work for its benefits and to avoid plant defence (Quentin, Abad, & Favery, 2013). Among others, these secretions modify the expression of cell wall-related enzymes and proteins of plant origin within the feeding site or in the surrounding cells (Figure 1). This, together with other morphological changes, triggers specific cell wall modifications that are essential for the feeding site formation and development of the nematode. They include increased modification and synthesis of cell wall, which results in its extension and massive thickening as well as formation of pronounced ingrowths at the interface with xylem vessels in both syncytia and giant cells induced by cyst and root-knot nematodes, respectively (Siddique, Sobczak, Tenhaken, Grundler, & Bohlmann, 2012; Wieczorek et al., 2006). On the other hand, extensive cell wall degradation occurs in syncytia (Goellner, Wang, & Davis, 2001; Sukno et al., 2006; Wieczorek et al., 2008). Moreover, the plant cell wall performance and integrity during the nematode infection is suggested to play an essential role in intense cross talk between the sensing and signalling of defective cell walls and the control of innate immune response against plant-parasitic nematodes (Figure 1) (Wieczorek & Seifert, 2012).

First, this chapter focuses on the proteins synthesized in the oesophageal gland cells and secreted through the stylet by plant-parasitic nematodes during the migration within host root and describes the role of nematode cellulolytic and pectolytic enzymes during different phases of parasitism process. Second, it summarizes recent knowledge of cell wall alterations that occur after the nematode has become sedentary and triggered the formation of the initial feeding cell and started to transform it into the feeding site. Based on microscopic observations and expression

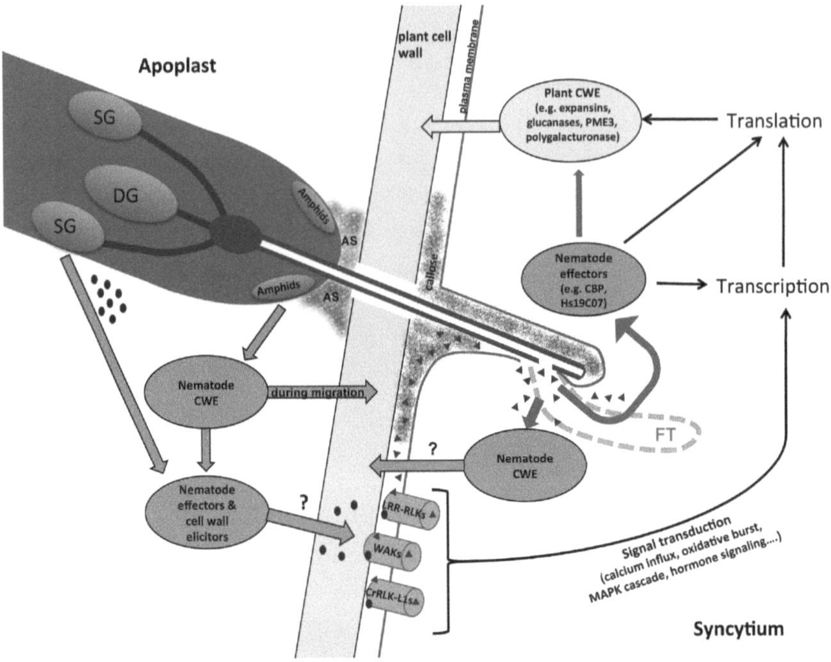

Figure 1 Cell wall modifying and degrading enzymes, of both nematode and plant origin, as well as nematode effector proteins and their functions in plant tissue and within the syncytia induced by cyst nematodes. During migration a cocktail of cell wall-related enzymes from nematode's subventral gland cells is released through the stylet facilitating the nematode movement through the root tissue. As consequence, damage-derived cell wall elicitors might be recognized by a set of different plant receptors mediating plant defence responses (e.g. LRR-RLKs, WAKs, CrRLK-L1s). In turn, the nematode might secrete effectors involved in the suppression of this induced host plant defence. During the sedentary stage, effectors from the dorsal gland are injected into the feeding site triggering substantial cellular changes including – amongst others – alterations in the expression of cell wall-related plant genes such as expansins or glucanases. These effectors may be deposited outside the plasma membrane and interact with membrane-bound receptors or further manipulate different pathways in the feeding site promoting its development. Various effector proteins (e.g. CBP, Hs19C07) may also directly affect plant cell wall enzymes (grey arrows). Blue dots – nematode's effectors and cell wall elicitors, red triangles – nematode's effectors injected into the feeding site, CWE – cell wall enzymes, SG – subventral gland, DG – dorsal gland, AS – amphidial secretions, FT – feeding tube. (See colour plate)

patterns of cell wall-related plant genes, it describes structural changes of the cell wall in feeding sites induced by sedentary cyst and root-knot nematodes.

2. MODIFICATIONS OF THE PLANT CELL WALL CAUSED BY CELL WALL MODIFYING AND DEGRADING PROTEINS SECRETED DURING MIGRATORY PHASE BY PLANT-PARASITIC NEMATODES

The nematode invasion into the root tissue requires a sequence of elaborate destructive and constructive modifications of plant cell wall (Bohlmann & Sobczak, 2014). Plant-parasitic cyst nematodes (*Heterodera* spp. and *Globodera* spp.) enter the root of host plant in the elongation zone and migrate intracellularly through the cortical cells in search of a single cell suitable for feeding site initiation within the central cylinder. During this rather harsh movement, the juveniles use the stylet to disrupt root cells causing a path of necrotized cells. In contrast, root-knot nematodes invade roots at the tip and then migrate within the central cylinder without causing a substantial damage of the root cells. For many years it was believed that during the migration stage the juveniles of plant-parasitic nematodes use only the stylet and mechanical force to break cell wall (Wyss, 1992; Wyss & Zunke, 1986). However, first evidences that plant-parasitic nematodes might additionally secrete cell wall-degrading enzymes were found more than four decades ago (reviewed in Deubert & Rohde, 1971). The origin of these proteins, however, was questionable, since the nematodes were often collected non-aseptically from the infected plant tissue. In the 1990s, when researchers working on proteins isolated from nematode subventral gland cells started to use tools of molecular biology, first reliable reports were published. Smant et al. (1998) found an endogenous coding sequence of an endo-1,4-beta-glucanase gene in the genome of two cyst nematodes, *Heterodera glycines* and *Globodera rostochiensis*. Further studies including immunolocalization showed that the expression of the isolated endoglucanase correlates with the migratory life stages of cyst nematodes (de Boer et al., 1999; Goellner, Smant, de Boer, Baum, & Davis, 2000) and the secreted protein accumulates along the migratory path of the juveniles (Wang et al., 1999). This was the ultimate evidence that the juveniles of plant-parasitic nematodes synthetize proteins in subventral gland cells and secrete them *in planta*. Based on the sequence similarity to bacterial glycosyl hydrolases (GHs) the potential horizontal gene transfer from prokaryotes to eukaryotes was suggested (Davis et al., 2000; Keen & Roberts, 1998; Smant et al., 1998; Yan et al., 1998). More recently, however, it was speculated that cyst and root-knot nematodes did not acquire these genes directly by lateral genes transfer from bacteria, but, which is more likely, they were passed on by ancestors of a family Pratylenchidae (Rybarczyk-Mydłowska et al., 2012). Last years brought gradually more information about endoglucanases as well as different pectolytic and hemicellulolytic enzymes from other plant-parasitic nematode species (reviewed in Davis, Haegeman, & Kikuchi, 2011; Kyndt, Haegeman, & Gheysen, 2008; Mitchum, Hussey, Davis, & Baum, 2007). All these reports indicate that the migratory juveniles are able to degrade not only cellulose but also other complex carbohydrate components of the cell wall and middle lamella, a strategy that considerably facilitates the nematode root invasion. A list of cell wall-related proteins secreted by plant-parasitic nematodes during the migratory phase is presented in Table 1.

2.1 Cellulose

The main component of the plant cell wall is cellulose, which consists of glucose residues. These sugar molecules form long (1,4)-β-linked glucan chains that are connected to adjacent glucans via hydrogen bonds forming crystalline cellulose microfibrils (Somerville, 2006) cross-linked by different hemicelluloses such as xyloglucan, glucomannan, xylan and mixed-linked glucans (Scheller & Ulvskov, 2010). The enzymes catalyzing the hydrolysis of glycosidic bonds in cellulose are glycoside hydrolyses, which are divided into different families according to their sequence similarity (EC 3.2.1.4; Henrissat & Bairoch, 1996). Several enzymes, mainly belonging to the family 5 (GHF5), have been found in many plant-parasitic nematodes from the order Rhabditida such as sedentary nematodes *Heterodera*, *Globodera* and *Meloidogyne* (Abad et al., 2008; Bera-Maillet, Arthaud, Abad, & Rosso, 2000; Gao, Allen, Davis, Baum, & Hussey,2004a; Gao, Allen, Maier, Davis, & Baum, 2002; Goellner et al., 2000; Ledger, Jaubert, Bosselut, Abad, & Rosso, 2006; de Meutter, Vanholme, Baun, Tytgat, & Gheysen, 2001; Rehman et al., 2009; Rosso et al., 1999; Smant et al., 1998; Yan, Smant, & Davis, 2001). Recent studies, however, report on endoglucanases from migratory nematodes such as *Rotylenchus*, *Pratylenchus* and *Ditylenchus* (Kyndt et al., 2008; Wubben, Callahan, & Scheffler, 2010). It is believed that these secreted cellulases soften and weaken the cellulose network during root invasion and the movement through the plant tissue to alleviate the mechanical breakdown of the cell wall. The structure of nematode GHF5 endoglucanases consists of several domains: signal peptide, GHF5 catalytic domain, a linker and sometimes a carbohydrate-binding module (CBM) of family 2 (CBM2) (Boraston, Bolam, Gilbert, & Davies, 2004; Gilkes, Henrissat, Kilburn, Miller, & Warren, 1991). It is suggested that the CBM

Table 1 Nematode Proteins Acting on Plant Cell Walls during the Migration through the Host Root

Gene/Protein	Nematode	Class	Function	References
Hg-pel-1	*Heterodera glycines*	Pectate lyase		de Boer et al. (2002)
Hspel1	*Heterodera schachtii*	Pectate lyase		Vanholme et al. (2007)
Hspel2	*H. schachtii*	Pectate lyase	Silencing resulted in low infection efficiency	Vanholme et al. (2007)
Gr-PEL1	*Globodera rostochiensis*	Pectate lyase		Popeijus et al. (2000)
Gr-PEL2	*G. rostochiensis*	Pectate lyase	Transient expression in *Nicotiana benthamiana* resulted in severe malformations of infiltrated tissues	Kudla et al. (2007)
GR-ENG-1	*G. rostochiensis*	Cellulase	Hydrolysis of carboxymethylcellulose	Smant et al. (1998)
GR-ENG-2	*G. rostochiensis*	Cellulase	Hydrolysis of carboxymethylcellulose	Smant et al. (1998)
HG-ENG-1	*H. glycines*	Cellulase		
HG-ENG-2	*H. glycines*	Cellulase		
GR-ENG1/2/3/4	*G. rostochiensis*	Cellulase	Targeting the mRNAs of secreted cellulases with dsRNA resulted in lower nematode infection	Chen et al. (2005)
Gr-EXPB1	*G. rostochiensis*	Expansin	Cell wall extension activity	Qin et al. (2004), Kudla et al. (2005)
Mi-pg-1	*Meloidogyne incognita*	Polygalacturonase		Jaubert et al. (2002)
Mi-xyl1	*M. incognita*	endo-1,4-β- xylanase		Mitreva-Dautova et al. (2006)
MI-ENG-2	*M. incognita*	1,4-β-endoglucanase		Ledger et al. (2006)
MI-CBP-1	*M. incognita*	Cellulose-binding protein		Ding et al. (1998)

Source: This table (here updated) was originally published in Bohlmann and Sobczak (2014).

helps the enzyme to bind to the substrate. In *Bursaphelenchus xylophilus*, endoglucanases belong to the GHF45 family, which suggests a different origin of these enzymes in the Aphelenchoidea superfamily (Kikuchi, Jones, Aikawa, Kosaka, & Ogura, 2004). The role of cellulases during the parasitism process was demonstrated by Chen, Rehman, Smant, and Jones (2005). They showed that invasive *G. rostochiensis* juveniles soaked in dsRNA targeting the mRNA of secreted cellulose were significantly less successful in invading the roots and inducing syncytia.

2.2 Hemicellulose

Hemicellulose together with pectins forms a matrix in which cellulose microfibrils are embedded within the plant cell wall. It is a complex polysaccharide network consisting of either xylose, glucose, mannose or mannose and glucose connected via (1,4)-β and (1,3)-β-glycosidic bonds (Bacic, Harris, & Stone, 1988). Xylan is a main component of hemicellulose and comprises (1,4)-β-linked xylopyranose units that are substituted with different sugar residues depending on the plant species (Collins, Gerday, & Feller, 2005). The activity of the nematode GHF5 endoglucanases that hydrolyze hemicellulose is limited to (1,4)-β-linked polysaccharides (Gao, Allen, Davis, Baum, & Hussey, 2004a). Some endoglucanases from *H. glycines* are able to hydrolyze xylan (Gao, Allen, Davis, Baum, & Hussey, 2004b). Several other endoxylanases were found in some *Meloidogyne* species (Abad et al., 2008; Mitreva-Dautova et al., 2006; Opperman et al., 2008) and in the migratory nematode *Radopholus similis* (Haegeman, Vanholme, & Gheysen, 2009). These GHs share domain structure with endoglucanases and consist of a signal peptide, a catalytic domain and sometimes CBM domain (Haegeman et al., 2009).

2.3 Pectin

Pectin is besides cellulose and hemicellulose one of the main structural polymers of the primary plant cell wall (McNeil, Darvill, Fry, & Albersheim, 1984). It is thought to be responsible for wall porosity, charge density and spacing of cellulose microfibrils. It is mainly located in the middle lamella between two adjacent plant cells and contributes to intercellular adhesion and controlled separation. Highly methyl esterified pectin is synthetized in the Golgi apparatus as three different polysaccharides: rhamnogalacturonan I, rhamnogalacturonan II and homogalacturonan (HG) whose backbone comprises α-D-galacturonic acid residues. Different enzymes, e.g. pectin esterases and pectate lyases, drive the degradation of pectin resulting in the maceration of plant tissue that is essential for many plant pathogens. Until now, in plant-parasitic nematodes two groups of enzymes were found which degrade the unmethylated form of pectin: polygalacturonases and pectate lyases. Pectin esterases that act on methylated pectin were not found in secretions of plant-parasitic nematodes so far.

Pectate lyases (EC 4.2.2.2) catalyze the eliminative cleavage of unmethylated pectin and are virulent factors of many bacterial and fungal plant pathogens that use these enzymes to degrade cell wall to facilitate the tissue penetration and colonization (Barras, Gijsegem, & Chatterjee, 1994). They were also found in several cyst nematode species such as *H. glycines* (de Boer et al., 2002), *H. schachtii* (Vanholme et al., 2007), *G. rostochiensis* (Kudla et al., 2007; Popeijus et al., 2000) and also migratory nematode *B. xylophilus* (Kikuchi, Shibuya, Aikawa, & Jones, 2006). The putative role of pectate lyases secreted by nematodes is the softening of the cell wall and weakening of the middle lamella to facilitate the migration through the root tissue. This assumption is supported by the study showing that knockdown of *H. schachtii* pectate lyase genes by RNAi results in fewer infections (Vanholme et al., 2007). This effect clearly demonstrates the importance of this enzyme for nematode invasion and migration inside the host root. Further, the transient expression of a pectate lyase from *G. rostochiensis* in *Nicotiana benthamiana* triggers severe malformations of the infiltrated tissue, which points at possible important role of this enzyme during the infection and parasitism (Kudla et al., 2007).

Polygalacturonases cleave the bonds between galacturonic acid in unmethylated HG and are divided into two classes according to their specificity: endo-polygalacturonases and exo-polygalacturonases (Jaubert, Laffaire, Abad, & Rosso, 2002). The exo-type enzyme was found in secretion of root-knot nematode *Meloidogyne incognita* (Jaubert et al., 2002). In expressed sequence tag (EST) libraries of other *Meloidogyne* species such as *M. javanica*, *M. arenaria*, *M. hapla* and *M. chitwoodi* several polygalacturonase-like sequences were found.

2.4 Cell Wall Modification Driven by Proteins without Enzymatic Activity

Many cell wall-related proteins that have been found in secretions of plant-parasitic nematodes do not possess hydrolytic activity. They contain often a cellulose binding protein (CBP), a domain that is predicted to bind specifically to cellulose. The first such protein from plant-parasitic nematodes was found in oesophageal gland cells and in secretions of the root-knot nematode *M. incognita* (Ding, Shields, Allen, & Hussey, 1998). Subsequently, similar CBP sequences were identified from other species of plant-parasitic nematodes (Kikuchi et al., 2009; Kyndt et al., 2008; Mitchum et al., 2007). Peptide sequence upstream of a cellulose-binding protein from *G. rostochiensis* shows similarity to the plant expansin (Qin et al., 2004). These cell wall-related proteins exhibit no enzymatic activity. They are widespread in the plant kingdom and function as cell wall loosening and weakening factors that disrupt the non-covalent bonds between cellulose microfibrils and hemicelluloses (Cosgrove, 2000; McQueen-Mason & Cosgrove, 1994). In secretions of sedentary and migratory plant-parasitic nematodes, expansins were found and it is suggested that they weaken the plant cell wall making its structure better accessible for cell wall degrading enzymes secreted by the nematodes. This makes the migration of the parasite through the root tissue much easier and facilitates the mechanical disruption of the cell wall with the help of the stylet and lips (Abad et al., 2008; Haegeman, Kyndt, & Gheysen, 2010; Kikuchi et al., 2007, 2009; Kudla et al., 2005; Opperman et al., 2008; Roze et al., 2008).

3. STRUCTURAL MODIFICATION OF THE CELL WALL IN FEEDING SITES INDUCED BY PLANT-PARASITIC NEMATODES

After the penetration and migration through the host root tissue, cyst and root-knot nematodes induce in the central cylinder sophisticated feeding sites, syncytia and giant cells embedded in gall tissue, respectively. These specialized structures function as strong sink organs within the plant and provide water and all nutrients, which are indispensable for the further development and reproduction of the nematodes.

In the case of cyst nematodes, the migrating juvenile chooses a single suitable root cell (initial syncytial cell (ISC)) and initiates the formation of a syncytium. It pierces its wall with the hollow stylet and inserts it through the cell wall into the plant protoplast without rupturing the plasma membrane. Secretions from amphids, nematode's chemoreceptor sense organs, are involved in formation of the feeding plug (Figure 2(H) and (J)). This structure prevents the leakage of the syncytial cytoplasm along the stylet and is embedded into the outer syncytial cell wall at the stylet insertion site (Sobczak, Golinowski, & Grundler, 1999). Feeding plug is probably composed of callose and is characteristic for cyst nematodes and was not found in giant cells. Subsequently, secretions are injected into the ISC to

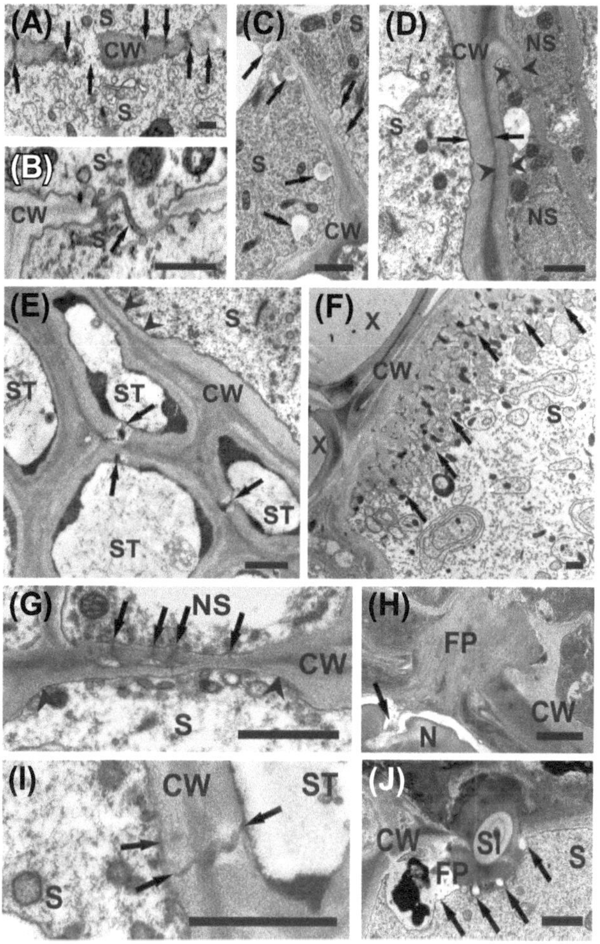

Figure 2 ***Structural features of cell walls of syncytia induced by*** **H. schachtii** ***in Arabidopsis roots.*** (A) Cell wall openings formed by widening of plasmodesmata (arrows). (B) Cell wall openings formed by local dissolution of cell wall without involvement of plasmodesmata. Arrow indicates middle lamella covered with plasma membrane. (C) Paramural bodies (arrows) formed at extensively digested part of internal cell wall. (D) Comparison of thickness of outer syncytial cell wall (between arrows) and cell wall of non-syncytial cells (between arrowheads). (E) Thin part of outer syncytial cell wall (arrowheads) facing sieve tube. Arrows indicate plasmodesmata between sieve tubes. (F) Well-developed system of cell wall ingrowths (arrows) formed at syncytial wall facing vessels. (G) A group of plasmodesmata (arrows) at thin part of outer syncytial cell wall (between arrowheads) facing non-syncytial parenchymatous cell. (H) Feeding plug in syncytial cell wall. Secretions emanating from nematode amphids are marked with arrows. (I) Plasmodesmata (arrows) between syncytial element and sieve tube. (J) Feeding plug with inserted cross-sectioned nematode stylet. Callose depositions are indicated with arrows. Transmission electron microscopy micrograms (A–J) of syncytia at 2 (C, J), 5 (B, H), 10 (D, E) and 13 (A, F, G, I) days post inoculation. CW, cell wall; FP, feeding plug; NS, non-syncytial cell; S, syncytium; SI, stylet; ST, sieve tube; X, xylem vessel. Bars = 1 μm. *This figure (here modified) was originally published in Bohlmann and Sobczak (2014).*

form the first feeding tube, which is a blind-ended structure, formed in the cytoplasm at the stylet orifice. This structure is composed of proteins and acts as a molecular sieve, allowing only molecules up to approximately 30 kDa to be withdrawn by the nematode (Eves van den Akker et al., 2014).

Triggered by largely unknown secretions injected into the feeding site through the stylet, the syncytium goes through many morphological changes. It strongly hypertrophies, becomes metabolically very active and enlarges via fusion of root cells by tightly controlled cell wall degradation (Figure 4(A) and (B) in Chapter 2) (Grundler, Sobczak, & Golinowski, 1998). During the first days this is achieved by widening of plasmodesmata occurring between the ISC and neighbouring root cells (Figure 2(A)) (Grundler et al., 1998; Hofmann, Wieczorek, Blöchl, & Grundler, 2007). Subsequently the young syncytium continuously enlarges via further cell wall dissolution without involvement of plasmodesmata (Figure 2(B)) (Grundler et al., 1998; Hofmann et al., 2007). The formation of such cell wall openings is accompanied by a number of paramural bodies (Figure 2(C)). Often parts of the syncytial cell wall remain rather thin such as cell wall regions facing sieve tubes (Figure 2(E)) or cells that are designated to be incorporated into the syncytium. In contrast, the outer syncytial wall thickens gradually (Figure 2(D)) and this process leads to occlusion of plasmodesmata. Therefore it was first suggested that syncytia are symplasmically isolated from adjacent root tissue. Recent studies, however, show that plasmodesmata in older syncytia are symplasmically connected to the phloem (Hofmann et al., 2007; Hofmann & Grundler, 2006; Hoth, Schneidereit, Lauterbach, Scholz-Starke, & Sauer, 2005). In outer syncytial cell wall plasmodesmata are formed occasionally *de novo* at the border to adjacent root cells (Figure 2(G)) or sieve tubes (Figure 2(I)). During several days after syncytium induction, more and more cells fuse with the feeding site that becomes a large hypertrophied conglomerate of root cell protoplasts (Wyss & Grundler, 1992).

The infective second-stage juveniles of root-knot nematodes penetrate the host root just behind the tip and subsequently migrate intercellularly between the cells. They move towards the root apex where they turn and enter the vascular cylinder. In the parenchyma, the nematodes trigger the formation of several, usually 5–7, multinucleate and strongly hypertrophied giant cells embedded in a gall tissue (Figure 3(A)) (Caillaud, Dubreuil, et al., 2008; Wyss, Grundler, & Münch, 1992). Their isotropic growth proceeds due to the additional synchronous nuclear divisions in the absence of cell division (Jones & Payne, 1978). Similar to syncytia, these cells are

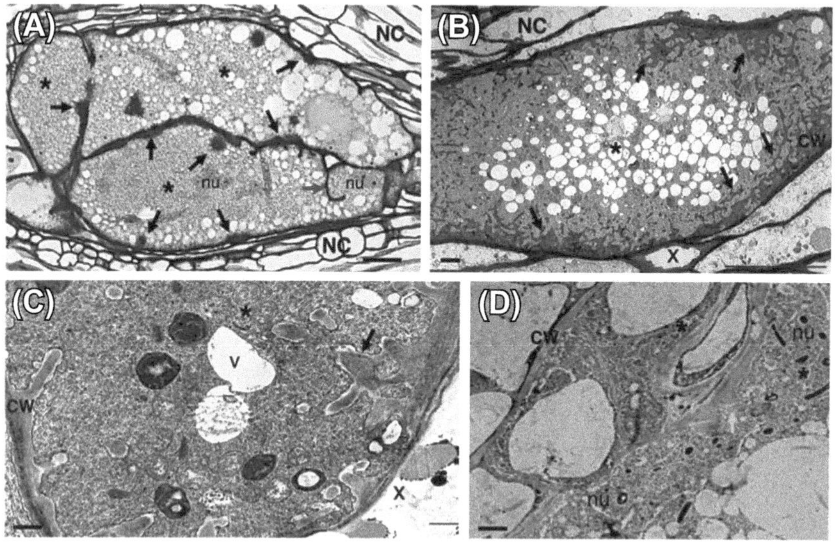

Figure 3 ***Anatomy of*** **Meloidogyne incognita*-induced giant cells in*** **Arabidopsis** ***roots.*** (A) Light microscopy image of sectioned giant cells embedded in a gall and stained with toluidine blue (black in print versions). Cell wall thickenings (black arrows) and a cell wall stub (red (grey in print versions) arrow) indicating arrest of cytokinesis. (B–D) Ultrastructure of giant cell sections showing cell wall ingrowths (black arrows) along regions predominantly flanking the vascular tissue. Note the xylem elements with thickened cell walls and dense cytoplasm containing numerous organelles including asymmetrically shaped nuclei and small vacuoles. (D) Detailed giant cells showing a plasmodesmata (green (light grey in print versions) arrow). Asterisk, giant cell; NC, neighbouring cells; x, xylem; CW, cell wall; V, vacuole; nu, nucleus. Bars = (A) 25 μm and (B–D) 5 μm. *This figure was originally published in Rodiuc et al. (2014).*

metabolically very active and act as a sink tissue that supplies the nematode with water and all nutrients (Caillaud, Dubreuil, et al., 2008). In addition, to ensure the nutrient transport, giant cells are connected with phloem cells via plasmodesmata (Figure 3(B)–(D)).

Both syncytia and giant cells act as nutrient transfer cells accumulating and synthesizing components necessary for survival and reproduction of the nematode (Rodiuc, Vieira, Banora, & de Almeida Engler, 2014). One of the structural traits of these unique cells is extended secondary cell wall ingrowths (Davis, Smith, & Cobb, 1990; Gunning & Pate, 1969). Similarly, in syncytia (Figure 2(F)) and giant cells (Figure 3(A)–(C)) at the interface with xylem vessels such elaborated ingrowths are formed (Golinowski, Grundler, & Sobczak, 1996; Jones & Northcote, 1972). First finger-like protrusions are visible in 5- to 7-day-old feeding sites. Subsequently, they branch and expand what results in formation of pronounced cell wall labyrinths that substantially increase the surface of plasma membrane allowing higher exchange of solutes between symplast and apoplast.

4. CELL WALL-RELATED EFFECTORS SECRETED BY PLANT-PARASITIC NEMATODES DURING SEDENTARY PHASE OF NEMATODE PARASITISM

The knowledge about nematode effectors and enzymes that act on the plant cell wall during the sedentary phase of parasitism is very limited (reviewed in Hewezi & Baum, 2013). However, in last years the number of reports showing that nematodes secrete such substances into already established feeding sites is slowly growing. Hewezi et al. (2008) showed that a cellulose-binding domain protein secreted by the juveniles of the sugar beet nematode *H. schachtii* interacts with a plant pectin methyl esterase (PME3) that gets thus activated and through direct interaction between both proteins may facilitate nematode parasitism. PME3 is expressed within the syncytia as well as in the adjacent root cells that will be fused with the expanding feeding structure, which suggests its important role for the growth of syncytium. Another recent study reported on the function of an oesophageal gland cell protein from *H. schachtii*, Hs19C07, during the development of syncytia (Lee et al., 2011). This novel effector interacts with *Arabidopsis* auxin influx transporter LAX3 that is expressed in lateral root primordia. It is thought to provide a hormonal signal that leads to the elevated expression of cell wall-modifying enzymes allowing the lateral roots to emerge. It was shown that increased activity of LAX3 triggers the auxin-dependent induction of a polygalacturonase. Therefore, it is suggested that Hs10C07 plays a role in LAX3-mediated auxin influx into the syncytium. It seems that this effector allows the nematode the regulation of the auxin flow in both feeding structure and adjacent root cells, thereby activating the hydrolysis of the plant cell wall and thus facilitating syncytium development.

Hamamouch et al. (2012) studied a novel 30C02 effector from *H. glycines* in *Arabidopsis* using its homologue from *H. schachtii*. The 30C02 protein interacts with a specific host plant 1,3-β-endoglucanase in both yeast and plant cells, possibly interfering with its role as a plant pathogenesis-related protein. Results from experiments with different mutant lines suggest that the 30C02 effector is essential for successful plant parasitism by cyst nematodes and that

Table 2 Nematode Proteins Acting on Plant Cell Walls during Sedentary Stage

Gene/ Protein	Nematode	Class	Function	References
Hs CBP	*Heterodera schachtii*	Cellulose binding protein	Interacts with a plant pectin methylesterase (PME3) to aid cyst nematode parasitism	Hewezi et al. (2008)
Hs19C07	*H. schachtii*		Interacts with *Arabidopsis* LAX3 triggering the expression of polygalacturonase	Lee et al. (2011)
30C02	*Heterodera glycines/H. schachtii*		Interacts with a specific host plant 1,3-β-endoglucanase	Hamamouch et al. (2012)

the host 1,3-β-endoglucanase plays a potential role in the defence response against *H. schachtii* infection.

A list of known cell wall proteins secreted by plant-parasitic nematodes, directly or indirectly, acting on plant cell wall during the sedentary phase is presented in Table 2.

5. EXPRESSION OF PLANT GENES ENCODING PROTEINS INVOLVED IN CELL WALL MODIFICATION AND BIOSYNTHESIS DURING SEDENTARY PHASE OF NEMATODE PARASITISM

The formation of the feeding sites induced by sedentary plant-parasitic nematodes requires different elaborated modifications of the plant cell wall. For both syncytia and giant cells induced by cyst and root-knot nematodes, respectively, it was shown that several plant cell wall-related proteins such as expansins, cellulases, pectate lyases and polygalacturonases are involved in their formation. The expression of most of these genes shows a specific temporal and special pattern within the feeding site or in adjacent root cells, a phenomenon that requires a tight and sophisticated regulation (reviewed in Sobczak, Fudali, & Wieczorek, 2011; Barcala et al., 2010; Caillaud, Lecomte, et al., 2008; Gheysen & Fenoll, 2002; Gheysen & Mitchum, 2009; Jammes et al., 2005; Szakasits et al., 2009; Wieczorek et al., 2006, 2008, 2014). Majority of the microscopic and molecular data concerning the cell wall modifications in nematode feeding sites was gained from the model plant *Arabidopsis*. For this reason the following parts of this chapter deal predominantly with the interaction between this plant and sedentary cyst and root-knot nematodes.

Arabidopsis genes involved in the cell wall modifications in syncytia induced by cyst nematodes and giant cells induced by root-knot nematodes are summarized in Tables 3 and 4, respectively.

5.1 Expansins

Expansins are cell wall proteins lacking hydrolytic activity that weaken the hydrogen bonds between cellulose and hemicellulose contributing to the expansion of the cell wall (McQueen-Mason & Cosgrove, 1994). They relax its structure making it more accessible to other cell wall-modifying and degrading enzymes. *Arabidopsis* expansin gene family contains 26 genes coding for alpha-expansins and five genes coding for beta-expansins. It was shown that these isoforms are differentially expressed in various plant organs and also in syncytia induced by cyst nematode *H. schachtii* (Wieczorek et al., 2006). Expression of some shoot-specific expansins such as *EXPA3* (At2g37640) and *EXPA16* (At3g55500), which are not expressed in the root of uninfected plants, was detected exclusively within the feeding site. This suggests their important role in the syncytium formation. The up-regulation of *EXPA1*, *EXPA6* and *EXPB1* in giant cells induced in *Arabidopsis* roots by *M. incognita* shows that these proteins have an important function also during the development of root-knot nematodes (Jammes et al., 2005).

5.2 Cellulases

Expansins weaken and relax the structure of the plant cell wall and make it more accessible to cell wall-related enzymes of plant origin that are involved in its degradation and further modification. One of the important groups comprises cellulases (EC 3.2.1.4; endo-1,4-β-glucanases) (Henrissat, Coutinho, & Davies, 2001). They are cellulolytic proteins that hydrolyze the 1,4-β-glucosidic linkages between glucose residues in cellulose. For instance, in *Arabidopsis* the endo-1,4-β-glucanases form a relatively large gene family with 25 members and these plant genes are specifically expressed in nematode feeding sites. Goellner et al. (2001) analyzed five endo-1,4-β-glucanases from tobacco (*Nicotiana tabacum*) and showed their up-regulation upon infection with root-knot and cyst nematodes. Wang, Replogle, Davis, and Mitchum (2007) isolated the *NtCel7* promoter and analyzed the up-regulation of the encoded gene in soybean (*Glycine max*), tomato and *Arabidopsis* by use of histochemical approaches and transgenic plants. Mitchum

Table 3 Plant Cell Wall-Related Genes Involved in Syncytium Formation in *Arabidopsis*

Gene	Name	Class	Function/Expression	References
At2g37640	*AtEXPA3*	Expansin	Loosening of cell walls, up-regulated in syncytia	Wieczorek et al. (2006)
At3g55500	*AtEXPA16*	Expansin	Loosening of cell walls, up-regulated in syncytia	Wieczorek et al. (2006)
At1g70710	*AtCel1*	Cellulase	Implicated in cell wall softening during early stages of syncytium development. Expressed in young syncytia	Mitchum et al. (2004) Wieczorek et al. (2008)
At4g24260	*AtKor3*	Cellulase	Degradation of cell walls. Decreased numbers of females develop on roots of T-DNA mutants	Wieczorek et al. (2008)
At3g14310	*PME3*	Pectin methylesterase	Interacts with a cellulose binding protein secreted by *Heterodera schachtii* during early phase of syncytium development	Hewezi et al. (2008)
At3g05910	*PAE (DiDi9C-12)*	Pectin acetylesterase	Up-regulated in and around young expanding syncytia induced by *H. schachtii*	Vercauteren et al. (2002)
At3g27400	*PLL18*	Pectate lyase-like	Up-regulated in syncytia. Important for proper development of the cyst nematode *H. schachtii*	Wieczorek et al. (2014)
At4g24780	*PLL19*	Pectate lyase-like	Up-regulated in syncytia. Important for proper development of the cyst nematode *H. schachtii*	Wieczorek et al. (2014)
At3g54920	*PMR6*	Pectate lyase-like	Up-regulated in syncytia. Known also as powdery mildew susceptibility gene	Vogel, Raab, Schiff, & Somerville (2002), Szakasits et al. (2009)
At1g14520	*MIOX1*	Myo-inositol oxygenase	Important for syncytium and nematode development, probably due to removal of excess myo-inositol to decrease the level of galactinol. Strong expression in syncytia.	Siddique et al. (2009), Siddique et al. (2014)

(Continued)

Table 3 Plant Cell Wall-Related Genes Involved in Syncytium Formation in *Arabidopsis*—cont'd

Gene	Name	Class	Function/Expression	References
At2g19800	*MIOX2*	Myo-inositol oxygenase	Important for syncytium and nematode development, probably due to removal of excess myo-inositol to decrease the level of galactinol. Strong expression in syncytia.	Siddique et al. (2009), Siddique et al. (2014)
At4g26260	*MIOX4*	Myo-inositol oxygenase	Important for syncytium and nematode development, probably due to removal of excess myo-inositol to decrease the level of galactinol. Strong expression in syncytia.	Siddique et al. (2009), Siddique et al. (2014)
At5g56640	*MIOX5*	Myo-inositol oxygenase	Important for syncytium and nematode development, probably due to removal of excess myo-inositol to decrease the level of galactinol. Strong expression in syncytia.	Siddique et al. (2009), Siddique et al. (2014)
At5g39320	*UGD1*	UDP-glucose dehydrogenase	Expressed in syncytia	Siddique et al. (2012)
At3g29360	*UGD2*	UDP-glucose dehydrogenase	Expressed in syncytia. In mutant plants cell wall ingrowths are not formed in syncytia	Siddique et al. (2012)
At5g15490	*UGD3*	UDP-glucose dehydrogenase	Expressed in syncytia. In mutant plants cell wall ingrowths are not formed in syncytia	Siddique et al. (2012)
At1g26570	*UGD4*	UDP-glucose dehydrogenase	Expressed in syncytia	Siddique et al. (2012)

Source: This table (here updated) was originally published in Bohlmann and Sobczak (2014).

Table 4 Plant Cell Wall-Related Genes Involved in Giant Cell Formation in *Arabidopsis*

Gene	Name	Class	Function/Expression	References
At3g05910	*PAE (DiDi9C-12)*	Pectin acetylesterase	Up-regulated in initiating giant cells and in cells surrounding the nematodes	Vercauteren et al. (2002)
At3g27400	*PLL18*	Pectate lyase-like	Up-regulated in galls	Wieczorek et al. (2014)
At4g24780	*PLL19*	Pectate lyase-like	Up-regulated in galls	Wieczorek et al. (2014)
At4g25810	*XTR6*	Xyloglucan-endotransglycosylase		Cabrera, Barcala, Fenoll, & Escobar (2014)
At1g69530	*AtEXP1*	Expansin	Up-regulated in galls/giant cells	Jammes et al. (2005) Barcala et al. (2010)
At2g28950	*ATEXP6*	Expansin	Up-regulated in galls/giant cells	Jammes et al. (2005) Barcala et al. (2010)
At2g20750	*AtEXPB1*	Expansin	Up-regulated in galls	Jammes et al. (2005)
At3g15720		Pectinase	Up-regulated in galls	Jammes et al. (2005)
At1g02730	*CSLD5*	Cellulose synthase-like d5	Up-regulated in giant cells	Barcala et al. (2010)
At2g06850	*EXGT-A1*	Endoxyloglucan transferase	Up-regulated in giant cells	Barcala et al. (2010)
At3g53190		Pectate lyase	Up-regulated in giant cells	Barcala et al. (2010)
At4g30270		Endoxyloglucan transferase	Up-regulated in giant cells	Barcala et al. (2010)
At4g32410	*AtCesA1*	Cellulose synthase	Primary cell wall biosynthesis. Up-regulation in giant cells.	Hudson (2009)
At4g39350	*AtCesA2*	Cellulose synthase	Up-regulation in giant cells.	Hudson (2009)
At5g05170	*AtCesA3*	Cellulose synthase	Primary cell wall biosynthesis. Up-regulation in giant cells.	Hudson (2009)
At5g44030	*AtCesA4*	Cellulose synthase	Secondary cell wall synthesis. Up-regulation in giant cells.	Hudson (2009)

(Continued)

Table 4 Plant Cell Wall-Related Genes Involved in Giant Cell Formation in *Arabidopsis*—cont'd

Gene	Name	Class	Function/Expression	References
At5g09870	*AtCesA5*	Cellulose synthase	Secondary cell wall synthesis. Up-regulation in giant cells.	Hudson (2009)
At5g64740	*AtCesA6*	Cellulose synthase	Primary cell wall biosynthesis. Up-regulation in giant cells.	Hudson (2009)
At5g17420	*AtCesA7*	Cellulose synthase	Secondary cell wall synthesis. Up-regulation in giant cells.	Hudson (2009)
At4g18780	*AtCesA8*	Cellulose synthase	Up-regulation in giant cells.	Hudson (2009)

et al. (2004) and Sukno et al. (2006) showed *AtCel1*-driven GUS (β-glucuronidase) expression in tobacco and *Arabidopsis* upon infection with *M. incognita*. Interestingly, infection with *Globodera tabacum* or *H. schachtii* did not lead to expression of the *AtCel1* promoter. Wieczorek et al. (2008) published a study on expression pattern of cellulase genes in *Arabidopsis* roots infected with *H. schachtii*. It shows that two endo-1,4-β-glucanases, *AtCel2* (*At1g02800*) and *AtKOR3* (*At4g24260*), are specifically up-regulated in syncytia induced by *H. schachtii*. Experiments with corresponding mutants showed that their involvement in the formation of the feeding site and subsequent development of the nematodes is essential as 50% less female nematodes developed on the mutant plants when compared to wild type plants.

5.3 Pectin Modifying and Degrading Enzymes

Pectin is another important component of the primary cell wall. During the nematode parasitism it has to be gradually modified and degraded to enable the expansion of the nematode feeding site. This process involves a set of different enzymes that often form large gene families in plants. For instance, in *Arabidopsis* there are 66 genes encoding polygalacturonases (Kim, Shiu, Thoma, Li, & Patterson, 2006), 66 genes coding for PMEs (Louvet et al., 2006), 12 genes encoding pectin acetylesterases (PAEs) (Gou et al., 2012), 26 pectate lyases or pectate lyase-like genes (PLLs) (Palusa, Golovkin, Shin, Richardson, & Reddy, 2007; Sun & van Nocker, 2010) and 67 pectin lyases (Cao, 2012). These different enzymes might play a crucial role during the intrusive growth of the feeding sites, for instance, allowing the proper formation of cell wall openings in syncytia (Figure 2(D)). Highly methyl esterified pectin is synthetized in the Golgi apparatus and only pectin lyases are able to degrade this form of pectin. There is, however, beside of some GeneChip data (Szakasits et al., 2009), lack of information concerning their expression and function in nematode feeding sites. After deposition of pectin in the apoplast PMEs and PAEs selectively remove the methyl and acetyl groups, respectively, making this polymer accessible to other enzymes. PMEs were shown to be differentially expressed in syncytia induced by *H. schachtii*. Szakasits et al. (2009) observed down-regulation of 10 isoforms and up-regulation of three other PMEs. Further, a CBP secreted by *H. schachtii* interacts with *Arabidopsis* PME3 facilitating the early steps of cyst-nematode infection (Hewezi et al., 2008). The authors showed the important role of this protein–protein interaction demonstrating that plants overexpressing PME3 are more susceptible while the knockdown mutants are more resistant to *H. schachtii*. According to Szakasits et al. (2009), two

PAEs were down-regulated and three up-regulated in syncytia induced by *H. schachtii*. Vercauteren, de Almeida Engler, DeGroodt, and Gheysen (2002) demonstrated up-regulation of PAE in both young giant cells induced by *M. incognita* and in syncytia of *H. schachtii*. After modification driven by PMEs and PAEs, pectin becomes more vulnerable to enzymes that are able to degrade the demethylated and deacetylated form. This is performed mainly by polygalacturonases and pectate lyases. The latter are encoded in *Arabidopsis* by 26 PLLs (Palusa, Golovkin, Shin, Richardson, & Reddy, 2007). According to GeneChip data presented by Szakasits et al. (2009), six PLLs were up-regulated and four down-regulated in syncytia induced by *H. schachtii* in *Arabidopsis* roots. Recently, the role of two of the up-regulated isoforms, *PLL18* (*At3g27400*) and *PLL19* (*At4g2470*), was studied during the development of syncytia and giant cells in more detail (Wieczorek et al., 2014). The authors showed that T-DNA mutants of both genes are more resistant to *H. schachtii* infection, suggesting their important role during the development of this cyst nematode. Analogous experiments with the root-knot nematode *M. incognita* did not reveal similar result, what might indicate a minor role of these enzymes in the development of root-knot nematodes.

5.4 Synthesis of New Cell Wall Material

In order to withstand the increased turgor pressure within the nematode feeding site the outer cell wall in syncytia and giant cells is considerably thickened (Figures 2(A, F) and 3(A)). Additionally, in both feeding sites elaborated ingrowths at the interface with xylem vessels are formed increasing the nutrient transport between symplast and apoplast (Figures 2(H) and 3(A–C)). To perform these structural changes (see Chapter III) and thus be able to fulfil these important functions, the cell wall in nematode feeding sites has not only be modified and partially digested but it has to be concomitantly synthetized *de novo*.

Cellulose biosynthesis in *Arabidopsis* involves a CesA gene family with eight enzymes responsible for formation of primary (*CesA1*, *-3* and *-6*) and secondary cell wall (*AtCesA4*, *-5* and *-7*). All *CesA* genes are temporally and quantitatively expressed in giant cells induced by *M. incognita*, with highest expression at five days post-inoculation (Hudson, 2009). Experiments indicated that up-regulated secondary cell wall *CesA* genes were localized within the giant cells and primary cell wall *CesA* genes were primarily localized to the surrounding gall tissue of the infection site. Further, the infection tests showed that the number of galls and female nematodes were decreased in mutants of eight of the *CesA* genes. This work indicates a critical role for plant *CesA* gene in generating the proper root morphology to promote root-knot nematode infection and for the development of feeding cells to support nematode growth and reproduction.

UDP-glucuronic acid (UDP-GlcA) is a major precursor of several cell wall polysaccharides and there are two alternative pathways that lead to its biosynthesis. UDP-GlcA is produced either by UDP-glucose dehydrogenase (UGD) through oxidation of UDP-glucose or by myo-inositol oxygenase (MIOX) that converts myo-inositol to glucuronic acid that can be further phosphorylated to produce glucuronic acid-1-phosphate which is then converted to UDP-GlcA. Both pathways seem to be affected in nematode feeding sites. Transcriptome analysis of syncytia induced by the cyst nematode *H. schachtii* in *Arabidopsis* roots (Szakasits et al., 2009) revealed that two MIOX genes are among the most strongly up-regulated genes in these feeding sites (MIOX4 and MIOX5) (Siddique et al., 2009). The other two genes are expressed in both control roots and syncytia. Nematode infection tests showed that the number of *H. schachtii* females was significantly reduced in root of double MIOX gene mutants. There was, however, no difference in the biochemical composition and ultrastructure of syncytial cell wall between mutants and wild type plants. Therefore, the authors suggested that the MIOX pathway in syncytia is not responsible for the production of cell wall precursors, but rather removes excess myo-inositol from syncytia (Siddique et al., 2014).

The second pathway involves four UDG genes (UDG1, UDG2, UDG3 and UDG4) that form a small family in *Arabidopsis* (Klinghammer & Tenhaken, 2007) and are expressed in syncytia induced by *H. schachtii* in *Arabidopsis* (Siddique et al., 2012). The experiments with mutants showed that *ugd2* and *ugd3* supported fewer and smaller female nematodes that were associated with smaller syncytia as compared with feeding sites in the wild type plants. Moreover, the analysis of the ultrastructure of syncytial wall revealed that in syncytia induced in the *ugd23* double mutant the cell wall ingrowths were absent (Siddique et al., 2012).

5.5 Chemical Composition of the Cell Wall in Nematode Feeding Structures

To date the knowledge of the chemical properties of the cell wall in, nematode feeding sites is rather limited mainly due to the very restricted amounts of material that can be sampled. For this reason, most information available now was obtained from simple histochemical staining experiments on feeding site sections or more recently gained from immunological studies that used monoclonal antibodies against different cell wall polymers. Sobczak (1996) showed for *H. schachtii* by use of different dyes that the syncytial wall contains pectin, whereas lignin and callose could not be detected. However, at the onset of parasitism translucent callose-like material was deposited around the nematode stylet in the ISC (feeding plug; Figure 2(H)), its cell wall as well as at walls of neighbouring cells. More recently, Davies, Lilley, Knox, and Urwin (2012) analyzed older syncytia induced in *Arabidopsis* by *H. schachtii* with a set of various monoclonal antibodies targeting different cell wall polysaccharides. The authors detected cellulose and hemicelluloses (xyloglucan and heteromannan), while xylan was absent which suggests that in syncytia the synthesis of the secondary cell wall does not occur. The staining with antibodies raised against pectin revealed that this polymer is present in the heavily methyl esterified form in older syncytia. On the one hand, these results suggest that the chemical composition of syncytial wall contributes to its strength, which is essential for withstanding the high turgor pressure within the feeding site. On the other hand, Davies et al. (2012) conclude that the high methylester status of pectin in older syncytia might contribute to the capacity for extension and flexibility of cell walls. In feeding sites, the latter trait is especially important for cell wall, which repeatedly contracts and expands during the active nutrient withdrawal performed by the nematode. Recently, Wieczorek et al. (2014) confirmed these results and additionally showed that there are larger amounts of unmethylated HG in cell walls of younger syncytia as compared to older ones. This is in line with the high expression levels of pectate lyases (Wieczorek et al., 2014) as well as PME3 (Hewezi et al., 2008) observed in syncytia at this stage. In contrast to syncytia, in young galls and walls of giant cells elevated levels of highly methylated pectin and lower levels of its unmethylated form were detected. The authors speculate that the reason could be the synthesis of new pectin that is necessary to enable cell walls of giant cells to expand and thicken. In gall tissue, a similar situation was found where parenchyma cells surrounding the giant cells undergo hyperplasia and hypertrophy.

REFERENCES

Abad, P., Gouzy, J., Aury, J. M., Castagnone-Sereno, P., Danchin, E. G., Deleury, E., et al. (2008). Genome sequence of the metazoan plant-parasitic nematode *Meloidogyne incognita*. *Nature Biotechnology, 26*, 909–915.

Bacic, A., Harris, P. J., & Stone, B. A. (1988). Structure and function of plant cell wall. In P. K. Stumpf, & E. E. Conn (Eds.), *The biochemistry of plants* (pp. 297–371). New York: Academic Press.

Barcala, M., García, A., Cabrera, J., Casson, S., Lindsey, K., Favery, B., et al. (2010). Early transcriptomic events in micro-dissected *Arabidopsis* nematode-induced giant cells. *Plant Journal, 61*, 698–712.

Barras, F., Gijsegem, F. V., & Chatterjee, A. K. (1994). Extracellular enzymes and pathogenesis of soft-rot Erwinia. *Annual Review of Phytopathology, 32*, 201–234.

Bera-Maillet, C., Arthaud, L., Abad, P., & Rosso, M. N. (2000). Biochemical characterization of MI-ENG1, a family 5 endoglucanase secreted by the root-knot nematode *Meloidogyne incognita. European Journal of Biochemistry, 267*, 3255–3263.

de Boer, J. M., Davis, E. L., Hussey, R. S., Popeijus, H., Smant, G., & Baum, T. J. (2002). Cloning of a putative pectate lyase gene expressed in the subventral esophageal glands of *Heterodera glycines. Journal of Nematology, 34*, 9–11.

de Boer, J. M., Yan, Y., Wang, X., Smant, G., Hussey, R. S., Davis, E. L., et al. (1999). Developmental expression of secretory beta-1,4-endoglucanases in the subventral esophageal glands of *Heterodera glycines. Molecular Plant-Microbe Interactions, 12*, 663–669.

Bohlmann, H., & Sobczak, M. (2014). The plant cell wall in the feeding sites of cyst nematodes. *Frontiers in Plant Science, 5*, 89.

Boraston, A. B., Bolam, D. N., Gilbert, H. J., & Davies, G. J. (2004). Carbohydrate-binding modules: fine-tuning polysaccharide recognition. *Biochemical Journal, 382*, 769–781.

Cabrera, J., Barcala, M., Fenoll, C., & Escobar, C. (2014). Transcriptomic signatures of transfer cells in early developing nematode feeding cells of *Arabidopsis* focused on auxin and ethylene signalling. *Frontiers in Plant Science, 5*, 107.

Caillaud, M. C., Dubreuil, G., Quentin, M., Perfus-Barbeoch, L., Lecomte, P., de Almeida Engler, J., et al. (2008). Root-knot nematodes manipulate plant cell functions during a compatible interaction. *Journal Plant Physiology, 165*, 104–113.

Caillaud, M. C., Lecomte, P., Jammes, F., Quentin, M., Pagnotta, S., Andrio, E., et al. (2008). MAP65-3 microtubule-associated protein is essential for nematode-induced giant cell ontogenesis in *Arabidopsis. Plant Cell, 20*, 423–437.

Cao, J. (2012). The pectin lyases in *Arabidopsis thaliana*: evolution, selection and expression profiles. *PLoS One, 7*, e46944.

Carpita, N. C., & Gibeaut, D. M. (1993). Structural models of primary cell walls in flowering plants-consistency of molecular structure with the physical properties of the wall during growth. *Plant Journal, 3*, 1–30.

Chen, Q., Rehman, S., Smant, G., & Jones, J. T. (2005). Functional analysis of pathogenicity proteins of the potato cyst nematode *Globodera rostochiensis* using RNAi. *Molecular Plant-Microbe Interactions, 18*, 621–625.

Collins, T., Gerday, C., & Feller, G. (2005). Xylanases, xylanase families and extremophilic xylanases. *FEMS Microbiology Reviews, 29*, 3–23.

Cosgrove, D. J. (2000). Loosening of plant cell walls by expansins. *Nature, 407*, 321–326.

Davies, L. J., Lilley, C. J., Knox, P. J., & Urwin, P. E. (2012). Syncytia formed by adult female *Heterodera schachtii* in *Arabidopsis thaliana* roots have a distinct cell wall molecular architecture. *New Phytologist, 196*, 238–246.

Davis, E. L., Haegeman, A., & Kikuchi, T. (2011). Degradation of the plant cell wall by nematodes. In J. Jones, G. Gheysen, & C. Fenoll (Eds.), *Genomics and molecular genetics of plant-nematode interactions* (pp. 255–272). Netherlands: Springer.

Davis, E. L., Hussey, R. S., Baum, T. J., Bakker, J., Schots, A., Rosso, M. N., et al. (2000). Nematode parasitism genes. *Annual Review of Phytopathology, 38*, 365–396.

Davis, R. W., Smith, J. D., & Cobb, B. G. (1990). A light and electron microscope investigation of the transfer cell region of maize caryopses. *Canadian Journal of Botany, 68*, 471–479.

Deubert, K. H., & Rohde, R. A. (1971). Nematode enzymes. In B. M. Zuckerman, W. F. Mai, & R. A. Rohd (Eds.), *Plant parasitic nematodes* (pp. 73–90). New York: Academic Press.

Ding, X., Shields, J., Allen, R., & Hussey, R. S. (1998). A secretory cellulose-binding protein cDNA cloned from the root-knot nematode (*Meloidogyne incognita*). *Molecular Plant-Microbe Interactions, 11*, 952–959.

Eves-van den Akker, S., Lilley, C. J., Ault, J. R., Ashcroft, A. E., Jones, J. T., & Urwin, P. E. (2014). The feeding tube of cyst nematodes: characterisation of protein exclusion. *PLoS One, 9*, e87289.

Gao, B. L., Allen, R., Davis, E. L., Baum, T. J., & Hussey, R. S. (2004a). Developmental expression and biochemical properties of a beta-1,4-endoglucanase family in the soybean cyst nematode, *Heterodera glycines*. *Molecular Plant Pathology, 5*, 93–104.

Gao, B., Allen, R., Davis, E. L., Baum, T. J., & Hussey, R. S. (2004b). Molecular characterisation and developmental expression of a cellulose-binding protein gene in the soybean cyst nematode *Heterodera glycines*. *International Journal for Parasitology, 34*, 1377–1383.

Gao, B., Allen, R., Maier, T., Davis, E. L., Baum, T. J., & Hussey, R. S. (2002). Identification of a new beta-1,4-endoglucanase gene expressed in the esophageal subventral gland cells of *Heterodera glycines*. *Journal of Nematology, 34*, 12–15.

Gheysen, G., & Fenoll, C. (2002). Gene expression in nematode feeding sites. *Annual Review of Phytopathology, 40*, 191–219.

Gheysen, G., & Mitchum, M. G. (2009). Molecular insights in the susceptible plant response to nematode infection. In R. H. Berg, & C. G. Taylor (Eds.), *Cell biology of plant nematode parasitism. Plant cell monographs* (pp. 45–81). Berlin: Springer-Verlag.

Gilkes, N. R., Henrissat, B., Kilburn, D. G., Miller, R. C., Jr., & Warren, R. A. (1991). Domains in microbial beta-1, 4-glycanases: sequence conservation, function, and enzyme families. *Microbiology Reviews, 55*, 303–315.

Goellner, M., Smant, G., de Boer, J. M., Baum, T. J., & Davis, E. L. (2000). Isolation of beta-1,4-endoglucanase genes from *Globodera tabacum* and their expression during parasitism. *Journal of Nematology, 32*, 154–165.

Goellner, M., Wang, X., & Davis, E. L. (2001). Endo-ß-1,4-glucanase expression in compatible plant nematode interactions. *Plant Cell, 13*, 2241–2255.

Golinowski, W., Grundler, F. M. W., & Sobczak, M. (1996). Changes in the structure of *Arabidopsis thaliana* during female development of the plant-parasitic nematode *Heterodera schachtii*. *Protoplasma, 194*, 103–116.

Gou, J. Y., Miller, L. M., Hou, G., Yu, X. H., Chen, X. Y., & Liu, C. J. (2012). Acetylesterase-mediated deacetylation of pectin impairs cell elongation, pollen germination, and plant reproduction. *Plant Cell, 24*, 50–65.

Grundler, F. M. W., Sobczak, M., & Golinowski, W. (1998). Formation of wall openings in root cells of *Arabidopsis thaliana* following infection by the plant-parasitic nematode *Heterodera schachtii*. *European Journal Plant Pathology, 104*, 545–551.

Gunning, B. E. S., & Pate, J. S. (1969). "Transfer cells" plant cells with wall ingrowths, specialized in relation to short distance transport of solutes- their occurrence, structure, and development. *Protoplasma, 68*, 107–133.

Haegeman, A., Kyndt, T., & Gheysen, G. (2010). The role of pseudo-endoglucanases in the evolution of nematode cell wall-modifying proteins. *Journal of Molecular Evolution, 70*, 441–452.

Haegeman, A., Vanholme, B., & Gheysen, G. (2009). Characterization of a putative endoxylanase in the migratory plant-parasitic nematode *Radopholus similis*. *Molecular Plant Pathology, 10*, 389–401.

Hamamouch, N., Li, C., Hewezi, T., Baum, T. J., Mitchum, M. G., Hussey, R. S., et al. (2012). The interaction of the novel 30C02 cyst nematode effector protein with a plant beta-1,3-endoglucanase may suppress host defence to promote parasitism. *Journal of Experimental Botany, 63*, 3683–3695.

Henrissat, B., & Bairoch, A. (1996). Updating the sequence-based classification of glycosyl hydrolases. *Biochemical Journal, 316*, 695–696.

Henrissat, B., Coutinho, P. M., & Davies, G. J. (2001). A census of carbohydrate-active enzymes in the genome of *Arabidopsis thaliana*. *Plant Molecular Biology, 47*, 55–72.

Hewezi, T., & Baum, T. J. (2013). Manipulation of plant cells by cyst and root-knot nematode effectors. *Molecular Plant-Microbe Interactions, 26*, 9–16.

Hewezi, T., Howe, P., Maier, T. R., Hussey, R. S., Mitchum, M. G., Davis, E. L., et al. (2008). Cellulose binding protein from the parasitic nematode *Heterodera schachtii* interacts with *Arabidopsis* pectin methylesterase: cooperative cell wall modification during parasitism. *Plant Cell, 20*, 3080–3093.

Hofmann, J., & Grundler, F. M. W. (2006). Females and males of root-parasitic cyst nematodes induce different symplasmic connections between their syncytial feeding cells and the phloem in *Arabidopsis thaliana*. *Plant Physiology and Biochemistry, 44*, 430–433.

Hofmann, J., Wieczorek, K., Blöchl, A., & Grundler, F. M. W. (2007). Sucrose supply to nematode-induced syncytia depends on the apoplasmic and symplasmic pathways. *Journal of Experimental Botany, 58*, 1591–1601.

Hoth, S., Schneidereit, A., Lauterbach, C., Scholz-Starke, J., & Sauer, N. (2005). Nematode infection triggers the de novo formation of unloading phloem that allows macromolecular trafficking of green fluorescent protein into syncytia. *Plant Physiology, 138*, 383–392.

Hudson, L.C. (2009). Analysis of cell wall synthesis genes in feeding cells formed by root-knot nematodes (Ph.D thesis). Plant Pathology, Raleigh: North Carolina State University.

Jammes, F., Lecomte, P., de Almeida Engler, J., Bitton, F., Martin-Magniette, M. L., Renou, J. P., et al. (2005). Genome wide expression profiling of the host response to root-knot nematode infection in *Arabidopsis*. *Plant Journal, 44*, 447–458.

Jaubert, S., Laffaire, J. B., Abad, P., & Rosso, M. N. (2002). A polygalacturonase of animal origin isolated from the root-knot nematode *Meloidogyne incognita*. *FEBS Letters, 522*, 109–112.

Jones, M. G., & Northcote, D. H. (1972). Nematode-induced syncytium – a multi-nucleate transfer cell. *Journal of Cell Science, 10*, 789–809.

Jones, M. G. K., & Payne, H. L. (1978). The early stages of nematode-induced giant cell formation in roots of *Impatiens balsamina*. *Journal of Nematology, 10*, 70–84.

Keen, N. T., & Roberts, P. A. (1998). Plant parasitic nematodes: digesting a page from the microbe book. *Proceedings of the National Academy of Sciences of the United States of America, 95*, 4789–4790.

Kikuchi, T., Aikawa, T., Kosaka, H., Pritchard, L., Ogura, N., & Jones, J. T. (2007). EST analysis of the pine wood nematode *Bursaphelenchus xylophilus* and *B. mucronatus*. *Molecular and Biochemical Parasitology, 155*, 9–17.

Kikuchi, T., Jones, J. T., Aikawa, T., Kosaka, H., & Ogura, N. (2004). A family of glycosyl hydrolase family 45 cellulases from the pine wood nematode *Bursaphelenchus xylophilus*. *FEBS Letters, 572*, 201–205.

Kikuchi, T., Li, H., Karim, N., Kennedy, M. W., Moens, M., & Jones, J. T. (2009). Identification of putative expansin-like genes from the pine wood nematode, *Bursaphelenchus xylophilus*, and evolution of the expansin gene family within the nematoda. *Nematology, 11*, 355–364.

Kikuchi, T., Shibuya, H., Aikawa, T., & Jones, J. T. (2006). Cloning and characterization of pectate lyases expressed in the esophageal gland of the pine wood nematode Bursaphelenchus xylophilus. *Molecular Plant-Microbe Interactions, 19*, 280–287.

Kim, J., Shiu, S. H., Thoma, S., Li, W. H., & Patterson, S. E. (2006). Patterns of expansion and expression divergence in the plant polygalacturonase gene family. *Genome Biology, 7*, R87.

Klinghammer, M., & Tenhaken, R. (2007). Genome-wide analysis of the UDP-glucose dehydrogenase gene family in *Arabidopsis*, a key enzyme for matrix polysaccharides in cell walls. *Journal of Experimental Botany, 58*, 3609–3621.

Kudla, U., Milac, A. L., Qin, L., Overmars, H., Roze, E., Holterman, M., et al. (2007). Structural and functional characterization of a novel, host penetration-related pectate lyase from the potato cyst nematode *Globodera rostochiensis*. *Molecular Plant Pathology, 8*, 293–305.

Kudla, U., Qin, L., Milac, A., Kielak, A., Maissen, C., Overmars, H., et al. (2005). Origin, distribution and 3D-modeling of Gr-EXPB1, an expansin from the potato cyst nematode *Globodera rostochiensis*. *FEBS Letters, 579*, 2451–2457.

Kyndt, T., Haegeman, A., & Gheysen, G. (2008). Evolution of GHF5 endoglucanase gene structure in plantparasitic nematodes: no evidence for an early domain shuffling event. *BMC Evolutionary Biology, 8*, 305.

Ledger, T. N., Jaubert, S., Bosselut, N., Abad, P., & Rosso, M. N. (2006). Characterization of a new beta-1,4-endoglucanase gene from the root-knot nematode *Meloidogyne incognita* and evolutionary scheme for phytonematode family 5 glycosyl hydrolases. *Gene, 382*, 121–128.

Lee, C., Chronis, D., Kenning, C., Peret, B., Hewezi, T., Davis, E. L., et al. (2011). The novel cyst nematode effector protein 19C07 interacts with the *Arabidopsis* auxin influx transporter LAX3 to control feeding site development. *Plant Physiology, 155*, 866–880.

Louvet, R., Cavel, E., Gutierrez, L., Guénin, S., Roger, D., Gillet, F., et al. (2006). Comprehensive expression profiling of the pectin methylesterase gene family during silique development in *Arabidopsis thaliana*. *Planta, 224*, 782–791.

McNeil, M., Darvill, A. G., Fry, S. C., & Albersheim, P. (1984). Structure and function of the primary cell walls of plants. *Annual Review of Biochemistry, 53*, 625–663.

McQueen-Mason, S., & Cosgrove, D. J. (1994). Disruption of hydrogen bonding between plant cell wall polymers by proteins that induce wall extension. *Proceedings of the National Academy of Sciences of the United States of America, 91*, 6574–6578.

de Meutter, J., Vanholme, B., Baun, G., Tytgat, T., & Gheysen, G. (2001). Preparation and sequencing of secreted proteins from the pharyngeal glands of the plant-parasitic nematode *Heterodera schachtii*. *Molecular Plant Pathology, 2*, 297–301.

Mitchum, M. G., Hussey, R. S., Davis, E. L., & Baum, T. J. (2007). Application of biotechnology to understand pathogenesis of nematode plant pathogens. In Z. K. Punja, S. de Boer, & H. Sanfacon (Eds.), *Biotechnology & plant disease management* (pp. 58–86). Oxford: CABI International.

Mitchum, M. G., Sukno, S., Wang, X., Shani, Z., Tsabary, G., Shoseyov, O., et al. (2004). The promoter of the *Arabidopsis thaliana* Cel1 endo-1,4-ßglucanase gene is differentially expressed in plant feeding cells induced by root-knot and cyst nematodes. *Molecular Plant Pathology, 5*, 175–181.

Mitreva-Dautova, M., Roze, E., Overmars, H., De Graaff, L., Schots, A., Helder, J., et al. (2006). A symbiont-independent endo-1,4-beta-xylanase from the plant-parasitic nematode *Meloidogyne incognita*. *Molecular Plant-Microbe Interactions, 19*, 521–529.

Opperman, C. H., Bird, D. M., Williamson, V. M., Rokhsar, D. S., Burke, M., Cohn, J., et al. (2008). Sequence and genetic map of *Meloidogyne hapla*: a compact nematode genome for plant parasitism. *Proceedings of the National Academy of Sciences of the United States of America, 105*, 14802–14807.

Palusa, S. P., Golovkin, M., Shin, S. B., Richardson, D. N., & Reddy, A. S. N. (2007). Organ-specific, developmental, hormonal and stress regulation of expression of putative pectate lyase genes in *Arabidopsis*. *New Phytologist, 174*, 537–550.

Popeijus, H., Overmars, H., Jones, J., Blok, V., Goverse, A., Helder, J., et al. (2000). Enzymology – degradation of plant cell walls by a nematode. *Nature, 406*, 36–37.

Qin, L., Kudla, U., Roze, E. H. A., Goverse, A., Popeijus, H., Nieuwland, J., et al. (2004). Plant degradation: a nematode expansin acting on plants. *Nature, 427*, 30.

Quentin, M., Abad, P., & Favery, B. (2013). Plant parasitic nematode effectors target host defense and nuclear functions to establish feeding cells. *Frontiers in Plant Science, 4*, 53.

Rehman, S., Postma, W., Tytgat, T., Prins, P., Qin, L., Overmars, H., et al. (2009). A secreted SPRY domain-containing protein (SPRYSEC) from the plant-parasitic nematode *Globodera rostochiensis* interacts with a CC-NB-LRR protein from a susceptible tomato. *Molecular Plant-Microbe Interactions, 22*, 330–340.

Rodiuc, N., Vieira, P., Banora, Y. M., & de Almeida Engler, J. (2014). On the track of transfer cells formation by specialized plant-parasitic nematodes. *Frontiers in Plant Science, 5*, 160.

Rosso, M. N., Favery, B., Piotte, C., Arthaud, L., de Boer, J. M., Hussey, R. S., et al. (1999). Isolation of a cDNA encoding a β-1,4-endoglucanase in the root-knot nematode *Meloidogyne incognita* and expression analysis during plant parasitism. *Molecular Plant-Microbe Interactions, 12*, 585–591.

Roze, E., Hanse, B., Mitreva, M., Van Holme, B., Bakker, J., & Smant, G. (2008). Mining the secretome of the root-knot nematode *Meloidogyne chitwoodi* for candidate parasitism genes. *Molecular Plant Pathology, 9*, 1–10.

Rybarczyk-Mydlowska, K., Maboreke, H. R., van Megen, H., van den Elsen, S., Mooyman, P., Smant, G., et al. (2012). Rather than by direct acquisition via lateral gene transfer, GHF5 cellulases were passed on from early *Pratylenchidae* to root-knot and cyst nematodes. *BMC Evolutionary Biology, 12*, 221.

Scheller, H. V., & Ulvskov, P. (2010). Hemicelluloses. *Annual Review of Plant Biology, 61*, 263–289.

Siddique, S., Endres, S., Atkins, J. M., Szakasits, D., Wieczorek, K., Hofmann, J., et al. (2009). Myo-inositol oxygenase genes are involved in the development of syncytia induced by *Heterodera schachtii* in *Arabidopsis* roots. *New Phytologist, 184*, 457–472.

Siddique, S., Endres, S., Sobczak, M., Radakovic, Z. S., Fragner, L., Grundler, F. M., et al. (2014). Myo-inositol oxygenase is important for the removal of excess myo-inositol from syncytia induced by *Heterodera schachtii* in *Arabidopsis* roots. *New Phytologist, 201*, 476–485.

Siddique, S., Sobczak, M., Tenhaken, R., Grundler, F. M., & Bohlmann, H. (2012). Cell wall ingrowths in nematode induced syncytia require UGD2 and UGD3. *PLoS One, 7*, e41515.

Smant, G., Stokkermans, J. P., Yan, Y., de Boer, J. M., Baum, T. J., Wang, X., et al. (1998). Endogenous cellulases in animals: isolation of ß-1,4-endoglucanase genes from two species of plant-parasitic cyst nematodes. *Proceedings of the National Academy of Sciences of the United States of America, 95*, 4906–4911.

Sobczak, M. (1996). Investigations on the structure of syncytia in roots of Arabidopsis induced by the beet cyst nematode *H. schachtii* and its relevance to the sex of the nematode (Ph.D thesis), Germany: Christian-Albrecht Universität.

Sobczak, M., Fudali, S., & Wieczorek, K. (2011). Cell wall modifications induced by nematodes. In J. Jones, G. Gheysen, & C. Fenoll (Eds.), *Genomics and molecular genetics of plant-nematode interactions* (pp. 395–422). Dordrecht: Springer.

Sobczak, M., Golinowski, W. A., & Grundler, F. M. W. (1999). Ultrastructure of feeding plugs and feeding tubes formed by *Heterodera schachtii*. *Nematology, 1*, 363–374.

Somerville, C. (2006). Cellulose synthesis in higher plants. *Annual Review of Cell and Developmental Biology, 22*, 53–78.

Somerville, C., Bauer, S., Brininstool, G., Facette, M., Hamann, T., Milne, J., et al. (2004). Toward a system approach to understanding plant cell walls. *Science, 306*, 2206–2211.

Sukno, S., Shimerling, O., McCuiston, J., Tsabary, G., Shani, Z., Shoseyov, O., et al. (2006). Expression and regulation of the *Arabidopsis thaliana* Cel1 endo 1,4 β glucanase gene during compatible plant-nematode interactions. *Journal of Nematology, 38*, 354–361.

Sun, L., & van Nocker, S. (2010). Analysis of promoter activity of members of the pectate lyase-like (PLL) gene family in cell separation in *Arabidopsis*. *BMC Plant Biology, 10*, 152.

Szakasits, D., Heinen, P., Wieczorek, K., Hofmann, J., Wagner, F., Kreil, D. P., et al. (2009). The transcriptome of syncytia induced by the cyst nematode *Heterodera schachtii* in *Arabidopsis* roots. *Plant Journal, 57*, 771–784.

Vanholme, B., Van Thuyne, W., Vanhouteghem, K., DeMeutter, J., Cannoot, B., & Gheysen, G. (2007). Molecular characterization and functional importance of pectate lyase secreted by the cyst nematode *Heterodera schachtii*. *Molecular Plant Pathology, 8*, 267–278.

Vercauteren, I., de Almeida Engler, J., DeGroodt, R., & Gheysen, G. (2002). An *Arabidopsis thaliana* pectin acetylesterase gene is upregulated in nematode feeding sites induced by root-knot and cyst nematodes. *Molecular Plant-Microbe Interactions, 15*, 404–407.

Vogel, J. P., Raab, T. K., Schiff, C., & Somerville, S. C. (2002). PMR6, a pectate lyase-like gene required for powdery mildew susceptibility in *Arabidopsis*. *Plant Cell, 14*, 2095–2106.

Wang, X., Meyers, D., Yan, Y., Baum, T., Smant, G., Hussey, R., et al. (1999). In *planta* localization of a ß-1,4-endoglucanase secreted by *Heterodera glycines*. *Molecular Plant-Microbe Interactions, 12*, 64–67.

Wang, X., Replogle, A., Davis, E. L., & Mitchum, M. G. (2007). The tobacco *Cel7* gene promoter is auxin-responsive and locally induced in nematode feeding sites of heterologous plants. *Molecular Plant Pathology, 8*, 423–436.

Wieczorek, K., Elashry, A., Quentin, M., Grundler, F. M. W., Favery, B., Seifert, G. J., et al. (2014). A distinct role of pectate lyases in the formation of feeding structures induced by both cyst and root-knot nematodes. *Molecular Plant-Microbe Interactions, 27*, 901–912.

Wieczorek, K., Golecki, B., Gerdes, L., Heinen, P., Szakasits, D., Durachko, D. M., et al. (2006). Expansins are involved in the formation of nematode-induced syncytia in roots of *Arabidopsis thaliana*. *Plant Journal, 48*, 98–112.

Wieczorek, K., Hofmann, J., Blochl, A., Szakasits, D., Bohlmann, H., & Grundler, F. M. (2008). *Arabidopsis* endo-1,4-ß-glucanases are involved in the formation of root syncytia induced by *Heterodera schachtii*. *Plant Journal, 53*, 336–351.

Wieczorek, K., & Seifert, G. J. (2012). Plant cell wall signaling in the interaction with plant-parasitic nematodes. In G. Witzany, & F. Baluska (Eds.), *Biocommunication of plants* (pp. 139–155). Berlin Heidelberg: Springer-Verlag.

Wubben, M. J., Callahan, F. E., & Scheffler, B. E. (2010). Transcript analysis of sedentary parasitic females of the semi-endoparasitic nematode *Rotylenchulus reniformis*. *Molecular and Biochemical Parasitology, 172*, 31–40.

Wyss, U. (1992). Observations on the feeding behaviour of *Heterodera schachtii* throughout development, including events during moulting. *Fundamental & Applied Nematology, 15*, 75–89.

Wyss, U., & Grundler, F. M. W. (1992). *Heterodera schachtii* and *Arabidopsis thaliana*, a model host-parasite interaction. *Nematologica, 38*, 488–493.

Wyss, U., Grundler, F., & Münch, A. (1992). The parasitic behaviour of second-stage juveniles of *Meloidogyne incognita* in roots of *Arabidopsis thaliana*. *Nematologica, 38*, 98–111.

Wyss, U., & Zunke, U. (1986). Observations on the behaviour of second stage juveniles of *Heterodera schachtii* inside host roots. *Revue Nématology, 9*, 153–165.

Yan, Y. T., Smant, G., & Davis, E. (2001). Functional screening yields a new beta-1,4-endoglucanase gene from *Heterodera glycines* that may be the product of recent gene duplication. *Molecular Plant-Microbe Interactions, 14*, 63–71.

Yan, Y. T., Smant, G., Stokkermans, J., Qin, L., Helder, J., Baum, T., et al. (1998). Genomic organization of four beta-1,4-endoglucanase genes in plant-parasitic cyst nematodes and its evolutionary implications. *Gene, 220*, 61–70.

CHAPTER FOUR

The Plant Cell Cycle Machinery: Usurped and Modulated by Plant-Parasitic Nematodes

Janice de Almeida Engler*[,1], Paulo Vieira[§], Natalia Rodiuc[¶], Maria Fatima Grossi de Sa[¶], Gilbert Engler*

*Institut National de la Recherche Agronomique, UMR 1355 ISA/Centre National de la Recherche Scientifique, UMR 7254 ISA/Université de Nice-Sophia Antipolis, UMR ISA, Sophia-Antipolis, France
[§]NemaLab/ICAAM – Instituto de Ciências Agrárias e Ambientais Mediterrânicas, Universidade de Évora, Núcleo da Mitra, Évora, Portugal
[¶]Laboratório de Interação Molecular Planta-Praga, Embrapa Recursos Genéticos e Biotecnologia, PqEB, Brasília, Distrito Federal, Brazil
[1]Corresponding author: E-mail: janice.almeida-engler@sophia.inra.fr

Contents

Abstract

Plant-parasitic nematodes, like root-knot and cyst nematodes, usurp and modulate the plant cell cycle machinery in their favour. Both, the plant mitotic cycle and the endocycle, are essential targets for a successful susceptible interaction between the host plant and these pathogens. Key cell cycle genes, as well as their inhibitor genes, are important components to allow the induction and maintenance of the nematode

Advances in Botanical Research, Volume 73
ISSN 0065-2296
http://dx.doi.org/10.1016/bs.abr.2014.12.003

feeding site (NFS) development. Giant cells undergo acytokinetic mitosis and DNA endoreduplication and are surrounded by rapidly dividing neighbouring cells. Nuclei in an initial syncytial cell do not divide and an increase in nuclei number results from cell fusion followed by endoreduplication. Cells neighbouring syncytia undergo active mitosis before their incorporation into the NFS. The increase of ploidy levels in nuclei of feeding cells most likely sustains their high metabolic activity needed for the nematode growth and reproduction. Functional analysis using adapted microscopy approaches allowed us to gain insight into the role of core cell cycle components in these intriguing feeding sites. In addition, we describe some imaging methods that were developed to investigate nuclear behaviour and how the cell cycle progress in the complex nematode feeding structures of plant host roots.

1. THE PLANT CELL CYCLE IN NEMATODE FEEDING SITES

Sedentary-endoparasitic nematodes are major agricultural worldwide pathogens causing huge annual yield loss (Chitwood, 2003). Among them, root-knot nematodes (RKN) and cyst nematodes (CN) are the most specialized plant-parasitic nematodes with the ability to induce significant changes in root host cells throughout the formation of feeding sites (Figure 1). RKN induce several multinucleate giant-feeding cells within a gall, whereas CN trigger the formation of a multinucleate syncytium. While feeding sites of both types of nematodes have the similar function to nourish the nematodes, their developmental programs differ.

RKN (*Meloidogyne* spp.) induce feeding sites within the host root (Figure 1(E)) by piercing and injecting secretions into five to eight giant-feeding cells generated within the root vascular cylinder (Figure 1(C) and (F); Caillaud, Dubreuil, et al., 2008). These giant cells (GCs) become hypertrophied and contain multiple peculiarly enlarged nuclei. Multinucleation in GCs is a result of numerous mitotic events (de Almeida Engler et al., 1999; Starr, 1993) of which cytokinesis is absent or interrupted (de Almeida Engler et al., 2004; Caillaud, Lecomte, et al., 2008; Jones & Payne, 1978). Throughout GC development and expansion, recurring synchronous mitotic events (Figure 1(A)) take place ending up in an enlarged multinucleate cell with outsized nuclei (de Almeida Engler, Engler, & Gheysen, 2011; Starr, 1993). Giant-feeding cells become highly polyploid most likely via successive endoreduplication cycles (Figure 1(B); de Almeida Engler et al., 1999, 2012; Starr, 1993; Vieira, Escudero, et al., 2013; Wiggers, Starr, & Price, 1990). Unconventional ways of DNA amplification, such as

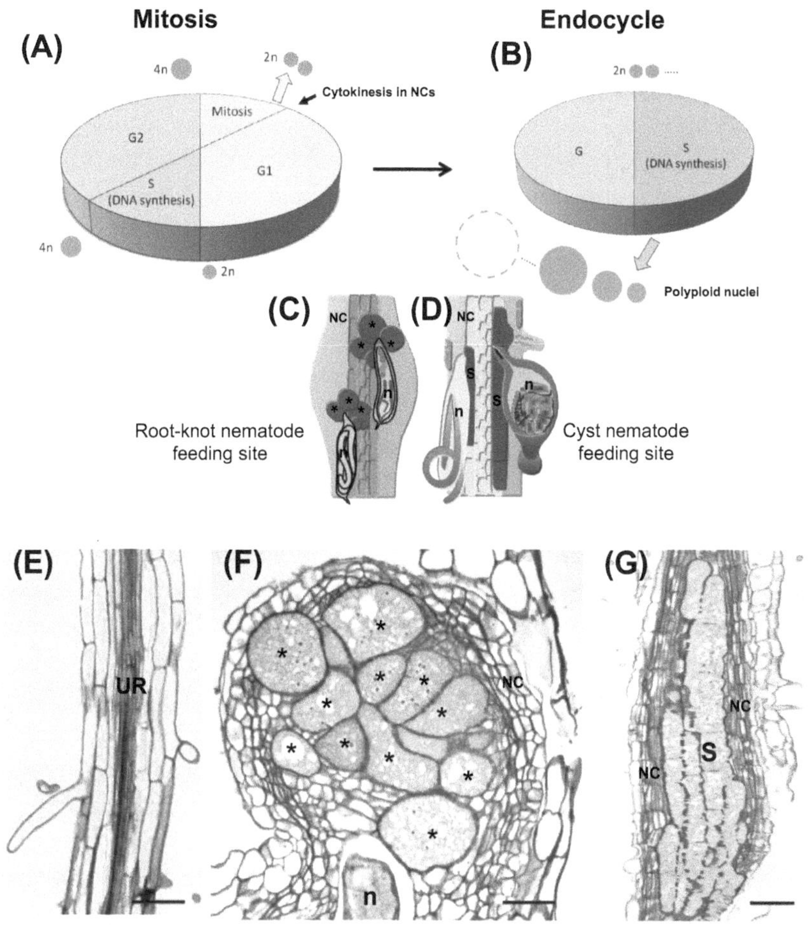

Figure 1 From mitosis to the endocycle (A–B) and root and nematode feeding sites morphology in the model host *Arabidopsis thaliana* (C–G). (A) The mitotic cycle in giant cells is acytokinetic and cytokinetic in neighbouring cells. (B) The endocycle occurring in galls and syncytia. (C–D) Schematic representations of a root-knot nematode-induced gall and a cyst nematode-induced syncytium. (E) Toluidine blue stained longitudinal section of an uninfected root, (F) of a gall and (G) of a syncytium. *, giant cell; n, nematode; NC, neighbouring cells; S, syncytium; UR, uninfected root. *Schemes (A) and (B) are modified from de Almeida Engler and Gheysen (2013) and (C) drawing by Mansour Karimi. Bars = 50 μm.* (See colour plate)

defective mitoses, nuclear fusion and events of endomitosis may also be occurring in GC nuclei. Nevertheless, endomitosis is less apparent to occur, given that chromosome condensation within the nuclear membrane has never been observed in GCs. Typical cellular features of these highly metabolically active giant-feeding cells are a dense cytoplasm filled with numerous organelles, small vacuoles and cell wall ingrowths (Figure 1(F); de Almeida Engler et al., 2011; Rodiuc, Vieira, Youssef Banora, & de Almeida Engler, 2014). Young GCs are surrounded by parenchymatic vascular tissue (VT) cells that are actively asymmetrically dividing in a disordered manner to form multifaceted root swellings named as root-knots or galls.

In contrast, CN (*Globodera* and *Heterodera* spp.) induce a multinucleate feeding site by puncturing and injecting secretions in one single vascular root cell that develop into a syncytium through incorporation of adjacent cells by extensive cell wall dissolution and protoplast fusion (Figure 1(D) and (G); Grundler, Sobczak, & Golinowski, 1998; Sijmons, Atkinson, & Wyss, 1994). Syncytium formation involves considerable reprogramming of root cell development. Neighbouring cells (NC) divide and are then progressively incorporated through cell wall dissolution along the plasmodesmata and later by de novo development of cell wall openings (Grundler et al., 1998; Hoth, Stadler, Sauer, & Hammes, 2008). A syncytium is highly metabolically active and has a dense cytoplasm with proliferating organelles. CN-induced feeding sites almost certainly become multinucleate solely due to the incorporation of nuclei from adjacent cells, following cell wall degradation and cell fusion rather than via mitotic activity. Expanding syncytia exhibit enlargement of nuclei and nucleoli (Endo, 1964; Golinowski, Sobczak, Kurek, & Grymaszewska, 1997), most likely as a result of onset endocycle mechanism (de Almeida Engler et al., 2012). It has been shown that the regulation of the cell cycle machinery in plant cells, as well as in nematode feeding site (NFS), involves changes in the expression and activation of a significant number of core cell cycle genes (de Almeida Engler et al., 1999, 2009, 2012; Vieira, Escudero, et al., 2013).

Development of both GCs and syncytia is a process still scarcely known and most likely are a consequence of the crosstalk between nematode-secreted proteins and various components of the host molecular pathways. It involves coordinated cell cycle programs resulting in the formation of multinucleate feeding cells (de Almeida Engler & Gheysen, 2013). This chapter will therefore focus on cell cycle genes involved in the endocycle of NFS and on the impact of cell cycle inhibitor genes (cyclin-dependent kinase inhibitors (CKI)) on RKN-induced gall development. Herein, we also overview diverse microscopic approaches that we have adapted, developed and applied to investigate plant cell cycle genes in both RKN-induced galls and CN-induced syncytia.

2. GENES INVOLVED IN THE ENDOCYCLE ARE CRITICAL FOR PLANT-PARASITIC NFS ESTABLISHMENT

The plant cell cycle is characterized by four sequential phases: DNA replication (S phase) followed by mitosis (M phase) and cytokinesis (cell division) (Figure 1(A)) which are separated by a first gap (G1 phase) between the end of mitosis and the beginning of DNA synthesis, and a second gap (G2 phase) between the end of DNA synthesis and the beginning of mitosis. Two main forms of incomplete cell cycles leading to genome multiplication are known in plants: mitotic polyploidization in which mitosis is blocked at or after metaphase (e.g. endomitosis or acytokinetic mitosis), and endoreduplication (also called endocycle), which is a variant of the standard mitotic cell cycle in which there are recurrent S-phase and G-phase without M-phase, resulting in an exponential increase in the cell ploidy levels (Figure 1(B); Gutierrez et al., 2002; Sugimoto-Shirasu & Roberts, 2003). Genome amplification, as seen in nematode feeding cells, is often associated with tissues implicated in growth and nutrition comparable to tapetal cells, endosperm or tomato fruits (Chevalier et al., 2011; D'Amato, 1964; Grafi & Larkins, 1995; Rodiuc et al., 2014).

Early cytological studies on NFS encouraged the idea that endoreduplication occurs in RKN-induced GCs and CN-induced syncytia (de Almeida Engler et al., 1999; Rohde & McClure, 1975; Rubinstein & Owens, 1964). Previous observations of ^{3}H-thymidine incorporation into the nuclei of GCs and syncytia have shown that intense DNA replication cycles take place in nematode feeding cells (see Section 4.1; de Almeida Engler et al., 1999). The relevance of DNA replication for GCs and syncytia has been also evidenced by using cell cycle blockers (see Section 4.4) and flow cytomertry measurements (Section 4.7).

To date, substantial progress has been made to understand the molecular basis of endoreduplication in plants, focussing on the model *Arabidopsis thaliana* (Boudolf et al., 2009; De Veylder, Larkin, & Schnittger, 2011; Lammens et al., 2008). Both the ordinary mitotic and the endoreduplication cycles essentially make use of the same universal cell cycle machinery (Bourdon et al., 2010), which involves cyclin-dependent kinases (CDK) activated by their counterpartners, the cyclins (CYC). Eukaryotes have distinct types of CDK and cyclins involved in different cell cycle phases. Since G1 and S phases are common to the mitotic cycle and the endocycle, similar proteins are critical for cell cycle progression of both mechanisms. Key components of the core cell cycle machinery, such as *CDKA*;1, *CDKB*1;1, *CYCA*2;1 and *CYCB*1;1, have been previously described to be early upregulated genes in NFS (see Sections 4.2 and 4.3; de Almeida Engler et al., 1999; Niebel et al., 1996).

Recently, a set of core cell cycle genes (*CCS52A1*, *CCS52B*, *DEL1* and *RHL1*) presently known to be implicated in the plant endoreduplication cycle was evaluated in both RKN- and CN-induced feeding cells (de Almeida Engler et al., 2012). Cell cycle switch 52 (CCS52) proteins (CCS52A and CCS52B) are part of the anaphase-promoting complex/cyclosome (APC/C) and play a essential role in cell cycle progression by targeting mitotic cyclins for degradation, promoting the conversion of mitotic cycles to endocycles (Fülöp et al., 2005; Tarayre, Vinardell, Cebolla, Kondorosi, & Kondorosi, 2004; Kondorosi & Kondorosi, 2004). High promoter activity (Favery et al., 2002), as well as transcript levels (de Almeida Engler et al., 2012) of *CCS52A1* was shown to be induced by biotrophic phytopathogens, such as in RKN-induced galls, or in nodules of beneficial plant symbionts (Cebolla et al., 1999). Elevated *CCS52B* promoter activity and mRNA levels have also been observed in young galls and syncytia (de Almeida Engler et al., 2012). In contrast, *CCS52A1* expression decreased in mature feeding sites, supporting the view that endoreduplication slows down during gall or syncytium maturation. Besides the high *CCS52* gene family expression and intense ^{3}H-labelling, both markers for ongoing DNA replication (see Section 4.1; de Almeida Engler et al., 1999), high transcript levels of additional cell cycle genes involved in DNA synthesis (de Almeida Engler et al., unpublished data) are also observed in NFS of both RKN or CN. Persisting high expression levels of genes typically expressed during S cell cycle phase in cells neighbouring the syncytia, suggests that cell division might be followed by cycles of endoreduplication in these cells prior to incorporation (de Almeida Engler et al., unpublished data). Additional endoreduplication cycles take place in nuclei after incorporation of NC into the feeding site. Flow cytometric analysis of syncytium nuclei supports these data showing peaks ranging from 2C to 32C (see Section 4.7; de Almeida Engler et al., unpublished data).

Overexpression, as well as knockdown, of *CCS52* and *DEL1* genes has an impact on GC and syncytium development. Upregulation of the different *CCS52* genes (*CCS52A2* and *CCS52B*), reported to be activators of the endocycle, speeds up gall and syncytium initiation and promotes the formation of both types of feeding sites. These observations suggest that cells at the root infection zone (elongation zone), due to their competence to dedifferentiate and enter the endocycle, are preferentially used for induction of root-knot and CN feeding sites. Accelerated induction of the endocycle forces a stop to the mitotic phase, also key for gall and syncytium development, as illustrated by their hampered development in the *CCS52*-overexpressing plants (de Almeida Engler et al., 2012). These findings support the idea that solely the endocycle phase is not sufficient for NFS development.

A transcription factor of the DP-E2F-like family (*E2Fe/DEL1*) (Vlieghe et al., 2005), which conserves the mitotic state of proliferating cells by repressing transcription of genes essential for cells to enter the DNA endoreduplication cycle, has been as well evaluated in NFS (de Almeida et al., 2012). *DEL1* transcripts are typically detected in mitotically dividing cells. Loss of *DEL1* function results in augmented ploidy levels, while ectopic expression diminishes endoreduplication (Vlieghe et al., 2005). One of the targets of DEL1 is the CCS52A2 (Lammens et al., 2008), which levels are controlled up to a particular threshold in the cell, as a result inhibiting endoreduplication (Lammens et al., 2008; Vlieghe et al., 2005). In addition to DEL1, several other transcription factors have been found to influence endoreduplication (Berckmans & De Veylder, 2009). As an example, the myb three repeat transcription factor AtMYB3R4 can act as an activator or repressor of endoreduplication, depending on its phosphorylation state (Wildermuth, 2010). It will be therefore appealing to find out if these are involved in NFS development.

Root hairless 1 (RHL1) is a homologue of the archaeal DNA topoisomerase VI, needed for successful progression of the endoreduplication cycle in *Arabidopsis* (Sugimoto-Shirasu et al., 2002; 2005). RKN could induce galls with tiny GCs in *rhl1* but unable to develop further. Once endoreduplication is triggered in GCs, repeated endocycle rounds possibly require DNA decatenation and disentanglement enzymes (de Almeida Engler et al., 2012). Moreover, syncytia could not be induced in *rhl1*.

Furthermore, additional components of the plant cell cycle are involved in post-transcriptional regulation influencing the endoreduplication cycle. For example, CDK inhibitors like ICKs/Kip-related proteins (KRPs) and SIAMESE-related genes that may inhibit mitosis and promote endoreduplication in *A. thaliana* (e.g. De Veylder et al., 2001; Kasili et al., 2010; Peres et al., 2007). The involvement of CKI (KRPs) during gall formation is discussed in Section 3.

3. THE INVOLVEMENT OF CKI DURING GALL FORMATION

Plants encode two plant-specific families of cyclin kinase inhibitors (CKIs), the ICK/KRPs (interactors/inhibitors of CDK, or also referred as KRP) and SIM/SMR (SIAMESE) families (Churchman et al., 2006;

De Veylder et al., 2001; Peres et al., 2007; Wang, Fowke, & Crosby, 1997; Van Leene et al., 2010). The use of cell cycle inhibitor genes of the KRP family has been investigated during NFS development and has been contemplated as a strategy to control gall development (Vieira, Engler, & de Almeida Engler, 2012; Vieira, Escudero, et al., 2013; Vieira, Kindt, Gheysen, & de Almeida Engler, 2013). In *Arabidopsis*, seven *KRP* family members have been identified and their encoded proteins localized to the nucleus (Bird, Buruiana, Zhou, Fowke, & Wang, 2007; Boruc et al., 2010; De Veylder et al., 2001; Vieira, Escudero, et al., 2013). The KRPs are small proteins with a C-terminal domain that are required for CDK- or cyclin-binding with an inhibitory function (De Veylder et al., 2001; Inzé & De Veylder, 2006). A biochemical property of CDK inhibitor proteins is its ability to interact with A-type CDKs and D-type cyclins (Wang, Zhou, Bird, & Fowke, 2008). The fine-tuning of KRP protein abundance in plant cells is a key factor to maintain the balance between cell proliferation and cell differentiation (Verkest et al., 2005; Weinl et al., 2005). It is recognized that elevated ectopic expression of *KRP* (e.g. *KRP2*) genes inhibits the mitotic as well as the endoreduplication cycles, while moderate overexpression of *KRP* genes favours endoreduplication over mitotic cell division (Verkest et al., 2005).

Promoter activity, mRNA in situ localization and detailed functional analyses of the *Arabidopsis KRP* gene family during RKN (*Meloidogyne incognita*) infection revealed their expression, and the involvement of three members of this family (*KRP2*, *KRP5* and *KRP6*) on gall ontogeny at different developmental stages (Coelho et al., unpublished data; Vieira et al., 2014; Vieira, Escudero, et al., 2013). The remaining *Arabidopsis KRP* family members (*KRP1*, *KRP3*, *KRP4* and *KRP7*) did not show any significant promoter activity and no significant transcript accumulation was observed in gall tissues. Differences in promoter activity and gene transcripts among *KRP2*, *KRP5* and *KRP6* within the gall tissues suggest diverse roles of these genes during plant-RKN interaction. Whereas *KRP2*, *KRP5* and *KRP6* expression was associated with early gall (7DAI) development, when high mitotic activity takes place, only *KRP2* and *KRP5* revealed GUS expression and transcript accumulation at subsequent stages of gall growth (>14DAI). This suggests that *KRP2* and *KRP5* expression might be also associated with the endoreduplication phase occurring during mid-stages of infection within the GCs (Vieira, Escudero, et al., 2013).

Although GUS promoter and transcript analyses illustrated continuous *KRP2* expression, in vivo GFP-KRP2 observations demonstrated fluctuations of the KRP2 protein levels throughout NFS expansion. Weak

GFP-KRP2 fluorescence was observed for GFP-KRP2 (2-14DAI), associated with the high mitotic activity phase observed in GCs (Vieira, Escudero, et al., 2013). In contrast, in older NFS (>14DAI) increased GFP-KRP2 levels in GCs nuclei suggest its association with the endoreduplication status in feeding cells (Vieira, Escudero, et al., 2013). Protein dynamics and functional analyses of the remaining gall expressed *KRP* genes (*KRP5* and *KRP6*) are still under scrutiny.

Investigation of the relevance of KRP2 for gall development has been attained by reverse genetics using single (*KRP2*) and double ($KRP2^{-/-}$ $KRP6^{-/-}$) mutant lines, as well as multisilenced RNAi lines for the seven *KRP* genes. Resulting data demonstrated that a mutation or attenuation of *KRP* genes in these lines resulted in ectopic activation of mitosis in RKN feeding-sites, illustrated by the increased number of nuclei in GCs, and abnormal proliferation of vascular cells surrounding these GCs (Vieira, Escudero, et al., 2013). This accelerated mitotic activity in galls is consistent with the loss-of-function of *KRP2* mutants in plant cells, which promotes the accumulation of active CDK/CYC complexes, facilitating and triggering cell proliferation (Anzola et al., 2010; Sanz et al., 2011). The increase of CDK levels above the normal endogenous cell threshold presumably triggers a shift toward a faster and longer mitotic phase in galls. Nevertheless, and despite this accelerated mitotic activity in gall tissues, development of the associated nematodes seems unchanged and these physiological alterations display no synergetic effect to RKN reproduction (Vieira, Escudero, et al., 2013).

On the other hand, RKN-infected *Arabidopsis* plants overexpressing *KRP2* lead to a strong inhibition of both mitotic and endoreduplication cycles, resulting in galls with an overall reduced size containing GCs with less nuclei and a severe reduction in the number of NC. The observed cutback in gall size was mostly associated with a blockage of mitotic activity of the normally proliferative NC, as well as the prominent inhibition of the mitotic activity within the GCs. KRP2 protein levels in GCs are tightly controlled during the establishment of the NFS (seen by GFP-KRP2 in vivo analysis), thus, an increase above endogenous levels (seen in the KRP2 overexpressing line) in GCs was sufficient to affect cell expansion, implying less availability of nutrients. Ultimately, most nematodes associated with these galls were unable to properly develop and reproduce (Vieira, Escudero, et al., 2013).

Likewise *KRP2*, two other *KRP* gene family members of *Arabidopsis*, named *KRP1* (Vieira, Escudero, et al., 2013) and *KRP4* (Vieira, Engler, et al., 2012), have been investigated during the RKN-plant host interaction.

The differential nuclear localization among *KRPs* in noninfected plants may reflect diverse functions or different effects on CDK/CYCs activity. While KRP1 and KRP4 proteins present both nuclear and sub-nuclear localization and co-localize with chromosomes during mitosis in plant cells (Boruc et al., 2010; Vieira, Engler, et al., 2012), KRP2 is evenly distributed in the interphase nucleus and apparently degraded during mitosis (Vieira, Escudero, et al., 2013). Unlike *KRP2*, both GUS promoter activity of *KRP1* and *KRP4* showed no expression in gall tissues. However, in a similar way as KRP2, ectopic expression of *KRP1* revealed a severe reduction of gall volume, inherent to the decreased GC size and proliferation of NC. A milder effect on gall development was observed when *KRP4* was overexpressed in galls as compared to that of *KRP1* and *KRP2*. Nevertheless, in both cases nematode development and offspring were significantly disturbed in comparison to control plants (Vieira, Engler, et al., 2012; Vieira, Escudero, et al., 2013).

It is conceivable that misexpression of *KRP1* and *KRP4* could target additional CDK complexes present within gall tissues, as both genes are normally not expressed during gall development. For example, the ectopic *KRP4* expression resulted in the aberrant segregation of nuclear DNA, disturbing GC expansion and nematode reproduction (Vieira, Engler, et al., 2012). Nonetheless, studies of the *KRP* gene family in RKN-induced galls suggest that these core cell cycle components follow a conserved cell cycle track in the same way as noninfected plant host (Van Leene et al., 2010; Vieira et al., 2013).

In summary, impinging on the expression levels of host KRPs can be envisaged as a strategy for disrupting NFS induction within their plant hosts. A plausible strategy is to ectopically express *KRP* genes in NFS, employing their inhibitory properties to the mitotic and endoreduplication cycles. Jointly, both cell cycle mechanisms might be a critical target for modulation and disruption of such specialized NFS.

4. UNDERSTANDING THE CELL CYCLE DURING PLANT–NEMATODE INTERACTIONS USING DIFFERENT MICROSCOPY APPROACHES

4.1 DNA Synthesis in Nematode Feeding Sites

The enhanced nuclear size in GCs and syncytia has intrigued researchers since the 1960s (de Almeida Engler et al., 1999; Rohde & McClure, 1975; Rubinstein & Owens, 1964). Therefore, tritiated thymidine incorporation studies have been performed in order to detect degrees of DNA synthesis needed for the successful parasitism achieved by RKN and CN (de Almeida Engler et al., 1999; Endo, 1971a,b). The cell cycle phase characterized by DNA synthesis is called S phase (Figure 1(A) and (B)). In order to visualize active DNA synthesis in individual nuclei of NFS the following microscopy approach is adopted: upon a 12-h incubation of nematode-infected plants with a medium containing ^{3}H-thymidine, NFS and root samples are processed for standard histology and tissue sections are covered with a photographic emulsion that generates contrasting silver grains upon ^{3}H exposure (de Almeida Engler et al., 1999). As such, DNA synthesis is visible as dark grains, under bright field optics, and as white grains seen by dark field optics, overlaying replicating nuclei. Increasing amounts of silver grains indicate more intense DNA replication. Weak label is present on the root VT as parenchyma cells only divide during secondary VT formation or as the result of new forming xylem and phloem elements (de Almeida Engler et al., 1999). Vascular parenchyma cells essentially retain a competence for cell division (Hemerly et al., 1993). In contrast, very strong ^{3}H labelling strikes GCs undergoing intense mitosis and less intense label is seen in neighbouring dividing cells (Figure 2(A)). Label gradually disappears at later time points in galls (>14DAI), indicating the end of the DNA replication phase and the exit of the mitotic cycle (see Section 4.6). Thus, a gall induced by RKN often

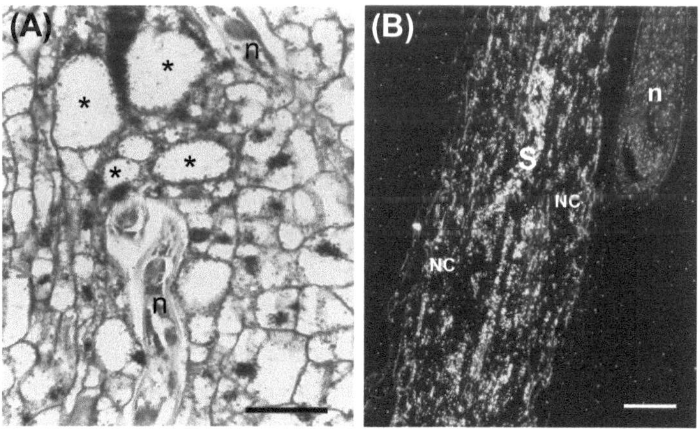

Figure 2 DNA synthesis in a gall and in a syncytium. (A) Dark grains seen by bright field optics illustrate high DNA synthesis in giant cells (arrow upper part) versus lower in neighbouring cells (arrow middle part). (B) White grains seen by dark field optics illustrate higher DNA synthesis in the feeding site (S) compared to neighbouring cells. *, giant cell; n, nematode; NC, neighbouring cells; S, syncytium. Bars = 50 μm.

shows synchronized mitotic GCs and dividing NC at early stages (<10DAI), but nuclear division and DNA synthesis seem not required once GCs are sufficiently developed. Differently, syncytia show intense DNA synthesis activity at early (around 5DAI) (Figure 2(B)) and later developmental stages (>21DAI). Intriguingly, NC close to the nematode's head persist in DNA synthesis. This localized DNA synthesis activity seems linked to the presence of the nematode and to the longitudinal expansion of the syncytium along the root host. This expansion will occur until a syncytium reaches a critical size for nematode reproduction, terminating its life cycle. Nuclear hypertrophy of NC seems to occur prior to their incorporation into the CN-induced syncytium (de Almeida Engler et al., 1999; Golinowski, Grundler, & Sobczak, 1996; Magnusson & Golinowski, 1991), therefore possibly contributing to resilient DNA synthesis surrounding the feeding site. No DNA synthesis activity has been detected in the endodermis, cortex or epidermal tissue layers of *M. incognita*-induced galls or of *Heterodera schachtii*-induced syncytia in *Arabidopsis* roots.

The data obtained support the idea that both types of nematodes are able to induce and form complex feeding structures by switching on a developmental program, which involves multiple cycles of DNA replication in their feeding cells. This increased DNA synthesis in nematode-infected plant roots is most likely needed to maintain the high metabolic activity in feeding sites (de Almeida Engler et al., 1999, 2012).

4.2 Promoter-GUS Fusion Analysis of Plant Cell Cycle Genes in Nematode Feeding Sites

During the last 25 years, the β-glucuronidase (*GUS*) gene has been widely used as a major reporter gene, mainly for the model plant *Arabidopsis*, due to its exceptional sensitivity to monitor promoter activity in all kind of cells, tissues and organs (Jefferson, Kavanagh, & Bevan, 1987). Although applied with success in various plant species, special care should be taken to distinguish endogenous glucuronidase activity from the proper transgene activity. Histochemical localization of GUS activity is performed using a chromogenic GUS substrate, 5-bromo-4-chloro-3-indolyl-β-d-glucuronide (X-Gluc). Although many dye variants including fluorescent ones, have been tagged to the glucuronide moiety, the indolyl-based substrate proved to be very sensitive, specific, nearly free of artefacts, and easy to use.

GUS reporter analysis on NFS revealed the promoter activity of a large number of cell cycle genes (e.g. Figure 3(A)–(D); de Almeida Engler et al., 1999, 2012; Vieira, Escudero, et al., 2013). Considering the large size of

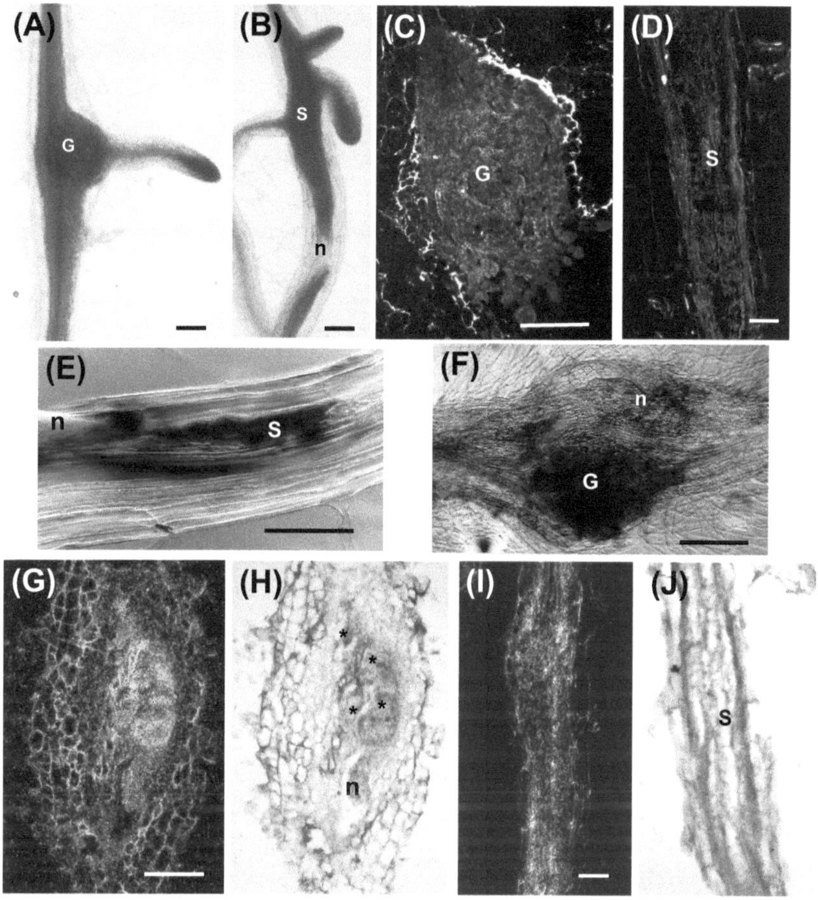

Figure 3 Expression analysis of cell cycle genes by promoter activity and mRNA in situ hybridization. *CKSpro:GUS* lines of whole (A and B) and sectioned (C and D) galls and syncytia induced by *Meloidogyne incognita* in *Arabidopsis* roots. Bright- (A and B) and dark-field (C and D) images illustrate GUS staining in blue and red respectively. (E and F) Whole mount mRNA in situ hybridization using a *CDKA;1* non-radioactive antisense probe (purple blue colour) on a whole syncytium 7 DAI and in a thick slice of a gall respectively. (G–J) In situ localization of *CCS52B* transcripts in a gall and a syncytium 7 DAI. White dots represent the hybridization signal from an antisense *CCS52B* radioactive probe. *, giant cell; G, gall; n, nematode; S, syncytium. Bars = 50 μm. (See colour plate)

nematode feeding cells and often the high levels of cell cycle gene expression, it is common to encounter serious hindrance of homogeneous substrate infiltration, as well as diffusion artefacts caused by the indigo precipitate close to the site of enzyme activity (De Block & Debrouwer, 1992). Although the use of ferri/ferrocyanide as an oxidation catalyst (Mascarenhas & Hamilton, 1992) strongly limits diffusion of the indigo precipitate, it does not eliminate all localization artefacts, mainly when working with large and heterogeneous tissue samples. Moreover, direct exposure of living tissues to a GUS assay does not block induction of gene expression and this can lead to false interpretation. Therefore, applying a brief fixation step, such as acetone pretreatment (90% at −20 °C) of tissues greatly eliminated differential component infiltration and consequently X-Gluc diffusion, still maintaining enzyme activity. Sample pretreatment with acetone instantaneously blocks gene expression avoiding misinterpretation of promoter activity (de Almeida Engler et al., 1999). Results described below were obtained using this adapted protocol.

Promoter activity of key cell cycle regulators like *CDKA*;1, *CYCB*1;1, *CDKB*2;1 and *CYCA*2;1 has been investigated (de Almeida Engler et al., 1999). These were the earliest expressed genes reported showing GUS activity within the first hours of nematode infection (Niebel et al., 1996). Infective steps were monitored in vitro where transparency of *Arabidopsis* roots allowed easy stereomicroscopic visualization of the pursuit of pre-parasitic J2 nematodes (ppJ2) within roots. Highest expression of all cell cycle markers (*CDKA*;1, *CYCB*1;1, *CDKB*2;1 and *CYCA*2;1) occurred in these initial infection phases with RKN as well as CN, and thereafter rapidly fading as NFS matured. Whole galls can be cleared and imaged by differential interference contrast optics (Figure 3(A) and (B); de Almeida Engler et al., 1999). However, a more precise GUS localization is achieved when imaging sectioned feeding sites using dark field optics as shown for *CKS*1 in Figure 3(C) and (D). The latter proved to be very valuable when dealing with less intense staining patterns. Indeed, there where transmitted bright field optics bathes your specimen in a sea of light, dark field optics only allows scattered light, generated in the specimen, (here mainly the indigo precipitate) to enter the objective lens. Weak signals can therefore be much better observed on a dark background. This analysis provided data illustrating that not only GCs but also syncytia underwent cell cycle activity mainly at initial phases of NFS development. At that moment, cell cycle activity in syncytial cells was an interesting observation, considering that mitotic activity as a rule is only observed in NC before their incorporation. Therefore, expression of a mitotic marker like *CYCB*1;1 suggested that the cell cycle in a syncytium progressed at least up to the G2 phase, prior mitosis. These data are supported by DNA synthesis assays described in Section 4.1. Analysis of promoter activity of other cell cycle genes, like *CCS52A1* and *KRPs* (Section 3), has been applied to study gene expression. These studies illustrate the relevance to incorporate this standard histochemical gene expression assay to examine genes involved in NFS development.

4.3 mRNA In situ Hybridization Performed on Paraffin Embedded, Freshly Sliced and Whole Tissue Preparations

Different mRNA in situ hybridization procedures were used to study cell cycle gene expression in nematode-induced feeding sites. Steady state mRNA levels of a large number of cell cycle genes in NFS have been studied by radioactive in situ hybridization on sections and, on whole-mounts and on-section in situ hybridization with radioactive and non-radioactive digoxigenin-labelled antisense probes. Results obtained confirmed and extended our promoter-GUS fusion analysis (de Almeida Engler et al., 1999, 2012; Vieira, Escudero, et al., 2013, unpublished data).

An in-house developed 'Whole Mount mRNA In situ Hybridization' was applied to analyze intact young syncytia and galls (Figure 3(E) and (F); de Almeida Engler, Van Montagu, & Engler, 1994, 1998). Large-sized mature galls were first Vibratome sliced (100–300 μm) to allow better probe penetration in GCs. This method permits a fast detection of transcripts without the need to go through laborious embedding and sectioning procedures (de Almeida Engler et al., 1994, 1998). However, to increase precision of transcript localization, the use of standard paraffin sectioned galls (10 μm) is recommended. For the latter, non-radioactive as well as radioactive probes can be used, the last being more sensitive for transcript detection (Figure 3(G)–(J); de Almeida Engler, de Groodt, Van Montagu, & Engler, 2001). Alternatively, mRNA in situ hybridizations can be carried out on metacrylate embedded sections (5 μm) resulting in a more precise cellular localization, but with lower sensitivity. A number of cell cycle genes have also been analyzed by radioactive in situ hybridization on sectioned galls and syncytia (e.g. de Almeida Engler et al., 2012; Vieira, Escudero, et al., 2013) as well as in *Arabidopsis* and radish seedlings (de Almeida Engler et al., 2009; Zhou et al., 2002). The method has been described in detail by de Almeida Engler et al. (2001).

Initially, four cell cycle gene markers (*CDKA*;1, *CDKB*2;1, *CDKB*1;1 and *CYCA*2.1) were investigated and reported to be early expressed in NFS by de Almeida Engler et al. (1999). *CDKA*;1, *CYCA*2.1 and *CDKB*2;1 are considered markers for competence for cell division in plants. The first is constitutively expressed throughout all cell cycle phases

(Hemerly et al., 1993), while *CDKB*1;1 (Segers et al., 1996; Van Leene et al., 2010) and *CYCA*2.1 (Burssens et al., 2000) are expressed at S to G2 phase and are correlated with active cell division. Expression analysis of genes involved in the endocycle control like *CCS52A*1 and *CCS52B* (Figure 3(G)–(J)) shows high expression in galls and syncytia (de Almeida Engler et al., 2012). Another endocycle control gene, *DEL*1, is weakly expressed in NFS.

Based on DNA synthesis, promoter activity and gene expression data on the cell cycle, we proposed that CN induce feeding cells to progress through the cell cycle, following the G1–S–G2–G1 cell cycle phases and shunting the M phase to enter into the endocycle. An initial mitotic activity might be noticed but blocked during nematode infection, possibly as a consequence of products secreted by the nematode or indirectly by manoeuvring cell cycle gene expression. On the other hand, GCs have been previously reported to go through all cell cycle phases including mitotic activity (G1–S–G2–M) (e.g. Starr, 1993; Vieira, Engler, et al., 2012) until the interruption of cytokinesis, as demonstrated by the formation of incomplete phragmoplast MT structures (de Almeida Engler et al., 2004; Youssef Banora et al., 2011) and aborted cell wall formation (Jones & Payne, 1978). The mitotic phase is then followed by a DNA synthesis phase using the plant endocycle machinery.

4.4 Effect of Pharmacological Cell Cycle Inhibitors on NFS Development

To verify whether DNA synthesis and mitosis are vital for galls and syncytial growth and development, we blocked plant cell cycle progression with phase-specific cell cycle inhibitors such as hydroxyurea (HU) and oryzalin (de Almeida Engler et al., 1999). HU inhibits ribonucleotide diphosphate reductase enzyme and thereby blocking DNA replication by depriving cells of newly synthesized deoxynucleotide triphosphates. Galls treated with HU can block cells at the G1/S transition (Young & Hodas, 1964). Oryzalin is a dinitroaniline herbicide that presents a strong binding affinity for plant tubulins, inhibiting microtubule polymerization blocking the cell cycle in G2. As such oryzalin-treated galls will be blocked cells just before mitosis. Other inhibitors have been previously applied to control nematode infection in various crop species (Davide & Triantaphyllou, 1968; Gershon, 1970; Griffin & Anderson, 1979; Orion & Minz, 1971; Orum, Bartels, & McClure, 1979; Romney, Anderson, & Griffin, 1974) but without any profound analysis on their effect on feeding site development.

DNA synthesis arrest in roots impedes the plant to respond to the presence of the infecting nematodes. Therefore, entry to S phase appears essential for the formation of both GCs and syncytia. S phase blockage applied after infection affects more GCs than syncytia, probably owing to continuous cell fusion in syncytia directing to cell-cycle-independent polynucleation. Ultimately, late treatments on established feeding sites do not affect nematode maturation and reproduction.

Low HU or oryzalin concentrations harmfully affect gall and to a lesser extend syncytia development with effect on nematode maturation and reproduction. Release of short treatments often resulted in recovery in gall development. Along these lines, experiments on galls and syncytia demonstrated that S phase block by HU affected both galls and syncytia, however, an oryzalin-induced G2 block mainly affected gall development and to a lesser extend syncytia expansion. This suggests that cycles of endoreduplication (see Section 2) or other ways of DNA amplification are not enough to propel GC expansion. Accordingly, DNA synthesis is critical for both galls and syncytium development, whereas progression to mitosis was crucial for galls and less for syncytia expansion.

Control experiments to evaluate the effect of chemical treatments on nematode viability suggest that the impermeable nematode cuticle prevents any detectable effect on nematode activity after cell cycle inhibitors incubation. Thus, blocking the plant cell cycle with drugs like HU or oryzalin presumably causes no meaningful effect on nematode development or maturation (de Almeida Engler et al., 1999). These studies show that the use of drugs disturbing the cell cycle can be useful to elucidate mechanisms controlling NFS development.

4.5 Immunocytochemical Detection of Proteins in Planta and in Nematodes

Even though most resins for tissue embedding have been effectively used to localize a wide array of proteins in plants, nearly all media prevent antibody penetrability. Localization of plant (de Almeida Engler et al., 2004; Gubler, 1989; Kronenberger, Desprez, Höfte, Caboche, & Traas, 1993; Youssef Banora et al., 2011) and nematode proteins (Figure 4(A)–(D); Danchin et al., 2010; Jaubert et al., 2005; Vieira, Youssef, et al., 2012), however, can be successfully achieved by immunolocalization using the butylmethyl methacrylate (BM) resin (Figure 4). BM-processed tissues offer a better resolution than paraffin-based embedding procedure or cryosections. Resolution improvement offered by BM and the good structural

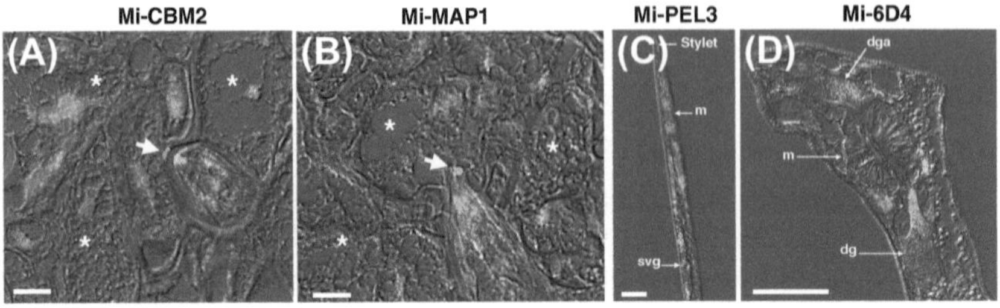

Figure 4 Immunolocalization of *Meloidogyne incognita* secreted proteins in methacrylate sections. (A) Gall sections showing the localization of Mi-CBM2 accumulated at the tip of the stylet and along the giant cell wall (arrow), and (B) Mi-MAP1 accumulated in the nematode amphids and secreted within the apoplasm (arrow). (C) Localization of Mi-PEL3 protein in the subventral glands of pre-parasitic J2, and (D) of Mi-6D4 protein in the dorsal gland (dg) and dorsal gland ampulla (dga) of an adult female. Each figure corresponds to the overlay of images showing the antibody localization (green), DAPI-satined nuclei (blue) and differential interference contrast (grey). *, giant cells; m, metacorpus. Bars = 20 μm. (See colour plate)

preservation of very fragile tissues like, e.g. encountered in *A. thaliana* allowed us to functionally analyze genes and proteins during plant development and nematode infection.

An adapted BM method has been effectively employed for the localization of a number of nematode-secreted proteins (Vieira, Youssef, et al., 2012), for example a Mi-CRT (Jaubert et al., 2005), Mi-CBM2-bearing proteins (Figure 4(A)) Mi-MAP-1 (Figure 4(B)), Mi-PEL3 (Figure 4(C)), Mi-6D4 (Figure 4(D)) and Mi-EFF1 (Danchin et al., 2010; Jaouannet et al., 2012; Vieira et al., 2011; Vieira, Youssef, et al., 2012). The protocol permits the simultaneous detection of nematode-secreted proteins, previously localized within nematode secretory glands, and plant cytoskeletal proteins such as actins, tubulins and an actin-binding protein (de Almeida Engler et al., 2004; Clément et al., 2009; Youssef Banora et al., 2011). A nematode microtubule-associated protein has also been localized in galls, as well as within pre-parasitic J2, parasitic J2 and females (Vieira, Youssef, et al., 2012).

Combined with other methods, like mRNA in situ hybridization, the BM method for protein localization contributes to uncover candidate effector genes responsible for a susceptible plant–nematode interaction.

4.6 Whole Mount Analysis of Fixed and Live Nuclei in Nematode Feeding Sites

Three dimensional (3D) reconstruction of individual images from botanical samples with a complex architecture like galls is possible by applying an

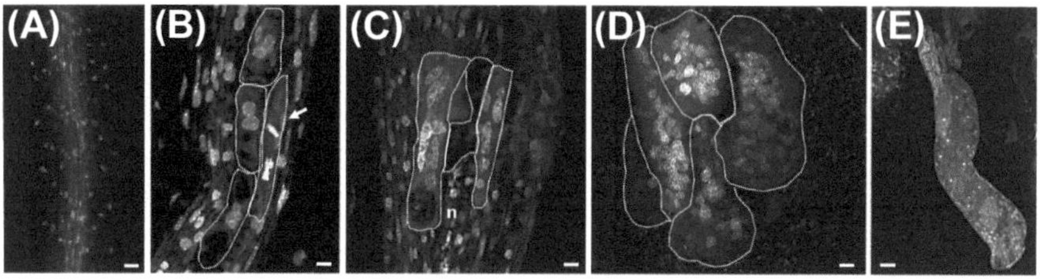

Figure 5 Maximum brightness projections of serial optical sections of whole-mount *Arabidopsis* roots stained with propidium iodide. (A) Noninfected root vascular tissue. (B) A gall (2 DAI) showing young giant cells containing two or more nuclei and mitotic divisions occurring within a giant cell (arrow). (C) A gall (4 DAI) showing an increased number of nuclei within expanding giant cells. (D) A gall (21 DAI) containing a large number of nuclei within giant cells. (E) Developing sedentary juvenile (J2) within the root tissue. Giant cells are marked with dashed lines. n, nematode. Bars = 10 μm.

adapted tissue clearing and propidium iodide staining (Figure 5) on whole fixed nematode-infected roots (Figure 5(B)–(D)) (Figure 5(A); Vieira, Engler, et al., 2012). Three dimensional and maximum brightness projections of a number of optical sections, generated by standard confocal microscopy of whole galls reveal unique information on the distribution of outsized nuclei within GCs from initial stages of infection to gall maturation (Figure 5(B)–(D)). The same method allows following nuclei at all stages of nematode development with high resolution (Figure 5(E); Vieira, Engler, et al., 2012). These observations illustrated that mitotic events are often restricted to young giant-feeding cells, that cell division phases are tightly synchronized and that in wild-type GCs nuclei are usually not connected (Figure 5(B)–(D); Vieira, Engler, et al., 2012).

The main strength of the whole mount clearing method is that it provides simultaneous information on the entire nuclear distribution in an intact gall as well as excellent resolution to discern nuclear morphology in the same sample. This largely facilitated the comparative analysis between RKN transgenic feeding sites of particular cell cycle genes and galls induced in the wild-type *Arabidopsis* roots. Obtaining similar data with standard histological procedures would be extremely time-consuming and very laborious. To get data from live-galls, we removed the cell layers surrounding GCs by performing thick Vibratome slices (100–300 μm) that were immediately imaged by confocal microscopy (de Almeida Engler et al., 2004). This strategy allowed us to in vivo visualize changes in nuclear organization and dynamics (de Almeida Engler et al., 2012) in KRP^{OE} lines (see Section 4.8; Vieira, Engler, et al., 2012; Vieira, Escudero, et al., 2013). Both approaches revealed that nuclei in the *KRP4*^{OE} line presented diverse shapes, interconnected by forming bridges, and often clustered possibly due to a mitotic instability (Vieira, Engler, et al., 2012). KRP4 degradation, during sister chromatid separation, appears to be a prerequisite for mitosis (Boruc et al., 2010), therefore suggesting a similar function for this cell cycle inhibitor protein in galls. As well, the enlarged nuclei containing bulky chromocenters in RKN-induced GCs, compared to wild-type, in mild *CCS52B*^{OE} lines supported the idea of the involvement of the three *Arabidopsis CCS52* genes in the endocycle (de Almeida Engler et al., 2012).

The possibility to visualize the nuclear organization in the entire gall from 3D reconstructed images offers an easy tool to quickly obtain an insight into the nuclear changes occurring not only in NFS but also in a more complex cellular context. This approach will help to better appreciate nuclei morphology in cells and other plant organs during altered gene expression.

However, confocal imaging has its limitations with as major drawback being the limited penetration depth mainly in more complex samples such as galls in thick roots. Recently a protocol has been developed based on classical histological tissue sectioning and 3D volumetric measurement analysis (Cabrera et al., 2015) which is not limited by specimen dimensions and that allows 3D morphological analysis of much bigger and less transparent galls. This approach is easy implementable in many laboratories and will be useful to gather 3D data from galls and GCs from anatomically more complex crops.

4.7 Flow Cytometry for Ploidy Level Studies in NFS

Application of fluorescence-based 4',6-diamidino-2-phenylindole (DAPI) flow cytometry technology allowed total DNA content measurements of NFS (de Almeida Engler et al., unpublished data; Vieira, Escudero, et al., 2013). Ploidy levels in GCs have been estimated in some RKN plant hosts via chromosome counting of squashed GCs (Starr, 1993). Galls induced by RKN in the model host *A. thaliana* show ploidy levels that varied from 2C up to 64C (Figure 6; Galbraith et al., 1991). These data strongly suggest that the 32C to 64C levels corresponds to the polyploidy of GCs considering that uninfected *Arabidopsis* roots ploidy levels go mainly from 2C to 16C (Figure 6). Recent analysis on tomato galls revealed ploidy levels going beyond 64C, like 128C to 256C (de Almeida Engler et al., unpublished data). Indeed, higher ploidy levels have been estimated for GCs on other plant hosts like pea, broad bean and lettuce (Starr, 1993). It will be interesting to further record ploidy levels of galls on other infected hosts using

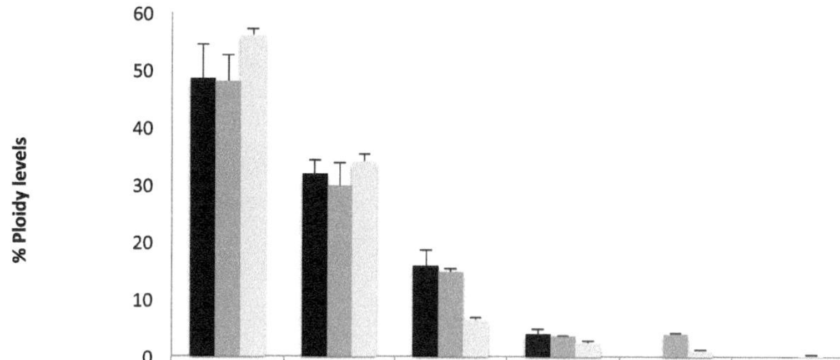

	2C	4C	8C	16C	32C	64C
■ Col-0 Non-infected	48.48	31.76	15.80	3.96	0.00	0.00
Col-0 Syncytia	47.95	29.79	14.73	3.60	3.93	0.00
Col-0 Galls	56.04	33.95	6.39	2.38	1.00	0.24

Figure 6 Flow cytometry analyses of noninfected, CN- and RKN-infected *Arabidopsis* roots. Each column represents the percentage of nuclei with various ploidy levels in noninfected roots, syncytia and galls at 40 days after inoculation. Two independent biological repetitions were performed using roots of six noninfected plants, and a pool of 30 syncytia and galls for each experiment, respectively.

flow cytometry, although interference of surrounding cells may hinder the accurate measurement of GCs ploidy. The limitation of the method is the need of large number of galls in order to extract sufficient GC nuclei (only 5–8 per gall), to perform accurate ploidy measurements. In addition, only a small fraction of the extracted nuclei actually turn out to be from GCs where the greater part of nuclei is from NC. Nuclei ploidy level measurements in mature syncytia of *Arabidopsis* revealed to be from 2C to 32C (Figure 6; de Almeida Engler, unpublished data). This suggests that highest DNA synthesis is seen in RKN-induced galls compared to CN-induced syncytia.

Flow cytometry measurements should be considered in the course of functional analysis of genes impairing gall or syncytia development.

4.8 Whole Mount and Fresh Sliced Galls Employed for In vivo Observations of Cell Cycle Proteins

The development of RKN and CN NFS clearly involves cell cycle activity. In vivo observations of feeding sites and nematode activity, combined with protein localization and dynamics, are now possible by confocal imaging of fresh, thick tissue slices (Figure 7). This strategy gives the possibility to sharply image giant-feeding cells lying deep inside galls (de Almeida Engler et al., 2004; Caillaud, Lecomte, et al., 2008; Clément et al., 2009; Vieira,

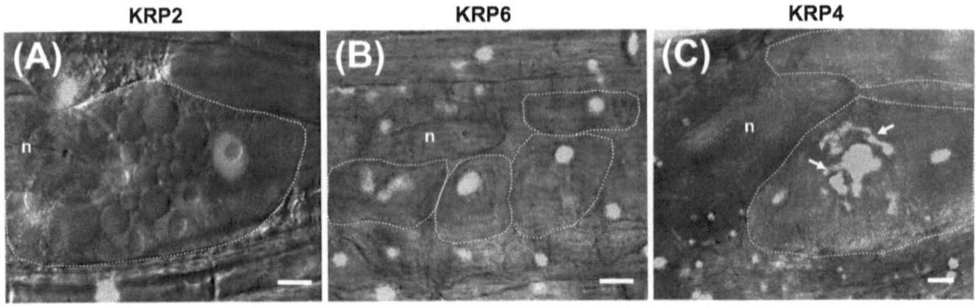

Figure 7 In vivo nuclear localization of different members of the *KRP* family in *Arabidopsis* galls induced by *Meloidogyne incognita*. (A) A gall (14 DAI) showing nuclear localization of GFP-KRP2 within giant cell nuclei. (B) A gall (4 DAI) showing nuclear localization of GFP-KRP6 within giant cell nuclei. (C) A gall (14 DAI) showing abnormal nuclear division in (A) and interconnected nuclei within giant cells (white arrows) in an *KRP4* (*35S:GFP-KRP4*) overexpressing line. Giant cells are marked with dashed lines. n, nematode. Bars = 10 μm. (See colour plate)

Escudero, et al., 2013). Therefore, infected roots (or freshly dissected galls) are mounted in agarose moulds and thick Vibratome slices are generated (100–300 μm) for immediate observation with a confocal microscope. Translucent and good to image GCs could only be obtained by taking care to eliminate a fraction of the NC while leaving GCs and the feeding nematode untouched. This allowed to image intact living GCs. Imaging proteins of interest in intact NC is less critical considering that they have a diameter of around 30 μm compared to GCs around 100–300 μm. Time laps imaging allowed us to follow fluctuations in protein levels and dynamics within living GCs, as well as nematode activity, like feeding (Vieira et al., 2012, 2013).

The use of vibro-sliced galls turned out to be a very attractive and fast method to follow protein dynamics within bulky samples imaged by confocal microscopy. This strategy may therefore contribute to the understanding of the role of cell cycle proteins in galls and to aid to dissect gene function overall during a susceptible plant nematode interaction.

5. CONCLUSIONS AND PERSPECTIVES

Herein, we discussed a view into the cell cycle usurped and modulated by plant-parasitic nematodes. We could show that the interplay of mitotic activity and gene amplification by endoreduplication are prerequisites for NFS maturation. We highlight different microscopic approaches applied to understand cell cycle control on RKN-induced galls and CN-induced syncytia during their ontogeny and maturation.

We illustrated that using diverse microscopy imaging methods in combination with functional gene analyses can greatly help to better understand the cell cycle machinery during plant–nematode interactions as well as in plant roots. The knowledge generated by these studies may well be useful to apply to other plant-pathogen interactions implicating the plant host cell cycle. In addition, the methodologies here described can be used for the functional analysis of any gene of interest from plant or pathogen origin.

Undoubtedly, developing novel strategies to fight against nematode parasitism befall an important concern when bearing in mind limitations on the use of chemical pesticides considered environmentally toxic. Therefore other approaches are required without delay (de Almeida Engler, Favery, Engler, & Abad, 2005) and alternative policies like manoeuvring the plant cell cycle and their regulators can be envisaged to control nematode damage in plant crops.

ACKNOWLEDGEMENTS

NR was supported with a grant of the National Council for Science and Technology (CNPq), JAE and GE had a 1 year foreign visiting professor grant from the Brazilian Federal Agency for Support and Evaluation of Graduate Education (CAPES) in 2013, three months in 2014 as foreign visiting researcher grant (PVE, CSF of CNPq), and work has also been supported by the CAPES/COFECUB program. PV has been founded by FEDER Funds through the Operational Programme for Competitiveness Factors – COMPETE and National Funds through FCT – Foundation for Science and Technology under the Strategic Projects PEst-C/AGR/UI0115/2011 and PEst-OE/AGR/UI0115/2014.

REFERENCES

de Almeida Engler, J., de Groodt, R., Van Montagu, M., & Engler, G. (2001). In situ hybridization to mRNA of *Arabidopsis* tissue sections. *Methods, 23*, 325–334.

de Almeida Engler, J., De Veylder, L., De Groodt, R., Rombauts, S., Boudolf, V., de Meyer, B., et al. (2009). Systematic analysis of cell cycle gene expression during *Arabidopsis* development. *The Plant Journal, 59*, 645–660.

de Almeida Engler, J., de Vleesschauwer, V., Burssens, S., Celenza, J. L., Inzé, D., Montagu, Van, et al. (1999). Molecular markers and cell cycle inhibitors show the importance of cell cycle progression in nematode-induced galls and syncytia. *The Plant Cell, 11*, 793–807.

de Almeida Engler, J., Engler, G., & Gheysen, G. (2011). Unravelling the plant cell cycle in nematode induced feeding sites. In J. Jones, G. Gheysen, & C. Fenoll (Eds.), *Genomics and molecular genetics of plant-nematode interactions* (pp. 349–368). Springer Science + Business Media.

de Almeida Engler, J., Favery, B., Engler, G., & Abad, P. (2005). Loss of susceptibility as an alternative for nematode resistance. *Current Opinion in Biotechnology, 16*, 112–117.

de Almeida Engler, J., & Gheysen, G. (2013). Nematode induced endoreduplication in plant host cells: why and how? *Molecular Plant–Microbe Interactions, 26*, 17–24.

de Almeida Engler, J., Kyndt, T., Vieira, P., Van Capelle, E., Boudolf, V., Sanchez, V., et al. (2012). *CCS52* and *DEL1* genes are key components of the endocycle in nematode-induced feeding sites. *The Plant Journal, 72*, 185–192.

de Almeida Engler, J., Van Montagu, M., & Engler, G. (1994). Whole-mount messenger RNA in situ hybridization in plants. *Plant Molecular Biology Reporter, 12*, 319–329.

de Almeida Engler, J., Van Montagu, M., & Engler, G. (1998). Whole-mount in situ hybridization in plants. In J. M. Martínez-Zapater, & J. Salinas (Eds.), *Arabidopsis protocols, methods in molecular biology* (Vol. 82, pp. 373–384). Totowa: Humana Press.

de Almeida Engler, J., Van Poucke, K., Karimi, M., De Groodt, R., Gheysen, G., Engler, G., et al. (2004). Dynamic cytoskeleton rearrangements in giant cells and syncytia of nematode infected roots. *The Plant Journal, 38*, 12–26.

Anzola, J. M., Sieberer, T., Ortbauer, M., Butt, H., Korbei, B., Weinofer, I., et al. (2010). Putative *Arabidopsis* transcriptional adaptor protein (PROPORZ1) is required to modulate histone acetylation in response to auxin. *Proceedings of the National Academy of Sciences of the USA, 107*, 10308–10313.

Berckmans, B., & De Veylder, L. (2009). Transcriptional control of the cell cycle. *Current Opinion in Plant Biology, 12*, 599–605.

Bird, D. A., Buruiana, M. M., Zhou, Y., Fowke, L. C., & Wang, H. (2007). *Arabidopsis* cyclin-dependent kinase inhibitors are nuclear-localized and show different localization patterns within the nucleoplasm. *Plant Cell Reports, 26*, 861–872.

Boruc, J., Mylle, E., Duda, M., De Clercq, R., Rombauts, S., Geelen, D., et al. (2010). Systematic localization of the *Arabidopsis* core cell cycle proteins reveals novel cell division complexes. *Plant Physiology, 152*, 553–565.

Boudolf, V., Lammens, T., Boruc, J., Van Leene, J., Van Den Daele, H., Maes, S., et al. (2009). CDKB1;1 forms a functional complex with CYCA2;3 to suppress endocycle onset. *Plant Physiology, 150*, 1482–1493.

Bourdon, M., Frangne, N., Mathieu-Rivet, E., Nafati, M., Cheniclet, C., Renaudin, J. P., et al. (2010). Endoreduplication and growth of fleshy fruits. *Progress in Botany, 71*, 101–132.

Burssens, S., De Almeida Engler, J., Beeckamn, T., Richard, C., Shaul, O., Ferreira, P., et al. (2000). Developmental expression of the *Arabidopsis thaliana CycA2*;1 gene. *Planta, 211*, 623–631.

Cabrera, J., Díaz-Manzano, F., Barcala, M., Arganda-Carreras, I., de Almeida Engler, J., Engler, G., et al. (2015). Phenotyping nematode feeding sites: three dimensional reconstruction and volumetric measurements of giant cells induced by root-knot nematodes in *Arabidopsis*. *New Phytologist*. http://dx.doi.org/10.1111/nph.13249.

Caillaud, M.-C., Dubreuil, G., Quentin, M., Perfus-Barbeoch, L., Lecomte, P., de Almeida Engler, J., et al. (2008). Root-knot nematodes manipulate plant cell functions during a compatible interaction. *Journal of Plant Physiology, 165*, 104–113.

Caillaud, M.-C., Lecomte, P., Jammes, F., Quentin, M., Pagnotta, S., Andrio, E., et al. (2008). MAP65-3 microtubule-associated protein is essential for nematode-induced giant cell ontogenesis in *Arabidopsis*. *The Plant Cell, 20*, 423–437.

Cebolla, A., Vinardell, J. M., Kiss, E., Olah, B., Roudier, F., Kondorosi, A., et al. (1999). The mitotic inhibitor ccs52 is required for endoreduplication and ploidy-dependent cell enlargement in plants. *The EMBO Journal, 18*, 4476–4484.

Chevalier, C., Nafati, M., Mathieu-Rivet, E., Bourdon, M., Frangne, N., Cheniclet, C., et al. (2011). Elucidating the functional role of endoreduplication in tomato fruit development. *Annals of Botany, 107*, 1159–1169.

Chitwood, D. J. (2003). Research on plant-parasitic nematode biology conducted by the United States Department of Agriculture Research Service. *Pest Management Science, 59*, 748–753.

Churchman, M. L., Brown, M. L., Kato, N., Kirik, V., Hulskamp, M., Inzé, D., et al. (2006). SIAMESE, a novel plant-specific cell cycle regulator controls endoreplication onset in *Arabidopsis thaliana*. *The Plant Cell, 18*, 3145–3157.

Clément, M., Ketelaar, T., Rodiuc, N., Youssef Banora, M. Y., Smertenko, A., Engler, G., et al. (2009). Actin-depolymerizing factor2-mediated actin dynamics are essential for root-knot nematode infection of *Arabidopsis*. *Plant Cell, 21*, 2963–2979.

Danchin, E. J. G., Rosso, M. N., Vieira, P., de Almeida Engler, J., Coutinho, P., Abad, P., et al. (2010). Multiple lateral gene transfers followed by duplications promoted plant parasitism ability in nematodes. *Proceedings of the National Academy of Sciences of the USA, 107*, 17651–17656.

Davide, R. G., & Triantaphyllou, A. C. (1968). Influence of the environment on development and sex differentiation of root-knot nematode. III. Effect of foliar application of maleic hydrazide. *Nematologica, 14*, 37–46.

De Block, M., & Debrouwer, D. (1992). In-situ enzyme histochemistry on plastic-embedded plant material. The development of an artefact-free β-glucuronidase assay. *The Plant Journal, 2*, 261–266.

De Veylder, L., Beeckman, T., Beemster, G. T., Krols, L., Terras, F., Landrieu, I., et al. (2001). Functional analysis of cyclin-dependent kinase inhibitors of *Arabidopsis*. *Plant Cell, 13*, 1653–1668.

De Veylder, L., Larkin, J. C., & Schnittger, A. (2011). Molecular control and function of endoreduplication in development and physiology. *Trends in Plant Science, 16*, 1360–1385.

D'Amato, F. (1964). Endopolyploidy as a factor in plant tissue development. *Caryologia, 17*, 41–52.

Endo, B. Y. (1964). Penetration and development of *Heterodera glycines* in soybean roots and related anatomical changes. *Phytopathology, 54*, 79–88.

Endo, B. Y. (1971a). Nematode-induced syncytia (giant cells): host/parasite relationships of Heteroderidae. In B. M. Zuckerman, W. F. Mai, & R. A. Rohde (Eds.), *Plant parasitic nematodes* (Vol. 2, pp. 91–117). New York: Academic Press.

Endo, B. Y. (1971b). Synthesis of nucleic acids at infection sites of soybean roots parasitized by *Heterodera glycines*. *Phytopathology, 61*, 395–399.

Favery, B., Complainville, A., Vinardell, J. M., Lecomte, P., Vaubert, D., Mergaert, P., et al. (2002). The endosymbiosis-induced genes ENOD40 and CCS52a are involved in endoparasitic–nematode interactions in *Medicago truncatula*. *Molecular Plant–Microbe Interactions, 15*, 1008–1013.

Fülöp, K., Tarayre, S., Kelemen, Z., Horváth, G., Kevei, Z., Nikovics, K., et al. (2005). *Arabidopsis* anaphase-promoting complexes: multiple activators and wide range of substrates might keep APC perpetually busy. *Cell Cycle, 4*, 1084–1092.

Galbraith, D. W., Harkins, K. R., & Knapp, S. (1991). Systemic endopolyploidy in *Arabidopsis thaliana*. *Plant Physiology, 96*, 985–989.

Gershon, D. (1970). Studies of aging in nematodes. I. The nematode as a model organism for aging research. *Experimental Gerontology, 5*, 7–12.

Golinowski, W., Grundler, F. M. W., & Sobczak, M. (1996). Changes in the structure of *Arabidopsis thaliana* during female development of the plant-parasitic nematode *Heterodera schachtii*. *Protoplasma, 194*, 103–116.

Golinowski, W., Sobczak, M., Kurek, W., & Grymaszewska, G. (1997). The structure of syncytia. In C. Fenoll, F. M. W. Grundler, & S. A. Ohl (Eds.), *Cellular and molecular aspects of plant–nematode interactions: Developments in plant pathology* (Vol. 10, pp. 80–97). Dordrecht, The Netherlands: Kluwer Academic Publishers.

Grafi, G., & Larkins, B. A. (1995). Endoreduplication in maize endosperm—involvement of m-phase-promoting factor inhibition and induction of s-phase-related kinases. *Science, 269*, 1262–1264.

Griffin, G. D., & Anderson, J. L. (1979). Effects of DCPA, EPTC, and chlorpropham on pathogenicity of *Meloidogyne hapla* to alfalfa. *Journal of Nematology, 11*, 32–36.

Grundler, F. M. W., Sobczak, M., & Golinowski, W. (1998). Formation of cell wall openings in root cells of *Arabidopsis thaliana* following infection by the plant-parasitic nematode *Heterodera schachtii*. *The European Journal of Plant Pathology, 104*, 545–551.

Gubler, F. (1989). Immunofluorescence localisation of microtubules in plant root tips embedded in butyl-methyl methacrylate. *Cell Biology International Reports, 13*, 137–145.

Gutierrez, C., Ramirez-Parra, E., Castellano, M. M., & del Pozo, J. C. (2002). G1 to S transition: more than a cell cycle engine switch. *Current Opinion in Plant Biology, 5*, 480–486.

Hemerly, A. S., Ferreira, P., de Almeida Engler, J., Van Montagu, M., Engler, G., & Inzé, D. (1993). *cdc2a* expression in *Arabidopsis* is linked with competence for cell division. *The Plant Cell, 5*, 1711–1723.

Hoth, S., Stadler, R., Sauer, N., & Hammes, U. Z. (2008). Differential vascularization of nematode-induced feeding sites. *Proceedings of the National Academy of Sciences of the USA, 105*, 12617–12622.

Inzé, D., & De Veylder, L. (2006). Cell cycle regulation in plant development. *The Annual Review of Genetics, 40*, 77–105.

Jaouannet, M., Perfus-Barbeoch, L., Deleury, E., Magliano, M., Engler, G., Vieira, P., et al. (2012). A root-knot nematode-secreted protein is injected into giant cells and targeted to the nuclei. *New Phytologist, 194*, 924–931.

Jaubert, S., Milac, A. L., Petrescu, A. J., de Almeida-Engler, J., Abad, P., & Rosso, M.-N. (2005). In planta secretion of a calreticulin by migratory and sedentary stages of root-knot nematode. *Molecular Plant–Microbe Interactions, 18*, 1277–1284.

Jefferson, R. A., Kavanagh, T. A., & Bevan, M. W. (1987). GUS fusions: β-glucuronidase as a sensitive and versatile gene fusion marker in higher plants. *The EMBO Journal, 6*, 3901–3907.

Jones, M. G. K., & Payne, H. L. (1978). Early stages of nematode-induced giant-cell formation in roots of *Impatiens balsamina*. *Journal of Nematology, 10*, 70–84.

Kasili, R., Walker, J. D., Simmons, A., Zhou, J., De Veylder, L., & Larkin, J. C. (2010). SIAMESE cooperates with the CDH1-like protein CCS52A1 to establish endoreplication in *Arabidopsis thaliana* trichomes. *Genetics, 185*, 257–268.

Kondorosi, E., & Kondorosi, A. (2004). Endoreduplication and activation of the anaphase-promoting complex during symbiotic cell development. *FEBS Letters, 567*, 152–157.

Kronenberger, J., Desprez, T., Höfte, H., Caboche, M., & Traas, J. (1993). A methacrylate embedding procedure developed for immunolocalization on plant tissues is also compatible with in situ hybridization. *Cell Biology International, 17*, 1013–1021.

Lammens, T., Boudolf, V., Kheibarshekan, L., Zalmas, L. P., Gaamouche, T., Maes, S., et al. (2008). Atypical E2F activity restrains APC/CCCS52A2 function obligatory for endocycle onset. *Proceedings of the National Academy of Sciences of the USA, 105*, 14721–14726.

Magnusson, C., & Golinowski, W. (1991). Ultrastructural relationships of the developing syncytium induced by *Heterodera schachtii* (Nematoda) in root tissues of rape. *Canadian Journal of Botany, 69*, 44–52.

Mascarenhas, J. P., & Hamilton, D. A. (1992). Artifacts in the localization of GUS activity in anthers of petunia transformed with a CaMV 35S-GUS construct. *The Plant Journal, 2*, 405–408.

Niebel, A., de Almeida Engler, J., Hemerly, A., Ferreira, P., Van Montagu, M., & Gheysen, G. (1996). Induction of *cdc2a* and *cyc1At* expression in *Arabidopsis* during early phases of nematode-induced feeding cell formation. *The Plant Journal, 10*, 1037–1043.

Orion, D., & Minz, G. (1971). The influence of morphactin on the root-knot nematode *Meloidogyne javanica* and its galls. *Nematologica, 17*, 107–112.

Orum, T. V., Bartels, P. G., & McClure, M. A. (1979). Effect of oryzalin and 1,1-dimethylpiperidinium chloride on cotton and tomato roots infected with the root-knot nematode, *Meloidogyne incognita*. *Journal of Nematology, 11*, 78–83.

Peres, A., Churchman, M. L., Hariharan, S., Himanen, K., Verkest, A., Vandepoele, K., et al. (2007). Novel plant-specific cyclin-dependent kinase inhibitors induced by biotic and abiotic stresses. *The Journal of Biological Chemistry, 282*, 25588–25596.

Rodiuc, N., Vieira, P., Youssef Banora, M., & de Almeida Engler, J. (2014). On the track of transfer cells formation by specialized plant-parasitic nematodes. *Frontiers in Plant Science, 5*, 1–14. epaper 160.

Rohde, R. A., & McClure, M. A. (1975). Autoradiography of developing syncytia in cotton roots infected with *Meloidogyne incognita*. *Journal of Nematology, 7*, 64–69.

Romney, R. K., Anderson, J. L., & Griffin, G. D. (1974). Effects of DCPA on seedling infection by root-knot nematode. *Weed Science, 22*, 51–54.

Rubinstein, J. H., & Owens, R. G. (1964). Thymidine and uridine incorporation in relation to the ontogeny of root-knot syncytia. *Contributions from Boyce Thompson Institute, 22*, 491–502.

Sanz, L., Dewitte, W., Forzani, C., Patell, F., Nieuwland, J., Wen, B., et al. (2011). The *Arabidopsis* D-type cyclin CYCD2;1 and the inhibitor ICK2/KRP2 modulate auxin-induced lateral root formation. *The Plant Cell, 23*, 641–660.

Segers, G., Gadisseur, I., Bergounioux, C., de Almeida Engler, J., Jacqmard, A., Van Montagu, M., et al. (1996). The *Arabidopsis* cyclin-dependent kinase gene *cdc2bAt* is preferentially expressed during S and G2 phases of the cell cycle. *The Plant Journal, 10*, 601–612.

Sijmons, P. C., Atkinson, H. J., & Wyss, U. (1994). Parasitic strategies of root nematodes and associated host cell responses. *Annual Review of Phytopathology, 32*, 235–259.

Starr, J. L. (1993). Dynamics of the nuclear complement of giant cells induced by *Meloidogyne incognita*. *Journal of Nematology, 25*, 416–421.

Sugimoto-Shirasu, K., & Roberts, K. (2003). 'Big it up': endoreduplication and cell-size control in plants. *Current Opinion in Plant Biology, 6*, 544–553.

Sugimoto-Shirasu, K., Roberts, G. R., Stacey, N. J., McCann, M. C., Maxwell, A., & Roberts, K. (2005). RHL1 is an essential component of the plant DNA topoisomerase VI complex and is required for ploidy-dependent cell growth. *Proceedings of the National Academy of Sciences of the USA, 102*, 18736–18741.

Sugimoto-Shirasu, K., Stacey, N. J., Corsar, J., Roberts, K., & McCann, M. C. (2002). DNA topoisomerase VI is essential for endo-reduplication in *Arabidopsis*. *Current Biology, 12*, 1782–1786.

Tarayre, S., Vinardell, J. M., Cebolla, A., Kondorosi, A., & Kondorosi, E. (2004). Two classes of the Cdh1-type activators of the anaphase-promoting complex in plants: novel functional domains and distinct regulation. *The Plant Cell, 16*, 422–434.

Van Leene, J., Hollunder, J., Eeckhout, D., Persiau, G., Van De Slijke, E., Stals, H., et al. (2010). Targeted interactomics reveals a complex core cell cycle machinery in *Arabidopsis thaliana*. *Molecular Systems Biology, 6*, 53.

Verkest, A., Manes, C. L., Vercruysse, S., Maes, S., Van Der Schueren, E., Beeckman, T., et al. (2005). The cyclin-dependent kinase inhibitor KRP2 controls the onset of the endoreduplication cycle during *Arabidopsis* leaf development through inhibition of mitotic CDKA;1 kinase complexes. *Plant Cell, 17*, 1723–1736.

Vieira, P., Danchin, E. G. J., Neveu, C., Crozat, C., Jaubert, S., Hussey, R. S., et al. (2011). The plant apoplasm is an important recipient compartment for nematode secreted proteins. *The Journal of Experimental Biology, 62*, 1241–1253.

Vieira, P., De Clercq, A., Stals, H., Van Leene, J., Van De Slijke, E., Van Isterdael, G., et al. (2014). The cyclin-depedent kinase inhibitor KRP6 induces mitosis and impairs cytokinesis in giant cells induced by plant-parasitic nematodes in *Arabidopsis*. *The Plant Cell, 26*, 2633–2647.

Vieira, P., Engler, G., & de Almeida Engler, J. (2012). Whole-mount confocal imaging of nuclei in giant feeding cells induced by root-knot nematodes in *Arabidopsis*. *New Phytologist, 195*, 488–496.

Vieira, P., Escudero, C., Rodiuc, N., Boruc, J., Russinova, E., Glab, N., et al. (2013). Ectopic expression of kip-related proteins restrains root-knot nematode-feeding site expansion. *New Phytologist, 199*, 505–519.

Vieira, P., Kindt, T., Gheysen, G., & de Almeida Engler, J. (2013). An insight into critical endocycle genes for plant-parasitic nematode feeding sites establishment. *Plant Signaling & Behavior, 8*, e24223.

Vieira, P., Youssef, M., Castagnone-Sereno, P., Rosso, M.-N., Engler, G., & de Almeida Engler, J. (2012). An immunocytochemical procedure for protein localization in nematode pre-parasitic and parasitic stages using methylacrylate-embedded tissues. *Phytopathology, 102*, 990–996.

Vlieghe, K., Boudolf, V., Beemster, G. T. S., Maes, S., Magyar, Z., Atanassova, A., et al. (2005). The DP-E2F-like gene DEL1 controls the endocycle in *Arabidopsis thaliana*. *Current Biology, 15*, 59–63.

Wang, H., Fowke, L. C., & Crosby, W. L. (1997). A plant cyclin-dependent kinase inhibitor gene. *Nature, 386*, 451–452.

Wang, H., Zhou, Y., Bird, D. A., & Fowke, L. C. (2008). Functions, regulation and cellular localization of plant cyclin-dependent kinase inhibitors. *Journal of Microscopy, 231*, 234–246.

Weinl, C., Marquardt, S., Kuijt, S. J. H., Nowack, M. K., Jakoby, M. J., Hülskamp, M., et al. (2005). Novel functions of plant cyclin-dependent kinase inhibitors, ICK1/KRP1, can act non-cell autonomously and inhibit entry into mitosis. *The Plant Cell, 17*, 1704–1722.

Wiggers, R. J., Starr, J. L., & Price, H. J. (1990). DNA content and variation in chromosome number in plant cells affected by *Meloidogyne incognita* and *M. arenaria*. *Phytopathology, 80*, 1391–1395.

Wildermuth, M. C. (2010). Modulation of host nuclear ploidy: a common plant biotroph mechanism. *Current Opinion in Plant Biology, 13*, 1–10.

Young, C. W., & Hodas, S. (1964). Hydroxyurea: inhibitory effect on DNA metabolism. *Science, 146*, 1172–1174.

Youssef Banora, Y. M., Rodiuc, N., Baldacci-Cresp, F., Smertenko, A., Bleve-Zacheo, T., Mellilo, M. T., et al. (2011). Feeding cells induced by phytoparasitic nematodes require γ-tubulin ring complex for microtubule reorganization. *PLoS Pathogens, 12*, e1002343.

Zhou, Y., Fowke, L. C., & Wang, H. (2002). Plant CDK inhibitors: studies of interactions with cell cycle regulators in the yeast two-hybrid system and functional comparisons in transgenic *Arabidopsis* plants. *Plant Cell Reports, 20*, 967–975.

CHAPTER FIVE

Metabolism in Nematode Feeding Sites

Shahid Siddique[1], Florian M.W. Grundler
INRES – Molecular Phytomedicine, Rheinische Friedrich-Wilhelms-University of Bonn, Bonn, Germany
[1]Corresponding author: E-mail: siddique@uni-bonn.de

Contents

Abstract

Plant-parasitic nematodes are dependent on their hosts for nutrient uptake. Whereas migratory nematodes feed on many different cells, sedentary nematodes induce hypermetabolic feeding sites. These feeding sites are the only source of nutrients throughout their life span of several weeks. The sink character of nematode feeding sites (NFSs) was established long ago by experiments with fluorescent dyes and isotope labelling in various plant species. However, until recently, we did not know much about the genes and mechanisms that drive the formation and maintenance of NFSs. Recent work in *Arabidopsis* has identified important players involved in NFS formation. In this chapter, we briefly review major findings related to metabolism in NFSs. Further we describe molecular data from *Arabidopsis* in detail to point out recent progress and to provide a framework for further research and molecular dissection of NFSs functioning.

The hypothesis that nematode feeding sites (NFSs) are hypermetabolic was assessed in experiments on tomato plants, which were infected with root-knot nematodes and were exposed to $^{14}CO_2$. The infected root segments accumulated significantly higher amounts of radioactivity as compared with uninfected segments. This led to the suggestion that infected areas

Advances in Botanical Research, Volume 73
ISSN 0065-2296
http://dx.doi.org/10.1016/bs.abr.2015.02.001

act as metabolic sinks to provide food for nematodes (Bird & Loveys, 1975; Mcclure, 1977). A number of phloem-loading experiments with fluorescent probes or sucrose demonstrated that solutes are translocated from leaves and accumulate in NFSs (Bockenhoff, Prior, Grundler, & Oparka, 1996; Dorhout, Gommers, & Kolloffel, 1993; Hofmann, Wieczorek, Blochl, & Grundler, 2007; Hoth, Schneidereit, Lauterbach, Scholz-Starke, & Sauer, 2005). Finally, all these observations were supported by transcriptomic analyses of NFSs from various plant species, which showed that there is a large increase in the transcript abundance for genes involved in primary metabolism (Ithal et al., 2007; Jammes et al., 2005; Puthoff, Nettleton, Rodermel, & Baum, 2003; Szakasits et al., 2009). In non-infected plants, seeds and pollens are particularly important nutrient sinks. Therefore, it is interesting that the transcriptome of syncytia is more closely related to the transcriptomes of seeds and pollens than to those of other parts of the roots (Szakasits et al., 2009).

1. METABOLISM IN NFSs

Most of the sinks in plants are connected with the source tissue through the phloem. Therefore, the composition and availability of nutrients in sinks is highly dependent on phloem transport and solute composition. Disaccharide sucrose, which is the most abundant sugar in phloem (Kursaanov, 1963), was expected to be the major available nutrient in NFSs. In fact, root exudates from tomato plants infected with the root-knot nematode *Meloidogyne incognita* were shown to contain twice the amount of sucrose as exudate from healthy roots (Wang & Bergeson, 1974). Similarly, experiments with the cyst nematode *Heterodera schachtii* showed that supplying sucrose in growth medium enhanced the development of this nematode in *Brassica rapa* roots (Grundler, Betka, & Wyss, 1991). Here we discuss the key metabolic changes with a particular emphasis on sugars in two different types of NFS, i.e. syncytia and giant cells (GCs), in the following sections.

1.1 Metabolism in Cyst Nematode-Induced Syncytia

First analyses to determine metabolite levels in syncytia were performed on soybean roots induced by *Heterodera glycines* (Gommers & Dropkin, 1977). Using Lowry's ultra-microanalytical technique (Lowry & Passonneau, 1972), they compared syncytia with actively growing root tips, and determined the metabolite concentration by quantitative conversion to pyridine nucleotides. The syncytia had similar amounts of ATP, glucose-6-phosphate and proteins compared to those of root tips but contained four times more glucose (Gommers & Dropkin, 1977). Due to technical restrictions, there was no further progress during the following decades. However, the establishment of *Arabidopsis thaliana* as a host provided a model system that allowed robust molecular genetic analysis of plant–nematode interactions (Sijmons, Grundler, Vonmende, Burrows, & Wyss, 1991). Considering the importance of sucrose for sink functioning, Hofmann et al. (2007) cut the syncytial root segments induced by *H. schachtii* in Arabidopsis and analyzed the sucrose content by HPLC-PAD. They found that the sucrose level was markedly higher in syncytia than in uninfected roots, confirming that it is an important source of carbohydrate for developing nematodes. The accumulation of sucrose in syncytia should have profound consequences in terms of storage and processing according to the demand of the developing nematode. In general, plants store excessive sucrose as starch in chloroplasts of photosynthetic tissue, which plays an important role in sustaining metabolism during the night (Smith & Stitt, 2007; Zeeman, Kossmann, & Smith, 2010). In comparison to photosynthetic tissues, nonphotosynthetic tissues may also convert sucrose to starch for long-term storage in specialized plastids called amyloplasts. Starch is remobilized by enzymes, such as amylases and glucosidases, to support various phases of growth such as seedling establishment (Fincher, 1989; Zeeman et al., 2010). Since syncytia contain large amounts of sucrose, it has long been suspected that starch granules are formed in syncytia and other NFSs. In fact, starch granules were detected in syncytia induced by *Nacobbus batatiformis* in sugar beet and *Nacobbus aberrans* in tomato (Jones & Payne, 1977; Schuster, Sandstedt, & Estes, 1964). *Nacobbus* spp. are false root-knot nematodes, and their secondary stage appears to have characteristics of both root-knot and cyst nematodes. However, until recently, it has not been possible to show a conclusive link between the occurrence of starch and functioning of syncytium. Therefore, the role of starch during the development and maintenance of syncytia induced by cyst nematode *H. schachtii* in Arabidopsis roots was recently investigated. Biochemical, microscopic and gene expression analyses showed that the amount of starch in syncytia was markedly higher than that in uninfected control roots (Figure 1) (Hofmann et al., 2008). Importantly, plants with impaired capacity of starch synthesis showed a significant decrease in susceptibility to nematodes (Hofmann et al., 2008). It was therefore suggested

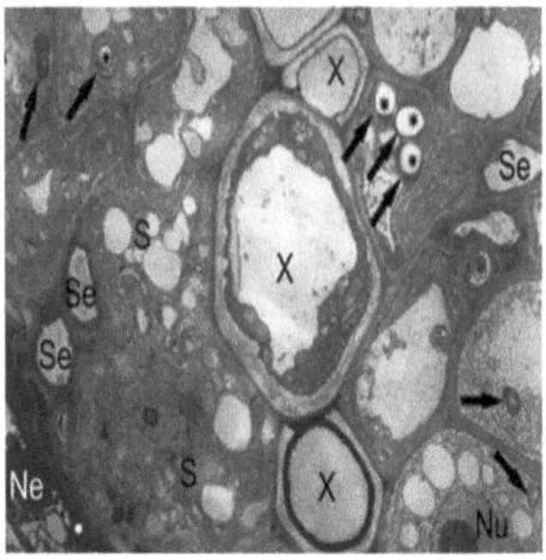

Figure 1 Cross-section of syncytium associated with J2 of *Heterodera schachtii* in Arabidopsis roots. S, syncytium; X, xylem; Ne, necrosis; Nu, nucleus; Se, sieve elements; Arrow, plastid; Asterisk, starch granules. Bars = 5 μm. *Hofmann et al. (2008).*

that starch serves as a carbohydrate buffer in syncytia and is required to cope with fluctuating sugar levels during various stages of nematode feeding (Hofmann et al., 2008). Alternatively, it could be that starch is produced in response to interruption in nematode feeding during moulting. This would lead to an excess of sugar in syncytia, thereby inducing the synthesis of starch.

Regardless of the reason for its synthesis, it is clear that nematodes are unable to take up large starch granules. It seems therefore plausible that starch is degraded to be available for cellular functions and nematode nutrient supply. However, not much is known about starch degradation in syncytium. Future research will aim to connect the role of starch degradation with nematode development. This will greatly help to understand the starch turnover within the syncytium linked to its function in sustaining nematode development.

As it is often the case, the importance of the discovery that syncytium is highly enriched in sucrose content lies in the fact that it raises more questions than it answers. One of the fundamental questions is how sucrose is processed in syncytium? In general, sucrose is processed in sink tissues to form glucose and fructose as precursor of further biochemical reactions. Enzymes from two families catalyze this reaction: invertases (INVs) and sucrose synthases (SUSs). Whereas INVs directly and irreversibly catalyze the conversion of sucrose to glucose and fructose, SUSs produce fructose and uridine diphosphate glucose (UDP)-glucose in a reversible reaction. To understand the sugar processing mechanisms at NFSs, a comprehensive analysis was performed by determining the role of INVs and SUSs in plant–nematode interactions (Cabello et al., 2014). Surprisingly, the development of both root-knot and cyst nematodes was enhanced in multiple INV and SUS mutants. Further analysis showed that the sink character of syncytia was enhanced in INV mutants, which in turn better supported the development of nematodes. The authors therefore concluded that the alteration of INVs and SUSs expression led to local and systemic changes in sugar processing and allocation, in the source—sink relationship, and in the availability of nutrition for nematodes (Cabello et al., 2014).

All of the above described results led to the conclusion that a remodelling of primary metabolism occurs at nematode-induced syncytia. However, published studies continued to lack a global, integrated analysis of syncytial metabolism. This was accomplished by performing transcriptomic and metabolomics profiling of syncytia induced by *H. schachtii* in Arabidopsis roots (Hofmann, El Ashry, et al., 2010; Szakasits et al., 2009). For transcriptome analysis, microaspiration was employed to isolate pure syncytial material. RNA was extracted and hybridized to Affymetrix GeneChips. Segments of the elongation zone of uninfected roots were used as a control. The results showed that out of a total of 21,138 genes, the transcripts of 3893 genes (18.4%) increased and 3338 (15.8) significantly decreased (Szakasits et al., 2009). A gene ontology enrichment of upregulated genes showed that categories associated with high metabolic activity were preferentially overrepresented. A more detailed description of the transcriptome data is provided in Chapter "Introductory Chapter on the Basic Biology of Cyst Nematodes". Therefore, we will focus on the study of the metabolome in this section. Hofmann, El Ashry, et al. (2010) performed a metabolite profiling study to obtain detailed insights into the metabolic changes in nematode-induced syncytia. The root segments containing syncytia were cut and metabolite profiling was performed using gas chromatography coupled to mass spectrometry. Corresponding segments from uninfected roots were used as a control. The results revealed a highly active and coordinated metabolism in infected syncytia. There was a strong local and systemic increase in the levels of various amino acids, phosphorylated metabolites, sugars and organic acids. Among sugars, a pronounced increase in sucrose, raffinose, trehalose and 1-kestose was observed in the syncytia and shoots of infected plants. Arabidopsis does not normally accumulate 1-kestose; therefore, the accumulation of 1-kestose in syncytia is indicative of a unique metabolic response and raises interesting questions regarding its role in the plant—nematode interaction.

A network analysis of syncytial metabolites showed that myoinositol phosphate forms a significant number of correlations with other metabolites in the syncytium, indicating an important role for inositol metabolism (Hofmann, El Ashry, et al., 2010). Myoinositol is a precursor for myoinositol phosphates, phytic acid, phosphatidylinositol phosphate, galactinol and sphingolipids, which have been implicated in a variety of cellular processes (Irvine & Schell, 2001). In addition, myoinositol is converted to UDP-glucuronic acid, which is a major precursor of cell wall polysaccharides. The conversion of myoinositol to UDP-glucuronic acid is catalyzed by the enzyme myoinositol oxygenase (MIOX), encoded by four genes in Arabidopsis (Kanter et al., 2005). As all four MIOX genes are highly upregulated in syncytia (Figure 2), their myoinositol content is significantly reduced compared with uninfected roots (Siddique et al., 2009, 2014; Szakasits et al., 2009). Detailed biochemical, genetic and molecular analyses showed that control of myoinositol metabolism through the expression of *MIOX* genes in the syncytium is required for the proper development of syncytia and to repress defence-signalling pathways via galactinol during parasitism (Siddique et al., 2014).

1.2 Metabolism in Root-Knot Nematode-Induced Giant Cells

GCs, like syncytia, are highly specialized feeding structures that are induced and maintained by root-knot nematodes. Root cells surrounding the infection site swell concomitant with the formation of GCs, leading to the formation of the typical galls. The GC differentiation requires extensive changes in cellular structure and metabolism. This was reflected in a series of experiments conducted by Owen and co-workers in tomato, in which radiotracers were used to demonstrate increased amounts of DNA, RNA and phosphorus in galls compared with healthy roots (Owens & Rubinstein, 1966; Owens & Specht, 1964, 1966). These authors also found that the rate of metabolism, especially in pathways leading to nucleic acid and protein synthesis, was much higher in galls (Owens & Rubinstein, 1966). Observations of the utilization of the hexose monophosphate pathway in nematode-infected tomato roots showed that the pathway is 1.4 to 1.8 times more active in galls compared with adjacent noninfected roots (De Mott, 1965, pp. 63). Endo and Veech performed a number of experiments to identify correlations between metabolite levels and enzyme activity in GCs on soybean roots induced by *M. incognita* (Veech & Endo, 1969). They observed that the activities of malate, isocitrate, succinate, glucose-6-phosphate, alkaline phosphatase, acid phosphatase, esterase, peroxidase, adenosine triphosphatase and cytochrome oxidase were much higher at the site of infection than in noninfected tissue. These

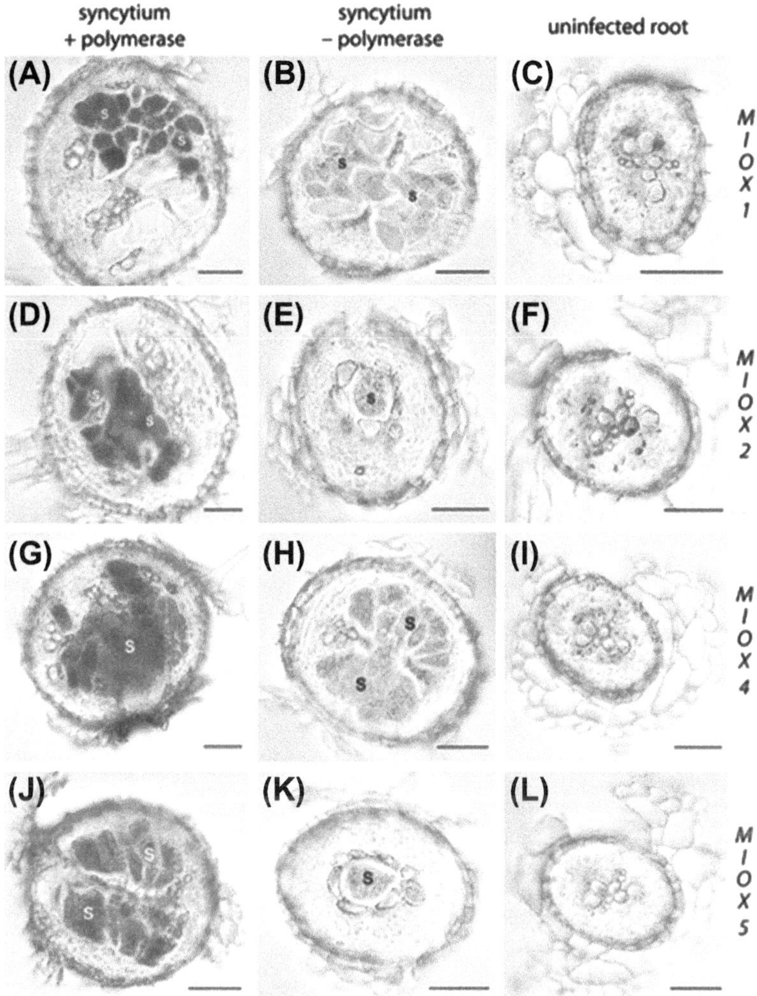

Figure 2 In situ reverse transcriptase chain reaction (RT-PCR) of *MIOX* gene expression in syncytia (s). (A–C) *MIOX1*; (D–F) *MIOX2*; (G–I) *MIOX4*; (J–L) *MIOX5*. (A,D,G,J) specific reaction; (B,E,H,K) control without polymerase; (C,F,I,L) uninfected roots. Bar, 50 μm. *Siddique et al. (2009)*. (See colour plate)

experiments further showed that cells affected by nematodes show a general increase in enzyme activity (Gommers & Dropkin, 1977; Veech & Endo, 1969). Gommers and Dropkin employed a microanalytic technique to demonstrate that GCs on garden balsam (*Impatiens balsamina*) contained higher concentrations of ATP, glucose-6-phospshate, glucose and amino acids (Gommers & Dropkin, 1977).

Microarray analysis of Arabidopsis roots infected with *M. incognita* and *Meloidogyne javanica* of hand-excised galls as compared to noninfected roots was performed in Arabidopsis and tomato along different stages of development (Jammes et al., 2005; Portillo et al., 2013). The functional categories with the highest number of genes were those related to metabolism, which is in accordance with the hypothesis that GCs act as strong sinks. Transcript abundance for most of the genes involved in cell cycle, energy metabolism, protein synthesis and DNA processing increased in galls as compared to control roots. Although transcriptome analysis of galls provided a detailed view of gene expression, they include GCs and the surrounding tissues, which might lead to a dilution of the specific mRNA population within GCs. Therefore, Barcala et al. (2010) used laser capture microdissection for microarray analysis of very young GCs at 3 dpi in Arabidopsis roots induced by *M. javanica* and Portillo et al. (2013), in tomato GCs at 3 and 7 dpi. Again, the functional categories with the highest number of upregulated genes included metabolism, RNA and protein. Similarly, isolation of GCs induced by *Meloidogyne graminicola* on rice roots and subsequent transcriptome analysis revealed a general induction of primary metabolism (Ji et al., 2013). More details can be found in Chapter "Recent Advances in Understanding Plant-Nematode Interactions in Monocots".

Finally, the major compounds of primary metabolism in roots and galls were quantified in galls induced in *Medicago truncatula* by *M. incognita*. Starch contents were also measured using an enzymatic assay (Baldacci-Cresp et al., 2012) and clear differences were observed between galls and uninfected roots. Out of 37 identified metabolites, six amino acids, glucose, sucrose, trehalose, malate and fumarate accumulated at high levels in galls compared with uninfected roots. Furthermore, the amount of starch increased three-fold in galls, suggesting that starch acts as a carbohydrate buffer during nematode development (Baldacci-Cresp et al., 2012).

These changes in the primary metabolism of galls are similar to those observed in syncytia and suggest that both feeding sites share functional similarities despite their different ontogeny (Kyndt, Vieira, Gheysen, & de Almeida-Engler, 2013). This is further supported by recent experiments on members of the sugar-processing enzyme families INV and SUS, which showed that a disruption of the function of these enzymes in Arabidopsis produced equivalent positive effect on susceptibility to cyst and root-knot nematodes (Cabello et al., 2014). In light of the major similarities regarding primary metabolism, investigating the formation and functioning of both NFSs, GCs and syncytia, among different plant species at the molecular level will be of critical importance in the coming years. This is not to ignore the variations in other plant species that have been reported in literature. It will probably lead to the identification of genes that are commonly regulated in syncytium and GCs in different plant species, which in turn will help to select additional sources of resistance in crop plants against nematodes.

2. VASCULARIZATION AND NUTRIENT DELIVERY

Phloem tissue consists of two cell types: **sieve elements (SEs)** and **companion cells (CCs)**. The SEs are specialized, elongated cells connected together at the interface by pores in the cell wall, which facilitate extensive solute transport. The CCs are parenchymatic cells having a large number of ribosomes and mitochondria. Because mature SEs do not contain a nucleus, vacuoles or certain other organelles, they depend on CCs for their maintenance. SEs and CCs are connected through an extensive network of plasmodesmatas (PDs), thus forming an SE/CC complex (Figure 3). Because

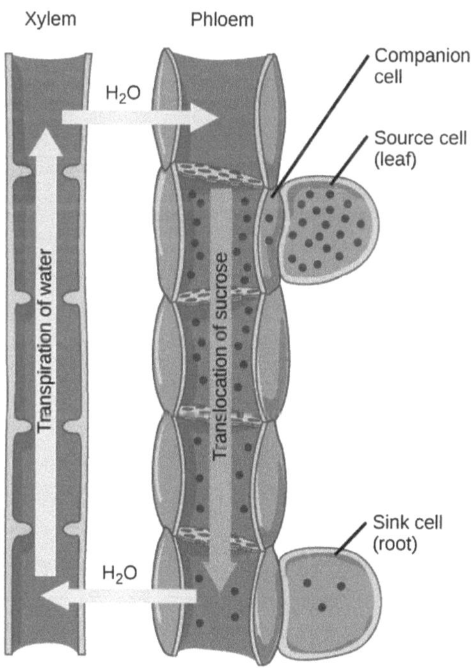

Figure 3 Diagram showing actual models of Phloem translocation in plants. *Reprinted from OpenStax College, Transport of Water and Solutes in Plants. OpenStax CNX. May 10, 2013. Download for free at http://cnx.org/contents/e5aabc6f-71d9-40d5-99f0-0fb2d8d47317@5@5.*

syncytia and GCs are both metabolic sinks, they must have a strong connection to phloem to ensure a supply of assimilates. Although similarities have been drawn between the functioning of syncytia and galls, there are several important differences in the manner that nutrients are transported towards and into these two different feeding sites.

2.1 Solute Supply to Syncytia

Studies of the connectivity between phloem and syncytia have been exclusively performed on syncytia induced in roots of Arabidopsis upon *H. schachtii* infection. Therefore, we will mainly focus on this interaction in this section. The Arabidopsis sucrose transporter AtSUC2 is expressed in the CC of the phloem and is essential for long-distance transport in plants. Using AtSUC2 as a marker for CCs and RS6 as an SE-specific monoclonal antiserum, it was shown that phloem surrounding syncytia is formed de novo and consists of a large number of SEs and few CCs (Hoth et al., 2005; Hoth, Stadler, Sauer, & Hammes, 2008). More recently, using additional markers for tissue identity, it was demonstrated that the phloem surrounding syncytia is a metaphloem (Absmanner, Stadler, & Hammes, 2013).

The manner by which nutrients enter the syncytium from the plant vasculature is a matter of ongoing debate. Initially, syncytia were thought to be symplasmically isolated. This hypothesis was based on experiments in which fluorescent dye microinjected into syncytia was unable to move into adjacent plant cells (Bockenhoff & Grundler, 1994). Further microscopic observations showed that PDs between syncytia and phloem are blocked from the syncytial side by the deposition of an unknown wall material (Grundler, Sobczak, & Golinowski, 1998). Symplasmic isolation of the syncytium was further supported by the microinjection of a range of low-molecular weight fluorescent probes into the syncytium that were unable to move out of the syncytium (Bockenhoff et al., 1996). Loading phloem with both fluorescent probe and ^{14}C-labelled sucrose in leaves resulted in the detection of much higher tracer levels in syncytia than in adjacent areas. Although a strong signal was detected in the syncytia, phloem transport occasionally continued past the syncytium towards the root apex but was often clearly restricted to the syncytium, indicating massive phloem unloading activity (Bockenhoff et al., 1996). Because syncytia were thought to be symplasmically isolated, these observations raised the question of how these nutrients are transported from phloem into syncytia. Sucrose transporters are found in phloem, where they facilitate the uptake of sucrose from the apoplast into the SE/CC complex (Stadler, Brandner, Schulz,

Gahrtz, & Sauer, 1995; Truernit & Sauer, 1995), and in sink tissues, where they catalyze the uptake of sucrose for storage purposes (Vanbel, 1993). Therefore, the detection of AtSUC2 transcripts in the syncytium induced by female nematodes led to the idea that assimilates are mainly imported into syncytia via the apoplast (Juergensen et al., 2003). Further, a detailed expression analysis of 90 annotated sugar transporter genes in Arabidopsis was performed in nematode-induced syncytia, and it was observed that 11 of those genes were significantly upregulated and 19 were significantly downregulated compared with control roots (Hofmann et al., 2009). Functional characterization using loss-of-function mutants demonstrated the importance of these transporters for proper development of nematodes and syncytia (Hofmann et al., 2009). However, experiments with plants expressing free or membrane-anchored green fluorescent protein (GFP) under the control of the AtSUC2 promoter (pAtSUC2:GFP) have suggested that AtSUC2 expression occurs exclusively in CCs surrounding syncytia. Subsequently, GFP moves into SEs and eventually into syncytia (Hoth et al., 2005). The authors of this study therefore concluded that solute transport between phloem and syncytia occurs via PDs (Hoth et al., 2005). The existence of symplasmic route was further studied using grafting experiments, in which Arabidopsis scions expressing pAtSUC2:GFP were grafted onto wild-type roots (Hofmann & Grundler, 2006). Two days after grafting, the roots that showed GFP within the phloem were infected with nematodes. No GFP signal was detected during early stages of syncytium development; however, GFP was detected at 8 days post inoculation (dpi) and later spread throughout the syncytium. To confirm the existence of symplasmic transport during syncytium development, the occurrence of PDs was studied during the infection process. It has been established that plant viruses exploit PD-mediated transport with the help of specialized movement proteins (MPs) to facilitate their DNA or RNA movement between plant cells. Several MPs are localized to PDs and, upon binding, significantly increase the size exclusion limits of PDs (Hofius et al., 2001; Lazarowitz & Beachy, 1999; Lucas, 2006; Scholthof, 2005; Waigmann, Ueki, Trutnyeva, & Citovsky, 2004). Therefore, transgenic plant lines expressing potato leafroll virus (p35S:MP^{17}-GFP) were inoculated with nematodes and monitored for the occurrence of PDs (Figure 4). Only few PDs could be detected at 4 dpi in syncytia and at the interface between syncytia and SEs. However, a higher number of PDs were detected at 7 dpi (Hofmann, Youssef-Banora, de Almeida-Engler, & Grundler, 2010; Hoth et al., 2008). On the other hand, contrasting results were observed in another study where numerous

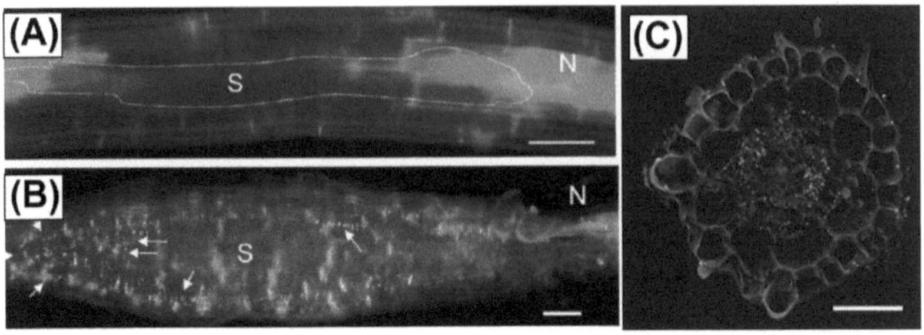

Figure 4 (A) and (B), Localization of plasmodesmatas (PDs) in the Arabidopsis cell walls of syncytium expressing p35S:MP[17]-GFP. Occasional PDs are present at 4 dpi (A) and numerous PDs (white arrows) are detected at 10 dpi (B). (C) p35S:MP[17]-GFP fluorescence in a cross-section of a syncytium at 5 dpi. Many PDs are found in the area of the developing syncytium. S = syncytium, N = nematode. White arrows in (C) indicate GFP fluorescence. Bars, (A) and (B) = 20 μm, (C) = 40 μm. *(A) and (B), Hofmann, Youssef-Banora, et al. (2010); (C), Hoth et al. (2008).* (See colour plate)

PDs were observed at 3-5 dpi (Hoth et al., 2008). Although these studies provided strong evidence for the presence of PDs between syncytium and SEs, it was not clear whether these PDs were functional and played an active role in the symplasmic transport of solute. Callose deposition has been shown to regulate the transport of symplasmic solute by modifying the size exclusion limit of PDs (Wolf, Deom, Beachy, & Lucas, 1991). Therefore, deposition of callose along PD in syncytia was investigated by using an anti-callose antibody. Specific callose deposition along PDs was detected in young syncytia 4 days after inoculation, indicating impaired solute transport. However, callose deposition decreased significantly by 7 days after inoculation.

Considering all the different studies, syncytia appear to be new organ-like root structures, which undergo a defined programme for differentiation. During the initial phase of development, the syncytium is symplasmically isolated which would facilitate cellular reorganization. Although, secondary PDs are formed during this early phase, they are not yet functional. Through the enhanced activity of transporters, the young syncytia are supplied with sugars that are on one hand necessary for the growth of the nematode and on the other hand for the cellular modifications occurring during syncytium formation. Secondary metaphloem is established around syncytia during its expansion. Later, secondary PDs connect the newly differentiated phloem with the expanding syncytium thus forcing the plant to cover the high demand for assimilates supporting growth of both nematode and syncytium.

2.2 Solute Supply to Giant Cells

In contrast to syncytia, CCs are only present during young stages of GCs development but are absent in mature root knots. Therefore, phloem consists of only SEs. It remained unclear whether CCs are consumed or their identity is lost during de novo phloem formation (Hoth et al., 2005, 2008). Using additional markers and hormone response elements, it has since been demonstrated that the phloem surrounding GCs is protophloem (Absmanner et al., 2013).

It is generally assumed that root-knot nematode-induced GCs are symplasmically isolated and that solute transport into feeding site occurs via transporters across membranes (Hoth et al., 2008; Jones, 1981; Jones & Dropkin, 1976). Indeed, a microarray study found that 26 transporter genes representing diverse transport processes were differentially upregulated in response to root-knot nematode infection of Arabidopsis roots (Hammes et al., 2005).

However, contrasting results were obtained in other reports, calling into question the symplasmic isolation of GCs. First, few PDs were present in microscopic observations of GCs and neighbouring cells in the roots of *Impatiens balsamina* (Jones & Dropkin, 1976). Second, it was found that the membrane-impermeable fluorescent dye carboxyfluorescein (CF) accumulated in GCs 2 days following application to tomato leaves (Dorhout et al., 1993). Given that CF requires a relatively long time to spread in tomato leaves (Dorhout et al., 1993), it is unlikely that specific and intense accumulation of CF in GCs could be attributed to slow diffusion across the membrane, as has been suggested elsewhere (Hoth et al., 2008; Wright, Horobin, & Oparka, 1996). Third, a clear MP^{17}-GFP signal used to localize PDs was observed in the walls between GCs and neighbouring cells at 13 dpi, further calling into question the symplasmic isolation of GCs from the surrounding cells (Figure 5). In conclusion, the route of assimilate transport to GCs is less clear than for syncytia. Therefore, additional experiments, for example, grafting transgenic plants expressing phloem-mobile visual markers on wild-type root-stocks infected with root-knot nematodes would help to shed light on this interesting question.

3. AMINO ACID METABOLISM IN NFSs

Amino acids play a vital role in protein synthesis and are precursors for large number of key metabolites. Given the fact that nematodes cannot synthesize all amino acids, it has been suggested that they obtain them from

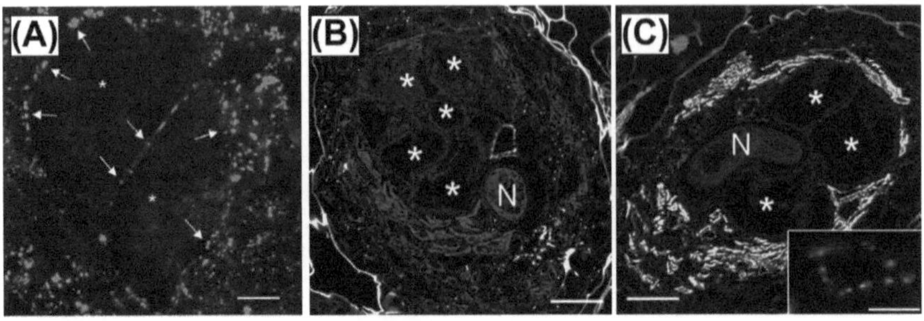

Figure 5 Localization of plasmodesmatas (PDs) in the Arabidopsis cell walls of giant cells (GCs) expressing MP^{17}-GFP. (A) Abundant PDs (white arrows) are present at 13 dpi in walls between GCs and surrounding cells. (B) Cross-section of galls with red fluorescence result from Cy3-labelled second antibody used to detect the sieve element (SE)-specific RS6 antibodies at 18 dpi. (C) Localization of PDs in the Arabidopsis cell walls of GCs expressing Mp^{17}-GFP. Same cross-section as in (B). Green fluorescent protein (GFP) was detected by using an antiserum against it in C. Green colour results from Cy2-labelled second antibody. PDs are primarily present in SEs. N = nematode, Asterisks = GCs. Bars, (A) = 20 μm, (B) and (C) = 150 μm. *(A), Hofmann, Youssef-Banora, et al. (2010); (B) and (C), Hoth et al. (2008).* (See colour plate)

their feeding sites. Indeed, distinct changes in amino acid concentrations were observed in different host plants in response to *Meloidogyne* spp. infection (Hanounik & Osborne, 1975; Hedin & Creech, 1998; Lewis & Mcclure, 1975; Meon, Fisher, & Wallace, 1978). Nonetheless, all these results were obtained by analysis from whole root homogenates including nematodes and therefore did not reflect the actual situation in NFSs. Krauthausen and Wyss (1982) used for the first time a microanalytical technique to get insight into the relative changes in the levels of free amino acid in feeding sites induced by *H. schachtii* on roots of oil radish (*Raphamus sativus*) and oilseed rape (*Brassica napus*). Profound changes were observed in the amino acid composition of both species during different developmental stages of nematode development. For example, relative amounts (%) of valine and gamma-aminobutyric acid increased significantly in syncytium as compared to growing root tips. Similarly, glutamine was shown to have a positive influence on development of *H. schachtii* in *Brassica rapa* (Betka, Grundler, & Wyss, 1991).

Although these studies provided useful information about amino acid metabolism in NFS, it is only recently that a clear accumulation of amino acids in NFS has been shown through large-scale metabolomics approaches (Baldacci-Cresp et al., 2012; Hofmann, El Ashry, et al., 2010). Amino acid transport and metabolism in GCs have been recently discussed in a detailed review (Bartlem, Jones, & Hammes, 2014). I therefore will focus on amino acid metabolism in syncytium here.

In syncytia induced by *H. schachtii* in Arabidopsis roots, levels of glutamate, glutamic acid, aspartic acid and several other amino acids were highly increased (Hofmann, El Ashry, et al., 2010). Among them, aspartic acid is of special interest, as it is a precursor of several other important amino acids including the essential amino acid methionine, which was also highly enriched in syncytia. Methionine is a sulphur-containing amino acid and is an importance source of sulphur in animal diets. Apart from its role as protein constituent and in initiation of mRNA translation, it is very important to control levels of several key metabolites such as ethylene, polyamines and biotin in plant cells. Among them, ethylene and polyamines have been shown to play an important role in nematode infection and development (Hewezi et al., 2010; Wubben, Su, Rodermel, & Baum, 2001). Another notable change was an increase in shikimic acid-based aromatic amino acids which are precursors for several key metabolites including auxin, salicylic acid and other important phenolic compounds. Considering the direct regulation of shikimic acid-dependent pathways by nematodes, it is plausible that aromatic amino acids pathways may play an effective role in the development and functioning of syncytium as sink for nematodes (Gao et al., 2003; Lambert, Allen, & Sussex, 1999). Albeit metabolite profiling using GC-MS provided valuable insights into the amino acid metabolism of syncytium, the heat-sensitive amino acids such as arginine were missed in this study. To cover this aspect, we recently carried out a detailed characterization for arginine metabolism during plant—nematode interaction (Anwar et al., unpublished). Arginine is a nonessential amino acid, which not only serves as an important source of nitrogen but also as precursor of polyamines in plants. Our analyses showed that arginine is highly enriched in syncytium and manipulating the arginine levels has unusual effects on nematode performance in Arabidopsis (Anwar et al., unpublished). These peculiar effects corroborate the earlier studies showing the importance of amino acid metabolism for optimal syncytium functioning (Hofmann, El Ashry, et al., 2010).

4. CONCLUSION AND PERSPECTIVE

The NFSs are intriguing structures, which enable long-term feeding associations of nematodes to the plants (Table 1). However, our understanding of their metabolism is still fragmentary and far from being clearly

Table 1 Specific Metabolic and Vascular Features of Syncytium and Giant Cells (GCs) Reviewed in This Chapter

Feature	Syncytium	GCs
Sucrose	Increase	Increase
Trehalose	Increase	Increase
Glucose	Increase	Increase
Starch	Increase	Increase
Myoinositol	Increase	Unknown
Solute supply	Transporters and plasmodesmatas (PDs)	Transporters
Phloem development	De novo synthesis, sieve elements (SEs) and companion cells	De novo synthesis, only SEs
Phloem identity	Metaphloem	Protophloem

understood. We do not yet know which signals induce the formation of NFSs? How a nematode is able to get continuous supply of nutrients eluding the defence responses and plant detection systems? Which signals induce the formation of numerous phloem elements around NFSs? And what is the contribution of apoplastic and symplasmic transport in nutrient supply to NFSs during different stages of development? Advances in understanding the answers for these questions would require new histological, genetic and biochemical tools in the coming years. To start with, we should identify additional molecular players that are involved in regulation of metabolism in NFSs. By doing so, we will be able to specifically interfere in feeding site metabolism and study the consequences of these manipulations.

ACKNOWLEDGEMENTS

We apologize to many authors whose works on NFSs have not been cited here because of length constraints. We would also like to acknowledge Julia Holbein for her help to improve manuscript language.

REFERENCES

Absmanner, B., Stadler, R., & Hammes, U. Z. (2013). Phloem development in nematode-induced feeding sites: the implications of auxin and cytokinin. *Frontiers in Plant Science, 4*, 241.

Baldacci-Cresp, F., Chang, C., Maucourt, M., Deborde, C., Hopkins, J., Lecomte, P., et al. (2012). (Homo)glutathione deficiency impairs root-knot nematode development in *Medicago truncatula*. *PloS Pathogens, 8*(1).

Barcala, M., Garcia, A., Cabrera, J., Casson, S., Lindsey, K., Favery, B., et al. (2010). Early transcriptomic events in microdissected Arabidopsis nematode-induced giant cells. *The Plant Journal, 61*, 698–712.

Bartlem, D. G., Jones, M. G. K., & Hammes, U. Z. (2014). Vascularization and nutrient delivery at root-knot nematode feeding sites in host roots. *Journal of Experimental Botany, 65*(7), 1789–1798.

Betka, M., Grundler, F., & Wyss, U. (1991). Influence of changes in the nurse cell system (Syncytium) on the development of the cyst Nematode *Heterodera schachtii* — single amino acids. *Phytopathology, 81*(1), 75–79.

Bird, A. F., & Loveys, B. R. (1975). Incorporation of photosynthates by *Meloidogyne javanica. Journal of Nematology,* 7(2), 111–113.

Bockenhoff, A., & Grundler, F. M. W. (1994). Studies on the nutrient uptake by the beet cyst nematode *Heterodera schachtii* by in situ microinjection of fluorescent probes into the feeding structures in *Arabidopsis thaliana. Parasitology, 109*, 249–254.

Bockenhoff, A., Prior, D. A. M., Grundler, F. M. W., & Oparka, K. J. (1996). Induction of phloem unloading in *Arabidopsis thaliana* roots by the parasitic nematode *Heterodera schachtii. Plant Physiology, 112*(4), 1421–1427.

Cabello, S., Lorenz, C., Crespo, S., Cabrera, J., Ludwig, R., Escobar, C., et al. (2014). Altered sucrose synthase and invertase expression affects the local and systemic sugar metabolism of nematode-infected *Arabidopsis thaliana* plants. *Journal of Experimental Botany, 65*(1), 201–212.

De Mott, H. E. (1965). *Observations on the utilization of the hexose monophosphate pathway in nematode infected roots of tomato* (Ph.D. thesis). University of Virginia.

Dorhout, R., Gommers, F. J., & Kolloffel, C. (1993). Phloem transport of carboxyfluorescein through tomato roots infected with *Meloidogyne incognita. Physiology and Molecular Biology of Plants, 43*(1), 1–10.

Fincher, G. B. (1989). Molecular and cellular biology associated with endosperm mobilization in germinating cereal-Grains. *Annual Review of Plant Physiology, 40*, 305–346.

Gao, B. L., Allen, R., Maier, T., Davis, E. L., Baum, T. J., & Hussey, R. S. (2003). The parasitome of the phytonematode *Heterodera glycines. Molecular Plant-Microbe Interactions, 16*(8), 720–726.

Gommers, F. J., & Dropkin, V. H. (1977). Quantitative histochemistry of nematode-induced transfer cells. *Phytopathology, 67*(7), 869–873.

Grundler, F., Betka, M., & Wyss, U. (1991). Influence of changes in the nurse cell system (syncytium) on sex determination and development of the cyst nematode *Heterodera schachtii*: total amounts of proteins and amino acids. *Phytopathology, 81*(1), 70–74.

Grundler, F. M. W., Sobczak, M., & Golinowski, W. (1998). Formation of wall openings in root cells of *Arabidopsis thaliana* following infection by the plant-parasitic nematode *Heterodera schachtii. European Journal of Plant Pathology, 104*(6), 545–551.

Hammes, U. Z., Schachtman, D. P., Berg, R. H., Nielsen, E., Koch, W., McIntyre, L. M., et al. (2005). Nematode-induced changes of transporter gene expression in Arabidopsis roots. *Molecular Plant-Microbe Interactions, 18*(12), 1247–1257.

Hanounik, S. B., & Osborne, W. W. (1975). Influence of *Meloidogyne incognita* on content of amino acids and nicotine in tobacco grown under gnotobiotic conditions. *Journal of Nematology,* 7(4), 332–336.

Hedin, P. A., & Creech, R. G. (1998). Altered amino acid metabolism in root-knot nematode inoculated cotton plants. *Journal of Agricultural and Food Chemistry, 46*(10), 4413–4415.

Hewezi, T., Howe, P. J., Maier, T. R., Hussey, R. S., Mitchum, M. G., Davis, E. L., et al. (2010). Arabidopsis spermidine synthase is targeted by an effector protein of the cyst nematode *Heterodera schachtii. Plant Physiology, 152*(2), 968–984.

Hofius, D., Herbers, K., Melzer, M., Omid, A., Tacke, E., Wolf, S., et al. (2001). Evidence for expression level-dependent modulation of carbohydrate status and viral resistance by the potato leafroll virus movement protein in transgenic tobacco plants. *The Plant Journal, 28*(5), 529–543.

Hofmann, J., El Ashry, A., Anwar, S., Erban, A., Kopka, J., & Grundler, F. (2010). Metabolic profiling reveals local and systemic responses of host plants to nematode parasitism. *The Plant Journal, 62*(6), 1058–1071.

Hofmann, J., & Grundler, F. M. W. (2006). Females and males of root-parasitic cyst nematodes induce different symplasmic connections between their syncytial feeding cells and the phloem in *Arabidopsis thaliana*. *Plant Physiology and Biochemistry, 44*(5–6), 430–433.

Hofmann, J., Hess, P. H., Szakasits, D., Blochl, A., Wieczorek, K., Daxbock-Horvath, S., et al. (2009). Diversity and activity of sugar transporters in nematode-induced root syncytia. *Journal of Experimental Botany, 60*(11), 3085–3095.

Hofmann, J., Szakasits, D., Blochl, A., Sobczak, M., Daxbock-Horvath, S., Golinowski, W., et al. (2008). Starch serves as carbohydrate storage in nematode-induced syncytia. *Plant Physiology, 146*(1), 228–235.

Hofmann, J., Wieczorek, K., Blochl, A., & Grundler, F. M. W. (2007). Sucrose supply to nematode-induced syncytia depends on the apoplasmic and symplasmic pathways. *Journal of Experimental Botany, 58*(7), 1591–1601.

Hofmann, J., Youssef-Banora, M., de Almeida-Engler, J., & Grundler, F. M. W. (2010). The role of callose deposition along plasmodesmata in nematode feeding sites. *Molecular Plant-Microbe Interactions, 23*(5), 549–557.

Hoth, S., Schneidereit, A., Lauterbach, C., Scholz-Starke, J., & Sauer, N. (2005). Nematode infection triggers the de novo formation of unloading phloem that allows macromolecular trafficking of green fluorescent protein into syncytia. *Plant Physiology, 138*(1), 383–392.

Hoth, S., Stadler, R., Sauer, N., & Hammes, U. Z. (2008). Differential vascularization of nematode-induced feeding sites. *Proceedings of the National Academy of Sciences USA, 105*(34), 12617–12622.

Irvine, R. F., & Schell, M. J. (2001). Back in the water: the return of the inositol phosphates. *Nature Reviews Molecular Cell Biology, 2*(5), 327–338.

Ithal, N., Recknor, J., Nettleton, D., Maier, T., Baum, T. J., & Mitchum, M. G. (2007). Developmental transcript profiling of cyst nematode feeding cells in soybean roots. *Molecular Plant-Microbe Interactions, 20*(5), 510–525.

Jammes, F., Lecomte, P., Almeida-Engler, J., Bitton, F., Martin-Magniette, M. L., Renou, J. P., et al. (2005). Genome-wide expression profiling of the host response to root-knot nematode infection in Arabidopsis. *The Plant Journal, 44*(3), 447–458.

Ji, Hongli, Gheysen, Godelieve, Denil, Simon, Lindsey, Keith, Topping, Jennifer F., Nahar, Kamrun, et al. (2013). Transcriptional analysis through RNA sequencing of giant cells induced by *M. graminicola* in rice roots. *Journal of Experimental Botany*. ert219v1–ert219.

Jones, M. G. K. (1981). Host cell responses to endoparasitic nematode attack: structure and function of giant cells and syncytia. *Annals of Applied Biology, 97*, 353–372.

Jones, M. G. K., & Dropkin, V. H. (1976). Scanning electron microscopy of nematode induced giant transfer cells. *Cytobios, 15*(58–59), 149–161.

Jones, M. G. K., & Payne, H. L. (1977). Structure of syncytia induced by phytoparasitic nematode *Nacobbus aberrans* in tomato roots, and possible role of plasmodesmata in their nutrition. *Journal of Cell Science, 23*(Feb), 299–313.

Juergensen, K., Scholz-Starke, J., Sauer, N., Hess, P., van Bel, A. J. E., & Grundler, F. M. W. (2003). The companion cell-specific Arabidopsis disaccharide carrier AtSUC2 is expressed in nematode-induced syncytia. *Plant Physiology, 131*(1), 61–69.

Kanter, U., Usadel, B., Guerineau, F., Li, Y., Pauly, M., & Tenhaken, R. (2005). The inositol oxygenase gene family of Arabidopsis is involved in the biosynthesis of nucleotide sugar precursors for cell-wall matrix polysaccharides. *Planta, 221*(2), 243–254.

Krauthausen, H. J., & Wyss, U. (1982). Influence of the cyst nematode *Heterodera schachtii* on relative changes in the pattern of free amino acids at feeding sites. *Physiology and Plant Pathology, 21*, 425–436.

Kursaanov, A. L. (1963). Metabolism and the transport of organic solutes. *Advances in Botanical Research, 1*, 209–278.

Kyndt, T., Vieira, P., Gheysen, G., & de Almeida-Engler, J. (2013). Nematode feeding sites: unique organs in plant roots. *Planta, 238*(5), 807–818.
Lambert, K. N., Allen, K. D., & Sussex, I. M. (1999). Cloning and characterization of an esophageal-gland-specific chorismate mutase from the phytoparasitic nematode *Meloidogyne javanica*. *Molecular Plant-Microbe Interactions, 12*(4), 328–336.
Lazarowitz, S. G., & Beachy, R. N. (1999). Viral movement proteins as probes for intracellular and intercellular trafficking in plants. *The Plant Cell, 11*(4), 535–548.
Lewis, S. A., & Mcclure, M. A. (1975). Free amino acids in roots of infected cotton seedlings resistant and susceptible to *Meloidogyne incognita*. *Journal of Nematology, 7*(1), 10–15.
Lowry, O. H., & Passonneau, J. V. (1972). *A flexible system of enzymatic analysis*. New York: Academic Press.
Lucas, W. J. (2006). Plant viral movement proteins: agents for cell-to-cell trafficking of viral genomes. *Virology, 344*(1), 169–184.
Mcclure, M. A. (1977). *Meloidogyne incognita* – metabolic sink. *Journal of Nematology, 9*(1), 88–90.
Meon, S., Fisher, J. M., & Wallace, H. R. (1978). Changes in free proline following infection of plants with either *Meloidogyne javanica* or *Agrobacterium tumefaciens*. *Physiology and Plant Pathology, 12*(3), 251–256.
Owens, R. G., & Rubinstein, J. H. (1966). Metabolic changes induced by root-knot nematodes in host tissues. *Contributions from Boyce Thompson Institute, 23*, 199–214.
Owens, R. G., & Specht, H. N. (1964). Root-knot histogenesis. *Contributions from Boyce Thompson Institute, 22*, 471–490.
Owens, R. G., & Specht, H. N. (1966). Biochemical alteration induced in host tissues by root-knot nematodes. *Contributions from Boyce Thompson Institute, 23*, 181–198.
Portillo, M., Cabrera J,Lindsey, K., Topping, J., Andres, M. F., Emiliozzi, M., Oliveros, J. C., et al. (2013). Distinct and conserved transcriptomic changes during nematode-induced giant cell development in tomato compared with Arabidopsis: a functional role for gene repression. *New Phytologist, 197*, 1276–1290.
Puthoff, D. P., Nettleton, D., Rodermel, S. R., & Baum, T. J. (2003). Arabidopsis gene expression changes during cyst nematode parasitism revealed by statistical analyses of microarray expression profiles. *The Plant Journal, 33*(5), 911–921.
Scholthof, H. B. (2005). Plant virus transport: motions of functional equivalence. *Trends in Plant Science, 10*(8), 376–382.
Schuster, M., Sandstedt, R., & Estes, L. W. (1964). Starch formation induced by a plant parasitic nematode. *Science, 143*, 1342–1343.
Siddique, S., Endres, S., Atkins, J. M., Szakasits, D., Wieczorek, K., Hofmann, J., et al. (2009). Myo-inositol oxygenase genes are involved in the development of syncytia induced by *Heterodera schachtii* in Arabidopsis roots. *New Phytologist, 184*(2), 457–472.
Siddique, S., Endres, S., Sobczak, M., Radakovic, Z. S., Fragner, L., Grundler, F. M. W., et al. (2014). Myo-inositol oxygenase is important for the removal of excess myo-inositol from syncytia induced by *Heterodera schachtii* in Arabidopsis roots. *New Phytologist, 201*(2), 476–485.
Sijmons, P. C., Grundler, F. M. W., Vonmende, N., Burrows, P. R., & Wyss, U. (1991). *Arabidopsis thaliana* as a new model host for plant-parasitic nematodes. *The Plant Journal, 1*(2), 245–254.
Smith, A. M., & Stitt, M. (2007). Coordination of carbon supply and plant growth. *Plant, Cell & Environment, 30*(9), 1126–1149.
Stadler, R., Brandner, J., Schulz, A., Gahrtz, M., & Sauer, N. (1995). Phloem loading by the PmSuc2 Sucrose carrier from plantago major occurs into companion cells. *The Plant Cell, 7*(10), 1545–1554.
Szakasits, D., Heinen, P., Wieczorek, K., Hofmann, J., Wagner, F., Kreil, D. P., et al. (2009). The transcriptome of syncytia induced by the cyst nematode *Heterodera schachtii* in Arabidopsis roots. *The Plant Journal, 57*(5), 771–784.

Truernit, E., & Sauer, N. (1995). The promoter of the *Arabidopsis thaliana* Suc2 Sucrose-H^+ symporter gene directs expression of beta glucuronidase to the phloem – evidence for phloem loading and unloading by Suc2. *Planta, 196*(3), 564–570.

Vanbel, A. J. E. (1993). Strategies of phloem loading. *Annual Review of Plant Physiology, 44*, 253–281.

Veech, J. A., & Endo, B. Y. (1969). The histochemical localization of several enzymes of soybeans infected with the root-knot nematode *Meloidogyne incognita acrita*. *Journal of Nematology, 3*, 265–276.

Waigmann, E., Ueki, S., Trutnyeva, K., & Citovsky, V. (2004). The ins and outs of nondestructive cell-to-cell and systemic movement of plant viruses. *Critical Reviews in Plant Sciences, 23*(3), 195–250.

Wang, E. L. H., & Bergeson, G. B. (1974). Biochemical changes in root exudate and xylem sap of tomato plants infected with *Meloidogyne incognita*. *Journal of Nematology, 6*(4), 194–202.

Wolf, S., Deom, C. M., Beachy, R., & Lucas, W. J. (1991). Plasmodesmatal function is probed using transgenic tobacco plants that express a virus movement protein. *The Plant Cell, 3*, 593–604.

Wright, K. M., Horobin, R. W., & Oparka, K. J. (1996). Phloem mobility of fluorescent xenobiotics in Arabidopsis in relation to their physicochemical properties. *Journal of Experimental Botany, 47*(304), 1779–1787.

Wubben, M. J. E., Su, H., Rodermel, S. R., & Baum, T. J. (2001). Susceptibility to the sugar beet cyst nematode is modulated by ethylene signal transduction in *Arabidopsis thaliana*. *Molecular Plant-Microbe Interactions, 14*(10), 1206–1212.

Zeeman, S. C., Kossmann, J., & Smith, A. M. (2010). Starch: Its metabolism, evolution, and biotechnological modification in plants. *Annual Review of Plant Biology, 61*, 209–234.

CHAPTER SIX

The Role of Lipid Signalling in Regulating Plant–Nematode Interactions

Sigal Brown Miyara*[,1], Iberkleid Ionit*[,§], Patricia Buki*, Mike Kolomiets[¶]

*Department of Entomology, Nematology and Chemistry Units, Agricultural Research Organization (ARO), the Volcani Center, Bet Dagan, Israel
[§]Department of Plant Pathology and Microbiology, Texas A&M University, College Station, TX, USA
[¶]Department of Plant Pathology and Microbiology, The Robert H. Smith Faculty of Agriculture, Food and Environment, The Hebrew University of Jerusalem, Jerusalem, Israel
[1]Corresponding author: E-mail: sigalhor@volcani.agri.gov.il

Contents

Abstract

Fatty acids (FAs) are important source of reserve energy and essential components of membrane lipids in all living organisms. In plants, FA-dependent signalling pathways play a key role in defense against pests and pathogens. Recent discoveries demonstrate more direct roles for FAs and their oxidation products in inducing various modes of plant defenses by modulating the basal, effector-triggered, and systemic immunity in plants. However, despite the widely recognized biological significance of FA-derived signals in regulating defense and pathogenicity processes in various plant–pathogen systems, the nature of lipid signals in a crosstalk between plants and nematodes has just began to be studied. This chapter summarizes the current knowledge of the involvement of FAs, FA-derived oxylipins and enzymes catalyzing FA metabolism in regulating plant response to nematode infection. The recent evidence implicating

Advances in Botanical Research, Volume 73
ISSN 0065-2296
http://dx.doi.org/10.1016/bs.abr.2014.12.004

oxylipins, especially the phytohormone jasmonic acid, one of the best studied oxylipins, in either successful defense or susceptibility following nematode inoculation is the focus of this discussion. To further demonstrate the existence of lipid-mediated signal crosstalk between plants and nematodes, we present the case of a group of the nematode's effectors that facilitate infection by modifying host lipid-based defenses. This review will focus on signals related to FA-signalling that govern both resistance and susceptibility in plants against parasitic nematodes as well as on a group of effector genes that might interfere with the host lipid-based defense.

1. INTRODUCTION

Plant parasitic nematodes (PPNs) represent one of the largest sources of biotic stress experienced by plants that are very difficult to control. Although precise economic evaluation is challenging, worldwide crop losses due to PPNs have been estimated at €78 billion annually (Chitwood, 2003; Koenning et al., 1999). All sedentary endoparasitic PPNs establish intimate association with host plant roots by manipulating fundamental aspects of plant cell biology (Ithal, Recknor, Nettleton, Hearne et al., 2007). Among the most devastating PPNs are the sedentary root-knot nematodes (RKN), *Meloidogyne* spp. and the cyst nematodes (*Heterodera* and *Globodera* spp.), which are all obligate biotrophs (Bird, 2004). These parasites establish a sophisticated dialogue with selected host cells mediated by nematode-secreted effectors, which alter host cell development and gene expression. Two forms of feeding sites are associated with both groups: large multinucleate cells, known as giant cells (GC) or fused feeding cells, known as syncytia which are the sole source of nutrition for the developing RKN or cyst nematode respectively (Figure 1); (Abad, Favery, Rosso, & Castagnone-Sereno, 2003; Dubreuil, Magliano, Deleury, Abad, & Rosso, 2007; Gheysen & Fenoll, 2002; Jones, 1981).

The RKNs are sedentary parasites whose development is strictly dependent on nematode feeding site (NFS) formation and maintenance, ensuring a continuous nutrient supply (Vanholme et al., 2004). Although it is not known how RKNs establish their GCs, increasing knowledge indicates that glandular secretions (effectors) injected into plant cells by the nematodes interact directly or indirectly with essential plant components, leading to the establishment and maintenance of NFS (Baum, Hussey, & Davis, 2007; Davis et al., 2000; Davis & Kemerait, 2009; Niebel et al., 1993; Sijmons, Atkinson, & Wyss, 1994; Vanholme et al., 2004). Effectors are secreted, parasite-produced proteins originating from the esophageal gland cells that facilitate plant infection by modifying host cell response and biology.

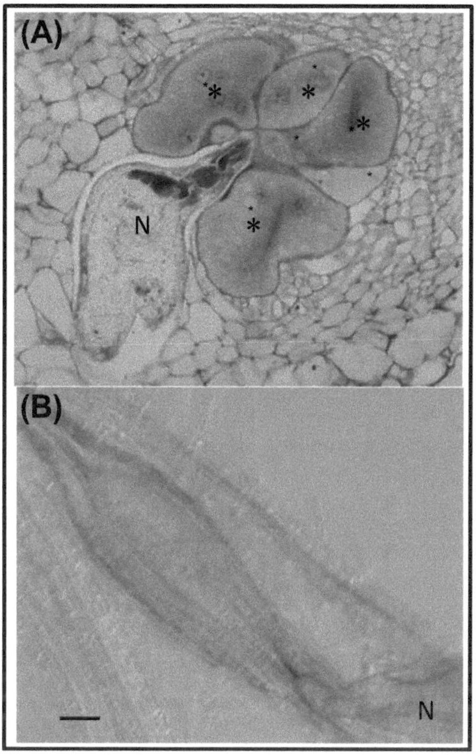

Figure 1 (A) Cross-section of RKN feeding site. Giant cells (*) induced by a root-knot nematode (N). Host cells are enlarged, contain multiple enlarged nuclei, and have highly enriched cytoplasm. (B) *Arabidopsis* root infected by the cyst nematode *Heterodera schachtii*. The multinucleated cell, the feeding site called syncytium, adjacent to the developed nematode (N) is observed.

Two esophageal gland types are involved in producing effectors: (1) two subventral glands, which are mainly active during the migratory phase of parasitism; and (2) one dorsal gland, which becomes more active during development and maintenance of the feeding site (Davis et al., 2000). Other organs in contact with the environment that produce secretory proteins include the amphids and the cuticle. During the last two decades several cuticle PPN proteins have been identified, including proteins important for parasitism (Jones, Reavy, Smant, & Prior, 2004; Robertson et al., 2000; Spiegel, Kahane, Cohen, & Sharon, 1997). Since cuticle proteins do not originate in the esophageal gland cells and, therefore, do not fit into the strict definition of nematode effectors, there is a need for an accurate identification of these proteins.

Even though, previous studies were successful in identifying putative effectors of PPNs (e.g., Abad et al., 2008; Bellafiore et al., 2008; Dautova et al., 2001; Doyle & Lambert, 2002; Elling et al., 2009; Gao et al., 2001, 2003; Goellner, Smant, De Boer, Baum, & Davis, 2000; Haegeman, Joseph et al, 2011; Huang et al., 2003, 2004; Jaubert et al., 2002; Opperman et al., 2008; Popeijus et al., 2000; Rosso et al., 1999; Roze et al., 2008; Smant et al., 1998; Vanholme et al., 2006; Yan et al., 1998) (see Chapters 11, 12, 13 for a detailed understanding), they did not provide enough evidence of their nature as effectors. This is complicated as the vast majority of known effectors found in PPNs do not share sequence similarity with any known genes including the large number of effector genes from bacterial, fungal and oomycete pathogens. This suggests that the fundamental mechanisms of nematode parasitism are probably distinct from those of other pathogens. Notable exceptions are nematode effectors with similarities to bacterial or fungal cell wall-degrading or modifying enzymes (Adam, Phillips, Jones, & Blok, 2008; Doyle & Lambert, 2002; Qin et al., 2004; Rosso et al., 1999; Smant et al., 1998; Stokkermans & Dunbar, 1998) and microbial chorismate mutases (Doyle & Lambert, 2003), all of which might have been acquired through horizontal gene transfer (Danchin et al., 2010; Haegeman, Jones et al., 2011; Scholl, Sosinski, Thomas, & Windham, 2008). In light of this, the PPN genes sharing homology to microbial effectors are some of the few phytonematode effector genes whose functions have been studied in detail. Additional examples of characterized nematode effector genes include CLAVATA3/ESR-like (CLE) signalling peptides (Huang, Allen, Davis, Baum, & Hussey, 2006; Lu et al., 2009; Replogle et al., 2011; Wang et al., 2010, 2005). However, with very few exceptions, the functions of effector genes that lack matches with known genes remain much unclear (Hewezi et al., 2008, 2010). Thus, in spite of a significant amount of data, the majority of PPN effectors remain uncharacterized, hampering the progress towards a comprehensive understanding of RKN parasitism.

1.1 Local and Systemic Plant Defense Mechanisms against Plant Parasitic Nematodes

Given that many PPN species invade their hosts and remain inside from several weeks to months, they inevitably expose themselves to being detected by pathogen-associated molecular patterns (PAMP) immune receptors. To date, no PAMP receptors have been identified that are directed against conserved cues of PPN (Wondafrash, Van Dam, & Tytgat, 2013). This could be due to several strategies developed by nematodes to evade or suppress PAMP-triggered immunity (PTI) (Davies & Curtis, 2011; Nobre & Evans, 1998). Once inside the plant root, the developing nematodes undergo three consecutive molts in each of which the new cuticle has a new composition, thereby creating a new challenge to the plant immune system (Spiegel, Inbar, Kahane, & Sharon, 1995).

Despite the strategies to avoid recognition, nematode invasion activates the host immune response. For instance, as early as 12 h after root penetration by the RKN *Meloidogyne incognita*, during root migration phase (Wyss & Grundler, 1992), genes encoding peroxidases, cell wall modification enzymes, lipoxygenases (LOX), and proteinase inhibitors were induced (Gheysen & Fenoll, 2002). Whether these represent a typical wounding response controlled by jasmonic acid (JA), or PTI induced by PAMP detection is unclear. Interestingly, studies of root inoculation with sedentary endoparasitic juveniles clearly show that only a fraction of them succeeds in penetrating the roots and inducing feeding sites (Tygat et al., 2002; Wyss & Grundler, 1992). Taken together, these observations point to a weak PTI that might generate priming signals for a stronger defense at later time points.

Several lines of evidence support the case for the activation of systemic-induced defense after PPN infection. Microarray analysis of *Arabidopsis thaliana* after infection with the cyst nematode *Heterodera schachtii* revealed a strong induction of *VSP2*, a marker gene for the MYC2 branch of the JA defense pathway, in the whole root system at 3 days after nematode inoculation (Puthoff, Nettleton, Rodermel, & Baum, 2003). Similarly, transcriptome analysis during time course experiments of soybean plants interaction with the cyst nematode *Heterodera glycines*, showed a clear induction of the JA pathway in the whole root system at all the time points inspected (Alkharouf et al., 2006). Likewise, systemic induction of the JA pathway in soybean roots was observed at 2, 5 and 10 days after *H. glycines* infection (Ithal, Recknor, Nettleton, Maier et al., 2007). However, locally, in the developing syncytia, JA-regulated defense was suppressed (Ithal, Recknor, Nettleton, Hearne et al.,2007). Genetic studies showed that *A. thaliana* SA-deficient mutants exhibit an increased susceptibility to *H. schachtii*, whereas exogenous application of SA renders wild-type plants less susceptible (Wubben, Jin, & Baum, 2008). A comparison of systemic defense signalling after rice infection with an RKN and a migratory endoparasitic nematode *Hirschmanniella oryzae* performed by Kyndt et al. (2012) showed that following inoculation with *H. oryzae*, systemic JA and ethephon (ET) signalling are induced at day 3, while the SA pathway was suppressed.

However, at 7 days, both the JA and ET signalling were repressed. A slight different response was observed for the RKNs during *Meloidogyne graminicola*-rice interaction where activation of both SA and JA but suppression of the ET pathway at 3 days was observed while at 7 days JA was also repressed.

Overall, published results suggest that sedentary endoparasitic nematodes initially induce the JA, ET and SA pathways. These three phytohormones induce unique defense responses individually, and they also interact synergistically or antagonistically to further orchestrate downstream signalling. For microbial pathogens, extensive studies on *A. thaliana* and other plants show that intricate signalling networks involving defense phytohormones facilitate the fine-tuning of signal transduction pathways, regulating most effective plant defense response (Glazebrook, 2001; Hammond-Kosack & Parker, 2003; Kachroo & Kachroo, 2009). In general, SA is known to suppress the JA pathway (Spoel et al., 2003). This can lead to trade-offs in the defense responses when plants are attacked simultaneously by different pathogens. However, this antagonistic interaction between SA and JA seems to be dependent on the concentration of both hormones, whereby a low concentration of both result in a synergistic effect, while high concentrations lead to antagonism (Mur, Kenton, Atzorn, Miersch, & Wasternack, 2006). The SA–JA antagonism seems also to be dependent on the time that has passed between the induction of both hormonal pathways (Koornneef et al., 2008). Moreover, ET signalling prevents SA-mediated suppression of the JA pathway (Leon-Reyes et al., 2010). Studies on nematodes indicate that, while nematode parasitism resulted in SA inhibition (Uehara, Sugiyama, Matsuura, Arie, & Masuta, 2010), JA/ET as well as the phytohormone auxin promote susceptibility to phytonematodes (Abad et al., 2003; Bhattarai et al., 2008; Goverse, Overmars et al., 2000; Grunewald, Cannoot, Friml, & Gheysen, 2009; Karczmarek, Overmars, Helder, & Goverse, 2004; Wubben, Rodermel, & Baum, 2004; Wubben, Su, Rodermel, & Baum, 2001). Evidence from these studies further suggests that signalling pathways dependent on hormones may be manipulated by virulent nematodes during the initiation and maintenance of feeding sites (see Chapter 7). It is likely that sedentary nematodes that establish an intimate interaction with their hosts manipulate the hormone balance in host cells through effectors. While previous studies focused on the interaction between nematode infection and the JA pathway (Bhattarai et al., 2008; Cooper, Jia, & Goggin, 2005), many details of JA signalling related to FA derivatives in regulating plant defense during RKN parasitism remained unclear.

1.2 Lipid Signals as Part of the General Plant Defense Signalling

During the last years, a number of studies have revealed the role of lipids and lipid metabolites during plant–pathogen interactions. One of the lipid-mediated defense pathways is initiated by oxidation of FAs by the LOX pathway resulting in the production of over 500 estimated molecular species of oxylipins. While the biological roles of the majority of oxylipins are poorly studied, the best-characterized oxylipin, JA, has been shown conclusively as a major signal for defense against diverse pathogens (Blée, 2002; Feussner & Wasternack, 2002; Yan, Borrego, & Kolomiets, 2013). Another defense-related function of lipids lies in the ability of unsaturated FA to remodel membrane lipid composition (Kachroo & Kachroo, 2009). Yet another group of lipids known as very long-chain FA (VLCFA) are produced in response to diverse environmental stress (Upchurch, 2008). The VLCFAs are FAs containing 20–36 carbons synthesized in the endoplasmic reticulum. They are crucial for a wide range of biological processes in plants such as biosynthesis of the plant cuticle and the production of sphingolipids, structurally diverse complex lipids that are important as membrane components and bioactive molecules (Samuels et al., 2008; Sperling & Heinz, 2003; Worrall, Ng, & Hetherington, 2003).

Among the signalling cascades, FA and oxylipin metabolic and signalling pathways regulate many defense and developmental pathways (Feussner & Wasternack, 2002). Oxylipins are produced by the lipoxygenase (LOX: 9-LOX or 13-LOX regio-specific) and α-dioxygenase (α-DOX) enzyme families (Chen et al., 2004; Feussner & Wasternack, 2002), which are structurally and functionally diverse and collectively or individually play fundamental roles in the physiological and pathological processes of plants and vertebrates (Figure 2, Vellosillo et al., 2007). Numerous studies aimed at defining the function of oxylipins have shown that in addition to the regulation of organism development and growth (Caldelari, Wang, Farmer, & Dong, 2011; Vellosillo et al., 2007), they act in defense as evidenced by differential expression of oxylipin-biosynthetic genes in response to inoculation with plant pathogens (Jalloul et al., 2002; Melan et al., 1993; Sanz, Moreno, & Castresana, 1998; Turner, Ellis, & Devoto, 2002). Among the best-characterized oxylipins are jasmonates including JA itself, its immediate precursor 12-oxo-phytodienoic acid (OPDA) and multiple JA derivatives that induce various plant defense responses, including the biosynthesis of protective secondary metabolites (Block, Schmelz, Jones, & Klee, 2005). Despite their widely recognized biological significance in defense against

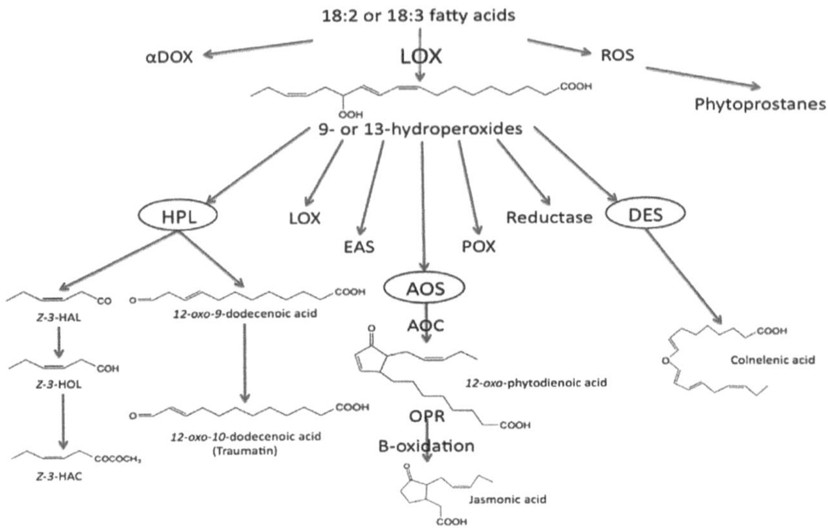

Figure 2 ***Oxylipin pathway.*** Peroxidation of PUFAs [Linoleic acid, (LA) 18:2 and Linolenic acid (LeA) 18:3] by LOXs leads to 9-LOX–derived and 13-LOX-derived hydroperoxy PUFAs that are the substrates of lipoxygenase (LOX), allene oxide synthase (AOS), divinyl ether synthase (DES), hydroperoxide lyase (HPL), α-dioxygenase (DOX), allene oxide cyclase (AOC), epoxy alcohol synthase (EAS), peroxygenase (PR) and reductase (RD) (Vellosillo et al., 2007).

microbial pathogens and insect herbivores, the functions of the JA- and other oxylipin-mediated signalling during nematode infection remain unknown.

Plant cells accumulate JA and MeJA in response to pathogens or pathogen-derived elicitors (Browse & Howe, 2008; Gundlach, Muller, Kutchan, & Zenk, 1992). Jasmonates are primarily derived from linolenic acid, 18-carbon FA containing three double bonds, α-18:3, which is released from plastid membrane lipids via the activity of phospholipase A1. A 13-lipoxygenase (13-LOX) adds molecular oxygen at C-13 of α-18:3. The 13-hydroperoxide generated from this reaction enters JA biosynthesis via one of the seven branches of the LOX pathway, the allene oxide synthase (AOS) branch (Figure 2). The first cyclic and biologically active compound produced is OPDA, which is reduced by OPDA reductase (OPR) followed by three β-oxidation cycles to generate the C12, JA (Creelman & Mulpuri, 2002). First genetic evidence for the role of JA in defense signalling came from research on JA-biosynthetic mutants. The *Arabidopsis fad3 fad7 fad8* triple mutant is unable to accumulate JA because of a deficiency in C18:3 and is hypersusceptible to invasion by insect larvae

(McConn, Creelman, Bell, Mullet, & Browse, 1997). The *fad3 fad7 fad8* mutant plants are also highly susceptible to root rot by *Pythium jasmonium* (previously *Pythium mastophorum*), and this susceptibility can be alleviated by the exogenous application of MeJA (Vijayan, Shockey, Levesque, Cook, & Browse, 1998).

LOX, AOS and OPR mutants also displayed altered plant responses to pathogens (Andersson, Kourtchenko, Dangl, Mackey, & Ellerstrom, 2006; Mei, Qi, Sheng, & Yang, 2006). In *Arabidopsis,* 13-LOX–derived oxylipin biosynthesis is induced upon recognition of the avirulence protein Avr-Rpm1 and is associated with hypersensitive response (HR) induced during race-specific resistance to the bacterial pathogen *Pseudomonas syringae* (Andersson et al., 2006). In potato cells, pathogen-derived elicitors preferentially stimulate a second LOX pathway catalyzed by a 9-LOX enzyme (Gobel et al., 2001). Silencing a pathogen-induced 9-LOX gene in tobacco enhances susceptibility to *Phytophthora parasitica* (Rance, Fournier, & Esquerre-Tugaye, 1998). The induction of 9-LOX activity during an incompatible interaction between cotton and *Xanthomonas campestris* is also associated with HR (Jalloul et al., 2002). In rice, *OsAOS2* expression is induced upon infection with *Magnaporthe grisea,* and overexpression of this gene in transgenic plants confers enhanced resistance to the blast fungus in rice (Mei et al., 2006). Correspondingly, silencing the *StAOS2* gene enhances susceptibility to *Phytophthora infestans* and *Erwinia carotovora* in potato. In maize, the *ZmOPR1/2* transcripts are strongly induced in resistant but not in susceptible lines infected with *Fusarium verticillioides* (Zhang et al., 2005). The *ZmOPR1/2* transcripts are also induced earlier during incompatible interactions with *Cochliobolus heterostrophus* (Zhang et al., 2005). Recent genetic evidence showed that global JA-deficiency in maize due to inactivation of the two JA-producing maize OPR isoforms, ZmOPR7 and 8, resulted in complete lack of immunity to *Pythium* (Yan et al., 2012) and *F. verticillioides* (Christensen et al., 2014). Maize seedlings lacking JA are unable to survive when grown in natural, nonsterile soils, further supporting the notion that JA is required against necrotrophic and/or hemibiotrophic soil-borne pathogens.

Besides JA-Ileu, the JA precursor OPDA and its C16 structural homologue dinor OPDA (dn-OPDA) also serve as signalling molecules in plant defense (Stintzi, Weber, Reymond, Browse, & Farmer, 2001). Similar to JA, OPDA and dn-OPDA accumulate to varying levels in response to wounding and pathogen infection (Weber, Vick, & Farmer, 1997). Exogenous application of OPDA induces the transcription of stress-related genes, which are distinct from those induced in response to JA (Stintzi et al., 2001). OPDA accumulating *opr3* mutant of *Arabidopsis* shows enhanced resistance to necrotrophic fungi in spite of its deficiency in JA, Stintzi et al., (2001) demonstrating distinct function of different molecules along the JA-biosynthetic pathway. Nonenzymatic oxidation of FAs also generates biologically active oxylipins, some of which are involved in plant-stress-related signalling, such as hydroxy FAs and phytoprostanes (Sattler et al., 2006). These are generated via the action of reactive oxygen species (ROS) that often accumulate in response to pathogen infection and abiotic stress. Although some biological activities overlap with those of JA and OPDA, they have other distinct biological activities and induce the transcription of a separate set of genes (Mueller et al., 2008).

2. THE ROLE OF LIPID SIGNALS IN REGULATING PLANT–NEMATODE INTERACTION

Recent findings implicate lipid metabolic pathways in the defense responses of plants to RKN (Abad et al., 2003; Gheysen & Fenoll, 2002; Veronico et al., 2006). For example, microarray-based expression profiling indicated that several FA metabolism genes, including those coding for LOX, patatin-like protein 1, and 12-oxophytodienoate reductase (OPR), are induced in susceptible soybean roots in response to soybean cyst nematode infection (Alkharouf et al., 2006). Similarly, LOX activity was increased in pea roots (Leone et al., 2001) and soybean roots (Alkharouf et al., 2006) upon infection with cyst nematodes. Moreover, degradation of GCs, indicative of an HR, is accompanied by an increased activity of LOXs in pea (Zacheo, Bleve-Zacheo, & Melillo, 1997). Genetic evidence of the involvement of 9-LOX biosynthetic maize pathway in resistance mechanisms to RKN was further provided by the analysis of the maize *lox3* knockout mutant, which displayed increased attraction and susceptibility to RKN compared with near-isogenic WT line (Gao et al., 2008). These and other studies implicated lipid metabolic pathways in the defense responses of plants to PPNs and provided conclusive genetic evidence of biological significance of lipid derivatives during plant–nematode interactions.

2.1 Specific Oxylipin Pathway Genes Play Vital Roles in Determining Host Status for RKN Infection

Investigation of the *A. thaliana* genome has led to the identification of six LOX genes encoding four 13-LOXs (*LOX2*, *LOX3*, *LOX4* and *LOX6*) and two 9-LOXs (*LOX1* and *LOX5*) (Bannenberg et al., 2009). Given the central regulatory roles of specific 9- and 13-LOX-derived oxylipins in regulating stressor-specific plant defense responses, understanding of the function of individual LOX isoforms in plant–nematode interaction is requested. Among the 13-LOXs, *A. thaliana* LOX2 (AtLOX2) appears to be distantly related to AtLOX3, AtLOX4 and AtLOX6, which are clustered together and have been suggested to be involved in plant's defense pathway (Bannenberg et al., 2009). LOX3 and LOX4 are highly similar, sharing 85% overall amino acid identity (Caldelari et al., 2011). They are being expressed in roots (Vellosillo et al., 2007). The function of these LOXs in response to PPNs was studied for both RKN and cyst nematode (Ozalvo et al., 2014). This study show that both LOX3 and LOX4 isoforms are induced in response to nematode infection as indicated by using *LOX promoter::GUS* lines (Ozalvo et al., 2014). An earlier study showed that LOX3 expression is associated with the initiation of lateral root formation (Vellosillo et al., 2007); (Figure 3(A–D)). *LOX3* expression was also detected during nematode migration and at later time points (Figure 3(E–J)). At 21 and 28 days after infection, a strong GUS signal was observed within the developing galls (Figure 3(H) and (I)). The *LOX4* expression pattern was notably different to that of *LOX3*. Yet, in noninoculated roots, *LOX4* was predominantly expressed at the stem–root juncture (Figure 4(A) and (B)). In contrast to *LOX3*, no GUS staining was detected in roots primordia, developing lateral roots or mature root tip (Figure 4(C) and (D)). Upon nematode inoculation, GUS expression was correlated with nematode parasitic stages: a weak GUS signal was observed during the migratory phase of *Meloidogyne javanica* J2 at 24 hours post inoculation (hpi) (Figure 4(E) and (F)), 48 hpi (Figure 4(G) and (H)) and at 72 hpi (Figure 4(I) and (J)). GUS signal remained faint in the galls produced by *M. javanica* at 21 and 28 days post inoculation (dpi) (Figure 4(K)and (L)). Histological analysis of roots expressing the *LOX3*-GUS construct at 28 dpi, where all GCs were mature and nematodes had developed into the fourth juvenile stage, clearly showed strong GUS expression inside the cells surrounding the female body and the hyperplastic cells surrounding the GCs inside the vascular cylinder, and within the developed GC systems (Figure 5(A–C)). *Arabidopsis* roots expressing *LOX4-GUS* construct showed a very weak signal in a few cells associated with the mature GC system (Figure 5(D–F)). Similar to the GUS signal induced by RKN, *LOX3* promoter-GUS and *LOX4* promoter-GUS plants inoculated by the cyst nematode *H. schachtii* demonstrated strong GUS signal upon infection (Figure 6). Functional analyzes of the *lox3* and *lox4* mutants

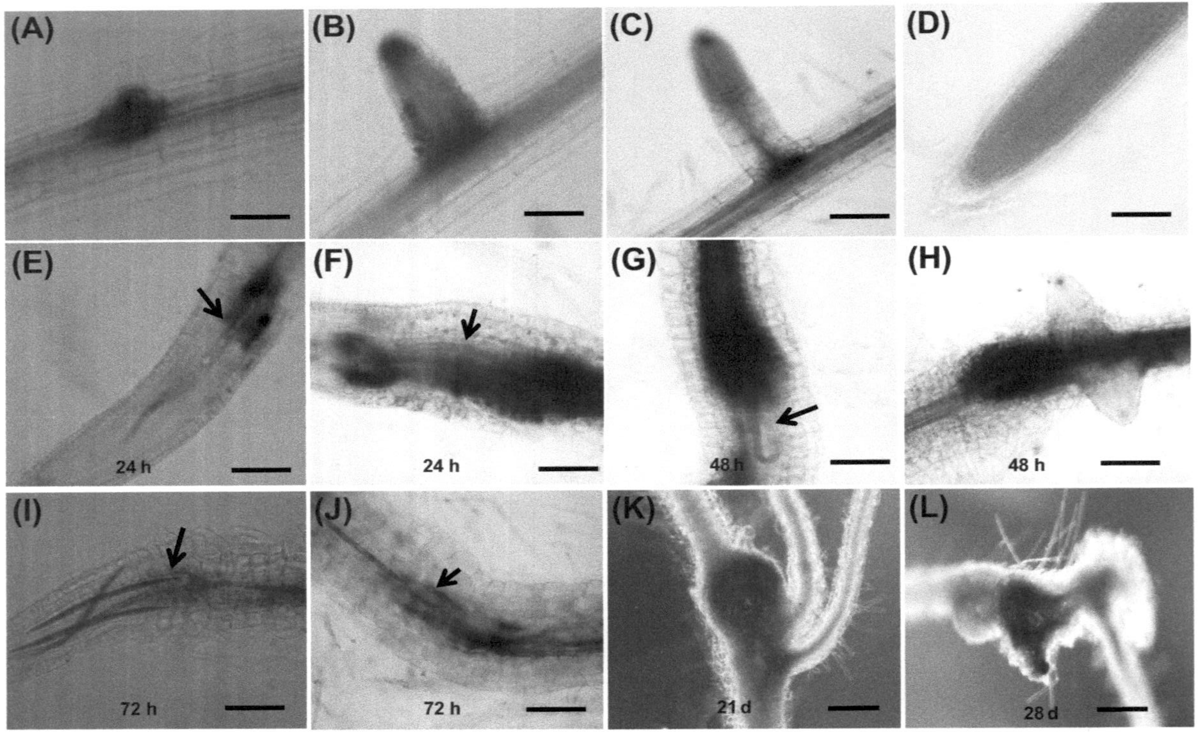

Figure 3 ***Microscopic analysis of* GUS*-expression patterns without inoculation and upon RKN infection in* Arabidopsis *plants harbouring the* LOX3 *promoter-GUS fusion construct.*** Noninfested control root harbouring the *LOX3*-GUS fusion construct demonstrated strong GUS staining of lateral root primordia (A) and emerging lateral root (B). GUS staining decreased with time for mature lateral root and was no longer detected in the mature root tip (C, D). In *Arabidopis* plants infested with *M. javanica*, strong GUS signal was observed during J2 migration along the vascular bundle at 24 hpi (E, F) and during entrance into the sedentary phase at 48 and 72 hpi (G–J). Strong GUS staining was observed in galls 28 dpi by *M. javanica*.(K, L). (A–J) Light micrographs as viewed under light microscope. (K, L), Bright-field image of galls photographed using a stereomicroscope. Bars: (A) = 50 μm, (B–J) = 100 μm, (K, L) = 500 μm (Ozalvo et al., 2014). (See colour plate)

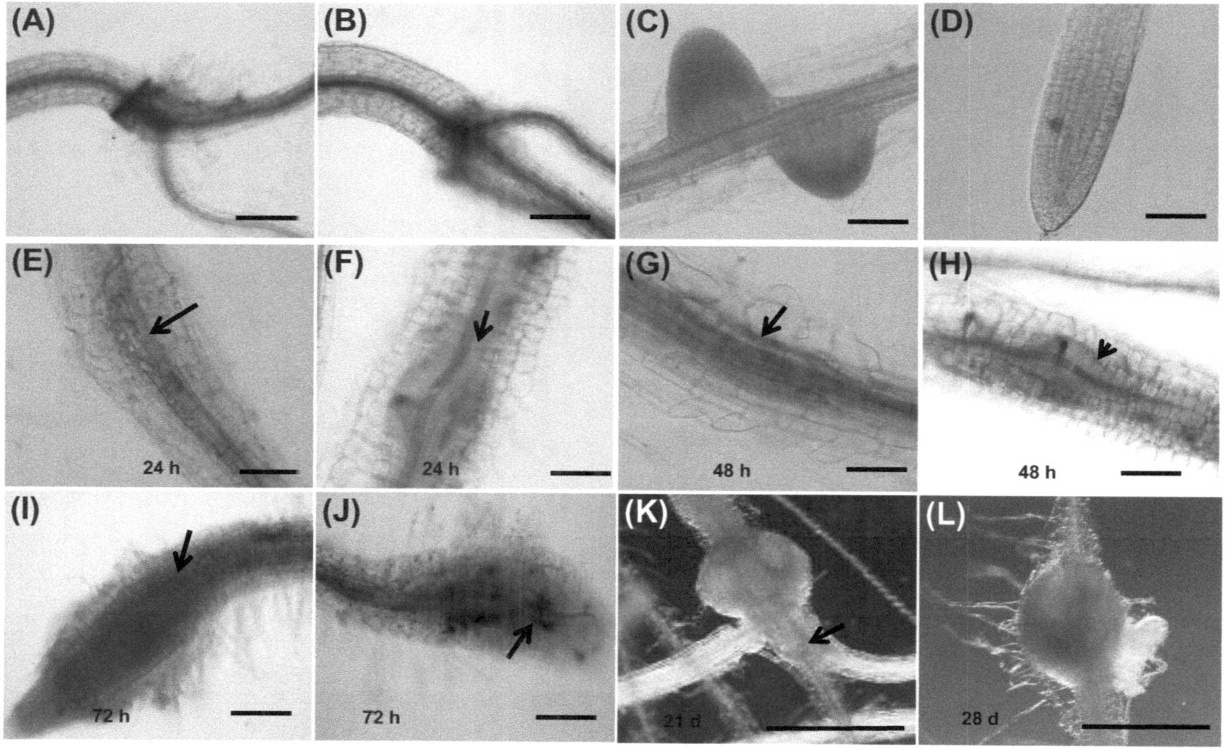

Figure 4 ***Microscopic analysis of GUS-expression patterns upon RKN infection in* Arabidopsis *plants harbouring the* LOX4 *promoter-GUS fusion.*** Noninfested control root harbouring the *LOX4*-GUS fusion construct demonstrated consistent GUS staining at the stem–root juncture (A, B), while no GUS staining was detected in roots primordia, developing lateral roots or mature root tip (C, D). Upon nematode inoculation, GUS expression was correlated with nematode parasitic stages: a weak GUS signal was observed during the migratory phase of *M. javanica* J2 at 24 hpi (E, F), 48 hpi (G, H) and at 72 hpi (I, J). GUS signal remained faint in galls 21 and 28 dpi (K, L) by *M. javanica*. Bars: (A–J) = 100 μm, (K, L) = 500 μm (Ozalvo et al., 2014). (See colour plate)

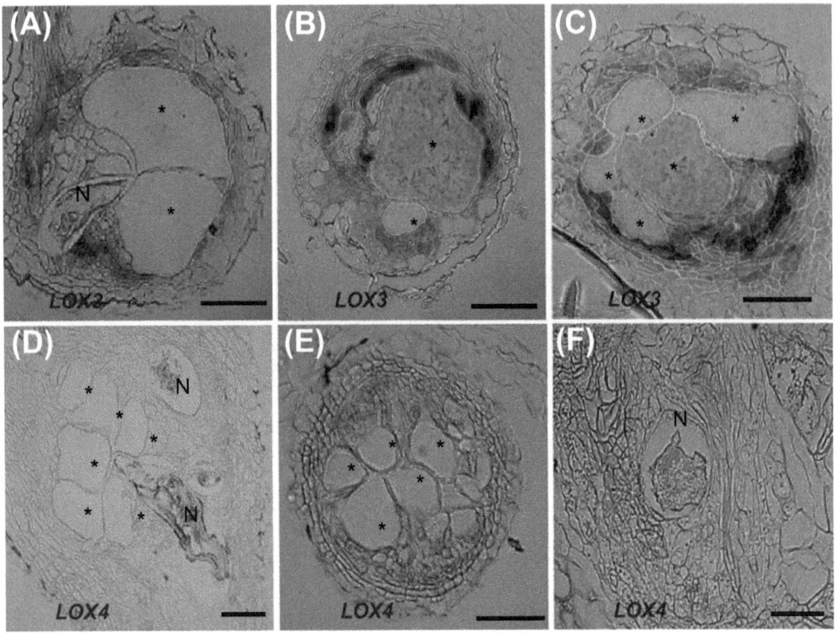

Figure 5 ***Microscopic analysis of GUS activity in cross-section of* Arabidopsis *root gall expressing* LOX3 *and* LOX4 *promoter-GUS construct.*** At 28 dpi, all giant cells (GCs) were mature and nematodes developed into the fourth juvenile stage. Histological analysis of roots expressing the *LOX3*-GUS construct clearly showed strong GUS expression inside the cells surrounding the female body and the hyperplastic cells surrounding the GCs inside the vascular cylinder (A), and within the developed GC systems (B, C). *Arabidopsis* roots expressing *LOX4*-GUS construct still showed a very weak signal in a few cells associated with the developed GC system (D–F). The female body of the nematode (N) can be seen at the edge of the GCs (*). Bars = 200 μm (Ozalvo et al., 2014). (See colour plate)

demonstrated that mutants lacking *LOX3* were less susceptible than wild-type plants to RKN infection, as reflected by a decrease in female counts. Conversely, deficiency in LOX4 function led to a marked increase in females and egg mass number and in the female to male ratio of *M. javanica* and *H. schachtii*, respectively (Ozalvo et al., 2014). The susceptibility of *lox4* mutants was accompanied by increased expression of the AOS and AOC genes as well as increased level of JA measured in roots. This response was not found in *lox3* mutants. Taken together, these surprising results reveal that despite having 13-LOX activity, LOX3 and LOX4 play different roles in the interaction with nematodes, the outcome likely due to their association with distinct metabolic and signalling pathways. These results conclusively show that LOX4 plays a major role in controlling plant defense

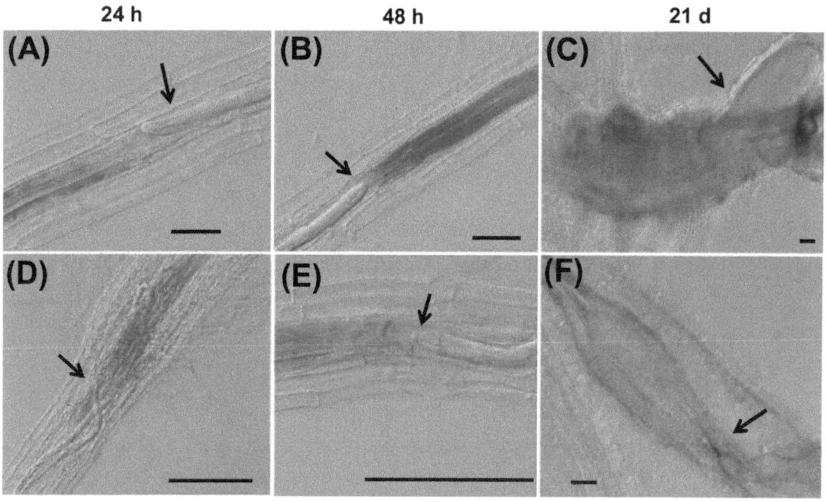

Figure 6 ***Microscopic analysis of GUS-expression patterns upon cyst nematode infection in* Arabidopsis *plants harbouring the* LOX3 *and LOX4 promoter-GUS fusion.*** Consistent GUS staining was observed as early as 24 hpi and the signal was centred inside the vascular cylinder close to the nematode head (A). The signal increased as the infection progressed to 48 hpi (B), and GUS signal remained at late infection stages, 21 dpi (C). In *Arabidopsis* plants harbouring the *LOX4* promoter-GUS fusion, LOX4 was activated at 24 hpi around the nematode body inside the vascular cylinder (D), and a clear signal was also observed at 48 hpi (E). The activity was maintained till the late infection stages but intensity markedly decreased at 21 dpi (F). Bars = 100 μm (Ozalvo et al., 2014). (See colour plate)

against nematode infection. It is reasonable to speculate that this interaction may be controlled by a suppressive signalling activity of as yet unknown LOX4-derived oxylipins. Similar to the latter case, Gobel, Feussner, & Rosahl (2003) reported that the 9-LOX-deficient potato plants showed increased levels of 13-LOX-derived oxylipins. Interestingly, plant roots and leaves are known to accumulate highly divergent levels of oxylipins, the so-called 'oxylipin signature' of a plant organ, as exemplified by the oxylipin profiling of potato organs (Stumpe et al., 2006). Similarly, Vellosillo et al. (2007) showed that *Arabidopsis* roots exhibit strong 9-LOX activity and the 9-LOX-defective mutants display an increased number of lateral roots. Interestingly, mutants insensitive to the 9-LOX-derived oxylipin, 9-HOT, exhibited increased susceptibility to a bacterial pathogen, *P. syringae* pv. *tomato* (Vellosillo et al., 2007). In summary, recent findings support the hypothesis that some oxylipins are not only endogenous modulators of root development but also signalling molecules for defense response.

2.2 The Role of the Jasmonate Pathway in Governing Plant–Nematode Interactions

While considerable genetic evidence suggests that JA is an important signalling molecule in plant defense (Takahashi et al., 2004), our current understanding of JA signalling in RKN–plant associations is very complex. Given that increased level of JA was observed in the *lox4* mutant line, the results suggest that increased level of JA acts as a susceptibility factor in the plant–nematode interaction (Ozalvo et al., 2014). These results where JA is associated with increased susceptibility are in a good agreement with those of Bhattarai et al. (2008) who analyzed the response of tomato mutants, *jai1* and *def1*, altered in JA signalling, and demonstrated that an intact JA-signalling pathway is required for tomato susceptibility to RKNs. In maize, *Mu*-insertional *lox3–4* mutants displayed increased attractiveness to RKNs, and an increased number of juveniles and eggs were accompanied by elevated levels of JA (Gao et al., 2008). Likewise, analysis of JA-deficient *opr7 opr8* double mutants of maize, OPR7 and OPR8, redundant paralogues responsible for JA biosynthesis, resulted in increased resistance to RKN (M.V. Kolomiets and J. Starr, Texas A&M University, College Station, TX, personal communication). As opposed to the accumulated results, Kyndt et al. (2011) found that in rice–RKN *M. graminicola* interaction, an intact JA biosynthesis pathway is a prerequisite for systemically induced resistance as indicated by increased susceptibility towards *M. graminicola* in the JA biosynthesis mutant *hebiba* than its control (Nahar, Kyndt, De Vleesschauwer, Hofte, & Gheysen, 2011). However, given that no measurements of JA were employed in planta it is difficult to speculate whether the absence of JA is the cause of the increased susceptibility observed. Nevertheless, the results in rice–RKN system demonstrate that in rice JA pathway is a key defense pathway involved in RKN interaction being efficiently activated and probably modulated by ET, while SA plays a minor role (Nahar et al., 2011). However it is possible that the differences occurring in these described systems with regard to the role of JA/ET and SA signalling pathway in the regulation of RKN disease development could be part of major difference between monocotyledonous and dicotyledonous plants. Interestingly, studying the effect of exogenous shoot application of JA on RKN disease development resulted in a similar effect for both studied plant systems. In rice–RKN *M. graminicola* system, exogenous ET and JA (methyl jasmonate) supplied to the shoots induced a strong systemic defense response in the roots, supported by a major up-regulation of pathogenesis-related genes *OsPR1a* and *OsPR1b*, while the salicylic acid analogue BTH (benzo-1,2,3-thiadiazole-7-carbothioic acid *S*-methyl ester) was a less potent systemic defense inducer from shoot to root (Nahar et al., 2011). Likewise, exogenous application of JA induces a systemic defense response that reduces avirulent nematode reproduction on susceptible tomato plants. Moreover, JA-dependent defenses proved to be heat-stable, whereas JA treatment enhanced Mi-mediated resistance at high temperature, but did not suppress either of the virulent nematodes tested (Cooper et al., 2005). Similarly, foliar application of JA to tomato resulted in a repressive effect on RKNs (Fujimoto et al., 2011). These results might reveal the opposite function for JA in regulation of plant responses to nematodes that were observed by Bhattarai et al. (2008) in tomato, by Gao et al. (2008) in maize and by Ozalvo et al. (2014) in *Arabidopsis*. However, it is likely that exogenous application of JA to shoots might induce the expression of various pathways in roots that are responsible for the increased resistance observed by Nahar et al. (2011) on rice and by Cooper et al. (2005) as well as by Fujimoto et al. (2011) on tomato.

To summarize, the accumulating results which indicate that maize *lox3–4* and *Arabidopsis lox4* mutants produce elevated levels of JA, and yet are more susceptible to RKN is rather intriguing and require careful consideration regarding the role of JA in plant–nematode interaction. It is more likely that host resistance mechanisms to nematodes may not rely solely on JA- and SA-mediated defense, perhaps some other oxylipins are involved. Thus, the function of JA in host–nematode interaction remains unclear for now, and obtaining conclusive evidence for its role will require the use of JA biosynthetic and analysis of signalling. Additionally, a careful quantification of JA and diverse other oxylipins in each mutant–nematode interaction will be also necessary.

2.3 Nematode Effectors Manipulate Lipid-Based Defense Signalling Pathways

Previous studies of the interactions between nematode parasitism and JA signalling, suggest that hormone signalling pathways may be manipulated by nematodes during the initiation and maintenance of their feeding sites (Bhattarai et al., 2008; Cooper et al., 2005; Goverse, Engler et al., 2000; Grunewald et al., 2009; Karczmarek et al., 2004; Soriano, Riley, Potter, & Bowers, 2004; Uehara et al., 2010; Wubben et al., 2001). Thus, it is likely that the hormone balance manipulation is mediated by nematode's secreted effector proteins. Secreted FA and retinol binding proteins family (FAR) have been proposed to facilitate animal parasitic nematodes infection by scavenging and transporting FAs, required for developmental processes and cellular differentiation of the parasite (Basavaraju et al., 2003; Fairfax et al., 2009; Garofalo, Kennedy et al., 2003; Garofalo et al., 2002; Kennedy et al., 1997). FAR proteins may also play a role in interfering with intracellular and intercellular lipid signalling related to host defenses (Basavaraju et al., 2003; Bath et al., 2009; Fairfax et al., 2009; Garofalo, Kennedy et al., 2003; Garofalo, Rowlinson et al, 2003; Garofalo et al., 2002; Kennedy et al., 1997; Prior et al., 2001). Up till the last few years, extensive studies were carried on FAR of animal and human parasites, particularly on their regulation and FA-binding characteristics (Basavaraju et al., 2003; Fairfax et al., 2009; Garofalo, Kennedy et al, 2003; Garofalo et al., 2002, Garofalo, Rowlinson et al, 2003; Kennedy et al., 1997). Analyzing the FAR of the potato cyst nematode, FAR protein (GpFAR1) of *Globodera pallida*, was the first insightful study that provided evidence for the role of these proteins in plant–nematode interactions. Being localized to the cuticle surface of the pre-parasitic J2 of this nematode species, GpFAR1 was shown to interfere with the plant LOX-mediated defense signalling by binding to LOX substrates and products (Prior et al., 2001). Although investigation on ligand-binding properties of recombinant FAR proteins have been reported, the in vivo role of FAR proteins during plant infection was not studied (Bath et al., 2009; Garofalo, Kennedy et al, 2003; Prior et al., 2001).

Recently, Iberkleid et al. (2014) identified and investigated the function of the first RKN *M. javanica* FAR protein (MjFAR-1) during the interaction of *M. javanica* with its tomato host. The *mj-far-1* mRNA was detected through *M. javanica* pre-parasitic J2s, migratory and sedentary parasitic stages by quantitative reverse transcriptase polymerase chain reaction (qRT-PCR). Further immunolocalization assays demonstrated that the FAR protein of *Meloidogyne* is secreted during sedentary stages, as evidenced by the accumulation of FAR at the nematode cuticle surface and along the adjacent host root tissues (Figure 7, (Iberkleid et al., 2014)). Tomato roots constitutively expressing *mj-far-1* demonstrated an increased susceptibility to RKN infection as observed by accelerated gall induction and expansion, accompanied by a higher percentage of nematodes developing into mature females compared to control roots. RNA interference assays that expressed double-stranded RNA complementary to *mj-far-1* in transgenic tomato lines specifically reduced nematode infection levels. The galls of overexpressing *mj-far-1* lines contained larger feeding cells that might support a faster nematode development and maturation. Roots overexpressing *mj-far-1* suppressed JA responsive genes such as the proteinase inhibitor (*Pin2*) and γ-*thionin*, illustrating the possible role of Mj-FAR-1 in manipulating the

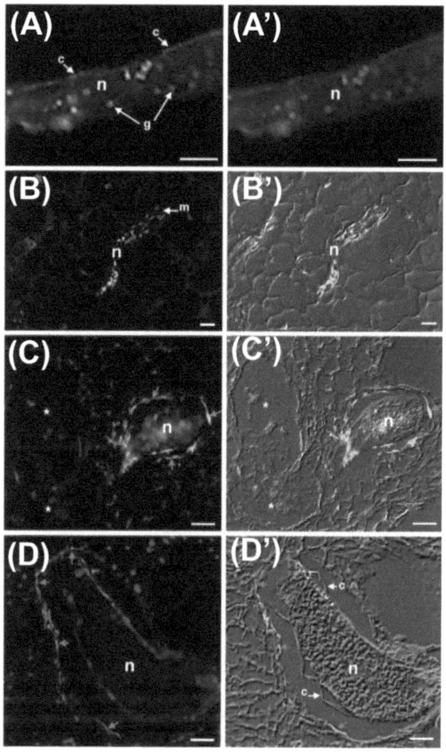

Figure 7 ***Immunodetection of FAR proteins in* Meloidogyne incognita *pre-parasitic J2 and during parasitism of* Arabidopsis thaliana *roots.*** (A) Cross nematode sections of pre-parasitic J2 displaying the protein localization at the nematode cuticle surface and circular granules structures within the posterior nematode body. (B) Localization of FAR proteins during nematode migration (B–B′), nematode sedentary stages at 10 DAI (C–C′) and 21 DAI (D–D′) within the roots of *A. thaliana*. Arrows point out the accumulation of FAR along the nematode cuticle and adjacent cells surrounding the nematode body at 10 and 21 DAI. Micrographs on the left are overlays of Alexa-488 fluorescence (green) and DAPI-stained nuclei (blue). Micrographs on the right are overlays of an Alexa-488 fluorescence (green), DAPI-stained nuclei (blue) and differential interference contrast (grey). **c**, cuticle, **g**, granules, **n**, nematode, **m**, metacorpus, * giant cell. Bars = 10 μm (Iberkleid et al., 2013). (See colour plate)

lipid-based signalling *in planta*. These data suggest that *Meloidogyne* FAR might have a strategic function during the interaction of the nematode with its host. Increased host susceptibility to nematode infection following the overexpression of nematode parasitism genes has been documented in the past (Hewezi et al., 2008; Patel et al., 2010), suggesting that an excess of some effector proteins can enhance a compatible host–parasite interaction via modulation of the plant stress (Patel et al., 2010) and defense (Hewezi et al., 2008) responses. The results presented by Iberkleid et al. (2014), provide evidence that Mj-FAR-1 facilitates infection most likely via the manipulation of host lipid-based defenses, as critical components for a successful parasitism by plant–parasitic nematodes.

3. CONCLUSIONS AND FUTURE ISSUES

Our understanding of lipid-based signals has only now being gained through the recent studies. The use of mutant lines altered in JA production or in FA metabolism is a promising strategy towards elucidating the role of lipid signals in regulating the parasitic interaction, partially explored in maize, tomato, rice and *Arabidopsis*. In future, the use of a wide battery of JA and other oxylipin biosynthetic and signalling mutants will be required in order to obtain conclusive evidence for the role of JA and other lipid signals in plant parasitic interaction. Studying the role of nematode's effectors in manipulating lipid-based signals and in interfering with intracellular plant signalling related to host defenses, will improve our understanding of nematode-induced manipulations of major plant defenses and signalling pathways. Aspects of FA metabolism and plant defense against nematode infection remain to be studied. It is still unknown how does the ability to import or export FA or FA-derived components contribute to plant defense. Another important question to be addressed is at what level(s) plant defense signalling is coupled to FA metabolism. Likewise it is important to understand at what rate does PPNs target host FA metabolism to ensure pathogenesis. To start answering these questions there is a strong need to integrate modern metabolomics lipidomics, protein–protein interaction assays and molecular genetics techniques and systems biology approaches allowing the exposure of further players regulating the parasitic interaction.

ACKNOWLEDGEMENTS

We appreciate all authors' contribution to this review as well as the financial support of grant obtained from the Israeli Chief of Scientist # 020-0687-02 (Ministry of Agriculture and Rural Development). We would like to thank Prof Carmen Fenoll and Prof Carolina Escobar Lucas for reviewing the chapter and bringing helpful suggestions.

REFERENCES

Abad, P., Favery, B., Rosso, M. N., & Castagnone-Sereno, P. (2003). Root-knot nematode parasitism and host response: molecular basis of a sophisticated interaction. *Molecular Plant Pathology, 4*, 217–224.

Abad, P., Gouzy, J., Aury, J. M., Castagnone-Sereno, P., Danchin, E. G., Deleury, E., et al. (2008). Genome sequence of the metazoan plant-parasitic nematode *Meloidogyne incognita*. *Nature Biotechnology, 26*, 909–915.

Adam, M. A. M., Phillips, M. S., Jones, J. T., & Blok, V. C. (2008). Characterisation of the cellulose-binding protein Mj-cbp-1 of the root knot nematode, *Meloidogyne javanica*. *Physiological and Molecular Plant Pathology, 72*, 21–28.

Alkharouf, N. W., Klink, V. P., Chouikha, I. B., Beard, H. S., MacDonald, M. H., Meyer, S., et al. (2006). Timecourse microarray analyses reveal global changes in gene expression of susceptible *Glycine max* (soybean) roots during infection by *Heterodera glycines* (soybean cyst nematode). *Planta, 224*, 838–852.

Andersson, M. X., Kourtchenko, O., Dangl, J. L., Mackey, D., & Ellerstrom, M. (2006). Phospholipase-dependent signalling during the AvrRpm1- and AvrRpt2-induced disease resistance responses in *Arabidopsis thaliana*. *Plant Journal, 47*, 947–959.

Bannenberg, G., Martinez, M., Rodriguez, M. J., Lopez, M. A., Ponce de Leon, I., Hamberg, M., et al. (2009). Functional analysis of alpha-DOX2, an active alpha-dioxygenase critical for normal development in tomato plants. *Plant Physiology, 151*, 1421–1432.

Basavaraju, S. V., Zhan, B., Kennedy, M. W., Liu, Y., Hawdon, J., & Hotez, P. J. (2003). Ac-FAR-1, a 20 kDa fatty acid- and retinol-binding protein secreted by adult *Ancylostoma caninum* hookworms: gene transcription pattern, ligand binding properties and structural characterisation. *Molecular and Biochemical Parasitology, 126*, 63–71.

Bath, J. L., Robinson, M., Kennedy, M. W., Agbasi, C., Linz, L., Maetzold, E., et al. (2009). Identification of a secreted fatty acid and retinol-binding protein (Hp-FAR-1) from *Heligmosomoides polygyrus*. *Journal of Nematology, 41*, 228–233.

Baum, T. J., Hussey, R. S., & Davis, E. L. (2007). Root-knot and cyst nematode parasitism genes: the molecular basis of plant parasitism. *Genetic Engineering (New York), 28*, 17–43.

Bellafiore, S., Shen, Z., Rosso, M. N., Abad, P., Shih, P., & Briggs, S. P. (2008). Direct identification of the *Meloidogyne incognita* secretome reveals proteins with host cell reprogramming potential. *PLoS Pathogens, 4*, e1000192.

Bhattarai, K. K., Xie, Q. G., Mantelin, S., Bishnoi, U., Girke, T., Navarre, D. A., et al. (2008). Tomato susceptibility to root-knot nematodes requires an intact jasmonic acid signaling pathway. *Molecular Plant-Microbe Interactions, 21*, 1205–1214.

Bird, D. M. (2004). Signaling between nematodes and plants. *Current Opinion in Plant Biology, 7*, 372–376.

Blée, E. (2002). Impact of phyto-oxylipins in plant defense. *Trends in Plant Science, 7*, 315–322.

Block, A., Schmelz, E., Jones, J. B., & Klee, H. J. (2005). Coronatine and salicylic acid: the battle between *Arabidopsis* and *Pseudomonas* for phytohormone control. *Molecular Plant Pathology, 6*, 79–83.

Browse, J., & Howe, G. A. (2008). New weapons and a rapid response against insect attack. *Plant Physiology, 146*, 832–838.

Caldelari, D., Wang, G., Farmer, E. E., & Dong, X. (2011). *Arabidopsis* lox3 lox4 double mutants are male sterile and defective in global proliferative arrest. *Plant Molecular Biology, 75*, 25–33.

Chen, G., Hackett, R., Walker, D., Taylor, A., Lin, Z., & Grierson, D. (2004). Identification of a specific isoform of tomato lipoxygenase (TomloxC) involved in the generation of fatty acid-derived flavor compounds. *Plant Physiology, 136*, 2641–2651.

Chitwood, D. J. (2003). Research on plant-parasitic nematode biology conducted by the United States Department of Agriculture-Agricultural Research Service. *Pest Management Science, 59*, 748–753.

Christensen, S. A., Nemchenko, A., Park, Y. S., Borrego, E. J., Huang, P. C., Schmelz, E., et al. (2014). The novel monocot-specific 9-lipoxygenase, ZmLOX12, is required to mount an effective jasmonate-mediated defense against Fusarium verticillioides in maize. *Molecular Plant-Microbe Interactions, 27*(11), 1263–1276.

Cooper, W. R., Jia, L., & Goggin, L. (2005). Effects of jasmonate-induced defenses on root-knot nematode infection of resistant and susceptible tomato cultivars. *Journal of Chemical Ecology, 31*, 1953–1967.

Creelman, R. A., & Mulpuri, R. (2002). The oxylipin pathway in *Arabidopsis*. Arabidopsis *Book, 1*, e0012.

Danchin, E. G., Rosso, M. N., Vieira, P., de Almeida-Engler, J., Coutinho, P. M., Henrissat, B., et al. (2010). Multiple lateral gene transfers and duplications have promoted plant parasitism ability in nematodes. *Proceedings of the National Academy of Sciences of the United States of America, 107*, 17651–17656.

Dautova, M., Rosso, M. N., Abad, P., Gommers, F. J., Bakker, J., & Smant, G. (2001). Single pass cDNA sequencing—a powerful tool to analyze gene expression in preparasitic juveniles of the southern root-knot nematode *Meloidogyne incognita*. *Journal of Nematology, 3*, 129–139.

Davies, K. G., & Curtis, R. H. C. (2011). Cuticle surface coat of plant-parasitic nematodes. *Annual Review of Phytopathology, 49*, 135–156.

Davis, R. F., & Kemerait, R. C. (2009). The multi-year effects of repeatedly growing cotton with moderate resistance to *Meloidogyne incognita*. *Journal of Nematology, 41*, 140–145.

Davis, E. L., Hussey, R. S., Baum, T. J., Bakker, J., Schots, A., Rosso, M. N., et al. (2000). Nematode parasitism genes. *Annual Review of Phytopathology, 38*, 365–396.

Doyle, E. A., & Lambert, K. N. (2002). Cloning and characterization of an esophageal-gland-specific pectate lyase from the root-knot nematode *Meloidogyne javanica*. *Molecular Plant-Microbe Interactions, 15*, 549–556.

Doyle, E. A., & Lambert, K. N. (2003). *Meloidogyne javanica* chorismate mutase 1 alters plant cell development. *Molecular Plant-Microbe Interactions, 16*, 123–131.

Dubreuil, G., Magliano, M., Deleury, E., Abad, P., & Rosso, M. N. (2007). Transcriptome analysis of root-knot nematode functions induced in the early stages of parasitism. *New Phytologist, 176*, 426–436.

Elling, A. A., Mitreva, M., Gai, X., Martin, J., Recknor, J., Davis, E. L., et al. (2009). Sequence mining and transcript profiling to explore cyst nematode parasitism. *BMC Genomics, 10*, 58.

Fairfax, K. C., Vermeire, J. J., Harrison, L. M., Bungiro, R. D., Grant, W., Husain, S. Z., et al. (2009). Characterization of a fatty acid and retinol binding protein orthologue from the hookworm *Ancylostoma ceylanicum*. *International Journal of Parasitology, 39*, 1561–1571.

Feussner, I., & Wasternack, C. (2002). The lipoxygenase pathway. *Annual Review of Plant Biology, 53*, 275–297.

Fujimoto, T., Tomitaka, Y., Abe, H., Tsuda, S., Futai, K., & Mizukubo, T. (2011). Expression profile of jasmonic acid-induced genes and the induced resistance against the root-knot nematode (*Meloidogyne incognita*) in tomato plants (*Solanum lycopersicum*) after foliar treatment with methyl jasmonate. *Journal of Plant Physiology, 168*, 1084–1097.

Gao, B., Allen, R., Maier, T., Davis, E. L., Baum, T. J., & Hussey, R. S. (2001). Identification of putative parasitism genes expressed in the esophageal gland cells of the soybean cyst nematode *Heterodera glycines*. *Molecular Plant-Microbe Interactions, 14*, 1247–1254.

Gao, B., Allen, R., Maier, T., Davis, E. L., Baum, T. J., & Hussey, R. S. (2003). The parasitome of the phytonematode *Heterodera glycines*. *Molecular Plant-Microbe Interactions, 16*, 720–726.

Gao, X., Starr, J., Gobel, C., Engelberth, J., Feussner, I., Tumlinson, J., et al. (2008). Maize 9-lipoxygenase ZmLOX3 controls development, root-specific expression of defense genes, and resistance to root-knot nematodes. *Molecular Plant-Microbe Interactions, 21*, 98–109.

Garofalo, A., Kennedy, M. W., & Bradley, J. E. (2003). The FAR proteins of parasitic nematodes: their possible involvement in the pathogenesis of infection and the use of *Caenorhabditis elegans* as a model system to evaluate their function. *Medical Microbiology and Immunology, 192*, 47–52.

Garofalo, A., Klager, S. L., Rowlinson, M. C., Nirmalan, N., Klion, A., Allen, J. E., et al. (2002). The FAR proteins of filarial nematodes: secretion, glycosylation and lipid binding characteristics. *Molecular and Biochemical Parasitology, 122*, 161–170.

Garofalo, A., Rowlinson, M. C., Amambua, N. A., Hughes, J. M., Kelly, S. M., Price, N. C., et al. (2003). The FAR protein family of the nematode *Caenorhabditis elegans*. Differential lipid binding properties, structural characteristics, and developmental regulation. *Journal of Biological Chemistry, 278*, 8065–8074.

Gheysen, G., & Fenoll, C. (2002). Gene expression in nematode feeding sites. *Annual Review of Phytopathology, 40*, 191–219.

Glazebrook, J. (2001). Genes controlling expression of defense responses in *Arabidopsis*–2001 status. *Current Opinion in Plant Biology, 4*, 301–308.

Gobel, C., Feussner, I., & Rosahl, S. (2003). Lipid peroxidation during the hypersensitive response in potato in the absence of 9-lipoxygenases. *Journal of Biological Chemistry, 278*, 52834–52840.

Gobel, C., Feussner, I., Schmidt, A., Scheel, D., Sanchez-Serrano, J., Hamberg, M., et al. (2001). Oxylipin profiling reveals the preferential stimulation of the 9-lipoxygenase pathway in elicitor-treated potato cells. *Journal of Biological Chemisty, 276*, 6267–6273.

Goellner, M., Smant, G., De Boer, J. M., Baum, T. J., & Davis, E. L. (2000). Isolation of beta-1,4-endo genes from *Globodera tabacum* and their expression during parasitism. *Journal of Nematology, 32*, 154–165.

Goverse, A., Engler, J. D., Verhees, J., van der Krol, S., Helder, J., & Gheysen, G. (2000). Cell cycle activation by plant parasitic nematodes. *Plant Molecular Biology, 43*, 747–761.

Goverse, A., Overmars, H., Engelbertink, J., Schots, A., Bakker, J., & Helder, J. (2000). Both induction and morphogenesis of cyst nematode feeding cells are mediated by auxin. *Molecular Plant-Microbe Interactions, 13*, 1121–1129.

Grunewald, W., Cannoot, B., Friml, J., & Gheysen, G. (2009). Parasitic nematodes modulate PIN-mediated auxin transport to facilitate infection. *PLoS Pathogens, 5*, e1000266.

Gundlach, H., Muller, M. J., Kutchan, T. M., & Zenk, M. H. (1992). Jasmonic acid is a signal transducer in elicitor-induced plant cell cultures. *Proceedings of the National Academy of Sciences of the United States of America, 89*, 2389–2393.

Haegeman, A., Jones, J. T., & Danchin, E. G. (2011). Horizontal gene transfer in nematodes: a catalyst for plant parasitism? *Molecular Plant-Microbe Interactions, 24*, 879–887.

Haegeman, A., Joseph, S., & Gheysen, G. (2011). Analysis of the transcriptome of the root lesion nematode *Pratylenchus coffeae* generated by 454 sequencing technology. *Molecular Biochemistry and Parasitology, 178*, 7–14.

Hammond-Kosack, K. E., & Parker, J. E. (2003). Deciphering plant-pathogen communication: fresh perspectives for molecular resistance breeding. *Current Opinion in Biotechnology, 14*, 177–193.

Hewezi, T., Howe, P., Maier, T. R., Hussey, R. S., Mitchum, M. G., Davis, E. L., et al. (2008). Cellulose binding protein from the parasitic nematode *Heterodera schachtii* interacts with *Arabidopsis* pectin methylesterase: cooperative cell wall modification during parasitism. *Plant Cell, 20*, 3080–3093.

Hewezi, T., Howe, P. J., Maier, T. R., Hussey, R. S., Mitchum, M. G., Davis, E. L., et al. (2010). *Arabidopsis* spermidine synthase is targeted by an effector protein of the cyst nematode *Heterodera schachtii*[1[W][OA]]. *Plant Physiology, 152*, 968–984.

Huang, G., Allen, R., Davis, E. L., Baum, T. J., & Hussey, R. S. (2006). Engineering broad root-knot resistance in transgenic plants by RNA silencing of a conserved and essential root-knot nematode parasitism gene. *Proceedings of the National Academy of Science of the United States of America, 103*, 14302–14306.

Huang, G., Dong, R., Maier, T., Allen, R., Davis, E. L., Baum, T. J., et al. (2004). Use of solid-phase subtractive hybridization for the identification of parasitism gene candidates from the root-knot nematode *Meloidogyne incognita*. *Molecular Plant Pathology, 5*, 217–222.

Huang, G., Gao, B., Maier, T., Allen, R., Davis, E. L., Baum, T. J., et al. (2003). A profile of putative parasitism genes expressed in the esophageal gland cells of the root-knot nematode *Meloidogyne incognita*. *Molecular Plant-Microbe Interactions, 16*, 376–381.

Iberkleid, I., Ozalvo, R., Feldman, L., Elbaz, M., Patricia, B., & Horowitz, S. B. (2014). Responses of tomato genotypes to avirulent and *Mi*-virulent *Meloidogyne javanica* Isolates occurring in Israel. *Phytopathology, 104*, 484–496.

Iberkleid, I. S., Vieira, P., de Almeida Engler, J., Spiegel, Y., & Brown Horowitz, S. (2013). Mj-FAR-1, a Fatty Acid-and Retinol-Binding protein secreted by Root-Knot Nematodes is required for tomato host susceptibility to *Meloidogyne javanica*. *PLoS ONE, 8*(5), e64586. http://dx.doi.org/10.1371/journal.pone.0064586.

Ithal, N., Recknor, J., Nettleton, D., Hearne, L., Maier, T. R., Baum, T. J., et al. (2007). Parallel genome-wide expression profiling of host and pathogen during soybean cyst nematode infection of soybean. *Molecular Plant-Microbe Interactions, 20*, 293–305.

Ithal, N., Recknor, J., Nettleton, D., Maier, T., Baum, T. J., & Mitchum, M. G. (2007). Developmental transcript profiling of cyst nematode feeding cells in soybean roots. *Molecular Plant-Microbe Interactions, 20*, 510–525.

Jalloul, A., Montillet, J. L., Assigbetse, K., Agnel, J. P., Delannoy, E., Triantaphylides, C., et al. (2002). Lipid peroxidation in cotton: xanthomonas interactions and the role of lipoxygenases during the hypersensitive reaction. *Plant Journal, 32*, 1–12.

Jaubert, S., Ledger, T. N., Laffaire, J. B., Piotte, C., Abad, P., & Rosso, M. N. (2002). Direct identification of stylet secreted proteins from root-knot nematodes by a proteomics approach. *Molecular and Biochemical Parasitology, 121*, 205–211.

Jones, M. G. K. (1981). Host cell responses to endoparasitic nematode attack: structure and function of giant cells and syncytia. *Annual Applied Biology, 97*, 353–372.

Jones, J. T., Reavy, B., Smant, G., & Prior, A. E. (2004). Glutathione peroxidases of the potato cyst nematode *Globodera rostochiensis*. *Gene, 324*, 47–54.

Kachroo, A., & Kachroo, P. (2009). Fatty acid-derived signals in plant defense. *Annual Review of Phytopathology, 47*, 153–176.

Karczmarek, A., Overmars, H., Helder, J., & Goverse, A. (2004). Feeding cell development by cyst and root-knot nematodes involves a similar early, local and transient activation of a specific auxin-inducible promoter element. *Molecular Plant Pathology, 5*, 343–346.

Kennedy, M. W., Garside, L. H., Goodrick, L. E., McDermott, L., Brass, A., Price, N. C., et al. (1997). The Ov20 protein of the parasitic nematode *Onchocerca volvulus*. A structurally novel class of small helix-rich retinol-binding proteins. *Journal of Biological Chemistry, 272*, 29442–29448.

Koenning, S. R., Overstreet, C., Noling, J. W., Donald, P. A., Becker, J. O., & Fortnum, B. A. (1999). Survey of crop losses in response to phytoparasitic nematodes in the United States for 1994. *Journal of Nematology, 31*, 587–618.

Koornneef, A., Leon-Reyes, R., Ritsema, T., Verhage, A., Den Otter, F. C., Van Loon, L. C., et al. (2008). Kinetics of salicylate-mediated suppression of jasmonate signaling reveal a role for redox modulation[1[OA]]. *Plant Physiology, 147*, 1358–1368.

Kyndt, T., Denil, S., Haegeman, A., Trooskens, G., Bauters, L., Van Criekinge, W., et al. (2012). Transcriptional reprogramming by root knot and migratory nematode infection in rice. *New Phytologist, 196*, 887–900.

Leon-Reyes, A., Du, Y., Koornneef, A., Proietti, S., Korbes, A. P., Memelink, J., et al. (2010). Ethylene signaling renders the jasmonate response of *Arabidopsis* insensitive to future suppression by salicylic Acid. *Molecular Plant-Microbe Interactions, 23*, 187–197.

Leone, A., Melillo, M. T., & Bleve-Zacheo, T. (2001). Lipoxygenase in pea roots subjected to biotic stress. *Plant Sci, 161*, 703–717.

Lu, S. W., Chen, S., Wang, J., Yu, H., Chronis, D., Mitchum, M. G., et al. (2009). Structural and functional diversity of CLAVATA3/ESR (CLE)-like genes from the potato cyst nematode *Globodera rostochiensis*. *Molecular Plant-Microbe Interactions, 22*, 1128–1142.

McConn, M., Creelman, R. A., Bell, E., Mullet, J. E., & Browse, J. (1997). Jasmonate is essential for insect defense in *Arabidopsis*. *Proceedings of the National Academy of Science of the United States of America, 94*, 5473–5477.

Mei, C., Qi, M., Sheng, G., & Yang, Y. (2006). Inducible overexpression of a Rice allene oxide synthase gene increases the endogenous jasmonic acid level, PR gene expression, and host resistance to fungal infection. *Molecular Plant-Microbe Interactions, 19*, 1127–1137.

Melan, M. A., Dong, X., Endara, M. E., Davis, K. R., Ausubel, F. M., & Peterman, T. K. (1993). An *Arabidopsis thaliana* lipoxygenase gene can be induced by pathogens, abscisic acid, and methyl jasmonate. *Plant Physiology, 101*, 441–450.

Mueller, S., Hilbert, B., Dueckershoff, K., Roitsch, T., Krischke, M., Mueller, M. J., et al. (2008). General detoxification and stress responses are mediated by oxidized lipids through TGA transcription factors in *Arabidopsis*. *Plant Cell, 20*, 768–785.

Mur, L. A. J., Kenton, P., Atzorn, R., Miersch, O., & Wasternack, C. (2006). The outcomes of concentration-specific interactions between salicylate and jasmonate signaling include synergy, antagonism and oxidative stress leading to cell death. *Plant Physiology, 140*, 249–262.

Nahar, K., Kyndt, T., De Vleesschauwer, D., Hofte, M., & Gheysen, G. (2011). The jasmonate pathway is a key player in systemically induced defense against root knot nematodes in rice. *Plant Physiology, 157*, 305–316.

Niebel, A., De Almeida Engler, J., Tire, C., Engler, G., Van Montagu, M., & Gheysen, G. (1993). Induction patterns of an extensin gene in tobacco upon nematode infection. *Plant Cell, 5*, 1697–1710.

Nobre, M. J. G., & Evans, K. (1998). Plant and nematode surfaces: their structure and importance in host-parasite interactions. *Nematologica, 44*, 103–124.

Opperman, C. H., Bird, D. M., Williamson, V. M., Rokhsar, D. S., Burke, M., Cohn, J., et al. (2008). Sequence and genetic map of *Meloidogyne hapla*: a compact nematode genome for plant parasitism. *Proceedings of the National Academy of Sciences of the United States of America, 105*, 14802–14807.

Ozalvo, R., Cabrera, J., Escobar, C., Christensen, S. A., Borrego, E. J., Kolomiets, M. V., et al. (2014). Two closely related members of *Arabidopsis* 13-lipoxygenases (13-LOXs), LOX3 and LOX4, reveal distinct functions in response to plant-parasitic nematode infection. *Molecular Plant Pathology, 15*, 319–332.

Patel, N., Hamamouch, N., Li, C., Hewezi, T., Hussey, R. S., Baum, T. J., et al. (2010). A nematode effector protein similar to annexins in host plants. *Journal of Experimental Botany, 61*, 235–248.

Popeijus, H., Overmars, H., Jones, J., Blok, V., Goverse, A., Helder, J., et al. (2000). Degradation of plant cell walls by a nematode. *Nature, 406*, 36–37.

Prior, A., Jones, J. T., Blok, V. C., Beauchamp, J., McDermott, L., Cooper, A., et al. (2001). A surface-associated retinol- and fatty acid-binding protein (Gp-FAR-1) from the potato cyst nematode *Globodera pallida*: lipid binding activities, structural analysis and expression pattern. *Biochemistry Journal, 356*, 387–394.

Puthoff, D. P., Nettleton, D., Rodermel, S. R., & Baum, T. J. (2003). *Arabidopsis* gene expression changes during cyst nematode parasitism revealed by statistical analyses of microarray expression profiles. *The Plant Journal, 33*, 911–921.

Qin, L., Kudla, U., Roze, E. H., Goverse, A., Popeijus, H., Nieuwland, J., et al. (2004). Plant degradation: a nematode expansin acting on plants. *Nature, 427*, 30.

Rance, I., Fournier, J., & Esquerre-Tugaye, M. T. (1998). The incompatible interaction between Phytophthora parasitica var. nicotianae race 0 and tobacco is suppressed in transgenic plants expressing antisense lipoxygenase sequences. *Proceedings of the National Academy of Sciences of the United States of America, 95*, 6554–6559.

Replogle, A., Wang, J., Bleckmann, A., Hussey, R. S., Baum, T. J., Sawa, S., et al. (2011). Nematode CLE signaling in *Arabidopsis* requires CLAVATA2 and CORYNE. *Plant Journal, 65*, 430–440.

Robertson, L., Robertson, W. M., Sobczak, M., Helder, J., Tetaud, E., Ariyanayagam, M. R., et al. (2000). Cloning, expression and functional characterisation of a peroxiredoxin from the potato cyst nematode *Globodera rostochiensis*. *Molecular and Biochemical Parasitology, 111*, 41–49.

Rosso, M.-N., Favery, B., Piotte, C., Arthaud, L., De Boer, J. M., Hussey, R. S., et al. (1999). Isolation of a cDNA encoding a beta-1,4-endoglucanase in the root-knot nematode *Meloidogyne incognita* and expression analysis during plant parasitism. *Molecular Plant-Microbe Interactions, 12*, 585–591.

Roze, E., Hanse, B., Mitreva, M., Vanholme, B., Bakker, J., & Smant, G. (2008). Mining the secretome of the root-knot nematode *Meloidogyne chitwoodi* for candidate parasitism genes. *Molecular Plant Pathology, 9*, 1–10.

Samuels, L., DeBono, A., Lam, P., Wen, M., Jetter, R., & Kunst, L. (2008). Use of *Arabidopsis* eceriferum mutants to explore plant cuticle biosynthesis. *J Vis Exp, 16*, 709.

Sanz, A., Moreno, J. I., & Castresana, C. (1998). PIOX, a new pathogen-induced oxygenase with homology to animal cyclooxygenase. *Plant Cell, 10*, 1523–1537.

Sattler, S. E., Mene-Saffrane, L., Farmer, E. E., Krischke, M., Mueller, M. J., & DellaPenna, D. (2006). Nonenzymatic lipid peroxidation reprograms gene expression and activates defense markers in *Arabidopsis* tocopherol-deficient mutants. *Plant Cell, 18*, 3706–3720.

Scholl, E., Sosinski, B. R., Thomas, V. P., & Windham, E. (2008). Sequence and genetic map of *Meloidogyne hapla*: a compact nematode genome for plant parasitism. *Proceedings of the National Academy of Sciences of the United States of America, 105*, 14802–14807.

Sijmons, P. C., Atkinson, H. J., & Wyss, U. (1994). Parasitic strategies of root nematodes and associated host cell responses. *Annual Review of Phytopathology, 32*, 235–259.

Smant, G., Stokkermans, J. P., Yan, Y., de Boer, J. M., Baum, T. J., Wang, X., et al. (1998). Endogenous cellulases in animals: isolation of beta-1, 4-endoglucanase genes from two species of plant-parasitic cyst nematodes. *Proceedings of the National Academy of Sciences of the United States of America, 95*, 4906–4911.

Soriano, I. R., Riley, I. T., Potter, M. J., & Bowers, W. S. (2004). Phytoecdysteroids: a novel defense against plant-parasitic nematodes. *Journal of chemical ecology, 30*, 1885–1899.

Sperling, P., & Heinz, E. (2003). Plant sphingolipids: structural diversity, biosynthesis, first genes and functions. *Biochimica et Biophysica Acta, 1632*, 1–15.

Spiegel, Y., Inbar, J., Kahane, I., & Sharon, E. (1995). Carbohydrate-recognition domains on the surface of phytophagous nematodes. *Experimental Parasitology, 80*, 220–227.

Spiegel, Y., Kahane, I., Cohen, L., & Sharon, E. (1997). *Meloidogyne javanica* surface proteins: characterization and lability. *Parasitology, 115*(Pt 5), 513–519.

Spoel, S. H., Koornneef, A., Claessens, S. M., Korzelius, J. P., Van Pelt, J. A., Mueller, M. J., et al. (2003). NPR1 modulates cross-talk between salicylate- and jasmonate-dependent defense pathways through a novel function in the cytosol. *Plant Cell, 15*, 760–770.

Stintzi, A., Weber, H., Reymond, P., Browse, J., & Farmer, E. E. (2001). Plant defense in the absence of jasmonic acid: the role of cyclopentenones. *Proceedings of the National Academy of Sciences of the United States of America, 98*, 12837–12842.

Stokkermans, T. J., & Dunbar, M. T. (1998). Solar retinopathy in a hospital-based primary care clinic. *Journal of the American Optometric Association, 69*, 625–636.

Stumpe, M., Gobel, C., Demchenko, K., Hoffmann, M., Klosgen, R. B., Pawlowski, K., et al. (2006). Identification of an allene oxide synthase (CYP74C) that leads to formation of alpha-ketols from 9-hydroperoxides of linoleic and linolenic acid in below-ground organs of potato. *Plant Journal, 47*, 883–896.

Takahashi, H., Kanayama, Y., Zheng, M. S., Kusano, T., Hase, S., Ikegami, M., et al. (2004). Antagonistic interactions between the SA and JA signaling pathways in *Arabidopsis* modulate expression of defense genes and gene-for-gene resistance to cucumber mosaic virus. *Plant and Cell Physiology, 45*, 803–809.

Turner, J. G., Ellis, C., & Devoto, A. (2002). The jasmonate signal pathway. *Plant Cell, 14*(Suppl.), S153–S164.

Tygat, T., De Meutter, J., Vanholme, B., Clayes, M., Verreijdt, L., & Gheysen, G. (2002). Development and pharyngeal gland activities of *Heterodera schachtii* infecting *Arabidopsis thaliana* roots. *Nematology, 4*, 899–908.

Uehara, T., Sugiyama, S., Matsuura, H., Arie, T., & Masuta, C. (2010). Resistant and susceptible responses in tomato to cyst nematode are differentially regulated by salicylic acid. *Plant and Cell Physiology, 51*, 1524–1536.

Upchurch, R. G. (2008). Fatty acid unsaturation, mobilization, and regulation in the response of plants to stress. *Biotechnology Letters, 30*, 967–977.

Vanholme, B., De Meutter, J., Tytgat, T., Van Montagu, M., Coomans, A., & Gheysen, G. (2004). Secretions of plant-parasitic nematodes: a molecular update. *Gene, 332*, 13–27.

Vanholme, B., Mitreva, M., Van Criekinge, W., Logghe, M., Bird, D., McCarter, J. P., et al. (2006). Detection of putative secreted proteins in the plant-parasitic nematode *Heterodera schachtii. Parasitology Research, 98*, 414–424.

Vellosillo, T., Martinez, M., Lopez, M. A., Vicente, J., Cascon, T., Dolan, L., et al. (2007). Oxylipins produced by the 9-lipoxygenase pathway in *Arabidopsis* regulate lateral root development and defense responses through a specific signaling cascade. *Plant Cell, 19*, 831–846.

Veronico, P., Giannino, D., Melillo, M. T., Leone, A., Reyes, A., Kennedy, M. W., et al. (2006). A novel lipoxygenase in pea roots. Its function in wounding and biotic stress. *Plant Physiology, 141*, 1045–1055.

Vijayan, P., Shockey, J., Levesque, C. A., Cook, R. J., & Browse, J. (1998). A role for jasmonate in pathogen defense of *Arabidopsis. Proceedings of the National Academy of Sciences of the United States of America, 95*, 7209–7214.

Wang, X., Lee, C., Replogle, A., Joshi, S., Korkin, D., Hussey, R., et al. (2010). Dual roles for variable domain in protein trafficking and host-specific recognition of *Heterodera glycines* CLE effector proteins. *New Phytologist, 187*, 1003–1007.

Wang, X., Mitchum, M. G., Gao, B., Li, C., Diab, H., Baum, T. J., et al. (2005). A parasitism gene from a plant-parasitic nematode with function similar to CLAVATA3/ESR (CLE) of *Arabidopsis thaliana. Molecular Plant Pathology, 6*, 187–191.

Weber, H., Vick, B. A., & Farmer, E. E. (1997). Dinor-oxo-phytodienoic acid: a new hexadecanoid signal in the jasmonate family. *Proceedings of the National Academy of Sciences of the United States of America, 94*, 10473–10478.

Wondafrash, M., Van Dam, N. M., & Tytgat, T. O. (2013). Plant systemic induced responses mediate interactions between root parasitic nematodes and aboveground herbivorous insects. *Frontiers in Plant Science, 4*, 87.

Worrall, D., Ng, C. K., & Hetherington, A. M. (2003). Sphingolipids, new players in plant signaling. *Trends in Plant Science, 8*, 317–320.

Wubben, M. J., 2nd, Rodermel, S. R., & Baum, T. J. (2004). Mutation of a UDP-glucose-4-epimerase alters nematode susceptibility and ethylene responses in *Arabidopsis* roots. *Plant Journal, 40,* 712–724.

Wubben, M. J., 2nd, Su, H., Rodermel, S. R., & Baum, T. J. (2001). Susceptibility to the sugar beet cyst nematode is modulated by ethylene signal transduction in *Arabidopsis thaliana. Molecular Plant-Microbe Interactions, 14,* 1206–1212.

Wubben, M. J. E., Jin, J., & Baum, T. J. (2008). Cyst nematode parasitism of *Arabidopsis thaliana* is inhibited by salicylic acid (SA) and elicits uncoupled SA-independent pathogenesis-related gene expression in roots. *Molecular Plant-Microbe Interactions, 21,* 424–432.

Wyss, U., & Grundler, F. M. W. (1992). Feeding behavior of sedentary plant parasitic nematodes. *Netherlands Journal of Plant Pathology, 98,* 165–173.

Yan, Y., Borrego, E., & Kolomiets, M. V. (2013). Jasmonate biosynthesis, perception and function in plant development and stress responses. In P. R. V. Baez (Ed.), *Lipid metabolism.* InTech.

Yan, Y., Christensen, S., Isakeit, T., Engelberth, J., Meeley, R., Hayward, A., et al. (2012). Disruption of OPR7 and OPR8 reveals the versatile functions of jasmonic acid in maize development and defense. *Plant Cell, 24,* 1420–1436.

Yan, Y., Smant, G., Stokkermans, J., Qin, L., Helder, J., Baum, T., et al. (1998). Genomic organization of four beta-1,4-endoglucanase genes in plant-parasitic cyst nematodes and its evolutionary implications. *Gene, 220,* 61–70.

Zacheo, G., Bleve-Zacheo, T., & Melillo, M. T. (1997). Biochemistry of plant resistance response to nematode infection. In C. Fenoll, S. Ohl, & F. Grundler (Eds.), *Cellular and molecular aspects of plant-nematode interaction* (pp. 201–213). Dordrecht, The Netherlands: Kluwer Academic Publishers.

Zhang, J., Simmons, C., Yalpani, N., Crane, V., Wilkinson, H., & Kolomiets, M. (2005). Genomic analysis of the 12-oxo-phytodienoic acid reductase gene family of Zea mays. *Plant Molecular Biology, 59,* 323–343.

CHAPTER SEVEN

Developmental Pathways Mediated by Hormones in Nematode Feeding Sites

Javier Cabrera, Fernando E. Díaz-Manzano, Carmen Fenoll, Carolina Escobar[1]
Laboratory of Plant Physiology, Department of Environmental Sciences, Universidad de Castilla-La Mancha, Toledo, Spain
[1]Corresponding author: E-mail: carolina.escobar@uclm.es

Contents

Abstract

Sedentary plant endoparasitic (root-knot and cyst) nematodes induce the formation of their feeding sites by directing the transdifferentiation of normal plant root cells into nematode feeding cells, namely giant cells (GCs) and syncytia. In the past years, transcriptomic analyses combined with molecular cell biology have revealed dramatic and specific changes in gene expression in syncytia and GCs. Among the genes whose expression is modified to establish feeding sites are those involved in hormone regulated developmental pathways in the roots, particularly those related to auxins and cytokinins. The high concentrations of auxins and cytokinins in galls and syncytia have been described in detail by the use of reporter genes driven by specific promoters as 'sensors' of both phytohormones, such as *DR5*, *ARR5* or *TCS*. Moreover, several molecular evidences link the formation of nematode feeding sites (NFSs) to developmental processes such as maintenance of the root apical meristem, lateral root initiation or vascular tissue development, in which the two hormones are involved. The mechanisms that nematodes use to interfere with plant developmental pathways are unclear, but some seem to involve nematode secreted molecules, such as the *CLE*-like and the *CEP* peptides. Only in a few cases, plant hormone transduction and developmental circuits hijacked by nematodes to induce and maintain feeding sites have been studied in detail. Analysis combining hormone genetic sensors, mutants and comparative transcriptomics lead to the identification of

Advances in Botanical Research, Volume 73
ISSN 0065-2296
http://dx.doi.org/10.1016/bs.abr.2014.12.005

relevant plant regulators that are exploited for NFS differentiation. We present the current knowledge connecting the hormonal-controlled developmental processes of the root with the development of the NFS, which seem to be different for GCs and syncytia. For instance, LBD16 and WRKY23, two key transcription factors in the signal transduction leading to lateral root formation mediated by auxins, play distinctive roles during gall/GC and syncytia formation, respectively. However, the expression of either gene in the feeding site is not strictly plant auxin-dependent, indicating that their regulation by nematodes differs in some aspects from the endogenous pathways operating in normal root development. We also highlight the evidences linking gall and GC ontogeny to the pericycle and discuss the transfer cell-like identity of feeding cells.

1. INTRODUCTION

Plant endoparasitic nematodes induce the formation of sophisticated feeding structures inside the root that operate as physiological sinks to supply nutrients to the nematode (Perry & Moens, 2011). Among the most damaging groups of plant parasitic nematodes are the root-knot nematodes (RKNs; *Meloidogyne* spp.) and the cyst nematodes (CNs; *Heterodera* spp. and *Globodera* spp.), representing major threats to agriculture (Bird et al., 2009; Moens, Perry, & Starr, 2009). One of the most remarkable changes regarding cell morphogenesis directly induced by nematode effectors (see Chapters 11, 12 and 13) is the formation of giant cells (GCs) by RKNs and syncytia by CNs. They are cells specifically induced by nematodes to sustain their feeding and their obligated development inside the plant. Together, both nematode groups are able to infect almost all species of agricultural crops, as RKNs show a polyphagous behaviour (Moens et al., 2009), suggesting that the nematodes interfere with biological processes shared by most plant species in order to develop their feeding sites. One possibility is that they may 'hijack', at least partially, fundamental mechanisms of plant development, necessary for the survival and appropriate plant performance, as described for molecular transducers common to lateral root (LR) and gall formation (Cabrera, Díaz-Manzano, et al., 2014). Understanding nematode feeding site (NFS) formation based on a deep knowledge of the developmental processes occurring in a noninfected root was proposed before the blast of the *omics* (Scheres et al., 1997). In the age of trancriptomics, when hundreds of genes have been identified as differentially expressed during the process of plant–nematode interaction, it becomes a prerequisite to connect those molecular evidences to the signalling cascades mediating developmental processes in a noninfected root. Several evidences show that plant parasitic nematodes develop their feeding sites through modulation or interloping of those developmental mechanisms present in the plant. This chapter summarizes the advances in this topic. The molecular mechanisms used by microorganisms to interfere with plant processes are surprisingly subtle but can effectively modify predefined plant developmental patterns. A recent example is the phytoplasma virulence protein SAP54 that promotes the degradation of flowering regulatory proteins, generating a short circuit in a developmental process that transforms flowers into leaves, helping attractiveness to leaf-hopper vectors for phytoplasma reproduction and propagation (MacLean et al., 2014). Similarly, effector molecules secreted by the nematodes seem to interfere with developmental pathways, although still there is a lack of clear evidence about the particular transduction cascades modified or perturbed by nematode effectors (this chapter and Chapters 11 and 12; Lee et al., 2011).

In order to modify the molecular pathways used by the plants to develop their basic functional structures, it is reasonable to presume that the nematode interferes with the upstream hormonal control of these particular transduction cascades. In this respect, early experiments already pointed to the importance of auxins and cytokinins in the development of NFSs induced within the root, as increased concentrations of both phytohormones or their precursors were detected in galls induced by RKNs (Balasubramanian & Rangaswami, 1962; Krupasagar & Barker, 1969). More recently, mass spectrometric analysis confirmed the presence of auxins and cytokinins in the secretions of *Heterodera schachtii* and *Meloidogyne incognita* (De Meutter et al., 2003, 2005), which suggests a role for both hormones during NFS development. Moreover, the use of 'hormone-sensor systems' based on reporter genes evidences the activation of both auxin and cytokinin signalling pathways in the formation/maintenance of GCs, galls and syncytia. Early experiments showing the activation of the auxin-responsive promoter *GH3* in the galls formed by RKNs in white clover (Hutangura, Mathesius, Jones, & Rolfe, 1999) have been further confirmed with the use of the synthetic auxin responsive promoter *DR5*. It showed a clear activation in the GCs, galls and syncytia induced by plant parasitic nematodes in *Arabidopsis* either with β-glucuronidase or Green Fluorescent Protein at early (Grunewald, Cannoot, Friml, & Gheysen, 2009; Karczmarek, Overmars, Helder, & Goverse, 2004) and late NFS developmental stages (Absmanner, Stadler, & Hammes, 2013; Cabrera, Díaz-Manzano, et al., 2014). The activation of the *DR5*-based sensor occurs very early during GC formation, as brief incubation with the GUS substrate in conditions where diffusion is minimized shows a strong signal specifically within the GCs (Figure 1(A)). However, the signal was also present in the surrounding vascular cells after longer incubation times, suggesting that, not surprisingly, auxins are also present in the vascular cell layers surrounding the GCs (Figure 1(B) and (C); Cabrera, Díaz-Manzano, et al., 2014). Similarly, the activation of cytokinin-regulated genes has been demonstrated by using the responsive promoter *ARR5*, induced during early stages of RKN establishment (Lohar et al., 2004), and the synthetic cytokinin-responsive promoter *TCS*, that is induced in syncytia formed by *H. schachtii* in *Arabidopsis* but not in galls (Absmanner et al., 2013). Thereby, the overexpression in *Lotus japonica* of *CKX*, the enzyme that catalyzes the degradation of cytokinins, resulted in a reduction in the infection by RKNs (Lohar et al., 2004).

The auxin–cytokinin crosstalk is considered as the main hormonal control system regulating the developmental processes occurring in the roots, such as, the root apical meristem (RAM) maintenance, LR emergence or vascular tissue development (Bielach, Duclercq, Marhavý, & Benková, 2012; Bishopp, Help, et al., 2011; Dello Ioio et al., 2007). There are several molecular evidences showing that both hormones act antagonistically,

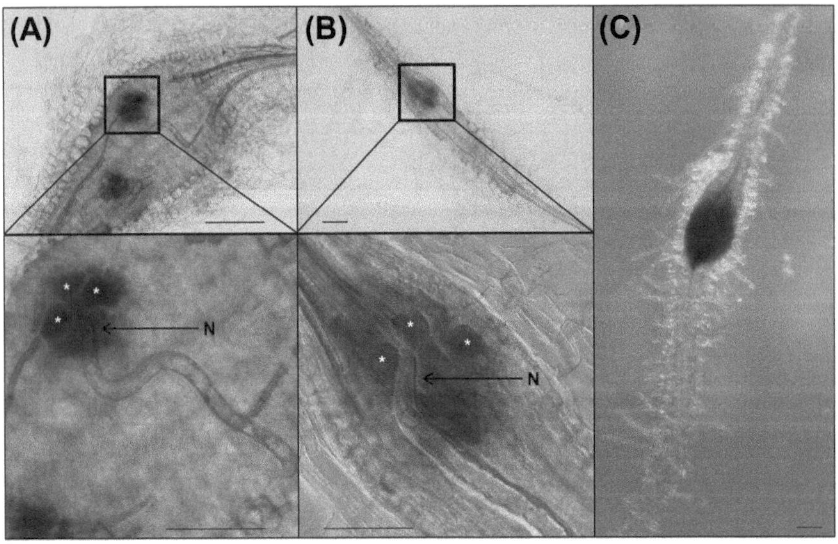

Figure 1 Transgenic *Arabidopsis* line *DR5::GUS* showing GUS activity in 4 dpi galls induced by *Meloidogyne javanica* at different incubation times in the GUS staining solution. (A) After 2 h incubation, GUS activity is centred in giant cells. (B) After 4 h incubation, staining spreads to the adjacent vascular tissue. (C) With overnight incubation, the signal covers most of the gall vascular tissues. N, nematode; *, giant cells. Zoom in images of (A) and (B) is shown in the panels below. Bars: 100 μm. (See colour plate)

i.e. auxin induces cell division in the meristems, while cytokinin stimulates the differentiation of these cells (reviewed in Bielach et al., 2012; Bishopp, Benková, & Helariutta, 2011; Moubayidin, Di Mambro, & Sabatini, 2009; Saini, Sharma, Kaur, & Pati, 2013).

Hormone signalling function in the plant–nematode interaction is not restricted to the interference with developmental pathways to establish the feeding site as they also play a key role in plant defences (review in Chapter 6). We present in this chapter the state of the art regarding our understanding of how nematodes interfere with hormone-regulated developmental pathways in the roots to establish their feeding sites, particularly those related to auxins and cytokinins. New data regarding morphometric parameters after GC reconstruction are also discussed in relation to their acquisition of transfer cell (TC)-like nature.

2. NEMATODE PEPTIDE HORMONES AS INTERCEPTORS OF PLANT DEVELOPMENT TO FORM FEEDING SITES

One of the major evidences demonstrating that nematodes actively interfere with the programmed development of the roots came from the discovery of a protein, *Hg-SYV46*, secreted by the CN *Heterodera glycines* that contains a structural motif of the *CLAVATA3/ESR-related* (*CLE*) family in *Arabidopsis* (Wang et al., 2001, 2005). *CLV3-like* peptide hormones have been shown to have multiple functions in many aspects of plant development and morphogenesis (Leasure & He, 2012). The similarities between *CLV3* and the nematode peptide *Hg-SYV46* are not merely structural. The overexpression of *Hg-SYV46* in *clv3* mutant plants rescues their phenotype, while *Hg-SYV46* overexpression in wild-type plants results in strong downregulation of *WUSCHEL* (*WUS*), similar to that occurring in the plants overexpressing *CLV3* (Wang et al., 2005). In *Arabidopsis*, membrane receptor kinases and several of the WUSCHEL-RELATED HOMEOBOX (WOX) transcription factors participate in *CLE* signalling (Leasure & He, 2012). While CLV3 acts in the shoot apical meristem repressing the expression of *WUS* and regulating the shoot apical meristem stem cell number through a negative feedback loop (Schoof et al., 2000), CLE40, another CLV3-like peptide, acts in the RAM regulating the expression of *WOX5* (Stahl, Wink, Ingram, & Simon, 2009) through the receptor-like kinase CRINKLY4 (ACR4). Overexpression of *CLE40* alters the expression of *WOX5* and promotes the differentiation of distal columella stem cells to columella cells (Stahl et al., 2009). Since the description of *Hg-SYV46*, more *CLE*-like genes have been described in *H. glycines* (Wang et al., 2010), *H. schachtii* (Patel et al., 2008; Wang et al., 2011) and *Globodera rostochiensis* (Guo, Ni, Denver, Wang, & Clark, 2011; Lu et al., 2009; see Chapters 11 and 12 for detailed explanation). Functional analysis via targeting these nematode *CLE*-like genes in plants by RNAi or by the infection of *CLE knockdown plants* showed a decrease in the infection rate and in the size of syncytia either in Arabidopsis or in soybean (Bakhetia, Urwin, & Atkinson, 2007; Patel et al., 2008; Guo et al., 2015). These constitute experimental evidences that confirm the role of *CLE*-like peptides in syncytia development (Replogle et al., 2011, 2013). All these data together suggest a role of the nematode peptide hormones in the development of the syncytia in *Arabidopsis*, possibly interfering with the aforementioned plant developmental pathways.

The *16D10* gene from *M. incognita* encoding a secretory peptide with a *CLE*-like sequence also showed functional characteristics of a component of a *CLE*-related pathway. In vivo expression of *16D10* dsRNA in *Arabidopsis* resulted in an increase in the resistance against RKN (Huang, Allen, Davis, Baum, & Hussey, 2006). In addition, the overexpression of 16D10, that directly interacts with SCARECROW-like transcription factors (SCR; Huang et al., 2006), does not rescue the *clv3* phenotype but stimulates root growth in *Arabidopsis* and tobacco, i.e. calli were formed in tips cut for subculturing (Huang et al., 2006). These results demonstrate 16D10 being an effector that substantially alters plant development. *SCR*, expressed specifically in the endodermis and cortex/endodermis initial cells of the root (Di Laurenzio et al., 1996), is a key regulator of radial patterning in the *Arabidopsis* root (Levesque et al., 2006) and is also directly activated by SHORT-ROOT (SHR; Levesque et al., 2006), therefore regulating root meristem identity and root development. Strikingly, transcripts from both *SCR* and *SHR* were downregulated in isolated GCs at 3 dpi (Barcala et al., 2010); however, transcriptomes specific of GCs are not available in earlier time points when developmental switches concerning cell development are probably crucial. Thus, the *16D10* gene constitutes another example of how nematodes could interfere with hormonal controlled developmental pathways of the root to generate their feeding sites.

The putative role of *CLE*-like nematode peptides might not be merely related to the meristems development. Yet, another group of *CLE* peptides such as tracheary element differentiation inhibitory factor (TDIF), a peptide hormone derived from *CLE41/44* (Ito et al., 2006) that induces the expression of *WOX4* in cambium cells (Hirakawa, Kondo, & Fukuda, 2010), participates in vascular development by promoting the division of cambium cells preventing their differentiation into xylem. Changes in auxin levels, such as those happening in nematode feeding cells, are necessary as the trigger signal for vascular development mediated by the *TDIF/WOX4* pathway (Donner, Sherr, & Scarpella, 2009; Scarpella, Marcos, Friml, & Berleth, 2006; Wenzel, Schuetz, Yu, & Mattsson, 2007). This agrees with the high degree of similarity found between the transcriptomes of 3 dpi GCs and suspension cells treated with brassinolide/boric acid that are differentiating into tracheids (Barcala et al., 2010; Kubo et al., 2005). These findings strongly point to provascular cells as putative precursors of the GCs, and are in agreement to initial data based on histological observations that proposed metaxylem, protoxylem or xylem parenchyma cells as the initial cells that develop into GCs (Bird, 1961; Bird & Koltai, 2000; Christie, 1936; Dropkin & Nelson, 1960; Niebel et al., 1993; Williamson & Hussey, 1996). In addition, proliferating tracheids and phloem elements have been described around GCs and syncytia (Absmanner et al., 2013; Bartlem, Jones, & Hammes, 2013; Hoth, Stadler, Sauer, & Hammes, 2008). The development of these vascular elements is governed by the balance between auxins and cytokinins, as those vascular cells from the galls that differentiate into phloem elements respond to auxins, but not to cytokinins, before differentiation (Absmanner et al., 2013). In contrast, the phloem around syncytia responded to both auxins and cytokinins (Absmanner et al., 2013). An example of the interference of the RKNs with the process of the vascularization in the galls is that overexpression of the enzyme chorismate mutase, secreted by *Meloidogyne javanica*, inhibits the final differentiation of root vascular cells (Doyle & Lambert, 2003), a phenotype that can be rescued by adding indole-3-acetic acid (IAA). This suggests that nematode-secreted chorismate mutase acts by reducing IAA levels. Interestingly, both CNs and RKNs encode in their genome proteins homologues to plant chorismate mutases (Bekal, Niblack, & Lambert, 2003; Chronis, Chen, Skantar, Zasada, & Wang, 2014; Huang et al., 2005; Jones et al., 2003; Vanholme, Haegeman, Jacob, Cannoot, & Gheysen, 2009). Therefore, the secretion of chorismate mutases by Plant Parasitic Nematodes (PPNs) might be a way to interfere with the auxin/cytokinin balance within the vascular cylinder in order to redirect the differentiation of vascular elements to NFCs.

Auxin gradients, generated by the PIN proteins, have been shown to be a common signal for the formation of different new organs in the plant (Benková et al., 2003; Vanneste & Friml, 2009). In this context, tomato plants treated with the polar auxin transport inhibitor NPA showed a reduction in the establishment of CNs and abnormal syncytia development (Goverse et al., 2000). Grunewald et al. (2009) studied the differential expression of *PIN* coding genes during early syncytia development in *Arabidopsis* confirming opposite regulation for different members of the family. A model was proposed in which PIN1 mediates the influx of auxin to the initial syncytia cells and PIN3 and PIN4 distribute the accumulated auxin laterally, allowing the expansion of the NFS (Grunewald et al., 2009). Thereby, CN infection rates and syncytia development are affected in *pin3* mutant plants (Grunewald et al., 2009). Together with PIN1, another auxin influx carrier, LAX3, that is expressed in the syncytium and in cells to be incorporated into the syncytium together with LAX1 (Lee et al., 2011), allow the syncytia growth by increasing the auxin levels in the neighbouring cells. Strikingly, an effector protein from *H. schachtii* (*Hs19C07*) can interact with LAX3 (Lee et al., 2011). In addition, *AUX1*, a closely related *AUX/LAX* family member, was upregulated in syncytia developed by *H. schachtii* and in galls induced by *M. incognita* in *Arabidopsis* (Mazarei, Lennon, Puthoff, Rodermel, & Baum, 2003). Both the *aux1/lax3* double mutant and the *aux1/lax1/lax2/lax3* quadruple mutant showed significant decrease in the number of female nematodes at both 14 and 30 dpi (Lee et al., 2011), suggesting that the *LAX3–Hs19C07* interaction could alter auxin levels to promote syncytia establishment. The analysis of *Arabidopsis* mutant lines *pin1/ttg-1* and *pin2* after *H. schachtii* infection, resulting in the reduction of nematode and syncytia development, confirmed the role of auxin transporters during the plant–nematode interaction (Goverse et al., 2000). However, most of the existing data are based on CNs and further analysis should be made in galls induced by RKNs to know the putative role of these proteins in this process.

Novel classes of peptide hormones as C-terminally encoded peptide (CEP) are emerging as regulators of the developmental process leading to gall formation. *CEP* genes have been identified in *M. incognita* and *Meloidogyne hapla* genomes (Bobay et al., 2013; Goverse & Bird, 2011) but not in the false RKN *Nacobbus aberrans* (Eves-van den Akker et al., 2014), nor in CNs (*H. glycines* and *G. rostochiensis*), migratory nematodes (*Radopholus similis* and *Pratylenchus coffeae*) or the free-living nematode (*Caenorhabditis elegans*; Bobay et al., 2013). In particular, MhCEP11 from *M. hapla* shows a significant sequence homology (Bobay et al., 2013) with the *Arabidopsis*-encoded CEP1 that is expressed in LR primordia. Overexpression of AtCEP1 with a constitutive promoter resulted in a reduced number of cells in the RAM (Ohyama, Ogawa, & Matsubayashi, 2008). In *Medicago truncatula*, the constitutive overexpression of MtCEP1 altered root development in several ways, by inhibition of LR formation, enhancement of nodulation and cortical, epidermal and pericycle cell divisions (Imin, Mohd-Radzman, Ogilvie, & Djordjevic, 2013). Future functional studies involving these molecules will elucidate their role as putative regulators of the gall and/or GC formation.

3. AUXINS, LATERAL ROOT FORMATION AND FEEDING SITES

CEPs are one of the most recently identified molecules that relate a root developmental process, LR formation, to gall development (Imin et al., 2013), but it is not the only connecting link. The aforementioned local increase in auxin levels favoured by PIN proteins is needed as well during the formation of LRs (Benková et al., 2003). There are several pieces of evidence that point to similarities and molecular connections between the processes of LR and NFS development. Among them, it was shown that the auxin-insensitive tomato mutant *diaegotropica*, *dgt* (Richardson & Price, 1982), which lacks LRs, was resistant to *M. incognita* and developed smaller syncytia upon CN infection (Goverse et al., 2000). In tomato and *M. truncatula*, two transcription factors, KNOX and PHAN, are induced in both GCs and LR meristems (Bird & Koltai, 2000; Koltai, Dhandaydham, Opperman, Thomas, & Bird, 2001). In *Arabidopsis* the downregulation of the Knotted1-like homeobox (KNOX) transcription factor KNAT6 yields an increment in the number of LRs (Dean, Casson, & Lindsey, 2004), in agreement to the suggested antagonistic action between auxin and KNOX transcription factors in organogenesis (reviewed in Scofield & Murray, 2006). Moreover, Barthels et al. (1997) used a promoter-tagging strategy to identify specific regulatory regions differentially activated in NFS as compared to uninfected roots. Surprisingly, among the 103 promoter tag lines that displayed a distinct activation response to nematode infection, 39 also exhibited induction at LR initiation sites. This has been further confirmed by in silico analysis of transcriptomes from galls and GCs in *Arabidopsis* that showed an enrichment of characteristic genes from LR initial cells in the transcriptome of 3 dpi GCs and galls (Cabrera, Díaz-Manzano, et al., 2014). LRs originate from divisions in the xylem pole pericycle (XPP) cells following and auxin-mediated signalling pathway. Two XPP marker lines, J0121 and J0192, showed strong and distinct GFP expression in the galls formed by *M. javanica* (Cabrera, Díaz-Manzano, et al., 2014;

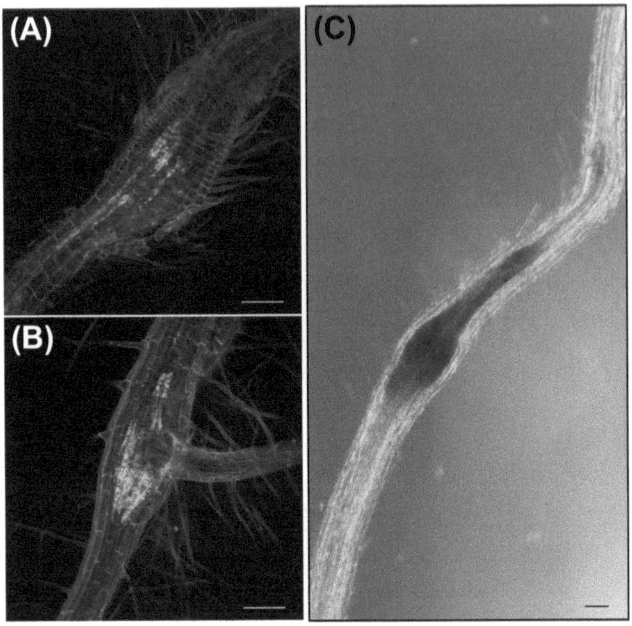

Figure 2 Transgenic lines of lateral root and XPP pericycle markers and *LBD16* expressing either GFP or GUS in galls formed by *Meloidogyne javanica*. (A) J0121 >> GFP at 4 dpi. (B) J0192 >> GFP at 4 dpi. (C) *pLBD16:GUS* at 7 dpi (Cabrera, Díaz-Manzano, et al., 2014). Bars: 100 μm. (See colour plate)

Figure 2(A) and (B)). Strikingly, in both lines GFP expression was mostly observed at both sides of the vascular cylinder and progressed inwards during gall development, which partially differs with the expression pattern found during LR formation (Cabrera, Díaz-Manzano, et al., 2014; Laplaze et al., 2005). In addition, the regular anticlinal and periclinal divisions observed during LR formation (Lavenus et al., 2013) were substituted by abnormal division planes in the cells proliferating inside the galls (Cabrera, Díaz-Manzano, et al., 2014). The expression pattern of *LBD16*, the gene whose promoter drives GFP expression in the J0192 enhancer trap line, mimics that of the J0192 line (Figure 2(C)). *LBD16* expression is detected in galls from *M. javanica* and *Meloidogyne arenaria* from 1 dpi up to 11–15 dpi and was regulated by auxins, as shown by its inhibition by α-(phenyl ethyl-2-one)-indole-3-acetic acid (PEO-IAA), an antagonist of IAA, similar to what happens in LR primordia (Cabrera, Díaz-Manzano, et al., 2014; Lee, Kim, Lee, & Kim, 2009; Okushima, Fukaki, Onoda, Theologis, & Tasaka, 2007). Although its expression at early stages appeared to correlate with the presence of auxins in the same cell types, at later stages the mere presence of auxins in the gall was not sufficient to activate *LBD16* expression (Cabrera, Díaz-Manzano, et al., 2014). These results may indicate the necessity for a threshold level of auxins in the gall to allow *LBD16* expression, which would mimic the scenario that occurs during the first divisions of LR development. However, the absence of signal in the syncytia formed in the *LBD16::GUS* line, where the 'auxin sensor' DR5 was also activated (Karczmarek et al., 2004), suggests that other signals apart from auxins could be contributing to the early activation of *LBD16* in galls and GCs. In this respect, secretions from *M. incognita* juveniles were also able to induce *LBD16* expression in *Arabidopsis* leaf protoplasts, suggesting an activation of the *LBD16* promoter by nematode secretions, in an autonomous manner. Furthermore, a reduction in gall formation of at least 20% was observed in *LBD16* loss of function lines as compared to wild type controls (Cabrera, Díaz-Manzano, et al., 2014). Interestingly, the expression pattern of the marker line *ProCycB1;1:CycB1;1(NT)-GUS*, active only during the G2/M transition, mimicked that of J0192 in XPP cells during early nematode establishment, and in most cells inside the vascular cylinder of the gall as the infection progressed. These results suggest that the founder cells contained in the XPP that divide to form a new LR (Péret, Larrieu, & Bennett, 2009) also divide during early gall formation (Cabrera, Díaz-Manzano, et al., 2014). *LBD16* loss of function lines showed abnormal GC development, pointing to a role of the pericycle during this process. The importance of XPP-specific genes during infection in *Arabidopsis* was further demonstrated by genetic ablation using a J0121 >> DTA line that showed a dramatic reduction in the infection and in the size of the GCs as compared to a control J0121 >> GFP line (Cabrera, Díaz-Manzano, et al., 2014).

The induction of *LBD16* during gall formation not only connects this process with LR formation but with the generation of calli (Cabrera, Díaz-Manzano, et al., 2014; Demeulenaere & Beeckman, 2014). *LBD* transcription factors and pericycle cells have been shown to be essential for the generation of calli from different organs through ectopic activation of an LR developmental program (Sugimoto, Jiao, & Meyerowitz, 2010). The ectopic expression of LBDs triggers spontaneous callus formation but their suppression inhibits the process (Fan, Xu, Xu, & Hu, 2012); whether or not gall development is somehow related to this process of callus formation through the activation of an LR initiation-like program remains to be elucidated. In silico data comparison supported this hypothesis, as those genes co-regulated with *LBD16* in different transcriptomes were integrated in signalling cascades mediated by auxins during LR and callus formation, as a particular feature of early developing RKN feeding sites (3 dpi) distinct to CNs. In contrast, cytokinin-induced genes were enriched in syncytia, whose transcriptomes hold a high similarity with the transcriptome of shoot regeneration from callus, modulated by cytokinins (Cabrera, Bustos, Favery, Fenoll, & Escobar, 2014; Cabrera, Fenoll, & Escobar, in press). In agreement with these analyses, subtle changes in the balance between cytokinin and auxin levels could be mediating the appearance of chloroplast-like structures inside 7 dpi GCs induced by *Meloidogyne graminicola* in rice, as showed by confocal microscopy (Ji et al., 2013; Kyndt, Vieira, Gheysen, & de Almeida-Engler, 2013) and in syncytia (Szakasits et al., 2009). Transcriptomic studies performed in isolated GCs and syncytia reflected the predominance of genes regulated by phytohormones among the differentially expressed genes as compared to noninfected tissues (reviewed in Cabrera, Bustos, et al., 2014; Cabrera, Fenoll, et al., in press; Escobar, Horowitz, & Mitchum, 2011). In this way, a direct in silico comparison of the transcriptomes of isolated GCs and syncytia at early developmental stages in *Arabidopsis* and the transcriptomes from seedlings treated with exogenous phytohormones contributed to increase the vast number of genes regulated by hormones that are also induced or repressed in NFCs (Cabrera, Bustos, et al., 2014; Cabrera, Fenoll, et al., in press). Clear differences between the hormone-related transcriptional balances of the two NFC types were found. While the percentage of auxin-induced genes stands out (26%) in GCs, in syncytia there are 21% of cytokinin-induced genes. On the contrary, in GCs there are more cytokinin-repressed genes as compared to those upregulated; in syncytia the number of cytokinin upregulated genes is higher than the number of downregulated ones. Interestingly, the number of genes repressed by auxins or cytokinins in GCs or syncytia was high, suggesting that gene repression driven by these hormones may be also crucial for the development of the NFS (Cabrera, Bustos, et al., 2014; Cabrera, Fenoll, et al., in press).

WRKY23 is another transcription factor regulated by auxins and induced by both RKN and CN. Loss of function lines showed an increased resistance to CNs (Grunewald et al., 2009). WRKY23 acts downstream of the signalling cascade (SLR/IAA14)-(auxin responsive factors) (ARF7/ARF19) during NFs and LR formation (Grunewald et al., 2008, 2012) and is needed for LR development through the stimulation of local flavonol biosynthesis (Grunewald et al., 2012). However, *WRKY23* expression is activated in NFS by an independent auxin pathway, suggesting the existence of other nematode-dependent signals in regulating *WRKY23* expression.

This could also be the case for *LBD16* expression that is activated by nematode secretions as previously mentioned (Cabrera, Díaz-Manzano, et al., 2014). Other members of the LBD transcription factor family, such as *LBD41*, are activated in both *M. incognita* and *H. schachtii* feeding sites (Fuller, Lilley, Atkinson, & Urwin, 2007), in agreement with transcriptomic analysis of GCs in *Arabidopsis* (Barcala et al., 2010).

The expression patterns of crucial molecular components of the auxin signalling pathway, encoded by genes of the *ARF*s family, have been addressed during syncytium development in *Arabidopsis* (Hewezi, Piya, Richard, & Rice, 2014). At early infection stages, 2–3 days after *H. schachtii* infection, *ARF3, 6, 10–12, 14, 15* and *20–22* were expressed inside the developing syncytium, while *ARF1, 2, 4, 5, 9, 18* and *19* were active in both syncytial and neighbouring cells. *ARF7* and *17* were mainly expressed at the edges of the syncytial and neighbouring cells and *ARF8* and *16* showed a weak response to *H. schachtii* infection (Hewezi et al., 2014). At 9–10 days after infection, *ARF1–3, 7, 17* and *20–22* were expressed in fully developed syncytium, whereas the expression of the other *ARFs* was restricted to the syncytial cells around the nematode head (Hewezi et al., 2014). Although still under investigation, the differential expression patterns of the *ARF* genes seem to be essential for the correct development of the CNs feeding cells. All these data suggest a subtle and complex regulation of auxin-mediated pathways based on a tight temporal and spatial control of molecular components such as ARFs in CNs feeding sites. A lack of knowledge of the regulation of these genes after RKN infection is still faced.

4. GIANT CELL MORPHOGENESIS AND TRANSFER CELL NATURE

In the previous sections, we described parallelisms between feeding site formation and developmental programs during the plant life cycle, such as LR formation. Although syncytia and GCs differ in their ontogeny and global transcriptional signatures, both develop cell wall ingrowths (CIs) to facilitate high rates of apoplastic/symplastic solute exchange. Both feeding site types also show similarities to the TCs that appear in different plant organs during plant development (reviewed in Offler, McCurdy, Patrick, & Talbot, 2003).

In GCs the amplification of the plasma membrane surface area could be up to 20-fold (reviewed in Jones & Goto, 2011). Syncytia induced by *H. schachtii* are symplastically isolated at 10–15 dpi (Hofmann, Wieczorek,

Blöchl, & Grundler, 2007) and the CIs are smaller in male- than in female-developed syncytia, suggesting that CI size control is based on the nutrient demand of the nematode (reviewed in Sobczak & Golinowski, 2008). Recently, three-dimensional reconstruction and volume measurements of GCs in *Arabidopsis* (Cabrera, Díaz-Manzano, et al., in press) brought some interesting findings that might explain, at least partially, their TC characteristics. The abnormally large size of the GCs implies a reduction in their surface area to volume ratio (S/V ratio; Cabrera, Díaz-Manzano, et al., in press), a factor that may compromise its functioning during nematode nourishing. Thus, the extensive formation of wall ingrowths lined with plasma membrane at certain developmental stages that defines them as TCs (Jones & Dropkin, 1976; Siddique, Sobczak, Tenhaken, Grundler, & Bohlmann, 2012) could be a response for a functional requirement to compensate the decrease of the S/V ratio as the GCs expand (reviewed in Cabrera, Díaz-Manzano, et al., in press; Rodiuc, Vieira, Banora, & de Almeida-Engler, 2014). Moreover, size regulation affects cell function in multiple ways, e.g. not only can nutrient and water movement be changed by altered surface/volume ratio, but also intercellular signalling might be influenced by changes in cell size and geometry. Recent studies in yeast revealed that transcription also changes specifically in response to cell dimensions (Sablowski & Dornelas, 2014; Wu, Rolfe, Gifford, & Fink, 2010).

Cell wall anatomy and composition in NFCs are quite similar to other TCs that develop in plants, and it is mainly composed of polysaccharides such as cellulose, hemicelluloses and pectin (reviewed in Rodiuc et al., 2014). Although regulatory signals for TC differentiation are not well known, the transcription factor *ZmMRP-1*, that has been described as a key component in the pathway leading to the formation of the TCs (Gómez, Royo, Guo, Thompson, & Hueros, 2002), is also induced in *Arabidopsis* galls (Barrero et al., 2009). Auxins and ethylene are the major phytohormones described as regulators of TC differentiation from several cell types (Dibley et al., 2009; Thiel et al., 2008; Thiel, Hollmann, et al., 2012; Thiel, Riewe, et al., 2012; Xiong, Li, Kang, & Chourey, 2011; Zhou et al., 2010). Common genes related to the auxin signalling cascades and transport from the *IAA/ARF/PIN* families play a role and/or show differential regulation in both TCs and NFCs (reviewed in Cabrera, Barcala, Fenoll, & Escobar, 2014; this chapter). Genes related with ethylene synthesis and signalling as those encoding membrane receptors, aminocyclopropane-1-carboxylic (ACC) acid oxidases or ACC acid synthase, are also to be induced in NFS (Cabrera, Barcala, et al., 2014; Tucker, Xue, & Yang, 2010). Increased concentration of ethylene in tomato galls was described long ago (Glazer, Orion, & Apelbaum, 1983). Moreover, functional analysis performed with ethylene-related mutants reinforce the importance of this hormone during plant–nematode interaction as ethylene-overproducing mutants *eto2* and *eto3* were more susceptible to CNs (Goverse et al., 2000; Wubben, Su, Rodermel, & Baum, 2001). Interestingly, ethylene overproduction in *eto2* mutants stimulated the formation of CIs or protuberances in syncytia along the vascular tissue, at late infection stages (Goverse et al., 2000), providing direct evidence for a putative role of ethylene in the stimulation of syncytia TC identity. On the other hand, mutants compromised in the ethylene signalling cascade (*etr1-1*, *ein2-1*, *ein3-1*, *eir1-1* and *axr2*) showed a lower susceptibility to infection by CNs (Goverse et al., 2000; Wubben et al., 2001). Interestingly, the most clarifying study of a functional implication in TCs characteristic of NFCs, such as the CIs formation, comes from the analysis of uridine diphosphate (UDP)-glucose dehydrogenase (UGD) coding genes. UGDs act through oxidation of UDP-glucose producing several cell wall polysaccharides. UGD2 and UGD3 are necessary for the production of CIs in syncytia and loss of function in double mutants severely affected nematode development (Siddique et al., 2012).

In conclusion, there are several evidences that nematodes may interfere or partially 'hijack' signal transduction pathways used by the plant to initiate and/or maintain developmental processes where auxins and cytokinins play a central role, such as SAM and RAM maintenance and LR and vascular tissue formation. In addition, nematodes may also interfere with transduction pathways leading to the differentiation of specialized plant cell types such as TCs. The understanding of the plant molecular components necessary to reprogramme normal plant cells into NFCs are an outstanding topic to be deciphered, what would contribute to the basic understanding of the plant–nematode interaction. This will also help to identify the interactions of nematode effectors with plant components, and ultimately, their mode of action while interfering with the plant developmental programs. Furthermore, this knowledge will constitute a powerful tool for engineering nematode resistance in plants and to direct the search for specific nematode control strategies.

ACKNOWLEDGEMENTS

This work was supported by the Spanish Government (AGL2010-17388 and AGL2013-48787-R to C. Escobar; and CSD2007-057 and PCIN-2013-053 to C. Fenoll). J. Cabrera and F. E. Díaz-Manzano were supported by predoctoral fellowships from the Spanish Government.

REFERENCES

Absmanner, B., Stadler, R., & Hammes, U. Z. (2013). Phloem development in nematode-induced feeding sites: the implications of auxin and cytokinin. *Frontiers in Plant Science, 4.*

Bakhetia, M., Urwin, P. E., & Atkinson, H. J. (2007). QPCR analysis and RNAi define pharyngeal gland cell-expressed genes of *Heterodera glycines* required for initial interactions with the host. *Molecular Plant–Microbe Interactions, 20*(3), 306–312.

Balasubramanian, M., & Rangaswami, G. (1962). Presence of indole compounds in nematode galls. *Nature, 194*, 774–775.

Barcala, M., García, A., Cabrera, J., Casson, S., Lindsey, K., Favery, B., et al. (2010). Early transcriptomic events in microdissected *Arabidopsis* nematode-induced giant cells. *The Plant Journal, 61*(4), 698–712.

Barrero, C., Royo, J., Grijota-Martinez, C., Faye, C., Paul, W., Sanz, S., et al. (2009). The promoter of *ZmMRP-1*, a maize transfer cell-specific transcriptional activator, is induced at solute exchange surfaces and responds to transport demands. *Planta, 229*(2), 235–247.

Barthels, N., van der Lee, F. M., Klap, J., Goddijn, O. J., Karimi, M., Puzio, P., et al. (1997). Regulatory sequences of *Arabidopsis* drive reporter gene expression in nematode feeding structures. *The Plant Cell, 9*(12), 2119–2134.

Bartlem, D. G., Jones, M. G., & Hammes, U. Z. (2013). Vascularization and nutrient delivery at root-knot nematode feeding sites in host roots. *Journal of Experimental Botany, 65*(7), 1789–1798.

Bekal, S., Niblack, T. L., & Lambert, K. N. (2003). A chorismate mutase from the soybean cyst nematode *Heterodera glycines* shows polymorphisms that correlate with virulence. *Molecular Plant–Microbe Interactions, 16*(5), 439–446.

Benková, E., Michniewicz, M., Sauer, M., Teichmann, T., Seifertová, D., Jürgens, G., et al. (2003). Local, efflux-dependent auxin gradients as a common module for plant organ formation. *Cell, 115*(5), 591–602.

Bielach, A., Duclercq, J., Marhavý, P., & Benková, E. (2012). Genetic approach towards the identification of auxin–cytokinin crosstalk components involved in root development. *Philosophical Transactions of the Royal Society B: Biological Sciences, 367*(1595), 1469–1478.

Bird, A. F. (1961). The ultrastructure and histochemistry of a nematode-induced giant cell. *The Journal of Biophysical and Biochemical Cytology, 11*(3), 701–715.

Bird, D. M., & Koltai, H. (2000). Plant parasitic nematodes: habitats, hormones, and horizontally acquired genes. *Journal of Plant Growth Regulation, 19*(2), 183–194.

Bird, D. M., Williamson, V. M., Abad, P., McCarter, J., Danchin, E. G., Castagnone-Sereno, P., et al. (2009). The genomes of root-knot nematodes. *Annual Review of Phytopathology, 47*, 333–351.

Bishopp, A., Benková, E., & Helariutta, Y. (2011). Sending mixed messages: auxin-cytokinin crosstalk in roots. *Current Opinion in Plant Biology, 14*(1), 10–16.

Bishopp, A., Help, H., El-Showk, S., Weijers, D., Scheres, B., Friml, J., et al. (2011). A mutually inhibitory interaction between auxin and cytokinin specifies vascular pattern in roots. *Current Biology, 21*(11), 917–926.

Bobay, B. G., DiGennaro, P., Scholl, E., Imin, N., Djordjevic, M. A., & Mck Bird, D. (2013). Solution NMR studies of the plant peptide hormone CEP inform function. *FEBS Letters, 587*(24), 3979–3985.

Cabrera, J., Barcala, M., Fenoll, C., & Escobar, C. (2014). Transcriptomic signatures of transfer cells in early developing nematode feeding cells of *Arabidopsis* focused on auxin and ethylene signalling. *Frontiers in Plant Science, 5.*

Cabrera, J., Bustos, R., Favery, B., Fenoll, C., & Escobar, C. (2014). NEMATIC: a simple and versatile tool for the in silico analysis of plant–nematode interactions. *Molecular Plant Pathology, 15*(6), 627–636.

Cabrera, J., Díaz-Manzano, F. E., Barcala, M., de Almeida-Engler, J., Engler, G., Fenoll, C., et al. Phenotyping nematode feeding sites: three dimensional reconstruction and volumetric measurements of giant cells induced by root-knot nematodes in Arabidopsis. New Phytologist, in press.

Cabrera, J., Díaz-Manzano, F. E., Sánchez, M., Rosso, M.-N., Melillo, T., Goh, T., et al. (2014). A role for *LATERAL ORGAN BOUNDARIES-DOMAIN 16* during the interaction *Arabidopsis – Meloidogyne* spp. provides a molecular link between lateral root and root-knot nematode feeding site development. *New Phytologist, 203*, 632–645.

Cabrera, J., Fenoll, C., & Escobar, C. Genes co-regulated with *LBD16* in nematode feeding sites inferred from in silico analysis show similarities to regulatory circuits mediated by the auxin/cytokinin balance in Arabidopsis. Plant Signaling and Behaviour, in press.

Christie, J. R. (1936). The development of root-knot nematode galls. *Phytopathology, 26*, 1–22.

Chronis, D., Chen, S., Skantar, A. M., Zasada, I. A., & Wang, X. (2014). A new chorismate mutase gene identified from *Globodera ellingtonae* and its utility as a molecular diagnostic marker. *European Journal of Plant Pathology, 139*(2), 239–246.

De Meutter, J., Robertson, L. E. E., Parcy, F., Mena, M., Fenoll, C., & Gheysen, G. (2005). Differential activation of *ABI3* and *LEA* genes upon plant parasitic nematode infection. *Molecular Plant Pathology, 6*(3), 321–325.

De Meutter, J., Tytgat, T., Witters, E., Gheysen, G., Van Onckelen, H., & Gheysen, G. (2003). Identification of cytokinins produced by the plant parasitic nematodes *Heterodera schachtii* and *Meloidogyne incognita*. *Molecular Plant Pathology, 4*(4), 271–277.

Dean, G., Casson, S., & Lindsey, K. (2004). *KNAT6* gene of *Arabidopsis* is expressed in roots and is required for correct lateral root formation. *Plant Molecular Biology, 54*(1), 71–84.

Dello Ioio, R., Linhares, F. S., Scacchi, E., Casamitjana-Martinez, E., Heidstra, R., Costantino, P., et al. (2007). Cytokinins determine *Arabidopsis* root-meristem size by controlling cell differentiation. *Current Biology, 17*(8), 678–682.

Demeulenaere, M. J., & Beeckman, T. (2014). The interplay between auxin and the cell cycle during plant development. In E. Zažímalová, J. Petrášek, & E. Benková (Eds.), *Auxin and its role in plant development* (pp. 119–141). Vienna: Springer.

Di Laurenzio, L., Wysocka-Diller, J., Malamy, J. E., Pysh, L., Helariutta, Y., Freshour, G., et al. (1996). The *SCARECROW* gene regulates an asymmetric cell division that is essential for generating the radial organization of the *Arabidopsis* root. *Cell, 86*(3), 423–433.

Dibley, S. J., Zhou, Y., Andriunas, F. A., Talbot, M. J., Offler, C. E., Patrick, J. W., et al. (2009). Early gene expression programs accompanying trans-differentiation of epidermal cells of *Vicia faba* cotyledons into transfer cells. *New Phytologist, 182*(4), 863–877.

Donner, T. J., Sherr, I., & Scarpella, E. (2009). Regulation of pre-procambial cell state acquisition by auxin signaling in *Arabidopsis* leaves. *Development, 136*(19), 3235–3246.

Doyle, E. A., & Lambert, K. N. (2003). *Meloidogyne javanica* chorismate mutase 1 alters plant cell development. *Molecular Plant–Microbe Interactions, 16*(2), 123–131.

Dropkin, V. H., & Nelson, P. E. (1960). The histopathology of root-knot nematode infections in soybeans. *Phytopathology, 50*(6), 442–447.

Escobar, C., Horowitz, S. B., & Mitchum, M. G. (2011). Transcriptomic and proteomic analysis of the plant response to nematode infection. In J. Jones, G. Gheysen, & C. Fenoll (Eds.), *Genomics and molecular genetics of plant–nematode interactions* (pp. 157–173). Dordrecht, Heidelberg, London & New York: Springer.

Eves-van den Akker, S., Lilley, C. J., Danchin, E. G., Rancurel, C., Cock, P. J., Urwin, P. E., et al. (2014). The transcriptome of *Nacobbus aberrans* reveals insights into the evolution of sedentary endoparasitism in plant-parasitic nematodes. *Genome Biology and Evolution, 6*(9), 2181–2194.

Fan, M., Xu, C., Xu, K., & Hu, Y. (2012). *LATERAL ORGAN BOUNDARIES DOMAIN* transcription factors direct callus formation in *Arabidopsis* regeneration. *Cell Research, 22*(7), 1169–1180.

Fuller, V. L., Lilley, C. J., Atkinson, H. J., & Urwin, P. E. (2007). Differential gene expression in *Arabidopsis* following infection by plant-parasitic nematodes *Meloidogyne incognita* and *Heterodera schachtii*. *Molecular Plant Pathology, 8*(5), 595–609.

Glazer, I., Orion, D., & Apelbaum, A. (1983). Interrelationships between ethylene production, gall formation, and root-knot nematode development in tomato plants infected with *Meloidogyne javanica*. *Journal of Nematology, 15*(4), 539.

Gómez, E., Royo, J., Guo, Y., Thompson, R., & Hueros, G. (2002). Establishment of cereal endosperm expression domains identification and properties of a maize transfer cell–specific transcription factor, *ZmMRP-1*. *The Plant Cell, 14*(3), 599–610.

Goverse, A., & Bird, D. (2011). The role of plant hormones in nematode feeding cell formation. In J. Jones, G. Gheysen, & C. Fenoll (Eds.), *Genomics and molecular genetics of plant–nematode interactions* (pp. 325–347). Dordrecht, Heidelberg, London & New York: Springer.

Goverse, A., Overmars, H., Engelbertink, J., Schots, A., Bakker, J., & Helder, J. (2000). Both induction and morphogenesis of cyst nematode feeding cells are mediated by auxin. *Molecular Plant–Microbe Interactions, 13*(10), 1121–1129.

Grunewald, W., Cannoot, B., Friml, J., & Gheysen, G. (2009). Parasitic nematodes modulate *PIN*-mediated auxin transport to facilitate infection. *PLoS Pathogens, 5*(1), e1000266.

Grunewald, W., De Smet, I., Lewis, D. R., Löfke, C., Jansen, L., Goeminne, G., et al. (2012). Transcription factor *WRKY23* assists auxin distribution patterns during *Arabidopsis* root development through local control on flavonol biosynthesis. *Proceedings of the National Academy of Sciences of the United States of America, 109*(5), 1554–1559.

Grunewald, W., Karimi, M., Wieczorek, K., Van de Cappelle, E., Wischnitzki, E., Grundler, F. M., et al. (2008). A role for *AtWRKY23* in feeding site establishment of plant-parasitic nematodes. *Plant Physiology, 148*(1), 358–368.

Guo, Y., Ni, J., Denver, R., Wang, X., & Clark, S. E. (2011). Mechanisms of molecular mimicry of plant *CLE* peptide ligands by the parasitic nematode *Globodera rostochiensis*. *Plant Physiology, 157*(1), 476–484.

Guo, X., Chronis, D., De La Torre, C. M., Smed, J., Wang, X., & Mitchum, M. G. (2015). Enhanced resistance to soybean cyst nematode Heterodera glycines in transgenic soybean by silencing putative CLE receptors. *Plant Biotechnol J.* http://dx.doi.org/10.1111/pbi.12313.

Hewezi, T., Piya, S., Richard, G., & Rice, J. H. (2014). Spatial and temporal expression patterns of auxin response transcription factors in the syncytium induced by the beet cyst nematode *Heterodera schachtii* in *Arabidopsis*. *Molecular Plant Pathology, 15*(7), 730–736.

Hirakawa, Y., Kondo, Y., & Fukuda, H. (2010). *TDIF* peptide signaling regulates vascular stem cell proliferation via the *WOX4* homeobox gene in *Arabidopsis*. *The Plant Cell, 22*(8), 2618–2629.

Hofmann, J., Wieczorek, K., Blöchl, A., & Grundler, F. M. (2007). Sucrose supply to nematode-induced syncytia depends on the apoplasmic and symplasmic pathways. *Journal of Experimental Botany, 58*(7), 1591–1601.

Hoth, S., Stadler, R., Sauer, N., & Hammes, U. Z. (2008). Differential vascularization of nematode-induced feeding sites. *Proceedings of the National Academy of Sciences of the United States of America, 105*(34), 12617–12622.

Huang, G., Allen, R., Davis, E. L., Baum, T. J., & Hussey, R. S. (2006). Engineering broad root-knot resistance in transgenic plants by RNAi silencing of a conserved and essential root-knot nematode parasitism gene. *Proceedings of the National Academy of Sciences of the United States of America, 103*(39), 14302–14306.

Huang, X., Tian, B., Niu, Q., Yang, J., Zhang, L., & Zhang, K. (2005). An extracellular protease from *Brevibacillus laterosporus G4* without parasporal crystals can serve as a pathogenic factor in infection of nematodes. *Research in Microbiology, 156*(5), 719–727.

Hutangura, P., Mathesius, U., Jones, M. G., & Rolfe, B. G. (1999). Auxin induction is a trigger for root gall formation caused by root-knot nematodes in white clover and is associated with the activation of the flavonoid pathway. *Functional Plant Biology, 26*(3), 221–231.

Imin, N., Mohd-Radzman, N. A., Ogilvie, H. A., & Djordjevic, M. A. (2013). The peptide-encoding *CEP1* gene modulates lateral root and nodule numbers in *Medicago truncatula. Journal of Experimental Botany, 64*(17), 5395–5409.

Ito, Y., Nakanomyo, I., Motose, H., Iwamoto, K., Sawa, S., Dohmae, N., et al. (2006). Dodeca-*CLE* peptides as suppressors of plant stem cell differentiation. *Science, 313*(5788), 842–845.

Ji, H., Gheysen, G., Denil, S., Lindsey, K., Topping, J. F., Nahar, K., et al. (2013). Transcriptional analysis through RNA sequencing of giant cells induced by *Meloidogyne graminicola* in rice roots. *Journal of Experimental Botany, 64*(12), 3885–3898.

Jones, M. G. K., & Dropkin, V. H. (1976). Scanning electron microscopy of nematode-induced giant transfer cells. *Cytobios, 15*(58–59), 149.

Jones, J. T., Furlanetto, C., Bakker, E., Banks, B., Blok, V., Chen, Q., et al. (2003). Characterization of a chorismate mutase from the potato cyst nematode *Globodera pallida. Molecular Plant Pathology, 4*(1), 43–50.

Jones, M. G., & Goto, D. B. (2011). Root-knot nematodes and giant cells. In J. Jones, G. Gheysen, & C. Fenoll (Eds.), *Genomics and molecular genetics of plant–nematode interactions* (pp. 83–100). Dordrecht, Heidelberg, London & New York: Springer.

Karczmarek, A., Overmars, H., Helder, J., & Goverse, A. (2004). Feeding cell development by cyst and root-knot nematodes involves a similar early, local and transient activation of a specific auxin-inducible promoter element. *Molecular Plant Pathology, 5*(4), 343–346.

Koltai, H., Dhandaydham, M., Opperman, C., Thomas, J., & Bird, D. (2001). Overlapping plant signal transduction pathways induced by a parasitic nematode and a rhizobial endosymbiont. *Molecular Plant–Microbe Interactions, 14*(10), 1168–1177.

Krupasagar, V., & Barker, K. R. (1969). Increased cytokinin concentrations in tobacco infected with the root-knot nematode *Meloidogyne incognita. Phytopathology, 56*, 885.

Kubo, M., Udagawa, M., Nishikubo, N., Horiguchi, G., Yamaguchi, M., Ito, J., et al. (2005). Transcription switches for protoxylem and metaxylem vessel formation. *Genes and Development, 19*(16), 1855–1860.

Kyndt, T., Vieira, P., Gheysen, G., & de Almeida-Engler, J. (2013). Nematode feeding sites: unique organs in plant roots. *Planta, 238*(5), 807–818.

Laplaze, L., Parizot, B., Baker, A., Ricaud, L., Martiniere, A., Auguy, F., et al. (2005). *GAL4-GFP* enhancer trap lines for genetic manipulation of lateral root development in *Arabidopsis thaliana. Journal of Experimental Botany, 56*(419), 2433–2442.

Lavenus, J., Goh, T., Roberts, I., Guyomarc'h, S., Lucas, M., De Smet, I., et al. (2013). Lateral root development in *Arabidopsis*: fifty shades of auxin. *Trends in Plant Science, 18*(8), 450–458.

Leasure, C. D., & He, Z. H. (2012). *CLE* and *RGF* family peptide hormone signaling in plant development. *Molecular Plant, 5*(6), 1173–1175.

Lee, C., Chronis, D., Kenning, C., Peret, B., Hewezi, T., Davis, E. L., et al. (2011). The novel cyst nematode effector protein *19C07* interacts with the *Arabidopsis* auxin influx transporter *LAX3* to control feeding site development. *Plant Physiology, 155*(2), 866–880.

Lee, H. W., Kim, N. Y., Lee, D. J., & Kim, J. (2009). *LBD18/ASL20* regulates lateral root formation in combination with *LBD16/ASL18* downstream of *ARF7* and *ARF19* in *Arabidopsis. Plant Physiology, 151*(3), 1377–1389.

Levesque, M. P., Vernoux, T., Busch, W., Cui, H., Wang, J. Y., Blilou, I., et al. (2006). Whole-genome analysis of the SHORT-ROOT developmental pathway in *Arabidopsis*. *PLoS Biology, 4*(5), e143.

Lohar, D. P., Schaff, J. E., Laskey, J. G., Kieber, J. J., Bilyeu, K. D., & Bird, D. M. (2004). Cytokinins play opposite roles in lateral root formation, and nematode and *Rhizobial* symbioses. *The Plant Journal, 38*(2), 203–214.

Lu, S. W., Chen, S., Wang, J., Yu, H., Chronis, D., Mitchum, M. G., et al. (2009). Structural and functional diversity of *CLAVATA3/ESR* (*CLE*)-like genes from the potato cyst nematode *Globodera rostochiensis*. *Molecular Plant–Microbe Interactions, 22*(9), 1128–1142.

MacLean, A. M., Orlovskis, Z., Kowitwanich, K., Zdziarska, A. M., Angenent, G. C., Immink, R. G., et al. (2014). Phytoplasma effector SAP54 hijacks plant reproduction by degrading MADS-box proteins and promotes insect colonization in a RAD23-dependent manner. *PLoS Biology, 12*(4), e1001835.

Mazarei, M., Lennon, K. A., Puthoff, D. P., Rodermel, S. R., & Baum, T. J. (2003). Expression of an *Arabidopsis* phosphoglycerate mutase homologue is localized to apical meristems, regulated by hormones, and induced by sedentary plant-parasitic nematodes. *Plant Molecular Biology, 53*(4), 513–530.

Moens, M., Perry, R. N., & Starr, J. L. (2009). *Meloidogyne* species – a diverse group of novel and important plant parasites. In R. N. Perry, M. Moens, & J. L. Starr (Eds.), *Root-knot nematodes* (pp. 1–13). United Kingdom: CAB International.

Moubayidin, L., Di Mambro, R., & Sabatini, S. (2009). Cytokinin – auxin crosstalk. *Trends in Plant Science, 14*(10), 557–562.

Niebel, A., de Almeida-Engler, J., Tire, C., Engler, G., Van Montagu, M., & Gheysen, G. (1993). Induction patterns of an extensin gene in tobacco upon nematode infection. *The Plant Cell, 5*(12), 1697–1710.

Offler, C. E., McCurdy, D. W., Patrick, J. W., & Talbot, M. J. (2003). TRANSFER cells: cells specialized for a special purpose. *Annual Review of Plant Biology, 54*, 431–454.

Ohyama, K., Ogawa, M., & Matsubayashi, Y. (2008). Identification of a biologically active, small, secreted peptide in *Arabidopsis* by in silico gene screening, followed by LC-MS-based structure analysis. *The Plant Journal, 55*(1), 152–160.

Okushima, Y., Fukaki, H., Onoda, M., Theologis, A., & Tasaka, M. (2007). *ARF7* and *ARF19* regulate lateral root formation via direct activation of *LBD/ASL* genes in *Arabidopsis*. *The Plant Cell, 19*(1), 118–130.

Patel, N., Hamamouch, N., Li, C., Hussey, R. S., Mitchum, M. G., Baum, T. J., et al. (2008). Similarity and functional analyses of expressed parasitism genes in *Heterodera schachtii* and *Heterodera glycines*. *Journal of Nematology, 40*(4), 299.

Péret, B., Larrieu, A., & Bennett, M. J. (2009). Lateral root emergence: a difficult birth. *Journal of Experimental Botany, 60*(13), 3637–3643.

Perry, R. N., & Moens, M. (2011). Introduction to plant-parasitic nematodes: modes of parasitism. In J. Jones, G. Gheysen, & C. Fenoll (Eds.), *Genomics and molecular genetics of plant–nematode interactions* (pp. 3–20). Dordrecht, Heidelberg, London & New York: Springer.

Replogle, A., Wang, J., Bleckmann, A., Hussey, R. S., Baum, T. J., Sawa, S., et al. (2011). Nematode *CLE* signaling in *Arabidopsis* requires *CLAVATA2* and *CORYNE*. *The Plant Journal, 65*(3), 430–440.

Replogle, A., Wang, J., Paolillo, V., Smeda, J., Kinoshita, A., Durbak, A., et al. (2013). Synergistic interaction of *CLAVATA1*, *CLAVATA2*, and RECEPTOR-LIKE PROTEIN KINASE 2 in cyst nematode parasitism of *Arabidopsis*. *Molecular Plant–Microbe Interactions, 26*(1), 87–96.

Richardson, L., & Price, N. S. (1982). Host–parasite relationships of *Meloidogyne incognita* and the *dia-geotropica* tomato mutant. *Journal of Nematology, 14*, 465–466.

Rodiuc, N., Vieira, P., Banora, M. Y., & de Almeida-Engler, J. (2014). On the track of transfer cell formation by specialized plant-parasitic nematodes. *Frontiers in Plant Science, 5*.

Sablowski, R., & Dornelas, M. C. (2014). Interplay between cell growth and cell cycle in plants. *Journal of Experimental Botany, 65*(10), 2703–2714.

Saini, S., Sharma, I., Kaur, N., & Pati, P. K. (2013). Auxin: a master regulator in plant root development. *Plant Cell Reports, 32*(6), 741–757.

Scarpella, E., Marcos, D., Friml, J., & Berleth, T. (2006). Control of leaf vascular patterning by polar auxin transport. *Genes and Development, 20*(8), 1015–1027.

Scheres, P., Sijmons, P. C., van den Berg, C., McKhann, H., de Vrieze, G., Willemsen, V., et al. (1997). Root anatomy and development, the basis for nematode parasitism. In C. Fenoll, F. M. Grundler, & S. A. Ohl (Eds.), *Cellular and molecular aspects of plant–nematode interactions* (pp. 25–38). Netherlands: Kluwer Academic Publishers.

Schoof, H., Lenhard, M., Haecker, A., Mayer, K. F., Jürgens, G., & Laux, T. (2000). The stem cell population of *Arabidopsis* shoot meristems is maintained by a regulatory loop between the *CLAVATA* and *WUSCHEL* genes. *Cell, 100*(6), 635–644.

Scofield, S., & Murray, J. A. (2006). *KNOX* gene function in plant stem cell niches. *Plant Molecular Biology, 60*(6), 929–946.

Siddique, S., Sobczak, M., Tenhaken, R., Grundler, F. M., & Bohlmann, H. (2012). Cell wall ingrowths in nematode induced syncytia require *UGD2* and *UGD3*. *PloS One, 7*(7), e41515.

Sobczak, M., & Golinowski, W. (2008). Structure of cyst nematode feeding sites. In R. H. Berg, & C. G. Taylor (Eds.), *Cell biology of plant nematode parasitism* (pp. 153–187). Berlin: Springer.

Stahl, Y., Wink, R. H., Ingram, G. C., & Simon, R. (2009). A signaling module controlling the stem cell niche in *Arabidopsis* root meristems. *Current Biology, 19*(11), 909–914.

Sugimoto, K., Jiao, Y., & Meyerowitz, E. M. (2010). *Arabidopsis* regeneration from multiple tissues occurs via a root development pathway. *Developmental Cell, 18*(3), 463–471.

Szakasits, D., Heinen, P., Wieczorek, K., Hofmann, J., Wagner, F., Kreil, D. P., et al. (2009). The transcriptome of syncytia induced by the cyst nematode *Heterodera schachtii* in *Arabidopsis* roots. *The Plant Journal, 57*(5), 771–784.

Thiel, J., Hollmann, J., Rutten, T., Weber, H., Scholz, U., & Weschke, W. (2012). 454 Transcriptome sequencing suggests a role for two-component signalling in cellularization and differentiation of barley endosperm transfer cells. *PloS One, 7*(7), e41867.

Thiel, J., Riewe, D., Rutten, T., Melzer, M., Friedel, S., Bollenbeck, F., et al. (2012). Differentiation of endosperm transfer cells of barley: a comprehensive analysis at the micro-scale. *The Plant Journal, 71*(4), 639–655.

Thiel, J., Weier, D., Sreenivasulu, N., Strickert, M., Weichert, N., Melzer, M., et al. (2008). Different hormonal regulation of cellular differentiation and function in nucellar projection and endosperm transfer cells: a microdissection-based transcriptome study of young barley grains. *Plant Physiology, 148*(3), 1436–1452.

Tucker, M. L., Xue, P., & Yang, R. (2010). 1-Aminocyclopropane-1-carboxylic acid (ACC) concentration and ACC synthase expression in soybean roots, root tips, and soybean cyst nematode (*Heterodera glycines*)-infected roots. *Journal of Experimental Botany, 61*(2), 463–472.

Vanholme, B., Haegeman, A., Jacob, J., Cannoot, B., & Gheysen, G. (2009). Arabinogalactan endo-1,4-β-galactosidase: a putative plant cell wall-degrading enzyme of plant-parasitic nematodes. *Nematology, 11*(5), 739–747.

Vanneste, S., & Friml, J. (2009). Auxin: a trigger for change in plant development. *Cell, 136*(6), 1005–1016.

Wang, X., Allen, R., Ding, X., Goellner, M., Maier, T., de Boer, J. M., et al. (2001). Signal peptide-selection of cDNA cloned directly from the esophageal gland cells of the soybean cyst nematode *Heterodera glycines*. *Molecular Plant–Microbe Interactions, 14*(4), 536–544.

Wang, J., Lee, C., Replogle, A., Joshi, S., Korkin, D., Hussey, R. S., et al. (2010). Dual roles for the variable domain in protein trafficking and host-specific recognition of *Heterodera glycines CLE* effector proteins. *New Phytologist, 187*(4), 1003–1017.

Wang, X., Mitchum, M. G., Gao, B., Li, C., Diab, H., Baum, T. J., et al. (2005). A parasitism gene from a plant-parasitic nematode with function similar to *CLAVATA3/ESR* (*CLE*) of *Arabidopsis thaliana. Molecular Plant Pathology, 6*(2), 187–191.

Wang, J., Replogle, A., Hussey, R. S., Baum, T. J., Wang, X., Davis, E. L., et al. (2011). Identification of potential host plant mimics of *CLAVATA3/ESR* (*CLE*)-like peptides from the plant-parasitic nematode *Heterodera schachtii. Molecular Plant Pathology, 12*(2), 177–186.

Wenzel, C. L., Schuetz, M., Yu, Q., & Mattsson, J. (2007). Dynamics of *MONOPTEROS* and *PIN-FORMED1* expression during leaf vein pattern formation in *Arabidopsis thaliana. The Plant Journal, 49*(3), 387–398.

Williamson, V. M., & Hussey, R. S. (1996). Nematode pathogenesis and resistance in plants. *The Plant Cell, 8*(10), 1735.

Wubben, M. J., Su, H., Rodermel, S. R., & Baum, T. J. (2001). Susceptibility to the sugar beet cyst nematode is modulated by ethylene signal transduction in *Arabidopsis thaliana. Molecular Plant–Microbe Interactions, 14*(10), 1206–1212.

Wu, C. Y., Rolfe, P. A., Gifford, D. K., & Fink, G. R. (2010). Control of transcription by cell size. *PLoS Biology, 8*(11), e1000523.

Xiong, Y., Li, Q. B., Kang, B. H., & Chourey, P. S. (2011). Discovery of genes expressed in basal endosperm transfer cells in maize using 454 transcriptome sequencing. *Plant Molecular Biology Reporter, 29*(4), 835–847.

Zhou, W., Wei, L., Xu, J., Zhai, Q., Jiang, H., Chen, R., et al. (2010). *Arabidopsis* tyrosylprotein sulfotransferase acts in the auxin/PLETHORA pathway in regulating postembryonic maintenance of the root stem cell niche. *Plant Cell Online, 22*(11), 3692–3709.

CHAPTER EIGHT

Recent Advances in Understanding Plant–Nematode Interactions in Monocots

Diana Fernandez*[,1], Anne-Sophie Petitot*, Maíra Grossi de Sá*[,a], Vũ Phong Nguyễn*[,§,a], Janice de Almeida Engler[¶], Tina Kyndt[||,b]

*IRD–Institut de Recherche pour le Développement, UMR 186 IRD-UM2-Cirad Résistance des Plantes aux Bioagresseurs, Montpellier, France

[§]Nông Lâm University, Hồ Chí Minh City, Việt Nam

[¶]Institut National de la Recherche Agronomique, UMR 1355 ISA/Centre National de la Recherche Scientifique, UMR 7254 ISA/Université de Nice-Sophia Antipolis, UMR ISA, Sophia-Antipolis, France

[||]Department Molecular Biotechnology, Ghent University (UGent), Gent, Belgium

[1]Corresponding author: E-mail: Diana.Fernandez@ird.fr

Contents

[a]NVP was recipient of a Vietnamese governmental PhD grant and MGS of a PhD grant from the Conselho Nacional de Desenvolvimento Cientifico e Tecnológico (CNPq-CSF) in Brazil.

[b]Tina Kyndt is supported by a postdoctoral fellowship from the Fonds Wetenschappelijk Onderzoek–Vlaanderen (FWO) in Belgium.

Advances in Botanical Research, Volume 73
ISSN 0065-2296
http://dx.doi.org/10.1016/bs.abr.2014.12.006

Abstract

Monocotyledonous crops including cereals, sugar cane, banana and cassava make an important tribute to agricultural commodities both in production and value worldwide. Diseases caused by nematodes in monocots are well documented and new data are emerging on interactions of these pathogens with some important monocotyledonous crop species. This review focuses on major monocotyledonous plant–nematode systems describing the biology and genetics of some interactions and highlighting efforts to identify and select resistant genetic resources. Molecular and histological data are presented having in the spotlight sedentary nematodes of the suborder Tylenchina. Histological observations are presented during nematode development and host resistance responses, comparing monocot and dicot plant hosts. Subsequently we focus on the recently published transcriptomics of monocotyledonous plant responses looking at specific metabolic changes and their role during plant–nematode interactions. Insights that may be learned from nematode interactions with dicotyledonous hosts are also discussed. As a final point, we look at nematode effectors during monocots–nematode interactions exploiting de novo transcriptome data produced from root-knot nematodes and cyst nematodes.

1. INTRODUCTION

Among flowering plants (Angiosperms) monocotyledons have been separated from dicotyledons based on morphological features particularly because the embryo is formed from one cotyledon in the former versus two cotyledons in the latter (Jussieu, 1789). Another morphological aspect differing among them is that monocots have adventitious roots that arise from stem nodes, after development of the seminal root system. Nodal roots are thicker, and are associated with tiller development. The root vascular system organization also differs in monocots, with the presence of a central metaxylem vessel in the stele, surrounded by alternate bands of xylem and phloem (Esau, 1953).

Based on molecular studies, monocotyledons form a well-separated monophyletic clade comprising 10 orders (and two families not yet assigned) (APG III system (Angiosperm Phylogeny Group III system)) of flowering plant classification (Angiosperm Phylogeny Group, 2009). The largest families of monocotyledons are the orchids, followed by the Poaceae, also called Gramineae or true grasses, being the fifth largest plant family. Other important monocots are the palm (Arecaceae), banana (Musaceae), ginger (Zingiberaceae) and the amaryllis (Amaryllidaceae) families. The latter includes ornamental plants cultivated for their blossom, notably lilies, amaryllis and tulips as well as ubiquitously consumed vegetables as onions and garlic.

Monocotyledons include economically important species, like sugar cane (*Saccharum officinarum*), and cultivated grains (cereals), including rice (*Oryza sativa*), wheat (*Triticum aestivum*; *Triticum durum*), maize (*Zea maïs*), sorghum (*Sorghum bicolor*) and others. Within the 20 most important food and agricultural commodities in the world (2012), rice is classified together with milk and meat (cattle, pig and chicken) (FAOSTAT, 2013, http://issuu.com/faooftheun/docs/syb2013issuu). Subsequently come wheat, sugar cane, maize, banana and cassava, listed together with other eight dicotyledonous crops. Monocot crops comprise 40% of the total cultivated area in the world, and cereals (e.g. wheat, maize, rice), starchy roots (cassava, yams) and fruits (banana and plantain) are the main source of carbohydrates for human and animal consumption. Cereals have been cultivated for their edible seeds and are an important part of the man diet worldwide. Sugar cane, an important source for sugar (sucrose) in the human diet, is now also extensively cultivated for biofuel (ethanol) production. Other monocotyledonous species providing carbohydrate or oils as food source include the date palm (*Phoenix dactylifera*), oil palm (*Elaeis guineensis*), coconut and other palm species. In addition to edible types monocots are cultivated as ornamentals, including turfgrasses, also widely used in many private and public lawns contributing to agro-economical sources. Finally, wild monocot species are weeds frequently acting as reservoirs for many plant pathogens.

Diseases caused by nematodes in monocots are well documented, and a series of exhaustive reviews about nematode parasites of bananas (Gowen, Quénéhervé, & Fogain, 2005), cereals (Bridge, Plowright, & Peng, 2005; Mc Donald & Nicol, 2005), palms (Griffith, Giblin-Davis, Koshy, & Sossama, 2005), pineapple (Sipes, Caswell-Chen, Sarah, & Apt, 2005) and sugar cane (Cadet & Spaull, 2005) have been compiled in a book focused on tropical and subtropical crop species (Luc, Sikora, & Bridge, 2005). However, in spite of their widespread occurrence and high abundance, only few data are available on monocot–nematode interactions (Kyndt, Fernandez, & Gheysen, 2014). Efforts have been made to control widespread pathogenic nematodes in banana and plantains (*Musa* spp.) severely affecting crop productivity and longevity (extensively reviewed in Quénéhervé, 2009). Among the most important root pathogens of banana are the burrowing nematode *Radopholus similis* and some species of the root lesion nematode *Pratylenchus* spp. These species attack primary roots disrupting the anchorage system resulting in host plant toppling or uprooting.

Production losses due to attacks by these nematode species can be high, also depending on other biotic and environmental factors (Quénéhervé, 2009). In addition, root-knot nematodes (RKN) belonging to the *Meloidogyne* genus, may also infect all banana varieties causing root deformations and stunting. Until the 1990s, control of nematode attacks on banana relied almost exclusively on the regular application of nematicides. However, as pointed out by Quénéhervé (2009), 'the golden age of chemical control with nematicides is definitely behind us' and different approaches to nematode management (e.g. cultural practices, plant resistance and biological control) must be used depending on each cropping systems. Owing to national and international efforts during the last 50 years, fundamental data have been generated on nematodes attacking important crops like bananas and plantains and new nematode management practices have been put in action.

It is therefore of crucial importance to identify novel sources of natural resistance to nematodes in crop species as well also to investigate more deeply plant–nematode interactions, including plant defence mechanisms and nematode manipulation of the host cell metabolism. Enhancing our basic knowledge on plant–nematode interactions may help defining future strategies for contributing to best plant health measures. Most of the available studies on the molecular interplay between plants and nematodes come from dicots (Kyndt, Vieira, Gheysen, & de Almeida-Engler, 2013). Therefore, the purpose of this review is to present up to date data and to identify major trends in plant–nematode interactions in monocots. As well we survey here and discuss in more detail existing histological and molecular data on the interaction of monocots and sedentary nematodes of the suborder Tylenchina.

2. MONOCOTYLEDONOUS PLANT–NEMATODE SYSTEMS: BIOLOGY AND GENETICS OF INTERACTIONS

RKN and cyst nematodes (CN) are both obligate sedentary endoparasites. In order to complete their life cycle, they must invade the roots of a susceptible plant where they induce a specialized nematode-feeding site, called syncytium for CN or giant cells for RKN. These specialized feeding structures serve as a constant food source to the nematode, allowing its development into a reproductive female (Kyndt et al., 2013). CN females mature into cysts that contain several hundred eggs, whereas RKN females exude egg masses directly on the roots or within the gall tissue. After hatching, new juveniles arising from these eggs will start a new reproduction cycle in another root from the same host plant, or will attack another plant.

Depending on the host plant and environmental conditions, the cycle lasts 15–45 days for RKN (Perry & Moens, 2011) and 14–65 days for CN (Cook & Noel, 2002). Several generations per cropping season are usually observed for RKN, but *Meloidogyne naasi* affecting temperate cereals has only one generation per cropping season (Rivoal & Cook, 1993). For CN, up to eight generations per season could be observed for *Heterodera trifolii* (Cook & Noel, 2002). On rice, two to three generations are usually observed for *Heterodera sacchari* (Bridge et al., 2005) but only one generation for *Heterodera avenae* on Gramineae.

RKN can cause swelling and galls throughout the root system. In some interactions, galls occur mostly at root tips. Severe infections result in a stunted root system preventing the development of the plant host (Bridge et al., 2005). CN infection in cereals often causes a 'bushy-knotted' appearance to the overall root system. This nomenclature is due to the proliferation of secondary roots, thought to be part of host's effort to compensate root damage. Lemon-shaped white females and brown cysts can be observed protruding from infected roots. Sometimes, root growth is suppressed (Mc Donald & Nicol, 2005).

Due to chemical pesticides prohibition, host natural resistance has become even more important for breeding aiming at nematode control (Molinari, 2011). Resistance to nematodes is usually defined as the plant capability to inhibit their reproduction (Roberts, 2002) and is determined by the nematode reproductive factor (Rf). Rf is the final nematode population at harvest (Pf) divided by the initial nematode population (Pi) at inoculation. Ideally, resistance should be combined with tolerance, which is the ability of the host plant to maintain yield potential in the presence of the nematode (Cook & Evans, 1987). It is also essential to determine the relative increase or decrease in nematode populations after a cropping season, which may impact the next crop production.

Plant resistance (*R*) genes or quantitative trait loci (QTL) conferring partial or high resistance to CN or RKN have been genetically mapped or characterized by map-based cloning, essentially in dicotyledonous plants (Kandoth & Mitchum, 2013; Williamson & Roberts, 2009). The *R* gene *Hs1pro-1* from sugar beet was the first plant gene cloned that conferred resistance to the CN *Heterodera schachtii* (Cai et al., 1997). It encodes a protein with a transmembrane domain and an extracellular leucine-rich repeat (LRR) domain. The first RKN *R* gene cloned was the *Mi-1* gene from tomato, encoding a nucleotide-binding (NB)-LRR protein (Milligan et al., 1998). Since then, the majority of nematode *R* genes cloned so far have been found to encode CC-NB-LRR or TIR-NB-LRR proteins from the superfamily of intracellular NB-LRR plant receptors (Kandoth & Mitchum, 2013). However, an unusual *R* gene has been recently identified in soybean encoding a serine hydroxymethyl-transferase, an enzyme involved in one-carbon metabolism conserved across kingdoms (Liu et al., 2012).

In monocotyledonous plants, despite the efforts to identify sources of resistance to nematodes, only a few *R* genes have been genetically mapped (Mc Donald & Nicol, 2005; Smiley & Nicol, 2009) and none of them, to our knowledge, have been cloned (Table 1). In parallel to the *R* gene mapping, breeders have identified plant accessions that present resistance or tolerance to nematodes. However, these sources of resistance are often complex traits inherited in a polygenic manner and the genes involved in these resistance mechanisms have not been identified (Fuller, Lilley, & Urwin, 2008). When possible, these sources of resistance have been transferred to crop cultivars through conventional breeding methods. We review hereafter the main models of monocotyledonous plant–nematode interactions and sources of resistance that have been identified (Table 1).

2.1 Wheat/Barley/Oat-*Heterodera avenae* Interactions

Heterodera avenae or cereal cyst nematode (CCN) is economically important in all temperate cereal-producing regions worldwide. *H. avenae*-infested plants appear initially as pale green seedlings that lack vigour and once mature are often severely stunted. Symptoms on roots are specific to the host species. Wheat and barley roots invaded by *H. avenae* branch excessively at locations where juveniles invade, resulting in a bushy or knotted appearance. Oat roots invaded by *H. avenae* are shortened and thickened but do not exhibit the typical knotted symptom (Smiley & Nicol, 2009).

Thirteen genes for resistance to *H. avenae* have been identified in cereals. Two genes (*Cre1* and *Cre8*) were identified in *T. aestivum,* the *Cre2–Cre7* genes were derived from *Aegilops* species, wild relatives of wheat, and the *CreR* gene was identified in *Secale cereale* (Smiley & Nicol, 2009). Four genes (*Rha1–4*) were identified in barley (Bakker, Dees, Bakker, & Goverse, 2006). However, many pathotypes occur within *H. avenae*, leading to difficulties to manage populations. For instance, the *Cre1* gene is highly effective against populations of *H. avenae* from Europe, North Africa and North America and moderately effective or ineffective to populations in Australia and Asia. The *Cre3* gene is effective against Australian populations but not

Table 1 Interaction Systems between Monocotyledonous Plants and Sedentary Nematodes

Family	Common Name (*Species*)	Nematode	Resistance Sources	Histological Data*	Molecular Data	References
Amaryllidaceae	Onion (*Allium cepa*)	*Meloidogyne* spp.; *Meloidogyne hapla*	*A. cepa*	—	—	Sikora & Fernandez (2005), Pang et al. (2009)
Arecaceae	Date palm (*Phoenix dactylifera*)	*Meloidogyne* spp.; *Meloidogyne incognita*	*P. dactylifera* ('Deglet-Noor', 'Samani')	S and R	—	Griffith et al. (2005), Eissa et al. (1998)
Bromeliaceae	Pineapple (*Ananas comosus*)	*Meloidogyne javanica*	—	—	—	Sipes et al. (2005)
Musaceae	Banana/Plantain (*Musa* spp.)	*Meloidogyne* spp.; *M. incognita*	*Musa* spp.	S	—	Gowen et al. (2005), Quénéhervé (2009), Davide & Marasigan (1985), Sudha & Prabhoo (1983)
Poaceae	Maize (*Zea mais*)	*Heterodera* spp.; *Heterodera zeae*	*Z. mais* inbred lines	—	—	Mc Donald & Nicol (2005), Hasmi et al. (1993)
		Meloidogyne spp.	*Z. mais* cultivars, inbred lines	S	*Z. mais* mutant *lox3-4*	Mc Donald & Nicol (2005), Windham & Williams (1994); Kagoda et al. (2011), Asmus, Ferraz, & Appezzato-da-Glória (2000)
	Oat (*Avena sativa*)	*Meloidogyne naasi*	—	S and R	—	Siddiqui (1971)

(*Continued*)

Table 1 Interaction Systems between Monocotyledonous Plants and Sedentary Nematodes—cont'd

Family	Common Name (*Species*)	Nematode	Resistance Sources	Histological Data*	Molecular Data	References
	Rice (*Oryza sativa*)	*Heterodera sacchari*	*Oryza glaberrima* (*Hsa-1Og*)	—	—	Reversat & Destombes (1995), Reversat & Destombes (1998), Lorieux et al. (2003)
		Meloidogyne graminicola	*O. glaberrima; Oryza longistaminata*	S and R	defence genes; root transcriptome; giant cells transcriptome; *M. graminicola* transcriptome	Soriano et al. (1999), Bimpong et al. (2010), Kyndt et al. (2012), Ji et al. (2013), Haegeman, Bauters, Kyndt, Rahman, & Gheysen (2013), Cabasan et al. (2014), Nguyễn et al. (2014)
		M. incognita	*O. glaberrima*	S	*M. incognita* genome	Abad et al. (2008), Plowright et al. (1999), Nguyễn et al. (2014)
	Sorghum/Barley (*Sorghum bicolor/Hordeum vulgare*)	*Meloidogyne naasi*	—	S	—	Ediz & Dickerson (1976)
	Sugar cane (*Saccharum officinarum*)	*Meloidogyne* spp.	—	—	—	Cadet & Spaull (2005)

	Turfgrass	*Meloidogyne* spp.	—	—	—	Mc Clure et al. (2012)
	Wheat/Barley (*Triticum aestivum*/ *H. vulgare*)	*Heterodera avenae*	*Triticum aestivum* (*Cre1*; *Cre8*), *Aegilops* spp. (*Cre2-7*), *H. vulgare* (*Rha1-4*)	S and R	defence genes; overexpression PIN2; *A. variabilis* transcriptome; *H. avenae* transcriptome	Oka, Chilt, & Spiegel (1997), Seah et al. (2000), Vishnudasan, Tripathi, Rao, & Khurana (2005), Bakker et al. (2006), Smiley & Nicol (2009), Simonetti et al. (2009), Simonetti, Alba, Montes, Delibes, & López-Braña (2010), Xu et al. (2012), Kumar et al. (2014)
		Meloidogyne spp.; *Meloidogyne artiellia*; *M. naasi*	*Aegilops* sp (*Rkn1-3*); *Hordeum chilense*	S and R	—	Siddiqui & Taylor (1970), Yu et al. (1990), Kaloshian et al. (1991), Balhadère & Evans (1995a), Balhadère & Evans (1995b), Mc Donald & Nicol (2005), Williamson et al. (2013)
Zingiberaceae	Ginger (*Zingiber* spp.)	*Meloidogyne* spp.; *Meloidogyne thailandica*	—	S	—	Handoo, Skantar, Carta, & Erbe (2005), Koshy et al. (2005), Shah & Raju (1977)

* S = susceptible; R = resistant; – no data found/available.

European populations of *H. avenae*. An 'International Root Disease Resistance Nursery' containing seven of the known *Cre* genes was coordinated by CIMMYT to establish the value of these genes in different regions of the world (Smiley & Nicol, 2009). Strategies for pyramiding resistance genes to CCN in wheat, seem more promising (Barloy et al., 2007).

2.2 Wheat/Barley-*Meloidogyne* spp. Interactions

Meloidogyne naasi and *Meloidogyne artiellia* have been shown to cause significant damage to wheat and barley. *M. naasi* is probably the most important RKN affecting grains in European countries and is present worldwide. It causes small-sized root galls on roots, especially the root tips. Symptoms of attack resemble those caused by *H. avenae* with patches of poorly growing and yellowing plants (Mc Donald & Nicol, 2005).

No genotype of bread wheat has been shown to be resistant to *M. naasi*. Sources of resistance to *Meloidogyne* species have been found in wild relative grasses, including *Aegilops variabilis* and *Hordeum chilense* (Yu, Person-Dedryver, & Jahier, 1990; Balhadère & Evans, 1995a). *Rkn1*, mediating resistance to *Meloidogyne javanica* was identified in the grass species *Aegilops tauschii* (Kaloshian, Waines, Roberts, & Thomason, 1991. *Rkn2*, conferring resistance to *M. naasi* is derived from *A. variabilis* (Yu et al., 1990). *Rkn3* was recently transferred from *Aegilops ventricosa* into wheat and confers resistance to *Meloidogyne incognita* and *M. javanica* (Williamson, Thomas, Ferris, & Dubcovsky, 2013). The resulting introgressed wheat line is tolerant to RKN infection, and shows potential to be used in crop rotation for nematode management. Combinations of *Rkn3* with *Rkn1* and *Rkn2* need to be tested to further increase the levels of wheat resistance to RKN.

2.3 Rice–Nematode Interactions

*2.3.1 Rice–*Heterodera sacchari *Interactions*

Heterodera sacchari was first found on rice in West Africa (Babatola, 1983) and later also in Asian countries (Reversat & Destombes, 1998). Production losses from *H. sacchari* infection on rice can be high under upland conditions, but are less severe under flooded conditions (Babatola, 1983). By causing root necrosis, chlorosis and reduced vigour and tillering, early *H. sacchari* infection, even with very low initial population densities, can result in severe grain losses (Coyne, 2000). Sources of resistance to *H. sacchari* were found in the African-cultivated rice species *Oryza glaberrima* TOG5681 (Reversat & Destombes, 1998). Characterization of *O. sativa* × *O. glaberrima* segregating populations allowed the identification of one major resistance gene, *Hsa-1Og*, with co-dominance of resistant and susceptible alleles (Lorieux et al., 2003).

2.3.2 Rice-*Meloidogyne* spp. Interactions

Meloidogyne graminicola, adapted to rice grown in irrigated and flooded systems, is a major pest of rice in Asia and in Central America. Infected rice root tips become swollen and hooked and plants become stunted and chlorotic with reduced growth. Important yield reductions are reported, up to 70% in an infected field (Bridge et al., 2005). In intensive rainfed rice systems, *M. incognita* which is not adapted to irrigated conditions, is prevalent (Plowright, Coyne, Nash, & Jones, 1999; Reversat, pers. com.). QTLs associated to quantitative resistance were identified in *O. sativa*, suggesting that it may be possible to breed tolerant *O. sativa* varieties (Shrestha, Uzzo, Wilson, & Price, 2007). Complete resistance to *M. graminicola* and *M. incognita* has been recorded in the African-related species, *Oryza longistaminata* and *O. glaberrima* (Plowright et al., 1999; Soriano, Schmitt, Brar, Prot, & Reversat, 1999; Bimpong et al., 2010; Cabasan, Kumar, Bellafiore, & De Waele, 2014). In addition, the *H. sacchari*-resistant TOG5681 accession presents resistance to *M. graminicola* (Bimpong et al., 2010) showing a significant reduction on nematode penetration and arrest of feeding site development, blocking further female development (Petitot, Haidar, Almeida Engler and Fernandez, unpublished data). Nematode resistance gene identification and mapping will be facilitated by *O. glaberrima* genome sequencing (Sakai et al., 2011). In addition, NERICA (New Rice for Africa) lines, having good yield capacity and adapted to lowland ecosystems have been developed. Some NERICAs were derived from TOG5681 and nematode-resistant lines are being characterized (Agnoun et al., 2012; Ndjiondjop et al., 2008).

2.4 Maize-*Meloidogyne* spp. and *Heterodera* spp. Interactions

Three CCN, *Heterodera zeae*, *H. avenae* and *Punctodera chalcoensis*, are major maize pests. *H. avenae* and *H. zeae* have been reported to occur worldwide whereas *P. chalcoensis* is limited in distribution to Mexico, where it causes severe damage on volcanic sandy soils (Mc Donald & Nicol, 2005). Five inbred lines, that contain resistance to an isolate of *H. zeae* from a Maryland population, were identified by Hashmi, Hashmi, Krusberg, and Huettel (1993).

Meloidogyne incognita and *Meloidogyne javanica* have been detected in almost all wide-reaching maize-growing regions. Galls and root tip branching are characteristic but typical gall symptoms may be totally absent, complicating diagnosis (Mc Donald & Nicol, 2005). Many maize cultivars have been reported to be resistant to *Meloidogyne* (Windham & Williams, 1994; Poerba, Windham, & Williams, 1990). Resistance to *M. incognita* in maize is primarily expressed by a slower nematode development rather than by differences in nematode penetration (Windham & Williams, 1994). More recently, RKN resistance sources were found in inbred lines from CYMMIT and their F1 progenies (Kagoda, Derera, Tongoona, Coyne, & Lorenzen, 2011).

2.5 Other Crop-Nematode Systems

Other important crops are affected by sedentary nematodes, especially by *Meloidogyne* spp., which causes, for instance, deformations and stunting on roots of all banana varieties (Quénéhervé, 2009). However, RKN mostly occur on banana roots in absence of the burrowing or lesion nematodes. Very little information is available on the existence of sources of resistance to RKN in *Musa*, except for nine cultivars which were described as 'resistant' to *M. incognita* in the Philippines (Davide & Marasigan, 1985). As well, several species of *Meloidogyne* such as *M. arenaria, M. incognita, M. javanica* and *M. chitwoodi* damage onion crops (Sikora & Fernandez, 2005), reducing onion bulb yield up to 70% (Pang, Hafez, & Sundararaj, 2009). The Mercury and R10T cultivars are tolerant and resistant to *Meloidogyne hapla* and to another parasitic nematode *Pratylenchus penetrans* being therefore used for crop rotations (Pang et al., 2009). In sugar cane, nematode species diversity is greater than in most cultivated crops (more than 300 species) and being RKN, like *M incognita* and *M. javanica* important pathogens. However, symptoms are not easily detected due to the small galls (Cadet & Spaull, 2005). More than 100 species of parasitic nematodes attack pineapple, and RKN are among the most important, especially *M. javanica* (Sipes et al., 2005). To date reported nematodes attacking palms are *Meloidogyne* spp. (Griffith et al., 2005) and two cultivars resistant to *M. incognita* have been characterized (Eissa, EI-Sherief, Abdel-Gawad, Ismail, & EI-Nagdi, 1998). Ginger is also affected by RKN, especially *M. arenaria, M. hapla, M. incognita* and *M. javanica*, inducing galling and rotting of roots and underground rhizomes (Koshy, Eapen, & Pandey, 2005). Finally, turfgrasses of golf courses and lawns in USA may be affected by several RKN, including *M. naasi* and *Meloidogyne marylandi* (Mc Clure, Nischwitz, Skantar, Schmitt, & Subbotin, 2012). Prospecting several turfgrasses in Israel showed that a range of them were non-hosts or resistant to RKN (Oka, Karssen, & Mor, 2003).

3. HISTOLOGICAL DESCRIPTIONS OF ROOTS DURING NEMATODE DEVELOPMENT AND HOST RESISTANCE RESPONSES

Histology of gall development during RKN infection in monocotyledonous plants has been investigated in banana, date palms, ginger, rice, sorghum, wheat and barley and in some of their relative wild species, and finally, turfgrasses (Table 1). However, only one study refers to CN in monocots that describes the development of the CCN (*H. avenae*) in susceptible and resistant wheat varieties (Seah, Miller, Sivasithamparam, & Lagudah, 2000). A large amount of reported data refer to *M. naasi*, an RKN species that attacks predominantly monocots and that is mostly threatening to barley and sorghum fields (Ediz & Dickerson, 1976).

3.1 Wheat/Barley–RKN Interactions

In wheat and barley, large numbers of *M. naasi* larvae penetrate root tips within 24 h and are found in the cortical tissue in the region of root elongation (Balhadère & Evans, 1995a; Siddiqui & Taylor, 1970). Three days after inoculation (DAI), nematodes are often observed in the stele and others are still migrating within the root cortex causing some damage. Larvae migration has been described to occur both inter- and intracellularly (Balhadère & Evans, 1995a; Siddiqui & Taylor, 1970). Five DAI, some giant cells appear in the stele, resulting from hypertrophy of protophloem and protoxylem cells. At this early stage of gall development, some common features with other plant hosts of *Meloidogyne*-induced giant cells are observed. These include cytoplasmic and nuclear enlargement, metabolically active cytoplasm (dense and granular) with two to four actively dividing nuclei (per giant cell section), each with several hypertrophied nucleoli (Balhadère & Evans, 1995a; Siddiqui & Taylor, 1970). Cortical cell proliferation is also observed around giant cells (Balhadère & Evans, 1995a; Siddiqui & Taylor, 1970), and atypical xylem cells containing wall reticulations developed around giant cells. These peculiar xylem cells surrounding giant cells are similar to wound-type xylem elements and may be part of the host's effort to maintain vascular continuity, following the disruption caused by giant cell induction (Bartlem, Jones, & Hammes, 2014). It is possible that xylem proliferation and the presence of hyperplastic parenchyma may be involved in containing giant cells development within this type of root tissues. Siddiqui & Taylor (1970) noticed that the giant cells induced by *M. naasi* in wheat roots are smaller than those reported in other plant hosts by other *Meloidogyne* spp. The giant cells display an irregular pattern of secondary wall thickening indicative of cell wall-reinforcing mechanisms to withstand the rising pressure in the root stele. This irregular wall thickening might be also involved in nutrient transfer since giant cells work like transfer cells too (Rodiuc, Vieira, Banora, & de Almeida-Engler, 2014). Giant cell size in infected wheat roots reaches its maximum at 15 DAI and undergoes not further increase. After 24–30 DAI, adult females are observed, an increasing number of small vacuoles appear in the cytoplasm and many giant cell nuclei degenerate (Balhadère & Evans, 1995a; Siddiqui & Taylor, 1970). Interestingly, in sorghum roots infected with race 5 of *M. naasi*, giant cells formed in the vascular tissue as well as within the root cortex (Ediz & Dickerson, 1976). Galls formed in the cortex tissue induced male nematode to develop indicating a less suitable condition for feeding cells development. The presence of nematodes trapped in cortex was reported in barley and wheat infected with *M. naasi*, but no successful galls has been established (Balhadère & Evans, 1995a; Siddiqui & Taylor, 1970).

No genotype of bread wheat has been shown to be resistant, although some wild grasses display complete resistance to *M. naasi*. In the *A. variabilis* and *H. chilense* accessions tested, resistance is being expressed at the cellular level by a hypersensitive-type reaction that quickly arrested *M. naasi* juvenile progression in roots. During migration within the cortex, cells in direct contact with nematodes degenerated rapidly, and walls are covered with non-lignified or lignified materials: callose was mobilized early in *A. variabilis*, whereas lignin accumulated in *H. chilense* (Balhadère & Evans, 1995a,b). Histo-localization of peroxidase in resistant roots showed a higher peroxidase activity around nematodes in cortical cells than in the susceptible roots (Balhadère & Evans, 1995b). In the stele, nuclear division in differentiating vascular cells was arrested where the nematode started to feed on; subsequently, the giant cell induction stopped. Within few days, resistant roots showed localized necrotic spots mainly dispersed in the cortex and no giant cell was formed. Interestingly, the expression of resistance by a rapid hypersensitive reaction traps the nematodes in the cortex and endodermis and does not seem to kill them. Unswollen nematodes are still seen within the host root and were still alive after 30 days (Balhadère & Evans, 1995a).

Wheat may as well be susceptible to other RKN species, like *M. marylandi* isolated from the grass *Zoysia japonica* occurring in Israel

(Oka et al., 2003). In contrast to other RKNs on cereals, *M. marylandi* infection in wheat leads to the formation of small root galls or no galls. There, larvae penetration seems to be restricted to the elongation zone without reaching the meristematic zone to enter the stele. The nematode just inserts its anterior end in the vascular cylinder, and induces giant cells from vascular parenchyma cells without much hypertrophy of the cortical cells around the feeding site (Oka et al., 2003). However, the endoderm cells near the nematode body and the giant cells collapse, which was not observed with other RKNs on cereals.

3.2 Rice–RKN Interactions

In rice, infection by *M. graminicola* and *M. incognita* was examined in a subset of *O. sativa* rice varieties, including ecotypes adapted to various growing conditions (Cabasan et al., 2014; Ibrahim, Ibrahim, & Rezk, 1972; Jena & Rao, 1977; Nguyễn et al., 2014). Giant cells are induced two DAI with *M. graminicola* but were visible only after six DAI in *M. incognita*-infected roots (Nguyễn et al., 2014) (Figure 1(A)). Cortical parenchyma proliferation and formation of multinucleated giant cells with dense cytoplasm are observed resulting in root galling. *Meloidogyne graminicola* induces the formation of typical hook-like galls at the root tip of young roots (Figure 1(B)). The number of giant cells at each feeding site varied from five to eight, and multiple feeding sites were observed in each root swelling (Cabasan et al., 2014; Nguyễn et al., 2014). Egg-laying females may be observed starting from 14 DAI for *M. graminicola* (Cabasan et al., 2014), but only from 28 DAI on for *M. incognita* (Nguyễn et al., 2014). *Meloidogyne graminicola* lay egg masses within the cortex whereas *M. incognita* females lay their eggs on the root surface (Figure 1(C)) (Nguyễn et al., 2014). Egg sac deposit and hatching of juveniles within the root cortex have been proposed to be specific adaptations for aquatic survival and infection ability of *M. graminicola* in rice grown in irrigated or flooded systems (De Waele & Elsen, 2007; Fortuner & Merny, 1979; Prot & Matias, 1995). The faster timing of infection and more efficient way of egg laying suggests that *M. graminicola* is more efficiently infecting rice plants.

Resistance against RKNs has been recorded in the African-related species *O. glaberrima* (Bimpong et al., 2010; Plowright et al., 1999; Soriano et al., 1999). Nematode reproduction is arrested in resistant varieties while some *O. glaberrima* accessions allow a residual reproduction of *M. graminicola* (Bimpong et al., 2010; Soriano et al., 1999). In resistant plants, nematodes may penetrate but infection is stopped at different stages

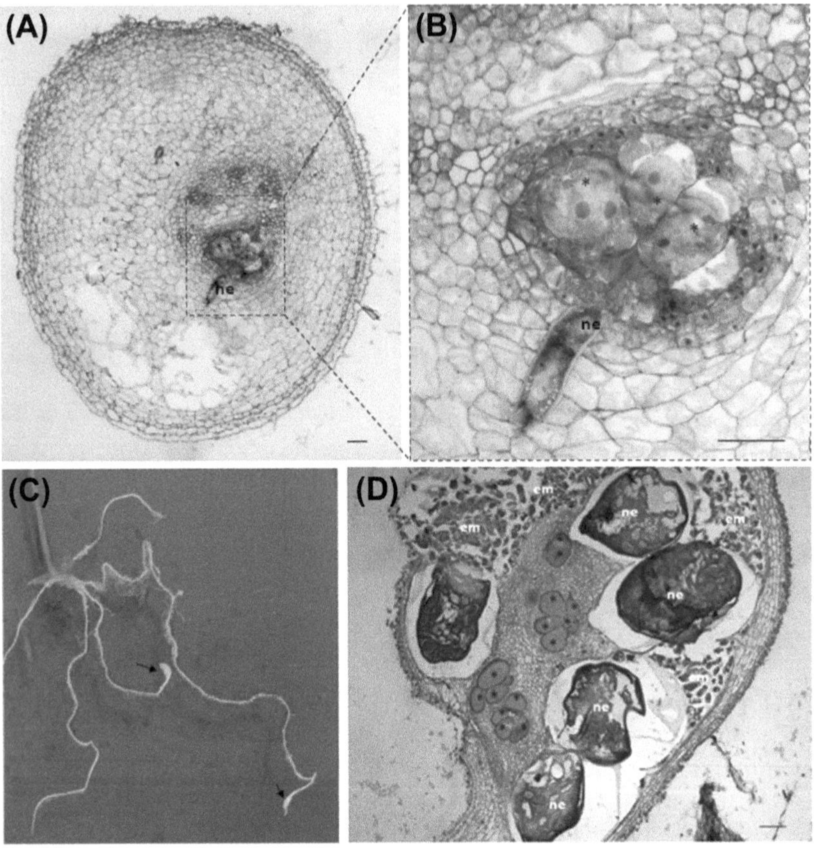

Figure 1 Galls induced by *Meloidogyne graminicola* in rice (*Oryza sativa*) roots. A, B and D: Histological analysis where root cross-sections (10 μm) obtained at 4 (A, B) and 31 (D) days after nematode infection were stained with toluidine blue. A: 4-days-old gall showing young giant cells containing two or more nuclei, with feeding nematode. B: close-up of A. C: photo of the root system of a 14-days-old rice plantlet 4 days after inoculation. Note the characteristic hook-type galls (arrows). D: 31-days-old galls containing giant cells, with several adult females with eggs embedded in the cortex tissue of a susceptible rice root. Asterisk: giant cell; ne: nematode; em: egg mass. Bars represent 100 μm. (See colour plate)

(Cabasan et al., 2014; A.S. Petitot, R. Haidar, J. de Almeida Engler, D. Fernandez, unpublished data). After some J2 had induced giant cell formation, the feeding sites developed poorly or degenerated after 2 weeks, preventing the nematode to develop into mature female. In some accessions, however, early defence response (HR-like) resulted in plant cell death around the nematode infection site. This occurred in some roots after

3–7 DAI, preventing the feeding site formation and leading to the nematode developmental arrest (Cabasan et al., 2014).

Overall, RKN life cycle and host resistance responses described for monocots do not differ substantially from dicotyledonous hosts. However, depending on the nematode species considered some features are different: juvenile migration within the root cortex both inter- and intracellularly for *M. naasi* (Balhadère & Evans, 1995a; Siddiqui & Taylor, 1970), and nematode body remaining within the cortex with its anterior end in the vascular cylinder for *M. marylandi* (Oka et al., 2003). Also typical of monocots are the development of giant cells within the cortex of turfgrass roots infected with *M. naasi* (Ediz & Dickerson, 1976) and the confinement of females and egg sacs entirely within the root for *M. graminicola* in rice (Fortuner & Merny, 1979). Egg sacs remaining within the root host have been also reported in banana infected by *M. incognita* (Sudha & Prabhoo, 1983). Another RKN species, *M. chitwoodi* may also produce eggs within potato host tissues. *Meloidogyne chitwoodi* has a wide host range attacking monocotyledonous and dicotyledonous crop plants and weed species. In potato (*Solanum tuberosum*) tubers, eggs are laid by the female in a gelatinous sac within 5–6 mm of the tuber surface. Interestingly, modified host cells form a protective layer around the egg mass and the juveniles as they hatch. This indicates that these adaptations may be due to the nematode species rather than the plant host.

4. TRANSCRIPTOMICS OF MONOCOTYLEDONOUS PLANT RESPONSES TO NEMATODES

Using transcriptome analyses, the response of a plant upon infection with nematodes can be efficiently monitored, leading to insights into the pathways that are (1) manipulated by the pathogen or (2) activated/suppressed by the plant as a defence response.

An RNA-Seq analysis of the incompatible interaction between *A. variabilis* and CCNs (Xu et al., 2013) analysed a pooled RNA sample to construct a de novo transcriptome of both infected and uninfected *Aegilops* roots at three time points, 30 hpi, 3 DAI and 9 DAIThe assembled transcriptome contained many unigenes associated with plant defence and resistance (Xu et al., 2013), but the RNA pooling made it impossible to draw conclusions about differential gene expression patterns affected by infection in these CCN-resistant monocot plants.

Up till now, the only transcriptome studies on the compatible interaction between monocots and nematodes are those of Kyndt et al. (2012) and Ji et al. (2013). The former study provides RNA-seq data obtained from (1) roots infected with root rot nematode *Heterodera oryzae* in comparison with uninfected roots and (2) root galls induced by *M. graminicola*, at three and seven DAI compared with control root tips (Kyndt et al., 2012). Ji et al. (2013) focus on the gene expression inside laser capture microdissected giant cells at 7 and 14 DAI and compared this profile with vascular cells from uninfected roots.

In case of infection with the migratory nematode *H. oryzae*, plant roots necrotize and die and this leads to a very strong defence response (Kyndt et al., 2012). For instance, infected roots showed a strong induction of the jasmonate biosynthesis pathway, many pathogenesis-related proteins, as well as a group of WRKY transcription factors. The early local response of rice roots upon migratory nematode infection shows a remarkable similarity with the early root response upon infection with the blast fungus *Magnaporthe oryzae* (Marcel, Sawers, Oakeley, Angliker, & Paszkowski, 2010). Most likely due to the trade-off between plant defence and plant growth, this strong defence induction is responsible for a suppression of primary metabolic pathways in the migratory nematode-infected root systems (Kyndt et al., 2012).

On the contrary, transcriptomics of galls and isolated giant cells formed by the biotrophic RKN in rice, revealed a remarkable induction of the primary metabolism and a suppression of secondary metabolism-based plant defence pathways. Similar metabolic changes were detected in galls and isolated giant cells in *Arabidopsis* (Barcala et al., 2010; Jammes et al., 2005). Repression of expression of the flavonoid, phenylpropanoid and jasmonate biosynthesis pathways, as well as many pathogenesis-related proteins was seen, mainly at the earliest studied time points after infection (Barcala et al., 2010; Ji et al., 2013; Kyndt et al., 2012). The transcriptome of galls showed a similar expression trends as the transcriptional responses observed in *Xanthomonas oryzae* pv. *oryzae*-infected shoots (GSE16793, Do et al. unpublished) as well as some marginal similarity to that of rice calli, 1–6 h after co-cultivation with *Agrobacterium* (Wang et al., 2010), while the transcriptome of *M. oryzae*-infected roots (Marcel et al., 2010) showed rather opposite expression profiles (Kyndt et al., 2012). In the data of RKN-infected tissue, it became clear that the primary metabolism is strongly induced, and therefore, here we provide a more detailed insight into these metabolic changes in isolated giant cells and galls. For a detailed review on the changes in secondary metabolism induced by nematode feeding in rice, we refer to Kyndt et al. (2014).

4.1 Induced Metabolite Production in Giant Cells and Galls

The primary metabolites produced by a plant are defined as those metabolites needed for plant growth and development, such as carbohydrates, lipids, proteins, and nucleic acids. A strong induction of this primary metabolism was detected in giant cells and galls in both monocot and dicot roots (Jammes et al., 2005; Ji et al., 2013; Kyndt et al., 2012).

The transcriptional profile of seven DAI giant cells, in comparison with control vascular cells of healthy plant roots, is exemplified in Figure 2. In Table 2, the metabolic pathways significantly different between giant cells and control cells are listed. Some pathways found to be significantly induced inside the rice giant cells at 7 and 14 DAI are, for instance, nucleotide synthesis (mainly purine biosynthesis), and phospholipid production, metabolites which are most likely needed for enhanced cellular activity. Protein and sucrose biosynthesis genes as well as glycolysis are also generally activated in galls and giant cells (Kyndt et al., 2012; Ji et al., 2013, Figure 2), suggesting that the heterotrophic RKN stimulates the cells within the gall to be more active and produce energy and nutrients to the benefit of the nematode. Although no metabolome studies have been done in rice gall tissue, transcriptome data suggest that starch is accumulating in the giant cells (Ji et al., 2013, Figure 2, Table 2). This carbohydrate was detected inside syncytia induced by CN in *Arabidopsis* (Hofmann et al., 2008) where it was hypothesized to act as a buffer and long-term storage in the nematode-feeding site to compensate for changing rates of solute uptake by the nematode.

Strikingly, the metabolic changes upon RKN infection are very similar to those seen in plants upon drought and cold stress (Kaplan et al., 2007; Rizhsky et al., 2004). In a study on the water status and growth of rice upon *H. sacchari* infection and drought, a reduction in leaf water potential and stomatal conductance was observed, symptoms of physiological drought in the plant (Audebert, Coyne, Dingkuhn, & Plowright, 2000). The uptake of nutrients and water by the feeding nematode apparently induces a response similar to dehydration in the infected root tissue, as exemplified, for instance, by the induction of many dehydration responsive genes (Kyndt et al., 2012) and the accumulation of ABA-responsive genes (Kyndt T., Nahar K., Gheysen G., unpublished data).

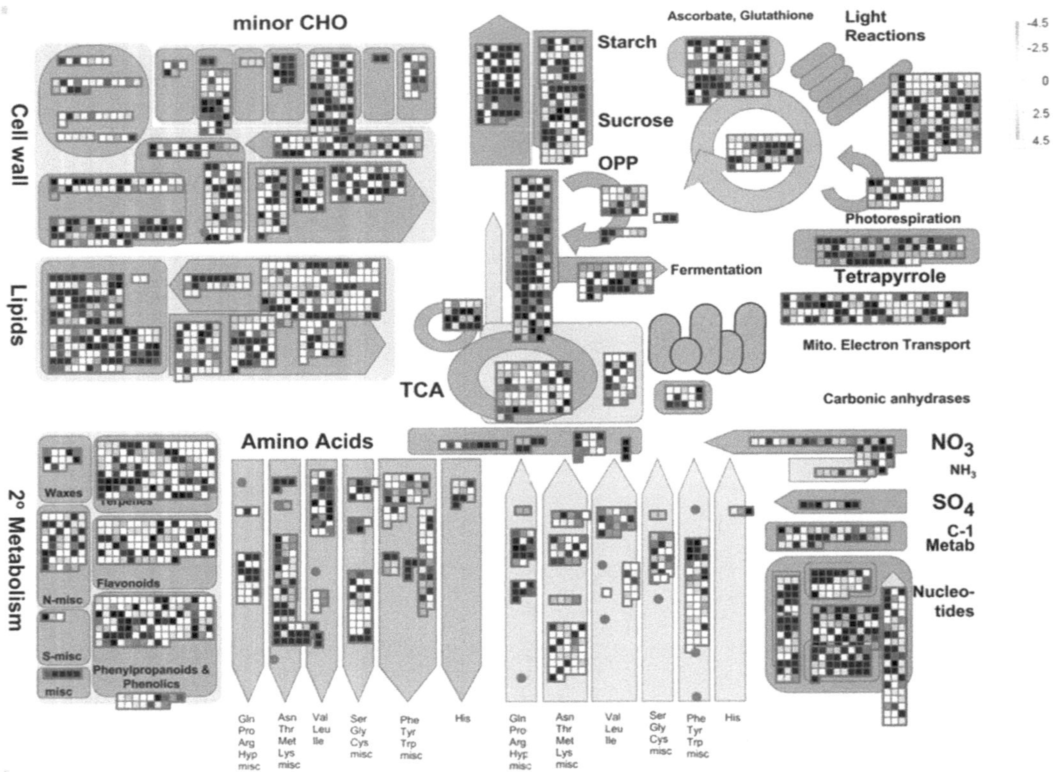

Figure 2 Visualization of transcriptome data of giant cells induced by *Meloidogyne graminicola* in rice (*Oryza sativa*) at 7 days after infection. The graph was obtained using MapMan (Thimm et al., 2004) and shows the expression profiles of genes involved in the general metabolism of the rice plant. The observed differential expression patterns are shown as the Log_2 fold changes of mRNA levels in giant cells versus control vascular root cells of uninfected plants. Red dots indicate that the gene is upregulated in giant cells versus the corresponding healthy control cells, while blue indicates downregulation. *Figure based on data provided by Ji et al. (2013).* (See colour plate)

Table 2 General Metabolic Pathways Significantly Affected ($p < 0.05$) in Seven DAI Giant Cells Induced by *Meloidogyne Graminicola* in *Oryza Sativa* Roots, in Comparison with Vascular Cells from Uninfected Roots, Based on the Transcriptome Data of Ji et al. (2013). The *p*-values were Calculated Using the Wilcoxon Signed Rank Test in MapMan (Thimm et al., 2004) and were Adjusted for Multiple Comparisons Using Benjamin and Hochberg Correction

Name of the Pathway	Number of Genes Included	*p*-value
Glycolysis	114	9.68E-09
Major CHO metabolism.synthesis.starch	64	4.57E-08
Secondary metabolism.phenylpropanoids	104	1.38E-05
Minor CHO metabolism.trehalose.potential TPS/TPP	19	1.58E-05
Amino acid metabolism.synthesis.aspartate family.misc.homoserine	17	2.57E-05
Tetrapyrrole synthesis	73	5.98E-04
Nucleotide metabolism.synthesis.purine	27	6.81E-04
Minor CHO metabolism.others	84	0.001971
Redox.ascorbate and glutathione	91	0.004207
Cell wall.precursor synthesis	59	0.006686
Lipid metabolism.Phospholipid synthesis	70	0.013896
Amino acid metabolism.synthesis.central amino acid metabolism.aspartate	6	0.018828
TCA/org. transformation.carbonic anhydrases	15	0.01937
Secondary metabolism.flavonoids	91	0.021608
Secondary metabolism.unspecified	5	0.022954
Amino acid metabolism.synthesis.aspartate family.asparagine	4	0.026806
Mitochondrial electron transport/ATP synthesis	97	0.026877
Lipid metabolism.FA desaturation	19	0.026877
Amino acid metabolism.degradation.central amino acid metabolism.aspartate	3	0.029755
Cell wall.pectin* esterases	37	0.03224
Amino acid metabolism.degradation.branched-chain group.shared	18	0.033044
PS.calvin cycle	47	0.03311
Secondary metabolism.simple phenols	16	0.033828
Amino acid metabolism.degradation.branched-chain group.leucine	11	0.040868
Amino acid metabolism.synthesis.serine-glycine-cysteine group.cysteine	30	0.041676
Amino acid metabolism.synthesis.glutamate family.arginine	21	0.048121
Amino acid metabolism.synthesis.serine-glycine-cysteine group.glycine	4	0.048121

*S = susceptible; R = resistant; - no data found/available.

4.2 A Specific Focus on Amino Acid Production

Amino acids form the building blocks for any protein produced inside a living cell. Changes in the amino acid composition of a cell can provide clues about the metabolic changes happening inside these cells, and in the case of nematode-feeding sites, amino acid changes can also reflect the nutritional needs of the nematodes along their developmental path.

While glycine biosynthesis genes are suppressed, genes involved in the biosynthesis of other amino acids, like asparagine, cysteine, arginine and homoserine are strongly induced inside giant cells in rice (Ji et al., 2013, Figure 2; Table 2). These amino acids might be simply ingested by the parasitic nematode, for the benefit of its growth and development. However, in addition, it has been shown that high endogenous homoserine and asparagine levels in pea act as inducers of infection by the necrotrophic root pathogen *Nectria hematococca* (Yang, Rogers, Song, Guo, & Kolattukudy, 2005). It was hypothesized that *N. hematococca* has evolved the ability to sense these amino acids and that this leads to the expression of virulence genes in plants (Yang et al., 2005). On the contrary, accumulation of homoserine rather leads to resistance against the obligate biotrophic leaf pathogen *H. arabidopsidis* in *Arabidopsis* (van Damme et al., 2009). Elucidating the role and importance of accumulating amino acids in a compatible rice–nematode interaction seems to be a worthwhile route for further investigation.

4.3 Transport of Nutrients into the Feeding Site

Investigations on phloem transport revealed that labelled solutes were able to enter CN-induced syncytia or RKN-induced giant cells (Bockenhoff, Prior, Grundler, & Oparka, 1996; Dorhout, Gommers, & Kolloffel, 1993) in tomato and *Arabidopsis*. Giant cells are embedded in a dense vascular network (Jones & Dropkin, 1976), from which they are generally believed to be symplastically isolated (Bartlem et al., 2014; Hoth, Stadler, Sauer, & Hammes, 2008). A strong differential regulation of various transporter genes has been revealed using a microarray analysis on galls formed in *Arabidopsis* roots (Hammes et al., 2005). Transmembrane lipid and protein transporter genes are also strongly reprogrammed in rice gall tissue (Kyndt et al., 2012), with a significant up-regulation of, for instance, amino acid and peptide transporters and different plant lipid transfer proteins. These changes suggest that uptake of nutrients from the phloem into giant cells is mostly mediated by transport across the plasma membranes of the sieve elements and the giant cells. However, Hofmann, Wieczorek, Blochl, and Grundler (2007) showed the presence of some plasmodesmata along the cell walls between giant cells and neighbouring cells at 10 and 13 DAI, indicating that symplastic transport of nutrients into the giant cells might also be possible.

4.4 Photosynthesis

An increase in leaf chlorophyll content, an indicator for induced nitrogen uptake by the plant (Peng et al., 1993) was found upon *H. sacchari* infection in rice (Audebert et al., 2000). It was hypothesized that this is a compensating mechanism for the nitrogen losses to the feeding nematode. Although this phenomenon has not been studied yet in leaves of RKN-infected rice plants, genes involved in photosynthesis were also locally induced in dark-grown giant cells, mainly at seven DAI (Figure 2 and Table 2: light reactions and tetrapyrrole), and the presence of chloroplast-like structures inside these root cells was confirmed by confocal imaging (Ji et al., 2013). Nevertheless, in *Arabidopsis* giant cells, photosynthesis was not found to be significantly modified (Barcala et al., 2010), but whether this is due to a physiological difference between monocots and dicots remains to be elucidated.

5. NEMATODE EFFECTORS IN MONOCOTS–NEMATODE INTERACTIONS

A key feature of sedentary plant-parasitic nematodes is the release of effector proteins secreted from their oesophageal gland cells through their stylet into host roots, so they can manipulate the cellular machinery and transform parenchymatic vascular cells in feeding sites (Hewezi & Baum, 2013). Transcriptomic and proteomic approaches identified putative secreted protein sequences from various nematode developmental stages or tissues, and from mixed samples of plant–nematode-infected tissues (reviewed in Haegeman et al., 2012). Comparative genomics approaches have facilitated a major breakthrough for identifying novel and highly specific *M. incognita* and *M. hapla* effector genes that constitute promising targets for controlling diseases caused by these RKN species (Danchin et al., 2013). *De novo* transcriptome analysis of *M. graminicola* and rice-infected tissues (Haegeman et al., 2012; A.S. Petitot, D. Fernandez, unpublished data) and of two developmental stages of the CCN *H. avenae* (Kumar et al., 2014) allowed identification of several hundreds of putative effector proteins in the genome of these nematode species. Genes encoding putative plant cell wall modifying enzymes (CAZymes), proteins involved in detoxification of reactive oxygen species, and homologs of known effectors in other nematode species (fatty acid and retinol-binding proteins, calreticulin, SPRYSEC etc.) were identified. However, most of these putative effectors have no annotation. Because homologous effectors were found in nematodes attacking dicotyledons or monocotyledons (Haegeman et al., 2012; Kumar et al., 2014), it is believed that some molecular mechanisms used in manipulating host defences may be similar. The efficiency of RNA interference (RNAi) to silence genes in plant-parasitic nematodes was demonstrated (Lilley, Davies, & Urwin, 2012) but, until recently had been limited to dicotyledonous host plants. RNAi soaking experiments targeting four *M. graminicola* genes showing strong homology to *Caenorhabditis elegans* genes with a lethal RNAi phenotype showed that these genes may be good targets for RNAi-based control strategies (Haegeman et al., 2012). In RNAi soaking experiments on *M. incognita*, suppression of a putative effector gene with unknown function affected the ability of the nematode to infect rice plants (Nguyễn, 2013). In dicotyledonous hosts, nematodes suppress the host defence signalling pathways to facilitate infection (Hewezi & Baum, 2013). For instance, in *Arabidopsis thaliana* the *M. incognita* calreticulin is able to suppress the basal defence responses induced by the bacterial elicitor *elf18* (Jaouannet et al., 2013). In rice, we showed that *M. incognita* and *M. graminicola* are able to suppress transcription of key immune regulatory genes (Kyndt et al., 2012; Nguyễn et al., 2014). Putative *M. incognita* and *M. graminicola* effector proteins expressed in oesophageal glands cells are currently being analysed for their role in RKN–rice interactions in our labs (Nguyễn, 2013; D. Fernandez & T. Kyndt, unpublished data). This new data should increase knowledge about the complex rice–nematode interaction, and could indicate possible candidates to control these nematodes.

6. CONCLUSIONS

Our understanding of the biological and molecular processes in monocot plant–nematode interactions is still fragmentary; however, new data show that common mechanisms act in monocot and dicot plant–nematode interactions. We highlighted some specific features in the developmental RKN cycle that are different from other RKN species attacking dicotyledons. New effector genes will soon be identified in some of the most studied RKN species adapted to graminaceous hosts. Transcriptome data have provided insights into the metabolic changes inside giant cells induced by RKN, but a functional characterization of many of these pathways remains to be done. This review also highlighted that molecular data on CN interactions with monocots are needed, as data were published so far only about the CCN *H. avenae*. Today, no *R* gene controlling nematodes in monocots has been cloned, but as an increasing number of cereal crops and their wild relative genomes are sequenced (Visendi, Batley, & Edwards, 2013), new data should become available to help identifying those found by genetic means and already mapped. A greater focus on crop breeding and improvement for nematode disease resistance may help to transfer these desirable traits into commercially acceptable cultivars. New data for the functional analysis of plant and nematode genes involved in host–parasite interaction may also arise from the monocot *Brachypodium distachyon*, a model species for grass genomics (Mur et al. 2011) that is susceptible to *M. graminicola* (Marcel et al., 2011). In this respect, the monocot, *M. graminicola* system may also be an excellent model for studying resistance to RKN.

REFERENCES

Abad, P., Gouzy, J., Aury, J. M., Castagnone-Sereno, P., Danchin, E. G., Deleury, E., et al. (2008). Genome sequence of the metazoan plant-parasitic nematode *Meloidogyne incognita*. *Nature Biotechnology, 26*, 909–915.

Agnoun, Y., Sie, M., Djedatin, G., Drame, K. N., Toulou, B., Ogunbayo, S., et al. (2012). Molecular profiling of interspecific lowland rice progenies resulting from crosses between TOG5681 and TOG5674 (*Oryza glaberrima*) and IR64 (*Oryza sativa*). *International Journal of Biology, 4*, 19–28.

Angiosperm Phylogeny Group. (2009). An update of the Angiosperm Phylogeny Group classification for the orders and families of flowering plants: APG III. *Botanical Journal of the Linnean Society, 161*, 105–121.

Asmus, G. L., Ferraz, L. C. C. B., & Appezzato-da-Glória, B. (2000). Alterações anatômicas em raízes de milho (*Zea mays* L.) parasitadas por *Meloidogyne javanica*. *Nematropica, 30*, 33–39.

Audebert, A., Coyne, D. L., Dingkuhn, M., & Plowright, R. A. (2000). The influence of cyst nematodes (*Heterodera sacchari*) and drought on water relations and growth of upland rice in Cote d'Ivoire. *Plant and Soil, 220*, 235–242.

Babatola, J. (1983). Pathogenicity of *Heterodera sacchari* on rice. *Nematologia Mediterranea, 1*, 21–25.

Bakker, E., Dees, R., Bakker, J., & Goverse, A. (2006). Mechanisms involved in plant resistance to nematodes. In S. Tuzun, & E. Bent (Eds.), *Multigenic and induced systemic resistance in plants* (pp. 314–334). US: Springer.

Balhadère, P., & Evans, A. A. F. (1995a). Histopathogenesis of susceptible and resistant responses of wheat, barley and wild grasses to *Meloidogyne naasi*. *Fundamental and Applied Nematology, 18*, 531–538.

Balhadère, P., & Evans, A. A. F. (1995b). Cytochemical investigation of resistance to root-knot nematode *Meloidogyne naasi* in cereals and grasses using cryosections of roots. *Fundamental and Applied Nematology, 18*, 539–547.

Barcala, M., Garcia, A., Cabrera, J., Casson, S., Lindsey, K., Favery, B., et al. (2010). Early transcriptomic events in microdissected Arabidopsis nematode-induced giant cells. *Plant Journal, 61*, 698–712.

Barloy, D., Lemoine, J., Abelard, P., Tanguy, A. M., Rivoal, R., & Jahier, J. (2007). Marker-assisted pyramiding of two cereal cyst nematode resistance genes from *Aegilops variabilis* in wheat. *Molecular Breeding, 20*, 31–40.

Bartlem, D. G., Jones, M. G. K., & Hammes, U. Z. (2014). Vascularization and nutrient delivery at root-knot nematode feeding sites in host roots. *Journal of Experimental Botany, 65*, 1789–1798.

Bimpong, I. K., Carpena, A. L., Mendioro, M. S., Fernandez, L., Ramos, J., Reversat, G., et al. (2010). Evaluation of *Oryza sativa* x *O. glaberrima* derived progenies for resistance to root knot nematode and identification of introgressed alien chromosome segments using SSR markers. *African Journal of Biotechnology, 9*, 3988–3997.

Bockenhoff, A., Prior, D. A. M., Grundler, F. M. W., & Oparka, K. J. (1996). Induction of phloem unloading in *Arabidopsis thaliana* roots by the parasitic nematode *Heterodera schachtii*. *Plant Physiology, 112*, 1421–1427.

Bridge, J., Plowright, R. A., & Peng, D. (2005). Nematode parasites of rice. In M. Luc, R. A. Sikora, & J. Bridge (Eds.), *Plant-parasitic nematodes in subtropical and tropical agricultures* (2nd ed.). (pp. 87–130). Wallingford, UK: CAB International.

Cabasan, M. T. N., Kumar, A., Bellafiore, S., & De Waele, D. (2014). Histopathology of the rice root-knot nematode, *Meloidogyne graminicola*, on *Oryza sativa* and *O. glaberrima*. *Nematology, 16*, 73–81.

Cadet, P., & Spaull, V. W. (2005). Nematode parasites of sugarcane. In M. Luc, R. A. Sikora, & J. Bridge (Eds.), *Plant-parasitic nematodes in subtropical and tropical agricultures* (2nd ed.). (pp. 645–674). Wallingford, UK: CAB International.

Cai, D., Kleine, M., Kifle, S., Harloff, H.-J., Sandal, N., Marcker, K., et al. (1997). Positional cloning of a gene for nematode resistance in sugar beet. *Science, 275*, 832–834.

Cook, R., & Evans, K. (1987). Resistance and tolerance. In R. H. Brown, & B. R. Kerry (Eds.), *Principles and practice of nematode control in crops* (pp. 179–231). Sydney: Academic Press.

Cook, R., & Noel, G. R. (2002). Cyst nematodes: *Globodera* and *Heterodera* species. In J. L. Starr, R. Cook, & J. Bridge (Eds.), *Plant resistance to parasitic nematodes* (pp. 71–105). Wallingford, UK: CABI Publishing.

Coyne, D. (2000). Pathogenicity of cyst nematode, *Heterodera sacchari*, on rice in sand and clay soil. *International Rice Research Notes, 25*, 17–18.

van Damme, M., Zeilmaker, T., Elberse, J., Andel, A., de Sain-van der Velden, M., & van den Ackerveken, G. (2009). Downy mildew resistance in *Arabidopsis* by mutation of homoserine kinase. *Plant Cell, 21*, 2179–2189.

Danchin, E. G., Arguel, M. J., Campan-Fournier, A., Perfus-Barbeoch, L., Magliano, M., Rosso, M. N., et al. (2013). Identification of novel target genes for safer and more specific control of root-knot nematodes from a pan-genome mining. *PLoS Pathogen, 9*(10), e1003745.

Davide, R. G., & Marasigan, L. Q. (1985). Yield loss assessment and evaluation of resistance of banana cultivars to the nematode *Radopholus similis* Thorne and *Meloidogyne incognita* Chitwood. *Phillipin Agricultural, 63*, 335–349.

De Waele, D., & Elsen, A. (2007). Challenges in tropical plant nematology. *Annual Review of Phytopathology, 45*(1), 457–485.

Dorhout, R., Gommers, F. J., & Kolloffel, C. (1993). Phloem transport of carboxyfluorescein through tomato roots infected with *Meloidogyne incognita*. *Physiological and Molecular Plant Pathology, 43*, 1–10.

Ediz, S. A., & Dickerson, O. J. (1976). Life cycle, pathogenicity, histopathology, and host range of race 5 of the barley root-knot nematode. *Journal of Nematology, 8*, 228–232.

Eissa, M. F. M., EI-Sherief, M. A., Abdel-Gawad, M. M., Ismail, A. A., & EI-Nagdi, W. M. A. (1998). Histological response of susceptible and resistant date-palm cultivars to the root-knot nematode, *Meloidogyne incognita* infection. *Pakistan Journal of Nematology, 16*, 103–109.

Esau, K. (1953). Plant anatomy. Chapman & Hall (Eds.). London, UK.

FAO Statistical Yearbook. (2013). World food and agriculture, FAO, Rome, http://issuu.com/faooftheun/docs/syb2013issuu.

Fortuner, R., & Merny, G. (1979). Root-parasitic nematodes of rice. *Nematology, 2*, 79–102.

Fuller, V. L., Lilley, C. J., & Urwin, P. E. (2008). Nematode resistance. *New Phytologist, 180*, 27–44.

Gowen, S. R., Quénéhervé, P., & Fogain, R. (2005). Nematode parasites of banana and plantains. In M. Luc, R. A. Sikora, & J. Bridge (Eds.), *Plant-parasitic nematodes in subtropical and tropical agricultures. 2nd Edition* (pp. 611–643). Wallingford, UK: CAB International.

Griffith, R., Giblin-Davis, R. M., Koshy, P. K., & Sossama, E. (2005). Nematode parasites of coconut and other palms. In M. Luc, R. A. Sikora, & J. Bridge (Eds.), *Plant-parasitic nematodes in subtropical and tropical agricultures* (2nd ed.). (pp. 493–527). Wallingford, UK: CAB International.

Haegeman, A., Mantelin, S., Jones, J. T., & Gheysen, G. (2012 Jan 15). Functional roles of effectors of plant-parasitic nematodes. *Gene, 492*(1), 19–31.

Haegeman, A., Bauters, L., Kyndt, T., Rahman, M. M., & Gheysen, G. (2013). Identification of candidate effector genes in the transcriptome of the rice root knot nematode *Meloidogyne graminicola. Molecular Plant Pathology, 14*, 379–390.

Hammes, U. Z., Schachtman, D. P., Berg, R. H., Nielsen, E., Koch, W., McIntyre, L. M., et al. (2005). Nematode-induced changes of transporter gene expression in Arabidopsis roots. *Molecular Plant-Microbe Interactions, 18*, 1247–1257.

Handoo, Z. A., Skantar, A. M., Carta, L. K., & Erbe, E. F. (2005). Morphological and molecular characterization of a new root-knot nematode, *Meloidogyne thailandica* n. sp. (Nematoda: Meloidogynidae), parasitizing ginger (*Zingiber* sp. *Journal of Nematology, 37*, 343–353.

Hashmi, G., Hashmi, S., Krusberg, L. R., & Huettel, R. N. (1993). Resistance in *Zea mays* to *Heterodera zeae*. *Journal of Nematology, 25*(4S), 820–823.

Hewezi, T., & Baum, T. J. (2013). Manipulation of plant cells by cyst and root-knot nematode effectors. *Molecular Plant-Microbe Interactions, 26*, 9–16.

Hofmann, J., Szakasits, D., Blochl, A., Sobczak, M., Daxbock-Horvath, S., Golinowski, W., et al. (2008). Starch serves as carbohydrate storage in nematode-induced syncytia. *Plant Physiology, 146*, 228–235.

Hofmann, J., Wieczorek, K., Blochl, A., & Grundler, F. M. W. (2007). Sucrose supply to nematode-induced syncytia depends on the apoplasmic and symplasmic pathways. *Journal of Experimental Botany, 58*, 1591–1601.

Hoth, S., Stadler, R., Sauer, N., & Hammes, U. Z. (2008). Differential vascularization of nematode-induced feeding sites. *Proceedings of the National Academy of Sciences of the United States of America, 105*, 12617–12622.

Ibrahim, I. A., Ibrahim, I. K. A., & Rezk, M. A. (1972). Host parasite relationship of *Meloidogyne incognita* (Kofoid & White) Chitw. on rice. *Nematologia Mediterranea, 1*, 8–14.

Jammes, F., Lecomte, P., Almeida-Engler, J., Bitton, F., Martin-Magniette, M. L., Renou, et al. (2005). Genome-wide expression profiling of the host response to root-knot nematode infection in *Arabidopsis*. *Plant Journal, 44*, 447–458.

Jaouannet, M., Magliano, M., Arguel, M. J., Gourgues, M., Evangelisti, E., Abad, P., et al. (2013). The root-knot nematode calreticulin Mi-CRT is a key effector in plant defense suppression. *Molecular Plant-Microbe Interactions, 26*, 97–105.

Jena, R. N., & Rao, Y. S. (1977). Nature of root-knot (*Meloidogyne graminicola*) resistance in rice (*Oryza sativa*). II. Histopathology of nematode infection in rice varieties. *Proceedings of the Indian Academy of Science, 86B*, 87–91.

Ji, H. L., Gheysen, G., Denil, S., Lindsey, K., Topping, J. F., Nahar, K., et al. (2013). Transcriptional analysis through RNA sequencing of giant cells induced by *Meloidogyne graminicola* in rice roots. *Journal of Experimental Botany, 64*, 3885–3898.

Jones, M. G. K., & Dropkin, V. H. (1976). Scanning electron-microscopy of nematode-induced giant transfer cells. *Cytobios, 15*, 149–161.

Jussieu, A.L. (1789). Genera plantarum: Secundum ordines naturales disposita, juxta methodum in Horto regio parisiensi exaratam, Herissant, Barrois (Eds.). Paris, France.

Kagoda, F., Derera, J., Tongoona, P., Coyne, D. L., & Lorenzen, L. (2011). Genetic analysis of resistance to nematodes in inbred maize *(Zea mays* L.) and maize hybrids. *Euphytica, 182*, 377–393.

Kaloshian, I., Waines, J. G., Roberts, P. A., & Thomason, I. J. (1991). Chromosomal location of root-knot-nematode-resistance gene in the D genome of wheat. *Journal of Heredity, 82*, 254–256.

Kandoth, P. K., & Mitchum, M. G. (2013). War of the worms: how plants fight underground attacks. *Current Opinion in Plant Biology, 16*, 457–463.

Kaplan, F., Kopka, J., Sung, D. Y., Zhao, W., Popp, M., Porat, R., et al. (2007). Transcript and metabolite profiling during cold acclimation of *Arabidopsis* reveals an intricate relationship of cold-regulated gene expression with modifications in metabolite content. *Plant Journal, 50*, 967–981.

Koshy, P. K., Eapen, S. J., & Pandey, R. (2005). Nematode parasites of spices, condiments and medicinal plants. In M. Luc, R. A. Sikora, & J. Bridge (Eds.), *Plant-parasitic nematodes in subtropical and tropical agricultures* (2nd ed.). (pp. 751–791). Wallingford, UK: CAB International.

Kumar, M., Gantasala, N. P., Roychowdhury, T., Thakur, P. K., Banakar, P., Shukla, R. N., et al. (2014). De novo transcriptome sequencing and analysis of the cereal cyst nematode, *Heterodera avenae. PLoS One, 9*(5), e96311, 6.

Kyndt, T., Denil, S., Haegeman, A., Trooskens, G., Bauters, L., Van Criekinge, W., et al. (2012). Transcriptional reprogramming by root knot and migratory nematode infection in rice. *New Phytologist, 196*, 887–900.

Kyndt, T., Fernandez, D., & Gheysen, G. (2014). Plant-parasitic nematode infections in rice: molecular and cellular insights. *Annual Review of Phytopathology, 52*, 7.1–7.19.

Kyndt, T., Vieira, P., Gheysen, G., & de Almeida-Engler, J. (2013). Nematode feeding sites: unique organs in plant roots. *Planta, 238*, 807–818.

Lilley, C. J., Davies, L. J., & Urwin, P. E. (2012). RNA interference in plant parasitic nematodes: a summary of the current status. *Parasitology, 139*, 630–640.

Liu, S., Kandoth, P. K., Warren, S. D., Yeckel, G., Heinz, R., Alden, J., et al. (2012). A soybean cyst nematode resistance gene points to a new mechanism of plant resistance to pathogens. *Nature, 492*(7428), 256–260.

Lorieux, M., Reversat, G., Garcia-Diaz, S. X., Denance, C., Jouvenet, N., Orieux, Y., et al. (2003). Linkage mapping of *Hsa-1Og*, a resistance gene of African rice to the cyst nematode, *Heterodera sacchari. Theorical and Applied Genetics, 107*, 691–696.

Luc, M., Sikora, R. A., & Bridge, J. (2005). *Plant parasitic nematodes in subtropical and tropical agriculture*. Wallingford, UK: CAB International.

Marcel, T. C., Niks, R. E., Wang, J. C., Antelme, S., Morel, J. B., Dufresne, M., et al. (2011). *Brachypodium distachyon*, a model grass to study plant-pathogen interactions. In *First european brachypodium workshop, October 19–21, Versailles, France.*

Marcel, S., Sawers, R., Oakeley, E., Angliker, H., & Paszkowski, U. (2010). Tissue-adapted invasion strategies of the rice blast fungus *Magnaporthe oryzae. Plant Cell, 22*, 3177–3187.

Mc Clure, M. A., Nischwitz, C., Skantar, A. M., Schmitt, M. E., & Subbotin, S. A. (2012). Root-knot nematodes in golf course greens of the western United States. *Plant Disease, 96*, 635–647.

Mc Donald, A. H., & Nicol, J. M. (2005). Nematode parasites of cereals. In M. Luc, R. A. Sikora, & J. Bridge (Eds.), *Plant-parasitic nematodes in subtropical and tropical agricultures* (2nd ed.). (pp. 131–191). Wallingford, UK: CAB International.

Milligan, S., Bodeau, J., Yaghoobi, J., Kaloshian, I., Zabel, P., & Williamson, V. (1998). The root knot nematode resistance gene *Mi* from tomato is a member of the leucine zipper, nucleotide binding, leucine-rich repeat family of plant genes. *Plant Cell, 10,* 1307–1319.

Molinari, S. (2011). Natural genetic and induced plant resistance, as a control strategy to plant-parasitic nematodes alternative to pesticides. *Plant Cell Reports, 30,* 311–323.

Mur, L. A. J., Allainguillaume, J., Catalán, P., Hasterok, R., Jenkins, G., Lesniewska, K., et al. (2011). Exploiting the *Brachypodium* Tool Box in cereal and grass research. *New Phytologist, 191,* 334–347.

Ndjiondjop, M. N., Semagn, K., Sie, M., Cissoko, M., Fatondji, B., & Jones, M. (2008). Molecular profiling of interspecific lowland rice populations derived from IR64 (*Oryza sativa*) and Tog5681 (*Oryza glaberrima*). *African Journal of Biotechnology, 7,* 4219–4229.

Nguyễn, V. P. (2013). *Caractérisation d'effecteurs de virulence du nématode à galles* Meloidogyne incognita *chez le riz (*Oryza sativa*).* PhD thesis (166 pp.). France: Université de Montpellier.

Nguyễn, V. P., Bellafiore, S., Petitot, A. S., Haidar, R., Bak, A., Abed, A., et al. (2014). *Meloidogyne incognita* - rice (*Oryza sativa*) interaction: a new model system to study plant-root knot nematode interactions in monocotyledons. *Rice,* 7(1), 23.

Oka, Y., Chilt, I., & Spiegel, Y. (1997). Are pathogenesis-related proteins induced by *Meloidogyne javanica* or *Hetorodera avenae* invasion? *Journal of Nematology, 29,* 501–508.

Oka, Y., Karssen, G., & Mor, M. (2003). Identification, host range and infection process of *Meloidogyne marylandi* from turf grass in Israel. *Nematology, 5,* 727–734.

Pang, W., Hafez, S. L., & Sundararaj, P. (2009). Screening of onion cultivars for resistance and tolerance to *Pratylenchus penetrans* and *Meloidogyne hapla. Nematropica, 39,* 47–55.

Peng, S., Garcia, F. V., Laza, R. C., & Cassman, K. G. (1993). Adjustment for specific leaf weight improves chlorophyll meter's estimate of rice leaf nitrogenconcentration. *Agron. J., 85,* 987–990.

Perry, R. N., & Moens, N. (2011). Introduction to plant-parasitic nematodes: modes of parasitism. In J. Jones, G. Gheysen, & C. Fenoll (Eds.), *Genomics and molecular genetics of plant-nematode interactions* (pp. 3–20). Dordrecht, the Netherlands: Springer.

Plowright, R. A., Coyne, D. L., Nash, P., & Jones, M. P. (1999). Resistance to the rice nematodes *Heterodera sacchari, Meloidogyne graminicola* and *M. incognita* in *Oryza glaberrima* and *O. glaberrima* × *O. sativa* interspecific hybrids. *Nematology, 1,* 745–751.

Poerba, Y. S., Windham, G. L., & Williams, W. P. (1990). Resistance of maize hybrids to *Meloidogyne javanica. Nematropica, 20,* 169–172.

Prot, J. C., & Matias, D. (1995). Effects of water regime on the distribution of *Meloidogyne graminicola* and other root-parasitic nematodes in a rice field toposequence and pathogenicity of *M. graminicola* on rice cultivar UPL R15. *Nematologica, 41,* 219–228.

Quénéhervé, P. (2009). Integrated management of banana nematodes. In A. Ciancio, & K. G. Mukerji (Eds.), *Integrated management of fruit crops and forest nematodes* (pp. 3–61). © Springer Science + Business Media B.V.

Reversat, G., & Destombes, D. (1995). Resistance to *Heterodera sacchari* in rice. *Nematologica, 41,* 333–334.

Reversat, G., & Destombes, D. (1998). Screening for resistance to *Heterodera sacchari* in the two cultivated rice species, *Oryza sativa* and *O. glaberrima. Fundamental and Applied Nematology, 21,* 307–317.

Rivoal, R., & Cook, R. (1993). Nematode pests of cereal. In K. Evans, D. L. Trudgill, & J. M. Webster (Eds.), *Plant parasitic nematodes in temperate agriculture* (pp. 259–303). Wallingford, UK: CAB International.

Rizhsky, L., Liang, H. J., Shuman, J., Shulaev, V., Davletova, S., & Mittler, R. (2004). When defense pathways collide. The response of *Arabidopsis* to a combination of drought and heat stress. *Plant Physiology, 134*, 1683–1696.

Roberts, P. A. (2002). Concepts and consequences of resistance. In J. L. Starr, R. Cook, & J. Bridge (Eds.), *Plant resistance to parasitic nematodes* (pp. 23–41). Wallingford, UK: CABI Publishing.

Rodiuc, N., Vieira, P., Banora, M. Y., & de Almeida-Engler, J. D. (2014). On the track of transfer cell formation by specialized plant-parasitic nematodes. *Frontiers in Plant Science, 5*, 1–14.

Sakai, H., Ikawa, H., Tanaka, T., Numa, H., Minami, H., Fujisawa, M., et al. (2011). Distinct evolutionary patterns of *Oryza glaberrima* deciphered by genome sequencing and comparative analysis. *Plant Journal, 66*, 796–805.

Seah, S., Miller, C., Sivasithamparam, K., & Lagudah, E. S. (2000). Root responses to cereal cyst nematode (*Heterodera avenae*) in hosts with different resistance genes. *New Phytologist, 146*, 527–533.

Shah, J. J., & Raju, E. C. (1977). Histopathology of ginger (*Zingiber officinale*) infested by soil nematodes *Meloidogyne* sp. *Phyton, 16*, 79–84.

Shrestha, R., Uzzo, F., Wilson, M. J., & Price, A. H. (2007). Physiological and genetic mapping study of tolerance to root-knot nematode in rice. *New Phytologist, 176*, 665–672.

Siddiqui, I. A. (1971). Histopathogenesis of galls induced by *Meloidogyne naasi* in oat roots. *Nematologica, 17*, 237–242.

Siddiqui, I. A., & Taylor, D. P. (1970). Histopathogenesis of galls induced by *Meloidogyne naasi* in wheat roots. *Journal of Nematology, 2*, 239–247.

Sikora, R. A., & Fernandez, E. (2005). Nematode parasites of vegetables. In M. Luc, R. A. Sikora, & J. Bridge (Eds.), *Plant-parasitic nematodes in subtropical and tropical agricultures* (2nd ed.). (pp. 319–392). Wallingford, UK: CAB International.

Simonetti, E., Alba, E., Montes, M. J., Delibes, A., & López-Braña, I. (2010). Analysis of ascorbate peroxidase genes expressed in resistant and susceptible wheat lines infected by the cereal cyst nematode, *Heterodera avenae*. *Plant Cell Reports, 29*, 1169–1178.

Simonetti, E., Veronico, P., Melillo, M. T., Delibes, A., Andrés, M. F., & López-Braña, I. (2009). Analysis of class III peroxidase genes expressed in roots of resistant and susceptible wheat lines infected by *Heterodera avenae*. *Molecular Plant Microbe Interactions, 22*, 1081–1092.

Sipes, B. S., Caswell-Chen, E. P., Sarah, J. L., & Apt, W. J. (2005). Nematode parasites of pineapple. In M. Luc, R. A. Sikora, & J. Bridge (Eds.), *Plant-parasitic nematodes in subtropical and tropical agricultures* (2nd ed.). (pp. 709–731). Wallingford, UK: CAB International.

Smiley, R. W., & Nicol, J. M. (2009). Nematodes which challenge global wheat production. In B. F. Carver (Ed.), *Wheat science and trade* (pp. 171–187). Ames, IA: Wiley-Blackwell.

Soriano, I., Schmitt, V., Brar, D. S., Prot, J.-C., & Reversat, G. (1999). Resistance to rice root-knot nematode *Meloidogyne graminicola* identified in *Oryza longistaminata* and *O. glaberrima*. *Nematology, 14*, 395–398.

Sudha, S., & Prabhoo, N. R. (1983). *Meloidogyne (Nematoda*: Meloidogynidae) induced root galls of the banana plant *Musa paradisiaca*–a study of histopathology. *Proceedings of the Indian Academy of Science (Animal Science), 92*, 467–473.

Thimm, O., Blasing, O., Gibon, Y., Nagel, A., Meyer, S., Kruger, P., et al. (2004). MAPMAN: a user-driven tool to display genomics data sets onto diagrams of metabolic pathways and other biological processes. *Plant Journal, 37*, 914–939.

Visendi, P., Batley, J., & Edwards, D. (2013). Next generation characterisation of cereal genomes for marker discovery. *Biology, 2*, 1357–1377.

Vishnudasan, D., Tripathi, M. N., Rao, U., & Khurana, P. (2005). Assessment of nematode resistance in wheat transgenic plants expressing potato proteinase inhibitor (PIN2) gene. *Transgenic Research, 14*, 665–675.

Wang, L., Xie, W. B., Chen, Y., Tang, W. J., Yang, J. Y., Ye, R. J., et al. (2010). A dynamic gene expression atlas covering the entire life cycle of rice. *Plant Journal, 61*, 752–766.
Williamson, V. M., & Roberts, P. A. (2009). Mechanisms and genetics of resistance. In R. N. Perry, M. Moens, & J. L. Starr (Eds.), *Root-knot nematodes* (pp. 301–325). Oxfordshire, UK: CABI.
Williamson, V. M., Thomas, V., Ferris, H., & Dubcovsky, J. (2013). An *Aegilops ventricosa* translocation confers resistance against root-knot nematodes to common wheat. *Crop Science, 53*, 1412–1418.
Windham, G. L., & Williams, W. P. (1994). Penetration and development of *Meloidogyne incognita* in roots of resistant and susceptible corn genotypes. *Journal of Nematology, 26*, 80–85.
Xu, D. L., Long, H., Liang, J. J., Zhang, J., Chen, X., Li, J. L., et al. (2013). *De novo* assembly and characterization of the root transcriptome of *Aegilops variabilis* during an interaction with the cereal cyst nematode. *BMC Genomics, 13*, 133.
Yang, Z. N., Rogers, L. M., Song, Y. D., Guo, W. J., & Kolattukudy, P. E. (2005). Homoserine and asparagine are host signals that trigger in planta expression of a pathogenesis gene in *Nectria haematococca*. *Proceedings of the National Academy of Sciences of the United States of America, 102*, 4197–4202.
Yu, M. Q., Person-Dedryver, F., & Jahier, J. (1990). Resistance to root-knot nematode *Meloidogyne naasi* (Franklin) transferred from *Aegilops variabilis* Eig to bread wheat. *Agronomie, 6*, 451–456.

CHAPTER NINE

Gene Silencing in Nematode Feeding Sites

Tarek Hewezi*[,1], Thomas J. Baum[§]
*Department of Plant Sciences, University of Tennessee, Knoxville, TN, USA
[§]Department of Plant Pathology and Microbiology, Iowa State University, Ames, IA, USA
[1]Corresponding author: E-mail: thewezi@utk.edu

Contents

Abstract

Sedentary plant-parasitic nematodes induce the formation of elaborate feeding sites in their host plants. The cell-specific gene expression analysis of nematode feeding sites yielded valuable information about the transcriptional regulation and gene regulatory networks associated with compatible and incompatible interactions of sedentary plant-parasitic nematodes and their host plants. Host endogenous small RNAs and RNA silencing pathways appear to play key roles in regulating gene expression changes and reprogramming infected root cells into specific syncytial or giant cell types. We highlight the wide gene silencing events that are associated with the initiation and formation of the nematode feeding sites. Also, we emphasize different types of transcriptional and post-transcriptional small RNA-mediated gene silencing in the feeding sites, highlighting their regulations and mode of actions. Finally, the implementation of two artificial RNA interference systems, namely host-induced gene silencing and virus-induced gene silencing as powerful tools for functional genomics of plant–nematode interactions is discussed.

1. INTRODUCTION

Small RNAs were first discovered in plants by David Baulcombe's group in 1999 (Hamilton & Baulcombe, 1999), and this discovery provided

Advances in Botanical Research, Volume 73
ISSN 0065-2296
http://dx.doi.org/10.1016/bs.abr.2014.12.007

the foundation of understanding the molecular basis of sequence-specific RNA interference (RNAi). The application of increasingly sophisticated next-generation sequencing technology led to elucidating the diversity and complexity of small RNA populations (Axtell, Jan, Rajagopalan, & Bartel, 2006; Fahlgren et al., 2007; Kasschau et al., 2007; Lu et al., 2005; Rajagopalan, Vaucheret, Trejo, & Bartel, 2006). Several classes of small RNAs have been identified and new classes are constantly being isolated from various organisms, which add additional depth and complexity to the small RNA populations. While various classes of small RNAs that mediate RNAi can function through different pathways, these pathways share one common and core feature: the formation of an effector complex, called the RNA-induced silencing complex (RISC) (Brodersen & Voinnet, 2006; Tomari & Zamore, 2005). These different pathways recognize and respond to a variety of aberrant RNAs and generally silence genes and genomic loci-sharing sequence similarity with the silencing trigger. Small RNAs have been known, early on, to impact a multitude of basic biological processes and developmental patterning (Chen, 2009). Recent studies provided clear evidence that host small RNA and RNA silencing pathways play fundamental roles in regulating the plant immune system against pathogen infection including plant-parasitic nematodes (Hewezi & Baum, 2013; Katiyar-Agarwal & Jin, 2010). The widespread and conserved functions of small RNA pathways led to the establishment of RNAi as a powerful functional genomics tool. Pioneering efforts have demonstrated the efficiency of using RNAi to diminish root-knot nematode and cyst nematode infection through host plant-induced gene silencing (reviewed by Rosso, Jones, and Abad (2009)) and this approach has proven effective in other pathosystems as well (Nunes & Dean, 2012). Also, a plant's ability to control viral infection through the triggering of RNAi by double-stranded viral replication intermediates can be exploited in a mechanism known as virus-induced gene silencing (VIGS) (Sudarshana, Roy, & Falk, 2007; Waterhouse & Fusaro, 2006). In this approach, recombinant virus genomes carry plant gene sequences and thus trigger the degradation of homologous endogenous transcripts. Recent studies have shown that VIGS can be used as a new tool to examine both host and nematode genes with potential functions in pathogenicity. Here we will discuss how plant small RNAs and RNA silencing pathways contribute to gene regulation in the nematode feeding sites and how RNAi mechanism can be utilized to identify genetic determinants of parasitism both from nematodes and host plants.

2. GLOBAL GENE DOWNREGULATION IN THE NEMATODE FEEDING SITES

Genome-wide expression profiling studies provided evidence that gene downregulation might be essential for proper feeding site formation. Microarray analysis of the compatible interaction between *Arabidopsis* and *Meloidogyne incognita* using giant cell-enriched root tissues revealed that more than half of the differentially expressed genes at various time points post infection were downregulated (Jammes et al., 2005). Similarly, whole genome microarrays were used to quantify plant gene changes of the fully developed *M. incognita*-induced galls in *Arabidopsis* (Fuller, Lilley, Atkinson, & Urwin, 2007). A set of 959 genes were identified as differentially expressed, and two-thirds of these genes were downregulated. This robust gene repression does not seem to be characteristic of mature giant cells in which crucial gene expression reprogramming events that lead to giant-cell differentiation are likely to have already occurred. A more recent microarray analysis of laser microdissected giant cells induced by *Meloidogyne javanica* at a very early stage of parasitism (i.e. 3 days after infection) revealed that more than 70% of the differentially expressed genes are downregulated (Barcala et al., 2010), indicating that sustained suppression is associated with giant-cell initiation and maturation. Interestingly, this remarkable gene repression during the early stage of giant-cell development was found to be conserved in tomato and *Arabidopsis*. The transcriptomes of tomato galls and laser capture microdissected giant cells were compared with those of *Arabidopsis*. Notably, 76.5% of the identified tomato–*Arabidopsis* homologues were found to be corepressed in both species, whereas only less than 1% were coinduced in the developing giant cells at 3 dpi (Portillo et al., 2013). It is well-known that successful establishment of cyst and root-knot nematode parasitism of host plants is associated with dynamic suppression of plant defence mechanisms. However, gene inactivation in the developing and mature giant cells seems to be a general trend of the transcription machinery of these cells and not restricted to genes related to defence response. In fact, almost all of the functional categories of the differentially expressed genes identified in these studies showed a higher number of downregulated genes than upregulated ones. Unlike other biotrophic interactions, it seems likely that repression of specific genes related to defence response and secondary metabolism, and several transcription factor gene families, as well as many genes of unknown function, are presumably the key processes associated with the differentiation of infected root cells into unique giant cells.

Similar to the extensive gene downregulation associated with giant-cell formation, the early stages of syncytium initiation and formation are associated with significant gene downregulation. More than 70% of the differentially expressed genes identified in 3 dpi microdissected incompatible syncytial cells induced by the soyabean cyst nematode, *Heterodera glycines*, were found to be suppressed (Klink, Overall, Alkharouf, MacDonald, & Matthews, 2007). In contrast, at later stages of infection (i.e. 8 dpi), when syncytium cells are formed and are undergoing maintenance, the majority of the syncytial differentially expressed genes were upregulated (Klink et al., 2007). This suggests that silencing of active transcriptional programmes may be required for syncytium initiation. This suggestion is further supported by comparing gene expression profiles of developing soyabean syncytia between compatible and incompatible interactions (Kandoth et al., 2011; Klink, Hosseini, Matsye, Alkharouf, & Matthews, 2009). In these comparisons, gene upregulation was found to be associated with the resistant interactions, while gene downregulation was associated with the susceptible interactions. While the molecular mechanisms underlying this dynamic gene expression change are not fully understood, recent studies shed light onto key roles of small RNA pathways in controlling gene expression of nematode-infected host cells both at transcriptional and posttranscriptional levels as discussed below.

3. INFLUENCE OF VARIOUS COMPONENTS OF SMALL RNA PATHWAYS ON NEMATODE PARASITISM

Small RNA-mediated gene silencing is regulated by various pathways and the main core of these pathways include dicer-like proteins (DCL), RNA-dependent RNA polymerases (RDR), Argonaute enzymes and RNA polymerase IV and V. Despite varying levels of functional specificities and redundancy among the pathways, the RNA silencing pathways generally involve generation of double-stranded RNA (dsRNA), cleavage of dsRNAs into small RNAs and finally targeting homologous sequences for silencing through the RISC (Brodersen & Voinnet, 2006; Tomari & Zamore, 2005). Small RNAs are involved in a wide range of biological processes that are essential for genome integrity, development and adaptive responses to biotic and abiotic stresses (Vaucheret, 2006). The modes of action of small RNAs are numerous and involve targeting mRNA for degradation and/or translational repression. Also, small RNAs are involved in DNA and histone methylation, and controlling the mobility and proliferation of transposable elements and exogenous nucleic acid. Small RNAs can be classified into two major classes; small interfering RNAs (siRNAs) and micro RNAs (miRNAs). siRNAs are generated from long dsRNAs resulting from antisense transcription or from the activity of RDR. In contrast, miRNAs are genetically defined at specific genomic loci and generated from single-stranded transcripts that form imperfectly paired stem-loop precursors.

The first evidence for the implication of small RNAs in regulating host gene expression in response to nematode infection was provided through the analysis of *dcl* and *rdr* mutants that are involved in the biogenesis and production of small RNAs. The *dcl* and *rdr* mutants exhibited reduced susceptibility to the beet cyst nematode *Heterodera schachtii* compared to the wild type (Hewezi, Howe, Maier, & Baum, 2008). These results suggested that small RNA accumulation is associated with gene regulation processes that are required for plant–nematode interactions. In accordance with the potential implication of siRNAs in regulating gene expression at the posttranscriptional level, siRNAs isolated from *Arabidopsis* roots after infection with *H. schachtii* showed perfect complementarity to protein-coding genes. In addition, a significant portion of the identified *Arabidopsis* siRNAs were mapped to transposons or retrotransposons of different types, including hAT-like, MuDR, LINE, LTR/Copia, LTR/Gypsy, RC/Helitron and Satellite (Hewezi, Howe, Maier, & Baum, 2008) and could have a role in controlling the movement and proliferation of transposable elements during infection. The potential regulatory function of these siRNAs during parasitism was further reinforced by qPCR analysis showing that many of these siRNAs are differentially expressed in response to *H. schachtii* infection.

The regulatory functions of siRNAs could be numerous as siRNAs can target imperfectly matched regions. Recently, a role of siRNAs in silencing specific genomic regions and the overlapping genes through DNA methylation has been reported (Dowen et al., 2012; Schmitz et al., 2013; Yu et al., 2013). It is well-known that de novo formation of DNA methylation in CG, CHG and CHH contexts where H refers to any nucleotide but G, is often established through a small RNA-directed DNA methylation pathway. This pathway involves the biosynthesis of 24-nt siRNAs that provide sequence-specific guides to silence genomic loci (Law & Jacobsen, 2010). Analysis of small RNA libraries prepared from *Arabidopsis* and soyabean roots infected with *H. schachtii* and *H. glycines*, respectively (Hewezi, Howe, Maier, & Baum, 2008; Li et al., 2012), revealed that 24-nt reads are the most abundant class relative to other small RNA populations.

Therefore, it is tempting to speculate that the reduced susceptibility phenotypes of *dcl* and *rdr* mutants, which are partially impaired in non-CG methylation, could be due to hypomethylation of specific genomic loci overlapping with genes that may be involved in defence response. Recently, Dowen et al. (2012) provided clear evidence of dynamic changes in DNA methylation in response to infection by the bacterial pathogen *Pseudomonas syringae* pv. *tomato* DC3000 (*Pst*). Using deep sequencing of bisulphite-treated DNA, they found that differentially methylated regions (DMRs) are preferentially associated with genes involved in defence response, and that hypomethylation in DMRs is frequently accompanied by activation of the proximal genes, specifically those with defence response function. Similarly, another recent study indicated that DNA demethylation restricts the multiplication and vascular propagation of *Pst* and, consequently, some immune response genes are repressed by DNA methylation (Yu et al., 2013). In soyabean, differential hypermethylation patterns have been identified in CG, CHG and CHH contexts at the genomic regions that contain multiple copies of soyabean cyst nematode (SCN) resistance gene *Rhg1* (Cook et al., 2014). In addition, mutant lines entirely defective in maintenance of CG methylation or non-CG methylation showed a notable level of resistance to *Pst* infection. Likewise, mutants partially impaired in CG methylation (decrease in DNA methylation1, *ddm1*) or non-CG methylation (*rdr1*, *rdr2* and *rdr6* and *dcl2/3/4* triple mutant) showed modest increases in *Pst* resistance (Dowen et al., 2012). Taken together, the finding that 24-nt small RNA reads are the most prevailing among other small RNA populations in nematode-infected root samples and that mutant lines partially compromised in DNA methylation are less susceptible to nematode infection suggest that nematodes exploit small RNA pathways to manipulate host gene expression not only at the posttranscriptional level but also at the transcriptional level to mediate susceptibility.

4. KEY REGULATORY ROLES OF miRNAs IN FEEDING SITE INITIATION AND FORMATION

In addition to their roles in developmental patterning, miRNAs are also essential components of plant responses to various biotic stress factors (Ruiz-Ferrer & Voinnet, 2009). Interestingly, several miRNAs showed differential expression patterns in response to *H. schachtii* (Hewezi, Howe, Maier, & Baum, 2008) or *H. glycines* infection (Li et al., 2012). In response to *H. schachtii* infection, 19 miRNA families, including miR156, miR157, miR158, miR159, miR160, miR161, miR164, miR165, miR166, miR167, miR168, miR169, miR171, miR172, miR390, miR396, miR398, miR403 and miR775, changed abundance following infection. One remarkable finding in this study is that at the early stage of infection (4 dpi), the majority of these miRNAs were significantly downregulated in inoculated roots compared with control roots. Similarly, in response to *H. glycines* infection, 43 soyabean miRNAs were identified as differentially expressed (Li et al., 2012). Out of these 43 miRNAs, 38 were downregulated 30 days after the emergence of soyabean seedlings. In a different pathosystem, several loblolly pine miRNAs were dramatically repressed in the stem galls induced by the rust fungus *Cronartium quercuum* f. sp. *fusiforme* (Lu, Sun, Amerson, & Chiang, 2007).

This downregulation could be the result of impairment of miRNA biogenesis as part of a common plant defence response against nematode invasion. It is well-known that the first phase of plant responses to pathogen infection during susceptible and resistant interactions is very similar and includes activation of general defence response pathways. In contrast, the second phase of responses is more specific and correlates well with the disease development. In this context, many of the nematode-responsive miRNAs showed upregulation at later stages of infection (Hewezi & Baum, 2012; Hewezi, Howe, Maier, & Baum, 2008). It may be important to mention that the expression levels of these miRNAs were quantified using whole root systems and that the expression levels of these miRNAs specifically in the nematode-induced feeding sites could be quite different.

The phytohormone auxin, as a key signalling molecule, plays key roles in the differentiation of infected root cells into the syncytium cell type. In addition, a role of auxin in the reprogramming of root cells adjacent to the developing syncytium prior to their incorporation within the developing syncytium has also been suggested (Goverse et al., 2000; Grunewald, Cannoot, Friml, & Gheysen, 2009; Hewezi, Piya, Richard, & Rice, 2014; Karczmarek, Overmars, Helder, & Goverse, 2004). Several miRNAs contribute to the regulation of auxin signal transduction by targeting different auxin response factors (ARFs) and NAC family transcription factors. ARF10, 16 and 17 are targeted by miR160 (Mallory, Bartel, & Bartel, 2005; Wang et al., 2005), while ARF6 and 8 are targeted by miR167 (Wu, Tian, & Reed, 2006). Both miR160 and miR167 were found to be strongly downregulated during the early stages of infection, while the five ARF targets showed a significant upregulation (Hewezi, Howe, Maier, & Baum, 2008). Very recently, using promoter activity assays, these ARFs were found to be expressed in the developing syncytium (ARF6, 8 and 10) and neighbouring cells (ARF16 and 17) (Hewezi et al., 2014). In addition, the expression level of miRNA164, which targets several NAC family transcription factors, was also downregulated in response to nematode infection (Hewezi, Howe, Maier, & Baum, 2008). Suppressing the expression of these miRNAs may be the mechanistic basis that enables high expression levels of ARFs in the nematode-induced feeding sites specifically during the initiation and formation stages.

Genes coding for transcription factors are abundant among the targets of miRNAs that were reported to change the expression in response to nematode infection. For example, miRNAs targeting growth regulating factors (GRFs) (miR396), Scarecrow-like transcription factor (miR171), MYB (miR159), HD-ZIP (miR166), AP2 (miR172), TCP (miR319) and SPL (miR159) in *Arabidopsis* or soyabean are regulated by nematode infection (Hewezi, Howe, Maier, & Baum, 2008; Hewezi, Maier, Nettleton, & Baum, 2012; Li et al., 2012). These transcription factors may be involved in the initiation of transcriptional cascades, thereby extending the effects of miRNAs to downstream signalling components. miR396 and its GRF target genes fit perfectly in this model. miR396 and its target genes *GRF1* and *GRF3* showed opposite expression patterns in the syncytium during formation and maintenance stages (Hewezi et al., 2012). miR396 was strongly downregulated during syncytium initiation/formation phases, and this downregulation was accompanied by increased *GRF1* and *GRF3* mRNA abundance. After the phase of syncytium formation was completed, miR396 expression increased to high levels in the syncytium, which post-transcriptionally silenced the expression of *GRF1* and *GRF3*. Thus, the expression changes of miR396 and *GRF1/3* define the syncytium induction/formation phase versus the maintenance phase. Suppressing the activity of GRFs through overexpression of miR396 or miR396-resistant versions of *GRF1/GRF3*, resulted in smaller syncytia and increased resistance to cyst nematode attack (Hewezi et al., 2012). These data clearly indicate that coordinated regulation of GRF1/3 and their negative regulator miR396 is required for proper syncytium development and that maintaining the homeostasis of miR396 and its two target genes at specific threshold levels seems to be critical for parasitic success. The homeostasis between miR396 and GRF1/GRF3 is established through a reciprocal feedback regulation, in which GRF1/GRF3 and miR396 negatively regulate each other's expression (Hewezi & Baum, 2012). Moreover, genome-wide expression profiling revealed that the miR396-GRF1/3 regulatory system can directly or indirectly modulate the expression of about half of the 7000 genes reported to change expression in the *Arabidopsis* syncytium by Szakasits et al. (2009) (Hewezi et al., 2012). Identification of potential targets of GRF1 and GRF3 suggested that both transcription factors contribute to the regulation of numerous biological processes associated with defence response and disease resistance (Liu, Rice, Chen, Baum, & Hewezi, 2014). More specifically, GRF1 and GRF3 contribute to the regulation of defence-related transcription factors, cytokinin signalling, cell wall modifications and secondary metabolite accumulation. Also, GRF1 and GRF3 appear to function as molecular links between miRNA signalling pathways by regulating the expression of several other miRNA target genes. The pleiotropic functions of miR396/GRF1/GRF3 regulatory system represent an attractive molecular target for the parasitic nematodes to modulate so many aspects of cell development and differentiation processes at once to force infected root cells into a syncytial cell fate.

5. SUPPRESSION OF NEMATODE GENES USING HOST-INDUCED GENE SILENCING

With the advent of RNAi in *Caenorhabditis elegans* (Fire et al., 1998) as a powerful gene silencing tool for reverse genetic analysis, this approach has been widely applied in a range of gene functional studies in a variety of organisms (Hannon, 2002; Tomari & Zamore, 2005). In 2006, Huang et al. described a novel RNAi-based approach to silence root-knot nematode effector genes. This approach relies on the production of stable transgenic plants expressing a dsRNA corresponding to targeted nematode genes. The dsRNA is recognized by the RNA silencing machinery and is cleaved into 21-nt siRNA molecules. siRNAs accumulate at high level in the feeding sites and are ingested as nematodes feed. In the nematode, the siRNAs presumably traverse membranes and are incorporated into the RISC and mediate sequence-specific mRNA degradation of the targeted nematode genes. This action results in depletion of the targeted transcripts and its translated protein, causing loss-of-function phenotypes. Overexpression of dsRNA of the root-knot nematode effector gene 16D10 in *Arabidopsis* resulted in resistance effective against the four major root-knot nematode species (Dinh, Brown, & Elling, 2014; Huang, Allen, Davis, Baum, & Hussey, 2006; Yang et al., 2013). Similarly, transgenic tobacco plants expressing dsRNAs targeting a splicing factor and an integrase gene in the root-knot nematode *M. incognita*-induced downregulation of the target genes in the nematode and subsequently a significant reduction in parasitic success (Yadav et al., 2006). Also, a consistent gene downregulation of the *M. javanica* transcription factor *Tis11* (*MjTis11*) was detected in nematodes feeding on transgenic tobacco plants overexpressing *MjTis11* dsRNA (Fairbairn et al., 2007). While this gene downregulation was associated with accumulation of siRNAs, no clear impact on nematode development was observed. One explanation of this result is that the embryo lethal phenotype of the homologous MjTis11 sequence observed in *C. elegans* could not be conserved in *M. javanica.* Otherwise, the levels of *MjTis11* downregulation may not have been high enough to compromise this gene's biological function. Recently, it has been shown that downregulation of *M. incognita* proteases by plant-delivered dsRNA negatively impacted nematode development as well as reproduction, egg viability and progeny virulence (Antonino de Souza Junior et al., 2013). Taken together, these studies demonstrate the feasibility of posttranscriptional silencing of endogenous root-knot nematode transcripts by expressing dsRNAs corresponding to the nematode genes in plant roots.

Similarly, several studies demonstrated the feasibility of this approach and its success in decreasing the abilities of plant-parasitic cyst nematodes to colonize host plants. For example, the major sperm protein (MSP) of *H. glycines* was selected as a target for in planta delivery of dsRNA (Steeves, Todd, Essig, & Trick, 2006). Expression analysis of transgenic soyabean plants demonstrated the processing of MSP dsRNA transcripts into siRNA molecules and subsequently a significant reduction in the reproductive potential of *H. glycines.* An RNAi construct targeted to silence the *H. glycines* gene encoding fructose-1, 6-diphosphate aldolase (HgALD) was transformed into soyabean roots of composite plants (Youssef, Kim, Haroon, & Matthews, 2013). The number of mature females on the transgenic hairy roots was reduced by more than 50%, suggesting a key role of HgALD in SCN metabolism.

Cyst and root-knot nematode effector genes have been shown to play key roles in mediating plant susceptibility (Hewezi & Baum, 2013), and thus they represent exciting targets for gene silencing. In this context, transgenic *Arabidopsis* plants generated to induce silencing of four *H. schachtii* effector genes caused mRNA reduction of the target genes in feeding nematodes and showed a significant reduction in parasitic success as evident by lower numbers of nematode females developing on the transgenic plants compared to those developing on wild-type plants (Sindhu et al., 2009). These results were further supported by the data showing that overexpression of two of these effector genes showed a dramatic increase in nematode susceptibility (Hewezi et al., 2010; Hewezi, Howe, Maier, Hussey et al., 2008). Together, these studies demonstrated the biological significance of targeting nematode effectors to interfere with nematode development and parasitic ability. Other studies provided additional evidence for the effectiveness of targeting nematode effector genes through host-induced RNAi approach. Transgenic *Arabidopsis* plants expressing dsRNA complementary to the *H. schachtii* annexin-like effector gene 4F01 showed a significant decrease in parasitic nematode 4F01 transcript levels and a dramatic reduction in nematode parasitism (Patel et al., 2010). In addition, *Arabidopsis*-derived RNAi targeted against the 30C02 effector gene from *H. glycines* strongly reduced nematode infection (Hamamouch et al., 2012). In that study, the production of siRNAs corresponding to the 30C02 transcripts was assayed in small RNA populations isolated from root tissues of the 30C02 RNAi plants. High numbers of siRNA reads generated from both strands were aligned to the 30C02 sequence with the 21-nt reads being the most abundant siRNA class (Hamamouch et al., 2012). The abundance of siRNA concentration was relatively uniform across the entire 30C02 sequence, indicating that siRNA generation was not associated with specific region.

Similarly, targeting root-knot nematode effector genes using host-induced gene silencing proved to be an efficient approach to interrupt nematode parasitism of host plants. Transgenic *Arabidopsis* plants expressing an RNAi construct of the *M. incognita* 8D05 effector gene showed up to 90% reduction in gall formation, indicating that the Mi-8D05 effector plays a critical role in the parasitic success of *M. incognita* (Xue et al., 2013). This critical functional role could be related to the regulation of water and solute transport in giant cells as 8D05 has been shown to specifically interact with a plant aquaporin (Xue et al., 2013). The *M. incognita* calreticulin (Mi-CRT) effector gene was similarly knocked down in feeding nematodes by expressing Mi-CRT dsRNA in transgenic *Arabidopsis* plants. About 75% depletion in the Mi-CRT transcript levels was observed and was found to be associated with reduced ability of the nematode to induce gall formation (Jaouannet et al., 2013). In contrast, stably transformed *Arabidopsis* plants overexpressing Mi-CRT were more susceptible to *M. incognita* infection. While the mode of action of Mi-CRT remains unknown, this effector could be involved in suppression of basal immune responses through yet unidentified mechanisms. Collectively, these studies demonstrate the potential in using host-induced gene silencing to control nematode infection. In this context, it may be important to mention that various studies produced variable results in terms of efficiency and consistency. Inconsistent efficiency could be due to low silencing efficiency associated with the construct size, integration site and number of transformation events. However, complete suppression of nematode reproduction as the ultimate goal could be achieved by increasing the level of silencing and by identifying nematode effector genes whose silencing results in a lethal phenotype. Currently, the number of nematode genes having dramatic impacts on nematode development and virulence remains limited and more screens are needed to identify these genes. Another possible strategy to attain complete suppression of nematode reproduction is to simultaneously target multiple nematode effectors that function in a sequential fashion during various stages of infection to maximize the efficiency and durability of silencing-associated phenotypes.

6. VIGS AS A TOOL FOR FUNCTIONAL GENOMICS OF PLANT–NEMATODE INTERACTIONS

Another approach of silencing nematode or host endogenous genes with small RNAs in the nematode feeding site is VIGS. This technology uses the virus as a tool to introduce small nucleotide sequences of the target gene as part of the virus genome into the plant to suppress the expression of this particular gene. Appropriate RNA viruses propagate systemically in the plant and generate dsRNA molecules as they replicate. Virus replication thus triggers plant defence mechanisms that detect and cleave dsRNA into siRNAs. These siRNAs also contain molecules homologous to the target gene, which then integrate into the RISC, bind to and degrade homologous RNA sequences, resulting in loss-of-function of targeted gene (Burch-Smith, Anderson, Martin, & Dinesh-Kumar, 2004; Purkayastha & Dasgupta, 2009). One of the main criteria of using VIGS in silencing nematode genes is the ability of the virus to replicate efficiently in the root system. Tobacco rattle virus (TRV) containing the *2b* gene has been shown to replicate within root tissues without inducing harmful symptoms (Valentine et al., 2004). Thus, a TRV-based vector was chosen to express dsRNA of the *H. schachtii* glyceraldehyde-3-phosphate dehydrogenase (GAPDH) sequence in *Arabidopsis* (Valentine et al., 2007). A reduction in the level of *GAPDH* mRNA in the nematodes feeding on *Arabidopsis* plants infected with the TRV was observed and found to be associated with a decrease in nematode size across the entire female population (Valentine et al., 2007).

However, the actual level of RNA downregulation observed in this system was marginal and the level of RNAi induced in individual nematodes as well as the proportion of nematodes subjected to the virus-encoded dsRNA was highly variable. In another study, TRV was used to target root-knot nematode genes in *Nicotiana benthamiana*, which is a good host for *M. incognita* and is more susceptible to virus infection compared to *Arabidopsis* (Dubreuil et al., 2009). TRV-inoculated plants were able to produce dsRNA and accumulate siRNAs corresponding to the nematode genes in the feeding cells. While downregulation of the targeted genes was observed in the progeny of the feeding nematodes, no clear effect on nematode development was detected. In addition, a high variability in gene silencing efficiency was observed among progeny collected from individual plants (Dubreuil et al., 2009). Variability in silencing efficiency could be due to the limited ability of TRV to propagate in root tissues and hence, to lead to siRNA accumulation. In addition, several factors including insert size, location of the target region and cloning orientation can also affect the silencing potential of VIGS vectors (Bennypaul et al., 2012; Liu & Page, 2008; Yuan et al., 2011; Zhang, Bradshaw, Whitham, & Hill, 2010) and these factors need to be optimized before such application. Recently, researchers used a bean pod mottle virus (BPMV)-based VIGS system in soyabean and accomplished uniform and steady gene silencing in various tissues, including roots for several weeks (Juvale et al., 2012). BPMV-based VIGS was used to silence a soyabean serine hydroxymethyltransferase (SHMT), the recently cloned SCN resistance gene *Rhg4* (Liu et al., 2012). Silencing of the SHMT gene in SCN-resistant plants by inoculation with BPMV-SHMT led to a significant increase in susceptibility to SCN relative to the BPMV empty vector (Liu et al., 2012). At the time of nematode inoculation, *SHMT* transcript levels were found to be decreased by more than 70% in the roots of plants inoculated with BPMV-SHMT compared with those inoculated with empty BPMV (Liu et al., 2012), confirming that the substantial increase in nematode susceptibility was the result of a significant decrease in the *SHMT* expression level. Also, TRV-mediated gene silencing in tomato was used to knock down the novel *M. javanica* effector gene *Mj-nulg1a* during parasitism (Lin et al., 2013). The expression of *Mj-nulg1a* in the nematodes was significantly downregulated and this downregulation diminished the parasitic ability of *M. javanica* as indicated by the low numbers of nematodes, galls and eggs observed on the silenced roots compared with the controls. Consistent with these findings, constitutive expression of the *Mj-nulg1a* gene in *Arabidopsis* produced plants that were more susceptible to *M. javanica* infection than the wild-type control, further supporting a key role of this effector in infection processes. Because *nulg1a* seems to be conserved in other root-knot nematode species, silencing this gene could provide broad-spectrum resistance to root-knot nematodes. The development of various viruses-based VIGS holds great promise to fully implement the VIGS system as a powerful reverse genetics tool suitable for high-throughput functional screening of a large number of genes, specifically those with lethal phenotypes.

7. CONCLUSIONS AND PERSPECTIVES

Clearly, there are massive gene expression changes manifested in nematode feeding sites, many of which actually represent downregulation events. Furthermore, it is increasingly clear also that small RNA molecules play critical roles in orchestrating these changes.

When considering the importance of small RNAs in the realm of plant–nematode interactions, it quickly becomes obvious that we have only scratched the surface so far. Knowledge of regulatory networks governing gene expression on the transcriptional and posttranscriptional levels will continue to increase in complexity and novel regulatory mechanisms and molecules will continue to be discovered. Studying the mechanisms and functions of individual RNA molecules continues to be of utmost importance, but to really understand a biological system like the formation and maintenance of nematode feeding sites such individual research projects need to be complemented by systems approaches that can harness and dissect the complexity researchers are dealing with.

In addition to a systems approach that is urgently needed, the importance of studying individual mechanistic facets of the complex system remains high. Particularly, deciphering the regulatory powers of individual miRNAs and siRNAs in the feeding sites is of utmost importance. Exploring how nematodes trigger these mechanisms in the first place, i.e. which effectors contribute which function to the overall development; remain as key discoveries that need to be made.

The more we learn, the more we have to marvel at the complexity that the coevolution of plant and sedentary nematode has produced. And the realization that the parasitic nematode has the tools and power to orchestrate these changes in a manner that leads to a stable and functioning status of the biological system created by the nematode is astonishing.

ACKNOWLEDGEMENTS

Research portrayed here was supported by grants from the soyabean check-off systems, USDA-NRI/AFRI, and NSF (Award #: 1145053), to mention a few. The authors also are supported by startup funds from the University of Tennessee, Institute of Agriculture as well as funds from the Iowa Agriculture Experiment Station, Hatch Act and the State of Iowa. The authors would like to thank their many colleagues for stimulating discussions and scientific exchange.

REFERENCES

Antonino de Souza Junior, J. D., Ramos Coelho, R., Tristan Lourenco, I., da Rocha Fragoso, R., Barbosa Viana, A. A., Lima Pepino de Macedo, L., et al. (2013). Knocking-down Meloidogyne incognita proteases by plant-delivered dsRNA has negative pleiotropic effect on nematode vigor. *PLoS One, 8*, e85364.

Axtell, M. J., Jan, C., Rajagopalan, R., & Bartel, D. P. (2006). A two-hit trigger for siRNA biogenesis in plants. *Cell, 127*, 565–577.

Barcala, M., Garcia, A., Cabrera, J., Casson, S., Lindsey, K., Favery, B., et al. (2010). Early transcriptomic events in microdissected *Arabidopsis* nematode-induced giant cells. *Plant Journal, 61*, 698–712.

Bennypaul, H. S., Mutti, J. S., Rustgi, S., Kumar, N., Okubara, P. A., & Gill, K. S. (2012). Virus-induced gene silencing (VIGS) of genes expressed in root, leaf, and meiotic tissues of wheat. *Functional and Integrative Genomics, 12*, 143–156.

Brodersen, P., & Voinnet, O. (2006). The diversity of RNA silencing pathways in plants. *Trends in Genetics, 22*, 268–280.

Burch-Smith, T. M., Anderson, J. C., Martin, G. B., & Dinesh-Kumar, S. P. (2004). Applications and advantages of virus-induced gene silencing for gene function studies in plants. *Plant Journal, 39*, 734–746.

Chen, X. (2009). Small RNAs and their roles in plant development. *Annual Review of Cell and Developmental Biology, 25*, 21–44.

Cook, D., Bayless, A., Wang, K., Guo, X., Song, Q., Jiang, J., et al. (2014). Distinct copy number, coding sequence and locus methylation patterns underlie *Rhg1*-mediated soybean resistance to soybean cyst nematode. *Plant Physiology, 165*, 630–647.

Dinh, P. T., Brown, C. R., & Elling, A. A. (2014). RNA interference of effector gene *Mc16D10L* confers resistance against *Meloidogyne chitwoodi* in *Arabidopsis* and potato. *Phytopathology, 104*, 1098–1106.

Dowen, R. H., Pelizzola, M., Schmitz, R. J., Lister, R., Dowen, J. M., Nery, J. R., et al. (2012). Widespread dynamic DNA methylation in response to biotic stress. *Proceedings of the National Academy of Sciences of the United States of America, 109*, E2183–E2191.

Dubreuil, G., Magliano, M., Dubrana, M. P., Lozano, J., Lecomte, P., Favery, B., et al. (2009). Tobacco rattle virus mediates gene silencing in a plant parasitic root-knot nematode. *Journal of Experimental Botany, 60*, 4041–4050.

Fahlgren, N., Howell, M. D., Kasschau, K. D., Chapman, E. J., Sullivan, C. M., Cumbie, J. S., et al. (2007). High-throughput sequencing of *Arabidopsis* microRNAs: evidence for frequent birth and death of MIRNA genes. *PLoS One, 2*, e219.

Fairbairn, D. J., Cavallaro, A. S., Bernard, M., Mahalinga-Iyer, J., Graham, M. W., & Botella, J. R. (2007). Host-delivered RNAi: an effective strategy to silence genes in plant parasitic nematodes. *Planta, 226*, 1525–1533.

Fire, A., Xu, S., Montgomery, M. K., Kostas, S. A., Driver, S. E., & Mello, C. C. (1998). Potent and specific genetic interference by double-stranded RNA in *Caenorhabditis elegans*. *Nature, 391*, 806–811.

Fuller, V. L., Lilley, C. J., Atkinson, H. J., & Urwin, P. E. (2007). Differential gene expression in *Arabidopsis* following infection by plant-parasitic nematodes *Meloidogyne incognita* and *Heterodera schachtii*. *Molecular Plant Pathology, 8*, 595–609.

Goverse, A., Overmars, H., Engelbertink, J., Schots, A., Bakker, J., & Helder, J. (2000). Both induction and morphogenesis of cyst nematode feeding cells are mediated by auxin. *Molecular Plant–Microbe Interactions, 13*, 1121–1129.

Grunewald, W., Cannoot, B., Friml, J., & Gheysen, G. (2009). Parasitic nematodes modulate PIN-mediated auxin transport to facilitate infection. *PLoS Pathogens, 5*, e1000266.

Hamamouch, N., Li, C., Hewezi, T., Baum, T. J., Mitchum, M. G., Hussey, R. S., et al. (2012). The interaction of the novel 30C02 cyst nematode effector protein with a plant beta-1,3-endoglucanase may suppress host defence to promote parasitism. *Journal of Experimental Botany, 63*, 3683–3695.

Hamilton, A. J., & Baulcombe, D. C. (1999). A species of small antisense RNA in posttranscriptional gene silencing in plants. *Science, 286*, 950–952.

Hannon, G. J. (2002). RNA interference. *Nature, 418*, 244–251.

Hewezi, T., & Baum, T. J. (2012). Complex feedback regulations govern the expression of miRNA396 and its GRF target genes. *Plant Signaling and Behavior, 7*, 749–751.

Hewezi, T., & Baum, T. J. (2013). Manipulation of plant cells by cyst and root-knot nematode effectors. *Molecular Plant–Microbe Interactions, 26*, 9–16.

Hewezi, T., Howe, P., Maier, T. R., & Baum, T. J. (2008). *Arabidopsis* small RNAs and their targets during cyst nematode parasitism. *Molecular Plant–Microbe Interactions, 21*, 1622–1634.

Hewezi, T., Howe, P., Maier, T. R., Hussey, R. S., Mitchum, M. G., Davis, E. L., et al. (2008). Cellulose binding protein from the parasitic nematode Heterodera schachtii interacts with *Arabidopsis* pectin methylesterase: cooperative cell wall modification during parasitism. *Plant Cell, 20*, 3080–3093.

Hewezi, T., Howe, P. J., Maier, T. R., Hussey, R. S., Mitchum, M. G., Davis, E. L., et al. (2010). *Arabidopsis* spermidine synthase is targeted by an effector protein of the cyst nematode *Heterodera schachtii*. *Plant Physiology, 152*, 968–984.

Hewezi, T., Maier, T. R., Nettleton, D., & Baum, T. J. (2012). The *Arabidopsis* microRNA396-GRF1/GRF3 regulatory module acts as a developmental regulator in the reprogramming of root cells during cyst nematode infection. *Plant Physiology, 159*, 321–335.

Hewezi, T., Piya, S., Richard, G., & Rice, J. H. (2014). Spatial and temporal expression patterns of auxin response transcription factors in the syncytium induced by the beet cyst nematode *Heterodera schachtii* in *Arabidopsis*. *Molecular Plant Pathology, 15*, 730–736.

Huang, G., Allen, R., Davis, E. L., Baum, T. J., & Hussey, R. S. (2006). Engineering broad root-knot resistance in transgenic plants by RNAi silencing of a conserved and essential root-knot nematode parasitism gene. *Proceedings of the National Academy of Sciences of the United States of America, 103*, 14302–14306.

Jammes, F., Lecomte, P., de Almeida-Engler, J., Bitton, F., Martin-Magniette, M. L., Renou, J. P., et al. (2005). Genome-wide expression profiling of the host response to root-knot nematode infection in *Arabidopsis*. *Plant Journal, 44*, 447–458.

Jaouannet, M., Magliano, M., Arguel, M. J., Gourgues, M., Evangelisti, E., Abad, P., et al. (2013). The root-knot nematode calreticulin Mi-CRT is a key effector in plant defense suppression. *Molecular Plant–Microbe Interactions, 26*, 97–105.

Juvale, P. S., Hewezi, T., Zhang, C., Kandoth, P. K., Mitchum, M. G., Hill, J. H., et al. (2012). Temporal and spatial bean pod mottle virus-induced gene silencing in soybean. *Molecular Plant Pathology, 13*, 1140–1148.

Kandoth, P. K., Ithal, N., Recknor, J., Maier, T., Nettleton, D., Baum, T. J., et al. (2011). The soybean *Rhg1* locus for resistance to the soybean cyst nematode *Heterodera glycines* regulates the expression of a large number of stress- and defense-related genes in degenerating feeding cells. *Plant Physiology, 155*, 1960–1975.

Karczmarek, A., Overmars, H., Helder, J., & Goverse, A. (2004). Feeding cell development by cyst and root-knot nematodes involves a similar early, local and transient activation of a specific auxin-inducible promoter element. *Molecular Plant Pathology, 5*, 343–346.

Kasschau, K. D., Fahlgren, N., Chapman, E. J., Sullivan, C. M., Cumbie, J. S., Givan, S. A., et al. (2007). Genome-wide profiling and analysis of *Arabidopsis* siRNAs. *PLoS Biology, 5*, e57.

Katiyar-Agarwal, S., & Jin, H. (2010). Role of small RNAs in host-microbe interactions. *Annual Review of Phytopathology, 48*, 225–246.

Klink, V. P., Hosseini, P., Matsye, P., Alkharouf, N. W., & Matthews, B. F. (2009). A gene expression analysis of syncytia laser microdissected from the roots of the *Glycine max* (soybean) genotype PI 548402 (Peking) undergoing a resistant reaction after infection by *Heterodera glycines* (soybean cyst nematode). *Plant Molecular Biology, 71*, 525–567.

Klink, V. P., Overall, C. C., Alkharouf, N. W., MacDonald, M. H., & Matthews, B. F. (2007). Laser capture microdissection (LCM) and comparative microarray expression analysis of syncytial cells isolated from incompatible and compatible soybean (*Glycine max*) roots infected by the soybean cyst nematode (*Heterodera glycines*). *Planta, 226*, 1389–1409.

Law, J. A., & Jacobsen, S. E. (2010). Establishing, maintaining and modifying DNA methylation patterns in plants and animals. *Nature Reviews Genetics, 11*, 204–220.

Lin, B., Zhuo, K., Wu, P., Cui, R., Zhang, L. H., & Liao, J. (2013). A novel effector protein, MJ-NULG1a, targeted to giant cell nuclei plays a role in *Meloidogyne javanica* parasitism. *Molecular Plant–Microbe Interactions, 26*, 55–66.

Liu, S., Kandoth, P. K., Warren, S. D., Yeckel, G., Heinz, R., Alden, J., et al. (2012). A soybean cyst nematode resistance gene points to a new mechanism of plant resistance to pathogens. *Nature, 492*, 256–260.

Liu, E., & Page, J. E. (2008). Optimized cDNA libraries for virus-induced gene silencing (VIGS) using tobacco rattle virus. *Plant Methods, 4*, 5.

Liu, J., Rice, J. H., Chen, N., Baum, T. J., & Hewezi, T. (2014). Synchronization of developmental processes and defense signaling by growth regulating transcription factors. *PLoS One, 9*, e98477.

Li, X., Wang, X., Zhang, S., Liu, D., Duan, Y., & Dong, W. (2012). Identification of soybean microRNAs involved in soybean cyst nematode infection by deep sequencing. *PLoS One, 7*, e39650.

Lu, S., Sun, Y. H., Amerson, H., & Chiang, V. L. (2007). MicroRNAs in loblolly pine (*Pinus taeda* L.) and their association with fusiform rust gall development. *Plant Journal, 51*, 1077–1098.

Lu, C., Tej, S. S., Luo, S., Haudenschild, C. D., Meyers, B. C., & Green, P. J. (2005). Elucidation of the small RNA component of the transcriptome. *Science, 309*, 1567–1569.

Mallory, A. C., Bartel, D. P., & Bartel, B. (2005). MicroRNA-directed regulation of *Arabidopsis* Auxin Response Factor17 is essential for proper development and modulates expression of early auxin response genes. *Plant Cell, 17*, 1360–1375.

Nunes, C. C., & Dean, R. A. (2012). Host-induced gene silencing: a tool for understanding fungal host interaction and for developing novel disease control strategies. *Molecular Plant Pathology, 13*, 519–529.

Patel, N., Hamamouch, N., Li, C., Hewezi, T., Hussey, R. S., Baum, T. J., et al. (2010). A nematode effector protein similar to annexins in host plants. *Journal of Experimental Botany, 61*, 235–248.

Portillo, M., Cabrera, J., Lindsey, K., Topping, J., Andres, M. F., Emiliozzi, M., et al. (2013). Distinct and conserved transcriptomic changes during nematode-induced giant cell development in tomato compared with *Arabidopsis*: a functional role for gene repression. *New Phytologist, 197*, 1276–1290.

Purkayastha, A., & Dasgupta, I. (2009). Virus-induced gene silencing: a versatile tool for discovery of gene functions in plants. *Plant Physiology and Biochemistry, 47*, 967–976.
Rajagopalan, R., Vaucheret, H., Trejo, J., & Bartel, D. P. (2006). A diverse and evolutionarily fluid set of microRNAs in *Arabidopsis thaliana*. *Genes and Development, 20*, 3407–3425.
Rosso, M. N., Jones, J. T., & Abad, P. (2009). RNAi and functional genomics in plant parasitic nematodes. *Annual Review of Phytopathology, 47*, 207–232.
Ruiz-Ferrer, V., & Voinnet, O. (2009). Roles of plant small RNAs in biotic stress responses. *Annual Review of Plant Biology, 60*, 485–510.
Schmitz, R. J., He, Y., Valdes-Lopez, O., Khan, S. M., Joshi, T., Urich, M. A., et al. (2013). Epigenome-wide inheritance of cytosine methylation variants in a recombinant inbred population. *Genome Research, 23*, 1663–1674.
Sindhu, A. S., Maier, T. R., Mitchum, M. G., Hussey, R. S., Davis, E. L., & Baum, T. J. (2009). Effective and specific in planta RNAi in cyst nematodes: expression interference of four parasitism genes reduces parasitic success. *Journal of Experimental Botany, 60*, 315–324.
Steeves, R. M., Todd, T. C., Essig, J. S., & Trick, H. N. (2006). Transgenic soybeans expressing siRNAs specific to a major sperm protein gene suppress *Heterodera glycines* reproduction. *Functional Plant Biology, 33*, 991–999.
Sudarshana, M. R., Roy, G., & Falk, B. W. (2007). Methods for engineering resistance to plant viruses. *Methods in Molecular Biology, 354*, 183–195.
Szakasits, D., Heinen, P., Wieczorek, K., Hofmann, J., Wagner, F., Kreil, D. P., et al. (2009). The transcriptome of syncytia induced by the cyst nematode *Heterodera schachtii* in *Arabidopsis* roots. *Plant Journal, 57*, 771–784.
Tomari, Y., & Zamore, P. D. (2005). Perspective: machines for RNAi. *Genes and Development, 19*, 517–529.
Valentine, T. A., Randall, E., Wypijewski, K., Chapman, S., Jones, J., & Oparka, K. J. (2007). Delivery of macromolecules to plant parasitic nematodes using a tobacco rattle virus vector. *Plant Biotechnology Journal, 5*, 827–834.
Valentine, T., Shaw, J., Blok, V. C., Phillips, M. S., Oparka, K. J., & Lacomme, C. (2004). Efficient virus-induced gene silencing in roots using a modified tobacco rattle virus vector. *Plant Physiology, 136*, 3999–4009.
Vaucheret, H. (2006). Post-transcriptional small RNA pathways in plants: mechanisms and regulations. *Genes and Development, 20*, 759–771.
Wang, J. W., Wang, L. J., Mao, Y. B., Cai, W. J., Xue, H. W., & Chen, X. Y. (2005). Control of root cap formation by MicroRNA-targeted auxin response factors in *Arabidopsis*. *Plant Cell, 17*, 2204–2216.
Waterhouse, P. M., & Fusaro, A. F. (2006). Plant science. Viruses face a double defense by plant small RNAs. *Science, 313*, 54–55.
Wu, M. F., Tian, Q., & Reed, J. W. (2006). *Arabidopsis* microRNA167 controls patterns of ARF6 and ARF8 expression, and regulates both female and male reproduction. *Development, 133*, 4211–4218.
Xue, B., Hamamouch, N., Li, C., Huang, G., Hussey, R. S., Baum, T. J., et al. (2013). The 8D05 parasitism gene of Meloidogyne incognita is required for successful infection of host roots. *Phytopathology, 103*, 175–181.
Yadav, B. C., Veluthambi, K., & Subramaniam, K. (2006). Host-generated double stranded RNA induces RNAi in plant-parasitic nematodes and protects the host from infection. *Molecular and Biochemical Parasitology, 148*, 219–222.
Yang, Y., Jittayasothorn, Y., Chronis, D., Wang, X., Cousins, P., & Zhong, G. Y. (2013). Molecular characteristics and efficacy of 16D10 siRNAs in inhibiting root-knot nematode infection in transgenic grape hairy roots. *PLoS One, 8*, e69463.

Youssef, R. M., Kim, K. H., Haroon, S. A., & Matthews, B. F. (2013). Post-transcriptional gene silencing of the gene encoding aldolase from soybean cyst nematode by transformed soybean roots. *Experimental Parasitology, 134*, 266–274.

Yuan, C., Li, C., Yan, L., Jackson, A. O., Liu, Z., Han, C., et al. (2011). A high throughput barley stripe mosaic virus vector for virus induced gene silencing in monocots and dicots. *PLoS One, 6*, e26468.

Yu, A., Lepere, G., Jay, F., Wang, J., Bapaume, L., Wang, Y., et al. (2013). Dynamics and biological relevance of DNA demethylation in *Arabidopsis* antibacterial defense. *Proceedings of the National Academy of Sciences of the United States of America, 110*, 2389–2394.

Zhang, C., Bradshaw, J. D., Whitham, S. A., & Hill, J. H. (2010). The development of an efficient multipurpose bean pod mottle virus viral vector set for foreign gene expression and RNA silencing. *Plant Physiology, 153*, 52–65.

CHAPTER TEN

Exploiting Solved Genomes of Plant-Parasitic Nematodes to Understand Parasitism

David M. Bird*[,1], Valerie M. Williamson[†], Charles H. Opperman*
*Department of Plant Pathology, NC State University, Raleigh, NC, USA
[†]Department of Plant Pathology, University of California-Davis, Davis, CA, USA
[1]Corresponding author: E-mail: bird@ncsu.edu

Contents

Abstract

The growing portfolio of sequenced plant-parasitic nematodes genomes is helping to drive the research agenda for the discipline of plant nematology. Comparative genomics has confirmed much of what had been deduced from expressed sequence tag sequencing, and expanded our understanding of the extent of horizontal gene transfer as a source of novelty during the evolution of parasitism. Full catalogues have been annotated for various phytolytic and other enzymes, and the evolutionary history of those genes deduced. Beyond these direct analyses, the genome sequences serve to underpin genetic, biochemical and physiological approaches, and we suspect that this will prove to be an invaluable legacy of genome sequencing.

Abbreviations

CEGMA Core eukaryotic genes mapping approach
CEP C-terminal encoded protein
EST Expressed sequence tag
GPCR G protein-coupled receptors
HGT Horizontal gene transfer
MIG *Meloidogyne incognita* group
PPN Plant-parasitic nematode
RKN Root-knot nematode
SCN Soybean cyst nematode

Advances in Botanical Research, Volume 73
ISSN 0065-2296
http://dx.doi.org/10.1016/bs.abr.2014.12.008

1. INTRODUCTION

It is no longer necessary to justify obtaining a complete genome sequence for any given plant-parasitic nematode (PPN); that case has been made elsewhere on numerous occasions (Bird, 2005; Bird & Opperman, 1998; Mitreva, Blaxter, Bird, & McCarter, 2005). Until very recently, the major barrier was cost. However, with advances in sequencing technology, it is now feasible to obtain sufficient raw sequence data to assemble a reasonably complete, whole genome of probably any PPN, typically with a cost of less than a few thousand US dollars, although subsequent bioinformatic analyses, including gene-finding and annotation may add to that cost. Thus it seems likely that many more PPN genomes will be solved as a method for gene discovery, and of course, this is to be welcomed. However, in this chapter we argue that the true power of a complete, well-annotated genome sequence lies in its role as the foundation upon which other information can be assembled, ideally derived from direct experimentation. More broadly, we argue that genomic data need to be considered in the context of genetics, biochemistry, physiology and evolutionary history. Indeed, the very word 'genome', a nearly century-old corruption of 'gene' and 'chromosome', points to the interdependency of genetics (information) and molecular biology (structure), and anticipates the need to consider the ensuing biology in a more holistic manner.

There is no doubt that the curated set of gene and protein models (WormPep), generated by the *Caenorhabditis elegans* Sequencing Consortium (1998) with enthusiastic community support, have proven invaluable in annotating the growing list of sequenced PPN genomes. But beyond that, we argue that tools need to be developed so that hypothesis-driven research can be conducted directly with PPN. Here we briefly discuss the current status of PPN genomes, with a particular emphasis on the root-knot nematode (RKN), *Meloidogyne hapla*. We emphasize *M. hapla* not simply because it is the species with which we are most familiar, but because features of the genome are clearly defined. The ~54-Mbp genome assembly (Opperman et al., 2008) is consistent with size prediction based independently on Cot analysis and flow cytometry (Leroy, Duperray, & Morand, 2003). Karyotypes are consistent with the linkage map (Figure 1) and the behaviour of chromosomes points to the genome as being diploid. Physical anchoring of the genome assembly with the linkage map confirms colinearity. We also argue that *M. hapla* has been developed into a genetic system sufficiently powerful to sustain hypothesis-driven research into the plant–parasite interaction.

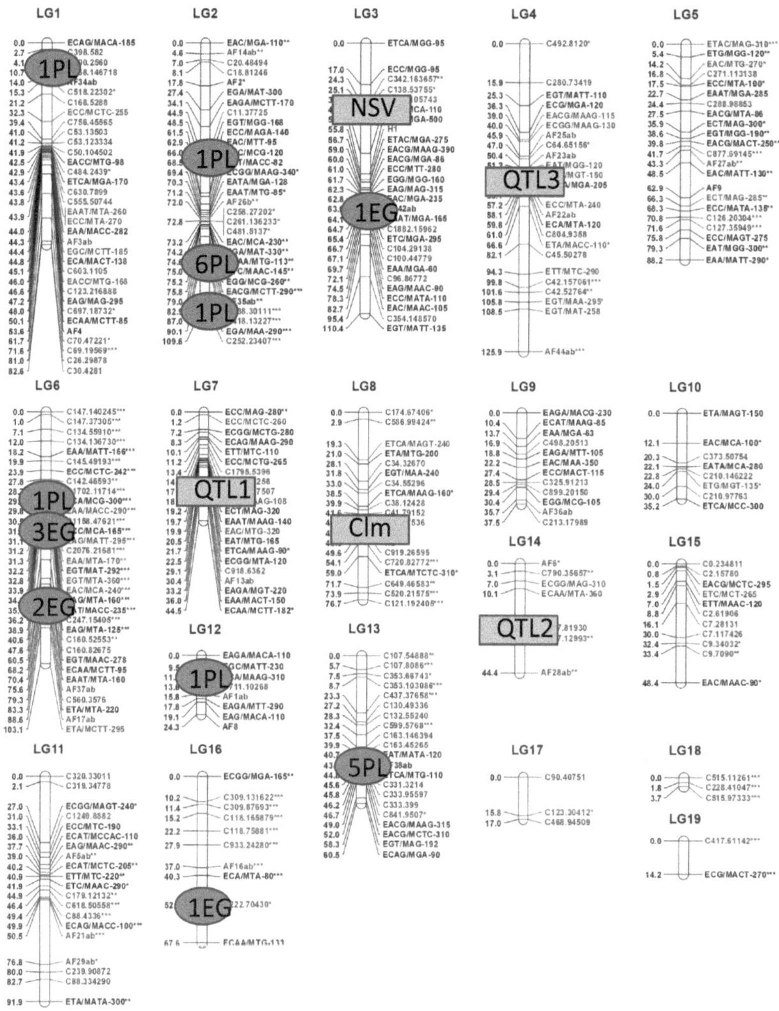

Figure 1 ***Genetic map of*** **Meloidogyne hapla.** This map is based on segregation of 414 polymorphic DNA markers in 183 F2 lines (Thomas et al., 2012). Green ovals indicate the approximate positions of genes encoding pectate lyase (PL) or endoglucanase (EG) with the numeral in front of the gene abbreviation indicating the number of closely linked copies at each locus. Yellow boxes indicate positions of phenotypic markers. *Clm* is a locus that modulates clumping, NSV determines ability to reproduce on common bean cultivar NemaSnap. QTL1, QTL2, QTL3, are three quantitative trait loci that modulate ability to reproduce on *Solanum bulbocastanum* SB22. (See colour plate)

Table 1 Genome Project Statistics

Species	Size	Contigs	N_{50}	Genes	Exons	Density
Caenorhabditis elegans	100.3	6	17.5	20,517	5	249
Ascaris suum	265.3	29,831	0.4	18,449	5	70
Meloidogyne incognita	82.1	9538	0.1	20,332	5	223
Meloidogyne hapla	54	3452	0.3	14,721	4	270
Pratylenchus coffeae	19.7	5821	0.01	6712	2	342
Radopholus similis	63.2	8176	0.02	11,315	4	172
Globodera pallida	124.7	6873	0.122	16,419	6	132

Terms and nematode species names are explained in the text. Values given are those at the time of public release and/or publication.

Additional genome sequences are available from public databases for *Meloidogyne incognita* (Abad et al., 2008) and *Meloidogyne chitwoodi*, as well as genome skims of *Meloidogyne floridensis* (Lunt, Kumar, Koutsovoulas, & Blaxter, 2014) and other *M. hapla* isolates (Thomas et al., 2012) permitting robust comparative genomics with the reference genome sequence of *M. hapla* (Opperman et al., 2008).

We do not intend this chapter to be a thorough review of PPN genomes. Not only are several comprehensive reviews already available (e.g. Bird et al., 2009; Bird et al., 2014), but the next few years will likely see much progress in this area. So, to avoid producing an article that might rapidly be rendered obsolete, we have tried to focus more broadly, with the hope that some of the points we discuss might become topics for active research in the future. Finally, as a consequence, some of our speculation will undoubtedly prove to be incorrect. Nonetheless, we feel that this article is timely, in large because sufficient genomes have now been obtained (Table 1) to permit some generalizations to be made, some of which, we hope, will prove useful.

2. THE EST Epoc

It has been more than 15 years since the complete genome sequence of the free-living nematode, *C. elegans* was obtained (*C. elegans* Sequencing Consortium, 1998). This much anticipated resource promised to be a powerful tool to probe, and hopefully understand the biology of PPN (Bird & Opperman, 1998). As famously noted by Mark Blaxter (1998): '*Caenorhabditis elegans* is a nematode'. Thus, it seemed reasonable that one nematode genome serves as an exemplar to guide understanding of another.

Indeed, even before the genome was available, the accumulation of *C. elegans* cDNA sequences served to nucleate parasitic nematode sequencing projects (Bird, Opperman, Jones, & Baillie, 1999; Blaxter & Bird, 1997). Importantly, early efforts to leverage the *C. elegans* genome annotation to curate data sets from parasitic nematodes proved highly effective for gene discovery. This is particularly true for RKN, which benefited from extensive expressed sequence tag (EST) sequencing projects in the USA and Europe (McCarter, Abad, Jones, & Bird, 2000; McCarter, Bird, Clifton, & Waterston, 2001; McCarter et al., 2003; Mitreva et al., 2005). ESTs correspond to the partial (sometimes complete) sequence of mRNAs and, as such, define the existence of the corresponding gene and confirm its expression. Beyond that, ESTs themselves proved to be amenable to semiautomated gene discovery (McCarter et al., 2003), and many genes with apparent roles in the host–parasite interaction were initially annotated using ESTs. Most of these so-called 'parasitism' genes encoded enzymes associated with degradation of plant cell walls (Smant et al., 1998), but other biochemical functions could be inferred from the data too. Beyond predicting biochemical function, the ESTs proved to be effective templates for strictly computational approaches to deduce evolutionary relatedness between species (Scholl & Bird, 2005), including identifying RKN genes with origins outside the phylum (Scholl, Thorne, McCarter, & Bird, 2003). Such genes are candidates for having been acquired from other organisms (particularly bacteria and fungi) by horizontal gene transfer (HGT), and may account for as much as 5% of the PPN gene space (Paganini et al., 2012).

However, despite the undeniable utility of ESTs to provide a glimpse of an organism's genetic repertoire, this approach suffers from certain shortcomings. Foremost, classical EST projects under-sample the diversity of any organism's transcriptome. Lowly expressed genes, genes with restricted spatial or temporal expression, or genes encoding small transcripts are typically underrepresented or absent in EST libraries. This is a particular problem with PPN, as is not feasible to sample individual cells or even entire organs, with the notable exception of the pharyngeal glands (Huang et al., 2003). Second, EST projects are not well suited to enumerating members of gene families. This is important because many genes with apparently clear roles in contributing to parasitic ability, such as the CarbohydrateActive Enzymes (CAzymes) encoded by RKN and other PPN, exist in families. A third limitation arises from how one defines a match (partly a consequence of how one sets various parameters in search algorithms). The initial examination of ESTs from the human-parasitic, filarial nematode *Brugia malayi*

(Bird et al., 1999) provides a good example. From a screen of 5555 random, unannotated *B. malayi* ESTs, it was established that 86% matched a *C. elegans* gene model. Alternatively stated, 14% failed to match. Are these genes uninteresting because they have no match? Or are they more interesting because, being present in the parasite but absent from the free-living *C. elegans*, they are candidates for playing a role in the parasitic interaction? These are important questions, but absent more data, are not easily answered.

In retrospect, it could be argued that the decade of EST-based approaches to genomics provided only limited insight into RKN biology, and the parasitic interaction in general. That is not to say that these data were of no value. In particular, previously controversial aspects of RKN biology, such as the secretion of cellulases (Bird, Downton, & Hawker, 1975), were validated (Smant et al., 1998). Similarly, the hypothesis that PPN acquired genes for cellulases and other phytolytic enzymes from bacteria (Keen & Roberts, 1998; Smant et al., 1998) has become firmly established. Perhaps the most useful contribution of the EST data was to train algorithms for gene-finding in genomic sequence, but that also has some limitations. For example, genes encoding very small proteins tend to be missed or mis-annotated. Nowadays, high-throughput RNA sequencing (RNA-Seq) technologies have superseded Sanger-based EST sequencing. Conceptually, RNA-Seq reads are equivalent to ESTs, but because RNA-Seq generates orders of magnitude more reads per run, many more primary transcripts are sampled, revealing new genes as well as splice variants. Examples of the power of the RNA-Seq approach for de novo gene discovery and the re-annotation of existing genomes are afforded by Eves van den Akker et al. (2014) and Guo, Bird, and Nielsen (2014) respectively.

3. WHOLE PPN GENOMES

A decade after the publication of the *C. elegans* genome, two RKN genomes were released: *M. incognita* (Abad et al., 2008) and *M. hapla* (Opperman et al., 2008), thus affording the opportunity for comparing their genomes to that of *C. elegans*. Projects to enhance the assemblies and annotation of both RKN genomes are ongoing, and for *M. hapla*, a re-annotation of the genome, based on mapping 2 billion RNA-Seq reads, has confirmed most of the published gene models and made mostly minor edits to others (Guo et al., 2014). Genome projects are in progress for many nematodes, and for some species or genera (especially the rhabditids) multiple, whole genome sequences have been obtained (Table 1). In the case of *M. hapla*, the genomes of strains VW8 and LM both have been sequenced, in addition to the reference sequence of the VW9 inbred line (Opperman et al., 2008; Thomas et al., 2012). Additionally, the genome of the pine wood nematode, *Bursaphelenchus xylophilus* was sequenced (Kikuchi et al., 2011), and recently the *Globodera pallida* (pale cyst nematode) genome was reported (Cotton et al., 2014). Besides these published genomes, sequences from several other PPN are available in advance of publication. Included in this list are *Pratylenchus coffeae, Radopholus similis* and *Globodera rostochiensis.* A draft, partially assembled but unannotated genome for the soybean cyst nematode *Heterodera glycines* also is available from the public DNA-sequence repositories.

With the notable exception of *C. elegans*, for which the number of scaffolds equals the number of chromosomes, which in turn equals the number of linkage groups, whole genome sequences typically have gaps, often thousands of gaps. Fortunately, as a consequence of the cloning/sequencing strategies used, it is possible to fairly accurately estimate the size of most of the gaps, which for *M. hapla* at least, are typically as small as 1 nt. However, until advances in technology permit gapless assembly (and this will likely occur in the near future), it is important to provide some metrics to give an estimate of the amount of the genome in an assembly, and also an estimate of the coverage. Commonly used to evaluate the assembly is a value termed the 'scaffold N_{50}' and, to a lesser extent, the number of contigs (Table 1). The value of N_{50} corresponds the position of the contig in a size-ranked list of contigs at which the sum of all larger contigs in the list equals 50% of the estimated total genome size. Larger values for N_{50} predict better assemblies. Recently, a new standard, termed the Core Eukaryotic Genes Mapping Approach (CEGMA) has been adopted to assess coverage completeness of the gene space and quality of the assembly (Parra, Bradnam, & Korf, 2007; Parra, Bradnam, Ning, Keane, & Korf, 2009). CEGMA is based on scoring the nascent genome assembly for 75% or greater coverage at the amino acid level of a panel of highly conserved genes that apparently are essential to all metazoans, presumably including all nematodes. The percentage of the total CEGMA compliment detected in a given genome assembly is a de facto representation of the assembly quality, including the number of gaps, and provides an indication of likelihood of finding any gene. CEGMA thus provides an estimate of completeness of genome coverage (Parra et al., 2007, 2009). For example, our recent assembly of the *M. chitwoodi* genome has a CEGMA score of 99%, indicating nearly complete genome coverage.

Examination of the genomes in Table 1 shows a more than 10-fold range of estimated sizes, from 265 Mbp for the human intestinal parasite, *Ascaris suum*, to 19.7 Mbp for *P. coffeae*. The latter value represents the smallest metazoan genome sequenced to date. With the caveat that the sample size is very small, the free-living nematode species appear to encode more genes than the animal parasites, which in turn have a larger compliment of genes than the plant parasites. Importantly there appears to be no relationship between gene number, protein size and overall genome size. However, there is a clear relationship between genome size and gene density. The largest genome (*Ascaris*) has the lowest gene density, and the smallest genome (*P. coffeae*) the highest. The *M. incognita* genome is median for both size and density. The average number of exons per gene is consistent with this trend. The larger genome size of *A. suum* is largely attributable to longer genes due to greatly expanded average intron size (mean = 1.1 kb vs 50 bp in *M. hapla*) and expanded intergenic space (Jex et al., 2011). Also contributing to 'simplicity' is a general lack of complex repeats that can hamper assembly. Similarly, %G + C is characteristically low, but not so low as to complicate assembly.

A prime motivation for solving the genome is gene discovery. Unlike ESTs, genome sequence presents each unique gene with equal stoichiometry. It also includes structural and regulatory elements (such as introns and the initiation codon), and these can be used as defining landmarks to delineate individual genes. As mentioned above, ESTs (from the same species as the genome) can serve as a training set for computer algorithms to discover those landmarks. Not surprisingly, the accuracy of these predictions is predicated on the accuracy and breadth of the training set. As a case in point, both automated gene-finding algorithms used to predict genes in the *M. hapla* genome reported a large region on contig 363 apparently devoid of protein-coding genes. However careful examination of this region shows that it encodes a 12-member family of C-terminal encoded protein (CEP) proteins (Bobay et al., 2013), which have been confirmed to encode transcripts (Figure 2). CEP are of interest because ectopic exposure to endogenous CEP-induced galls on roots in the same manner as does RKN. RKN-encoded CEP is likely recognizing the same receptors as native CEP (Bobay et al., 2013).

4. COMPARATIVE GENOMICS

In the simplest analysis, a whole genome sequence permits the genetic repertoire of that species to be deduced. For a PPN, genes hypothesized to encode functions directly related to the parasitic interaction can be tallied,

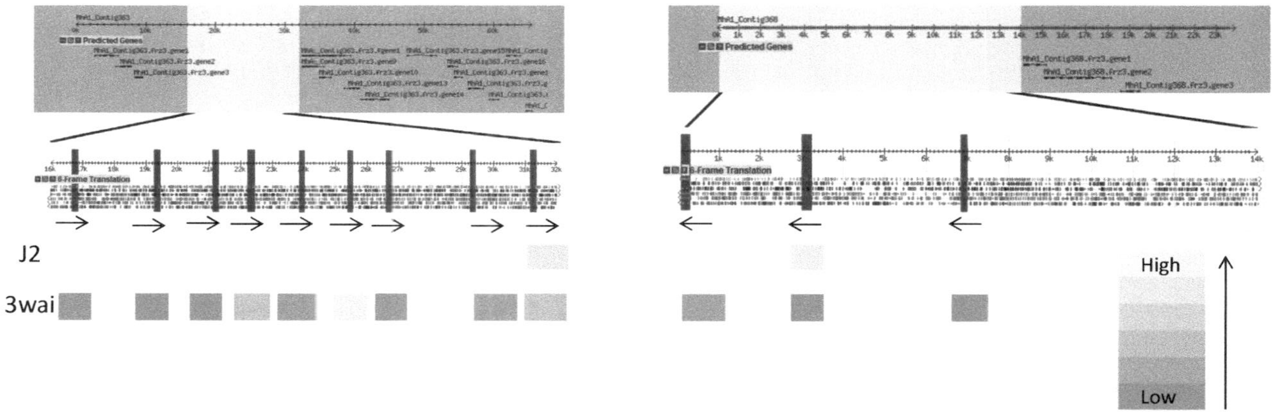

Figure 2 ***G-Browse view of the CEP super-locus in*** **Meloidogyne hapla.** Two contigs (olive boxes) exhibit typical gene density, except in the yellow boxes, which initially were annotated as being absent of genes, but which spans 12 CEP genes (arrows, which also indicate the direction of transcription) contigs over 122 kb. The heat map indicates transcriptional status in the pre-penetration J2, and *in planta* 3 weeks after infection. (See colour plate)

and indeed numerous such lists have been compiled. Beyond simple gene lists, however, it is possible to infer the presence (or absence) of biochemical pathways based on the presence of two or more component proteins although such inferences need to be made judiciously, with biological reality exerting priority over computational prediction. Interestingly, the findings are not always as anticipated. For example, many pathogenic bacteria have lost genes from pathways for biosynthesis of nutrients or other products that they can obtain from their host and therefore have genomes smaller than their free-living counterparts. Indeed, *M. hapla* encodes ~14,500 genes, compared to ~20,000 in *C. elegans*. However, there is no systematic loss of metabolic or catabolic pathways in the parasite compared to the free-living species. Rather, differences in gene number show a gene-by-gene pattern. Some gene families, such as the G protein-coupled receptors (GPCR), are greatly expanded in *C. elegans* compared to the RKN genomes (Abad et al., 2008; Opperman et al., 2008), possibly suggesting a greater requirement for these functions in the *C. elegans* niche. In nematodes, GPCR largely serve as olfactory receptors, so perhaps the hunter (*C. elegans*) needs to perceive a more diverse repertoire of environmental signals than does the sedentary plant endoparasite. Except for genes unique to the genus, gene families all seem to be smaller in *M. hapla* than in *C. elegans*, including those encoding ubiquitous products such as cuticular collagen. One of the powers of comparative genomics is that it allows genes, gene families and biochemical pathways to be compared across these parasites with distinct niches. Further, because genomic sequence provides relative positional information for genes, historical patterns of evolution can be deduced. For example, the establishment of gene families by duplication and inversion, can in many cases be reconstructed, as can the patterns of acquisition of new genes by HGT (Danchin et al., 2010).

5. DIVERSE REPRODUCTIVE MODES OF PPN IMPACT GENOME ANALYSIS

PPNs reproduce by multiple, distinct mechanisms ranging from obligate outcrossing (amphimixis) to mitotic parthenogenesis (Triantaphyllou & Hirschmann, 1980). Knowing which particular reproductive mode is employed by a species is prerequisite both for forward genetics and genome analysis. Cyst nematodes including *Heterodera schachtii*, *H. glycines* and *G. pallida* reproduce by outcrossing in which the egg and sperm nucleus are from different individual parents (Figure 3(A)). Consequently, even with extensive inbreeding it can be difficult or impossible to obtain strains

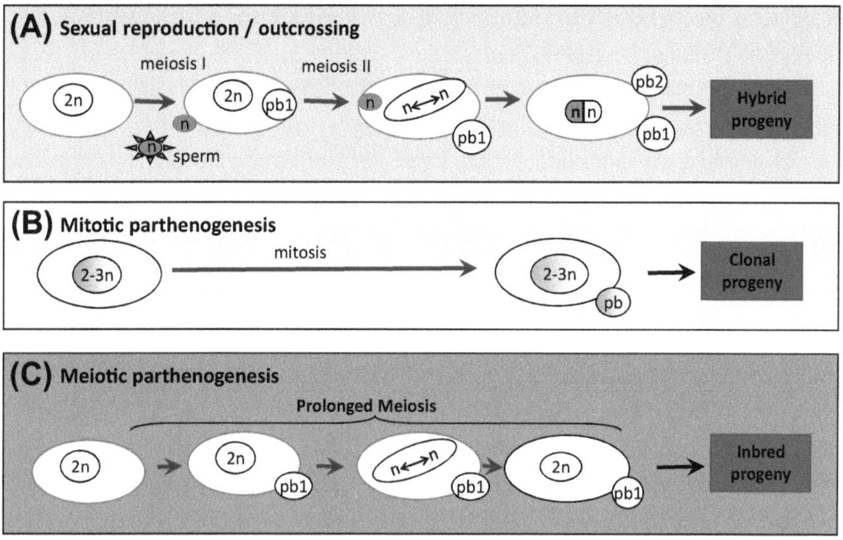

Figure 3 ***Reproductive mechanisms of root-knot nematodes and their outcomes.*** White ovals represent oocytes as they progress though the female gonad. (A) In sexual reproduction the sperm (red, dark grey in print version) fuses with the haploid (*n*) product of meiosis to form a hybrid zygote. (B) In mitotic parthenogenesis there is no reduction in ploidy or incorporation of genetic material from the male resulting in clonal progeny. (C) In meiotic parthenogenesis in *Meloidogyne hapla* meiotic the reduction division occurs, but in the absence of the male, the products of the second meiotic division rejoin to produce inbred progeny.

homozygous for all regions of the genome; heterozygosity can compromise genome assembly. In stark contrast to the cyst nematodes are the tropical RKN species belonging to a closely related species complex of *M. incognita*, *Meloidogyne javanica* and *Meloidogyne arenaria*, termed collectively the *M. incognita* group (MIG) (Hunt and Handoo, 2009). Most of these species reproduce exclusively by mitotic parthenogenesis (Figure 3(B)). Yet despite this lack of sexual reproduction, these species are widely distributed, have remarkably broad host ranges, and are amenable to selection and adaptive variation (Blok et al., 2008). As noted by Triantaphyllou (1985), one clue to their success may be in the variable chromosomal complement of isolates of a given species. For example, the chromosome number of *M. incognita* ranges can range from 32 to 46, depending on the strain, suggests that these species are aneuploid. Changes in ploidy of a specific chromosome could lead to phenotypic differences, especially if the individual is heterozygous for the relevant alleles. Gene conversion, transposable element movement or other specialized mechanisms of genetic change could also have a role in generating genetic variability (Castagnone-Sereno & Danchin, 2014; Gross & Williamson, 2011).

Comparisons of sequences of rDNA and of single-copy nuclear genes of MIG species fail to form phylogenetic lineages concordant with recognized species, leading to speculation that these species have reticulate origin, i.e. they each arose by hybridization of sexual species, possibly with one common parent, followed by loss of ability to reproduce sexually (Holterman et al., 2009; Lunt, 2008). In support of this, whole genome sequence analysis of *M. incognita* has revealed that its genome is apparently present as homologous but diverged, segment pairs, with on average 7% mean sequence divergence between aligned regions (Abad et al., 2008). Whole genome scans of additional species support the model of hybrid origin for MIG species and are beginning to provide clues to their evolutionary history, and perhaps to their parasitic success (Lunt et al., 2014). The presence of two or more diverged copies of most genomic regions severely compromises genome assembly, as neither assembly programs nor human annotators are especially adept at distinguishing alleles from homeologues.

Even within a single species (e.g. *M. hapla*), a range of reproductive mechanisms can be observed. For the majority of *M. hapla* isolates, reproduction is by facultative *meiotic* parthenogenesis (Triantaphyllou, 1985). Other isolates reproduce by obligate *mitotic* parthenogenesis, and even a hermaphroditic form has been reported (Triantaphyllou, 1993). For RKN species, sex is environmentally determined (Triantaphylou, 1973), with the default being oocyte-producing females. Simplistically (because there are many nuanced aspects to this process; see Figure 3(C)), in the facultative meiotic parthenogenetic strains, diploidy is restored and development initiated by fertilization (if sperm are present), or by fusion of the oocyte nucleus with fusion of the sister oocyte nuclei of the second meiotic division if sperm are absent if sperm are absent (Liu, Thomas, & Williamson, 2007; Triantaphyllou, 1966). Because the *M. hapla* genome selected for sequencing (VW9) was homozygous for most loci, its assembly provides a comparative reference for the more complex genomes, including that of *M. incognita* (Lunt et al., 2014).

6. INTEGRATION OF GENETICS WITH GENOMICS FOR PHENOTYPE-BASED IDENTIFICATION OF PARASITISM GENES

Because it permits genes to be identified based on the phenotype produced without any preconceived notion of their function, forward genetic analysis is a powerful experimental tool to causally relate biological processes. However, developing the necessary genetic resources has been challenging for parasitic nematodes. The typical starting point for a genetic experiment is to introduce genetic diversity by mutagenesis, but the obligate parasitic lifestyle deployed by PPN raises a barrier. Complex movement and behaviour is required for the parasite to complete the life cycle. Mutations that affect this orchestrated process, even for essentially trivial reasons, will disrupt the life cycle. Fortunately, considerable natural variation that is directly germane to the parasitic interaction exists in field populations.

Genetic analysis of PPN was pioneered in cyst nematodes, where screening progeny of controlled crosses with molecular markers permitted the construction of genetic (i.e. linkage) maps (Atibalentja et al., 2005; Rouppe van der Voort et al., 1999). Prior to that, traits contributing to parasitism on specific hosts had been found to segregate for some cyst nematode species (Dong & Opperman, 1997; Janssen, Bakker, & Gommers, 1991). Using highly inbred soybean cyst nematode (SCN) lines, Dong and Opperman (1997) demonstrated that parasitism is controlled by single, Mendelian loci in the parasite, termed *ror* (reproduce on resistant host). Different *ror* loci are required to overcome different host resistance genes. The major sources of resistance to SCN are QTLs on chromosome 8 (*Rhg4*) and 18 (*rhg1*). Susceptible cultivars of soybean carry a single copy of the *rhg-1* allele, but resistant cultivars carry 10 tandem copies. The *rhg1b* resistance locus derived from soybean PI88788 has been observed to function in a manner different from the classic R-gene mediated resistance to pathogens in that there is not a typical HR response, but rather a lack of development of feeding sites. In SCN, virulence to PI88788 is controlled by a dominant locus, *Ror-1*, and multiple copies of this locus act additively to increase aggressiveness of SCN on PI88788 sources of resistance (Dong & Opperman, 1997). It is interesting that the gene in SCN is dominant, while *rhg1b* is recessive in soybean and indicates a different kind of interaction between host and pathogen compared to the more classic models. In the SCN inbred lines, there are two recessive genes, *ror-2* and *ror-3*, that control virulence on PI90763 and Peking, respectively, and in contrast to *Ror-1-rhg1*, the *Rhg4-ror-2* interaction for SCN virulence on PI90763 resistance represents a classical gene-for-gene interaction, as has been observed for the *G. rostochiensis* potato system. The observation of additive effects by multiple copies of *ror* genes is one possible explanation for differing of aggressiveness between nematode populations on the same host genotype.

In contrast to cyst nematodes, the facultative meiotic parthenogenic *M. hapla* strains can reproduce by both selfing and by outcrossing; this is optimal for genetic analysis. Importantly, isolates of *M. hapla* display extensive phenotypic variability germane to the host–parasite interaction. For example, *M. hapla* isolates differ in their host range and pathogenicity on specific hosts (Chen & Roberts, 2003; Griffin & McKenry, 1989; Janssen, van Norel, Verkerk-Bakker, & Janssen, 1997; Liu & Williamson, 2006; Mitkowski, van der Beek & Abawi, 2002; Van der Beek, Los, & Pijanacker, 1998). Thus, rather than attempt to introduce genetic diversity by mutagenesis, it was decided to exploit the natural variability observed in field populations.

As noted above, the reproductive mode of *M. hapla* results in production of both selfed and outcrossed progeny, which is a situation ideal for genetic experimentation. Initially, a cross was made between two highly inbred strains, VW8 and VW9, which differ in the ability to reproduce on specific host plants (Liu & Williamson, 2006; Liu et al., 2007). DNA polymorphisms between parental lines were used to identify hybrid females, which, following parthenogenetic reproduction, produced a set of 183 F2 lines, each descended from the brood of a single F2 female. Molecular markers predominantly derived from SNPs identified between the sequenced and annotated VW9 genome and de novo sequence of VW8 were assessed in the F2 lines. Based on 182 markers distributed over 16 linkage groups corresponding in number to the haploid complement of chromosomes predicted by karyotyping (Figure 1; Thomas et al., 2012), a linkage map was constructed. Comparison of the marker order in the genetic map to that on sequence scaffolds revealed a correspondence of the genetic and physical maps and allows other genes on the contigs to be positioned genetically. For example, the positions and distribution of genes acquired by HGT have been mapped, providing insight into the acquisition and expansion of these genes. Similar pectate lyase and endoglucanse genes are clustered together, consistent with the hypothesis that expansion of the gene families occurred preferentially on the same linkage group (Thomas et al., 2012). Comparison of the genetic and physical maps of *M. hapla* also reveals an unusually high recombination rate, the highest for any animal described so far. In addition, a dearth of heterozygotes is found, suggesting that recombination resolution occurs by an unusual mechanism (Liu et al., 2007). As a consequence, the F2 lines are homozygous for most loci and thus resemble recombinant inbred lines (RILs).

The integrated genetic/genomic map and RIL-like F2 lines developed for *M. hapla* are powerful tools for identifying and cloning genes contributing to phenotypic differences in parasitism and other traits of interest. For example, these lines were utilized to investigate inheritance for a trait modulating clumping behaviour (*Clm*) (Wang, Lower, Thomas, & Williamson, 2010). *Meloidogyne hapla* strains differ in their propensity to form tight clumps in an in vitro assay. The parental strains, VW8 and VW9, used to produce the F2 lines are non-clumping and clumping strains, respectively. Assays of F2 lines for clumping ability revealed that this trait segregated as expected for a monogenic trait, and mapped to a locus on linkage group 8 (Wang et al., 2010; Figure 1). Similarly, this same set of F2 lines was used to assess segregation of ability to reproduce on *Solanum bulbocastanum* SB22, an isolate of a wild potato species that carries a nematode resistance gene (Thomas & Williamson, 2013). Comparison to the genetic map identified three quantitative trait loci (QTL) that could be positioned on linkage groups. A combination of two QTL alleles from one parent and one from the other appeared to be responsible for the transgressive phenotype of some F2 lines. Similarly a monogenic trait, NSV (NemaSnap Virulence), conferring ability to reproduce on common bean carrying a nematode-resistant gene has been positioned on a linkage group (Chen & Roberts, 2003; Gimeno & Williamson, unpublished data). While none of these mapped pathogenicity genes have yet been cloned, one has been localized to a deeply sequenced and well-annotated 14 kb region (Thomas & Williamson, 2013).

These examples demonstrate that *M. hapla* has emerged as a powerful system to integrate genetics and genomics and serves as a basis for ongoing efforts to clone the genes responsible for the parasitic interaction. We argue that beyond that, this platform is amenable to support structural analyses (e.g. Bobay et al., 2013), and we are pursuing this avenue.

REFERENCES

Abad, P., Gouzy, J., Aury, J.-M., Castagnone-Sereno, P., Danchin, E. G. J., Deleury, E., et al. (2008). Genome sequence of the metazoan plant-parasitic nematode *Meloidogyne incognita*. *Nature Biotechnology, 26*(8), 909–915.

Atibalentja, N., Bekal, S., Domier, L. L., Niblack, T. L., Noel, G. R., & Lambert, K. N. (2005). A genetic linkage map of the soybean cyst nematode *Heterodera glycines*. *Molecular Genetics and Genomics, 273*, 273–281.

Bird, D. McK. (2005). Model systems in agriculture: lessons from a worm. *Annals of Applied Biology, 146*, 147–154.

Bird, A. F., Downton, W. J. S., & Hawker, J. S. (1975). Cellulase secretion by second stage larvae of the root knot nematode (*Meloidogyne javanica*). *Marcellia, 38*, 165–169.

Bird, D. McK., & Opperman, C. H. (1998). *Caenorhabditis elegans*: a genetic guide to parasitic nematode biology. *Journal of Nematology, 30*, 299–308.

Bird, D. McK., Opperman, C. H., Jones, S. J. M., & Baillie, D. L. (1999). The *Caenorhabditis elegans* genome: a guide in the post genomics age. *Annual Review of Phytopathology, 37*, 247–265.

Bird, D. McK., Williamson, V. M., Abad, P., McCarter, J., Danchin, E. G. J., Castagnone-Sereno, P., et al. (2009). The genomes of root-knot nematodes. *Annual Review of Phytopathology, 47*, 333–351.

Bird, D. M., Jones, J. T., Opperman, C. H., Kikuchi, T., & Danchin, E. G. J. (2014). Signatures of adaptation to plant parasitism in nematode genomes. *Parasitology, 141*, 1–14.

Blaxter, M. A. (1998). *Caenorhabditis elegans* is a nematode. *Science, 282*, 2041–2046.

Blaxter, M., & Bird, D. McK. (1997). Parasitic nematodes. In D. L. Riddle, T. Blumenthal, B. Meyer, & J. Priess (Eds.), C. elegans *II* (pp. 851–878). Cold Spring Harbor, NY: Cold Spring Harbor Press.

Blok, V. C., Jones, J. T., Phillips, M. S., & Trudgill, D. L. (2008). Parasitism genes and host range disparities in biotrophic nematodes: the conundrum of polyphagy versus specialisation. *Bioessays, 30*, 249–259.

Bobay, B., DiGennaro, P., Scholl, E., Imin, N., Djordjevic, M., & Bird, D. M. (2013). Solution NMR studies of the plant peptide hormone CEP inform function. *FEBS Letters, 587*, 3979–3985.

Caenorhabditis elegans Sequencing Consortium. (1998). Genome sequence of the nematode *Caenorhabditis elegans*: a platform for investigating biology. *Science, 282*, 2012–2018.

Castagnone-Sereno, P., & Danchin, E. G. (2014). Parasitic success without sex - the nematode experience. *J Evol Biol, 27*, 1323–1333.

Chen, P., & Roberts, P. A. (2003). Virulence in *Meloidogyne hapla* differentiated by resistance in common bean (*Phaseolus vulgaris*). *Nematology, 5*, 39–47.

Cotton, J. A., Lilley, C. J., Jones, L. M., Kikuchi, T., Reid, A. J., Thorpe, P., et al. (2014). The genome and life-stage specific transcriptomes of *Globodera pallida* elucidate key aspects of plant parasitism by a cyst nematode. *Genome Biology, 15*, R43.

Danchin, E. G., Rosso, M. N., Vieira, P., de Almeida-Engler, J., Coutinho, P. M., Henrissat, B., et al. (2010). Multiple lateral gene transfers and duplications have promoted plant parasitism ability in nematodes. *Proceedings of the National Academy of Sciences, 107*, 17651–17656.

Dong, K., & Opperman, C. H. (1997). Genetic analysis of parasitism in the soybean cyst nematode *Heterodera glycines*. *Genetics, 146*, 1311–1318.

Eves van den Akker, S., Lilley, C. T., Danchin, E. G. J., Rancurel, C., Cock, P. J. A., Urwin, P. E., et al. (2014). The transcriptome of Nacobbus aberrans reveals insights into the evolution of sedentary endoparasitism in plant-parasitic nematodes. *Genome Biology and Evolution, 6*, 2181–2194.

Griffin, G. D., & McKenry, M. V. (1989). Susceptibility of Nevada synthetic XX germplasm to a California race of *Meloidogyne hapla*. *Journal of Nematology, 21*, 292–293.

Gross, S. M., & Williamson, V. M. (2011). Tm1: A Mutator/Foldback transposable element family in root-knot nematodes. *Plos One, 6*, e24534.

Guo, Y., Bird, D. M., & Nielsen, D. M. (2014). Improved structural annotation of protein-coding genes in the *Meloidogyne hapla* genome using RNA-Seq. *Worm, 3*, e29158.

Holterman, M., Karssen, G., van den Elsen, S., van Megen, H., Bakker, J., & Helder, J. (2009). Small subunit rDNA-based phylogeny of the Tylenchida sheds light on relationships among some high-impact plant-parasitic nematodes and the evolution of plant feeding. *Phytopathol, 99*, 227–235.

Huang, G., Gao, B., Maier, T., Allen, R., Davis, E. L., Baum, T. J., et al. (2003). A profile of putative parasitism genes expressed in the esophageal gland cells of the root-knot nematode *Meloidogyne incognita*. *Molecular Plant-Microbe Interactions, 16*, 376–381.

Hunt, D. J., & Handoo, Z. A. (2009). Taxonomy, identification and principal species. *Root-knot Nematodes, 1*, 55–88.

Janssen, R., Bakker, J., & Gommers, F. J. (1991). Mendelian proof for a gene-for-gene relationship between *Globodera rostochiasis* and the H1 resistance gene from *Solanum tuberosum* ssp. *andigena* CPC 1673. *Reveue de Nematology, 14*, 207–212.

Janssen, G., van Norel, A., Verkerk-Bakker, B., & Janssen, R. (1997). Intra- and interspecific variation of root-knot nematodes, *Meloidogyne* spp. with regard to resistance in wild tuber-bearing *Solanum* species. *Fundamental and Applied Nematology, 20*, 449–457.
Jex, A. R., Liu, S., Li, B., Young, N. D., Hall, R. S., Li, Y., et al. (2011). *Ascaris suum* draft genome. *Nature, 479*, 529–533.
Keen, N. T., & Roberts, P. A. (1998). Plant parasitic nematodes: digesting a page from the microbe book. *Proceedings of the National Academy of Sciences of the United States of America, 95*, 4789–4790.
Kikuchi, T., Cotton, J. A., Dalzell, J. J., Hasegawa, K., Kanzaki, N., McVeigh, P., et al. (2011). Genomic insights into the origin of parasitism in the emerging plant pathogen *Bursaphelenchus xylophilus*. *PLoS Pathog, 7*, e1002219.
Leroy, S., Duperray, C., & Morand, S. (2003). Flow cytometry for parasite nematode genome size measurement. *Molecular Biochemistry of Parasitology, 128*, 91–93.
Liu, Q. L., & Williamson, V. M. (2006). Host-specific pathogenicity and genome differences between inbred strains of *Meloidogyne hapla*. *J. Nematol., 38*, 158–164.
Liu, Q. L., Thomas, V. P., & Williamson, V. M. (2007). Meiotic parthenogenesis in a root-knot nematode results in rapid genomic homozygosity. *Genetics, 176*, 1483–1490.
Lunt, D. H. (2008). Genetic tests of ancient asexuality in Root Knot Nematodes reveal recent hybrid origins. *BMC Evolutionary Biology, 8*, 194.
Lunt, D. H., Kumar, S., Koutsovoulas, G., & Blaxter, M. L. (2014). The complex hybrid origins of the root knot nematodes revealed through comparative genomics. *PeerJ, 2*, e356. http://dx.doi.org/10.7717/peerj.356.
McCarter, J. A., Abad, P., Jones, J., & Bird, D. McK. (2000). Rapid gene discovery in plant parasitic nematodes via expressed sequence tags. *Nematology, 2*, 719–731.
McCarter, J., Bird, D. McK., Clifton, S., & Waterston, R. (2001). Nematode gene sequences; update for December 2000. *Journal of Nematology, 32*, 331–333.
McCarter, J. P., Mitreva, M. D., Martin, J., Dante, M., Wylie, T., Rao, U., et al. (2003). Analysis and functional classification of transcripts from the root-knot nematode *Meloidogyne incognita*. *Genome Biology, 4*, R26.1–R26.19.
Mitkowski, N. A., Van der Beek, J. G., & Abawi, G. S. (2002). Characterization of root-knot nematode populations associated with vegetables in New York State. *Plant Disease, 86*, 840–847.
Mitreva, M., Blaxter, M. L., Bird, D. McK., & McCarter, J. P. (2005). Comparative genomics in Nematodes. *Trends in Genetics, 21*, 573–581.
Opperman, C. H., Bird, D. McK., Williamson, V. M., Rokhsar, D. S., Burke, M., Cohn, J., et al. (2008). Sequence and genetic map of *Meloidogyne hapla*: a compact nematode genome for plant parasitism. *Proceedings of the National Academy of Sciences, 105*, 14802–14807.
Paganini, J., Campan-Fournier, A., Da Rocha, M., Gouret, P., Pontarotti, P., Wajnberg, E., et al. (2012). Contribution of lateral gene transfers to the genome composition and parasitic ability of root-knot nematodes. *PLoS One, 7*, e50875.
Parra, G., Bradnam, K., & Korf, I. (2007). CEGMA: a pipeline to accurately annotate core genes in eukaryotic genomes. *Bioinformatics, 23*, 1061–1067.
Parra, G., Bradnam, K., Ning, Z., Keane, T., & Korf, I. (2009). Assessing the gene space in draft genomes. *Nucleic Acids Research, 37*, 297–298.
Rouppe van der Voort, J. N. A. M., van Eck, H. J., van Zandvoort, P. M., Overmars, H., Helder, J., & Bakker, J. (1999). Linkage analysis by genotyping of sibling populations: a genetic map for the potato cyst nematode constructed using "pseudo-F2" mapping strategy. *Molecular Genetics and Genomics, 261*, 1021–1031.
Scholl, E. H., & Bird, D. McK. (2005). Resolving tylenchid evolutionary relationships through multiple gene analysis. *Molecular Phylogenetics and Evolution, 36*, 536–545.

Scholl, E. H., Thorne, J. L., McCarter, J. P., & Bird, D. McK. (2003). Horizontally transferred genes in plant-parasitic nematodes: a high-throughput genomic approach. *Genome Biology, 4*, R39.1–R39.12.

Smant, G., Stokkermans, J. P. W. G., Yan, Y., de Boer, J. M., Baum, T. J., Wang, X., et al. (1998). Endogenous cellulases in animals: Isolation of β-1,4-endoglucanase genes from two species of plant-parasitic cyst nematodes. *Proceedings of the National Academy of Sciences, 95*, 4906–4911.

Thomas, V. P., Fudali, S. L., Schaff, J. E., Liu, Q., Scholl, E. H., Opperman, C. H., et al. (2012). A sequence-anchored linkage map of the plant-parasitic nematode *Meloidogyne hapla* reveals exceptionally high genome-wide recombination. *G3: Genes, Genomics, Genetics, 2*, 815–824.

Thomas, V. P., & Williamson, V. M. (2013). Segregation and mapping in the root-knot nematode *Meloidogyne hapla* of quantitatively inherited traits affecting parasitism. *Phytopathology, 103*(9), 935–940.

Triantaphyllou, A. C. (1966). Polyploidy and reproductive patterns in the root-knot nematode *Meloidogyne hapla*. *Journal of Morphology, 118*, 403–413.

Triantaphyllou, A. C. (1973). Environmental sex differentiation of nematodes in relation to pest management. *Annual Review of Phytopathology, 11*, 441–462.

Triantaphyllou, A. C. (1993). Hermaphroditism in *Meloidogyne hapla*. *Journal of Nematology, 25*, 15.

Triantaphyllou, A. C., & Hirschmann, H. (1980). Cytogenetics and morphology in relation to evolution and speciation of plant-parasitic nematodes. *Annual Review of Phytopathology, 18*, 333–359.

Triantaphyllou, A. C. (1985). Cytogenetics, cytotaxonomy and phylogeny of root-knot nematodes. In J. N. Sasser, & C. C. Carter (Eds.), *An Advanced Treatise on Meloidogyne*. Raleigh: North Carolina State University.

Van der Beek, J., Los, J., & Pijanacker, L. (1998). Cytology of parthenogenesis of five *Meloidogyne* species. *Fundamental and Applied Nematology, 21*, 393–399.

Wang, C., Lower, S., Thomas, V. P., & Williamson, V. M. (2010). Root-knot nematodes exhibit strain-specific clumping behavior that is inherited as a simple genetic trait. *PLoS One, 5*, e15148.

CHAPTER ELEVEN

Emerging Roles of Cyst Nematode Effectors in Exploiting Plant Cellular Processes

Michael Gardner[a], Anju Verma[a], Melissa G. Mitchum[1]
Division of Plant Sciences and Bond Life Sciences Center, University of Missouri, Columbia, MO, USA
[1]Corresponding author: E-mail: goellnerm@missouri.edu

Contents

Abstract

Diverse phytonematodes deliver effector proteins into the cells of their plant hosts using a hollow, protrusible mouth spear called a stylet. Collectively, these stylet-secreted effector (SSE) proteins function to modulate cellular processes that favour nematode growth and development. We discuss the functions of SSEs, with an emphasis on cyst nematodes. Identified plant targets of several cyst nematode effectors have revealed their putative function in parasitism. Several cyst nematode effectors modulate host developmental pathways and alter phytohormone balance to favour feeding site establishment. Other effectors are deployed to various plant compartments to subvert plant stress and defence responses. We also discuss genome-aided discovery of the entire complement of cyst nematode effectors that promises to uncover the full range of effector functions and enable comparative analyses that address outstanding questions in plant–cyst nematode interactions such as the molecular basis of host specificity.

[a]These authors contributed equally to this work.

Advances in Botanical Research, Volume 73
ISSN 0065-2296
http://dx.doi.org/10.1016/bs.abr.2014.12.009

1. INTRODUCTION

Cyst nematodes represent a highly specialized group of phytoparasitic nematodes. Consisting primarily of species contained within the genera *Globodera* and *Heterodera*, they are characterized by their narrow host range, often consisting of a mere handful of plant species per nematode species, as well as a complex method of parasitizing host plants. Upon hatching from eggs within the soil, the infective second-stage juvenile (J2) nematode will migrate through the soil towards plant roots, attracted by a currently unknown signal. Once at the root surface, the nematode penetrates into the root and through the use of secreted cell wall degrading enzymes and mechanical punctures by the stylet migrates intracellularly towards the vascular tissue, where it selects a cell and begins establishment of a feeding site called a syncytium (Mitchum et al., 2013). Migration in this manner causes considerable damage to plant root tissue, potentially triggering early defence responses. The syncytium is comparable to the giant cells formed by the root-knot nematode in function, but is formed by dissolution of surrounding cell walls and fusion of hundreds of cells, instead of rapid cellular expansion of several cells around the head of the nematode. In addition, the augmented cell becomes highly metabolically active, integrated nuclei are enlarged, and dramatic changes in gene expression take place (Ithal et al., 2007; Szakasits et al., 2009). This feeding site is integral to survival of the nematode and serves as its sole nutrient source for the remainder of the sedentary stages of the nematode's life cycle. The sedentary nematode feeds on the plant, swelling up and molting before differentiating into sexes. The female nematodes remain in place while feeding whereas the males regain their vermiform shape and migrate to the exterior of the roots, where they fertilize eggs within the protruding females. After fertilization the female eventually dies, her body hardening into a protective casing for the eggs called a cyst that can break off and remain dormant in the soil for many years before stimulation triggers the cycle to begin again. In order to establish and maintain its unique feeding site, a cyst nematode must manipulate cellular processes within the host cell. This is accomplished through the use of secreted effectors, which serve as an interface between the pathogen and host. The best studied effectors are proteins originating in three oesophageal glands within the nematode, two subventral and one dorsal, which alter their expression as the nematode progresses through its life cycle (Davis, Hussey, Mitchum, & Baum, 2008; Hussey, Davis, & Baum, 2002). Following production in these glands they are next secreted, primarily through the stylet. These effectors have been shown to function both within the interior of the host cell as well as in the apoplast, and therefore could possibly be secreted to either location. Once inside the host plant these effectors must be responsible for manipulating a wide variety of cellular processes ranging from signaling and hormone balance to defence responses and developmental regulation (Gheysen & Mitchum, 2011). To date, the functional characterization of cyst nematode effectors has revealed emerging themes in suppression of plant defences and augmentation of host development, two vital functions the nematode must perform in order to survive. In this chapter, we review our current understanding of cyst nematode effectors with a particular focus on stylet-secreted effector proteins that have been subjected to some degree of functional characterization providing insight into their potential roles in feeding site formation (Figure 1, Table 1).

2. AUGMENTATION OF PLANT DEVELOPMENTAL PROCESSES

2.1 Peptide Mimicry

One of the most remarkable pathogen effector discoveries is the finding that cyst nematodes use CLAVATA3/endosperm surrounding region-related (CLE) ligand mimics to modulate host developmental programs for feeding cell formation (Wang et al., 2001, 2005; Gao et al., 2001; Olsen & Shriver, 2003). CLEs represent a class of pathogen effectors unique to plant–parasitic cyst nematodes, the only CLEs identified outside the plant kingdom to date, and the first example of a pathogen secreting a protein mimic of a plant peptide hormone (Mitchum, Wang, Wang, & Davis, 2012). CLE-like genes have been identified from several cyst nematode species belonging to the genera *Heterodera* and *Globodera*, including the soybean cyst nematode (SCN, *Heterodera glycines*, *HgCLE1* and *HgCLE2*) (Wang et al., 2005; Wang, Lee, et al., 2010), the beet cyst nematode (BCN, *Heterodera schachtii*, *HsCLE1* and *HsCLE2*) (Patel et al., 2008; Wang et al., 2011) and potato cyst nematode (PCN, *Globodera rostochiensis*, *GrCLE1*, *GrCLE4-A*, *GrCLE4-B*, *GrCLE4-C* and *GrCLE4-D*) (Lu et al., 2009). Similar to their plant counterparts (Ohyama, Shinohara, Ogawa-Ohnishi, & Matsubayashi, 2009), cyst nematode *CLEs* belong to gene families encoding small proteins with an N-terminal secretion signal peptide (SP) that targets these proteins for extracellular secretion, a variable domain (VD) and either a single (*Heterodera* spp.) or multiple (*Globodera* spp.)

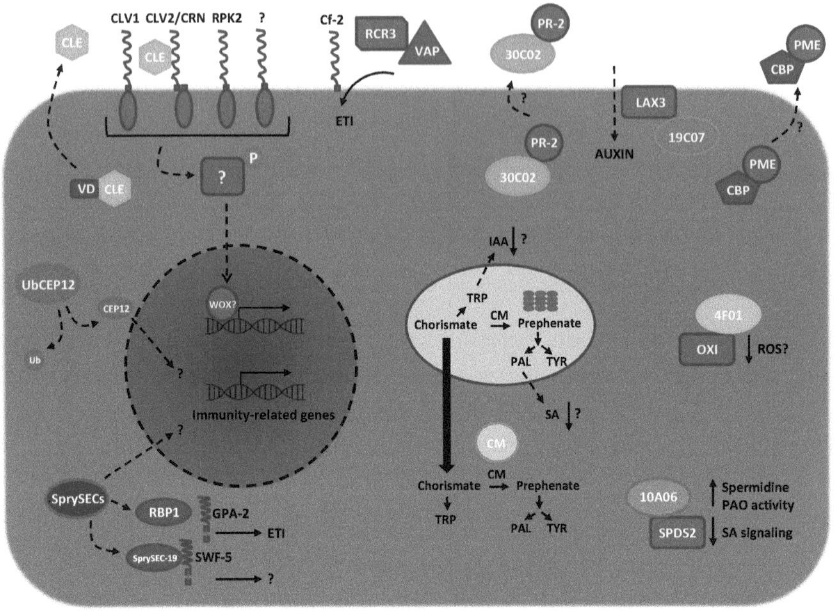

Figure 1 A diagram depicting the various plant targets and activities of stylet-secreted effectors (SSEs) from cyst nematodes. Cyst nematodes secrete a suite of SSEs into plant cells via a hollow, protrusible mouth spear or stylet. 19C07 interacts with LAX3 to possibly modulate auxin transport. CLAVATA3/endosperm surrounding region-related (CLE) peptides are peptide hormone mimics that interact with the extracellular leucine-rich repeat domain of plant receptor-like kinases with roles in stem cell maintenance. Cellulose-binding protein (CBP) interacts with pectin methylesterase (PME3) to likely modify cell wall properties. Chorismate mutase (CM) affects the plant shikimate pathway to modulate aspects of cellular metabolism. 10A06, 4F01 and 30C02 are novel effectors that interact with spermidine synthase (SPDS2), oxidoreductase (OXI) and β-1,3-endoglucanase (PR-2), respectively and may modulate plant defense responses. GrUBCEP12 is processed in planta to release free ubiquitin and the 12 amino acid GrCEP12 peptide to possibly affect the 26S proteasome and modulate plant defense. SPRYSECs are a large class of secreted SPRY-domain containing proteins that localize to different subcellular compartments of host cells. The R protein GPA-2 mediates effector-triggered immunity (ETI) by directly or indirectly recognizing the SPRYSEC effector RBP-1. SPRYSEC-19 interacts with SWF-5; however, the biological significance of this interaction is unknown. The R protein Cf-2 causes ETI by recognizing perturbations to the active site of the apoplastic protease Rcr3 by VAP-1. Host protein targets are green; cyst effectors are coloured. VD, variable domain; VAP, venom allergen-like protein; PR, pathogenesis related; PAO, polyamine oxidase; IAA, indole-3-acetic acid; UBCEP12, ubiquitin carboxyl extension protein 12; SA, salicylic acid; ROS, reactive oxygen species; WOX, WUSCHEL-like homeobox. (See colour plate)

Table 1 Predicted site of action, function, and known host targets of cyst nematode stylet-secreted effectors

Effector	Homology	Site of action	Host function	Interacting host proteins	References
Globodera spp.					
Globodera rostochiensis					
GrCM	Chorismate mutase	Cytoplasm	Hormone balance and defense suppression	—	Lu et al. (2008)
GrVAP1	Venom allergen-like protein	Apoplast	Defense suppression	Tomato RCR3 (apoplastic cysteine protease)	Lozano-Torres et al. (2012)
GrSPRYSEC-19	SPRYSEC domain-containing	Nucleus	Defense suppression	Tomato SWF-5 (CC-NB-LRR)	Rehman et al. (2009), Postma et al. (2012)
GrCLE1	CLE-like peptide	Apoplast	Developmental signaling	Arabidopsis CLV2, BAM1, BAM2 (LRR-RLKs); Potato CLV2	Lu et al. (2009); Guo, Ni, Denver, Wang, and Clark (2011); Chen et al. (2015)
GrCLE4	CLE-like peptide	Apoplast	Developmental signaling	—	Lu et al. (2009)
GrUBCEP12	Ubiquitin carboxyl extension protein	Nucleus	Protein degradation and defense suppression	—	Chronis et al. (2013)
Globodera pallida					
GpRBP-1	SPRYSEC domain-containing	Nucleus	Defense suppression	Potato GPA-2 (CC-NB-LRR)	Sacco et al. (2009)
GpCM	Chorismate mutase	Cytoplasm	Hormone balance and defense suppression	—	Jones et al. (2003), Yu et al. (2011)
Other Globodera spp.					
GtCM	Chorismate mutase	Cytoplasm	Hormone balance and defense suppression	—	Yu et al. (2011)
GeCM	Chorismate mutase	Cytoplasm	Hormone balance and defense suppression	—	Chronis et al. (2014)

(Continued)

Table 1 Predicted site of action, function, and known host targets of cyst nematode stylet-secreted effectors—cont'd

Effector	Homology	Site of action	Host function	Interacting host proteins	References
Heterodera spp.					
Heterodera glycines					
HgCLE1	CLE-like peptide	Apoplast	Developmental signaling	Soybean CLV1A, CLV2A, CLV2B, RPK2A, RPK2B, (LRR-RLPs)	Gao et al. (2003), Wang, Lee, et al. (2010), Guo et al. (2015)
HgCLE2	CLE-like peptide	Apoplast	Developmental signaling	Soybean CLV1A, CLV2A, CLV2B, RPK2A, RPK2B, (LRR-RLPs)	Wang et al. (2001, 2005), Wang, Lee, et al. (2010), Gao et al. (2003), Guo et al. (2015)
Hg30C02	Novel	Apoplast	Defense suppression	Arabidopsis PR-2 (β-1,3-endoglucanase)	Gao et al. (2003)
HgCM	Chorismate mutase	Cytoplasm	Hormone balance and defense suppression	—	Bekal et al. (2003)
HgCBP	Cellulose-binding protein	Apoplast	Cell wall modification	Arabidopsis PME3 (pectin methylesterase 3)	Gao et al. (2003)
Hg19C07	Novel	Cytoplasm	Hormone balance	Arabidopsis LAX3 (auxin influx transporter)	Gao et al. (2003)
Hg10A06	Novel	Cytoplasm	Defense suppression	Arabidopsis SPDS2 (spermidine synthase 2)	Gao et al. (2003)
Hg4F01	Annexin-like	Cytoplasm	Defense suppression	Arabidopsis oxidoreductase (2OG-Fe(II)oxygenase family)	Gao et al. (2003)
HgUBI1	Ubiquitin carboxyl extension protein	Nucleolus	Protein degradation	—	Gao et al. (2003)

HgVAP1/2	Venom allergen-like protein	—	—	—	Gao et al. (2001)
Heterodera schachtii					
HsCLE1	CLE-like peptide	Apoplast	Developmental signaling	—	Wang et al. (2011)
HsCLE2	CLE-like peptide	Apoplast	Developmental signaling	Arabidopsis CLV1, CLV2, BAM1, BAM2 (LRR-RLPs)	Wang et al. (2011), Guo et al. (2010)
Hs30C02	Novel	Apoplast	Defense suppression	Arabidopsis BG (β-1,3-endoglucanase)	Hamamouch et al. (2012)
HsCM	Chorismate mutase	Cytoplasm	Hormone balance and defense suppression	—	Vanholme et al. (2009)
HsCBP	Cellulose-binding protein	Apoplast	Cell wall modification	Arabidopsis PME3 (pectin methylesterase 3)	Hewezi et al. (2008)
Hs19C07	Novel	Cytoplasm	Hormone balance	Arabidopsis LAX3 (auxin influx transporter)	Lee et al. (2011)
Hs10A06	Novel	Cytoplasm	Defense suppression	Arabidopsis SPDS2 (spermidine synthase 2)	Hewezi et al. (2010)
Hs4F01	Annexin-like	Cytoplasm	Defense suppression	Arabidopsis oxidoreductase (2OG-Fe(II)oxygenase family)	Patel et al. (2010)
HsUBI1	Ubiquitin carboxyl extension protein	Nucleolus	Protein degradation	—	Tytgat et al. (2004)

CB-LB-LRR, coiled-coil nucleotide-binding leucine-rich repeat; LRR-RLKs, leucine-rich repeat–receptor-like kinases; CLE, CLAVATA3/endosperm surrounding region-related.

conserved 12 amino acid C-terminal CLE domain(s) (Mitchum et al., 2012). In the case of plant CLEs, the proteins are targeted to the plant cell secretory pathway for secretion to the extracellular space, whereas the nematode CLEs are targeted for secretion from the dorsal gland cell via the stylet into plant tissues. *CLE* genes are critical to the parasitic success of these nematodes as was demonstrated by post-transcriptional gene silencing via RNA interference (RNAi) (Bakhetia, Urwin, & Atkinson, 2007; Patel et al., 2008). Synthesis of the 12-aa CLE motif peptides and exogenous application to plant roots or overexpression of cyst nematode *CLE* genes in plants (Lu et al., 2009; Wang et al., 2005; Wang, Lee, et al., 2010; Wang et al., 2011) causes a range of phenotypes, including premature termination of shoot, root and floral meristems similar to plant CLEs. Cyst nematode CLEs were also shown to partially or fully complement the phenotypes of an Arabidopsis *clavata3* (*clv3*) mutant (Lu et al., 2009; Wang et al., 2005; Wang, Lee, et al., 2010; Wang et al., 2011), a plant CLE which regulates the balance of stem cell proliferation versus differentiation in the central zone of the shoot apical meristem. Together these data indicate that cyst nematode CLEs may function as ligand mimics of endogenous plant CLE peptides.

CLEs are the most abundantly produced effector in the dorsal oesophageal gland cell of parasitic life stages and their detection in the host cell (syncytium) cytoplasm (Lu et al., 2009; Wang et al., 2005; Wang, Lee, et al., 2010) is consistent with a role in feeding cell formation and maintenance. The VD of mature CLE proteins is sufficient for trafficking of the CLE peptides from the cytoplasm back to the apoplast to exert their function (Wang, Joshi, Korkin, & Mitchum, 2010). It has been hypothesized that nematode CLEs are likely trafficked into the host cell secretory pathway to acquire post-translational modifications (i.e. glycosylation) similar to endogenous plant CLE peptides that are important for their function in planta (Ohyama et al., 2009). Using ectopic in planta expression coupled with nanoLC-MS/MS analysis, it was recently determined that the in vivo bioactive forms of GrCLE1 (Lu et al., 2009), a multidomain CLE protein secreted by PCN during plant parasitism, are 12-aa glycopeptides with striking structural similarity to mature plant CLE peptides (Ohyama et al., 2009; Chen et al., 2015), adding direct evidence that cyst nematode CLE effectors can be modified and processed by host cellular machinery to become bioactive mimics of endogenous plant CLE signals. In eukaryotes, it is well established that proteins carrying a secretion signal cross the endoplasmic reticulum (ER) membrane via co-translational translocation mechanisms; however, more recent findings in yeast and mammalian systems indicate that short secretory proteins can also be targeted and inserted into the ER post-translationally (Johnson, Powis, & High, 2013). Such pathways, while they likely exist, have not yet been characterized in plants underscoring the potential significance of these findings. Thus, the VD function highlights a novel concept in pathogen effector trafficking by illustrating the potential for movement of effectors from the cytoplasm to the apoplast following delivery to host cells. This idea is significant as it ascribes a new role to undiscovered post-translational trafficking pathways in plants as a subcellular host target of pathogens to facilitate infection.

It is well established that plant CLE domains are processed into 12 or 13 amino acid peptides, several of which serve as ligands for receptors to mediate plant signaling important for shoot, floral, root and vascular stem cell maintenance by regulating the expression of members of the WUSCHEL-like homeobox (WOX) transcription factor family. Several receptor–CLE ligand pairs have been characterized and direct physical interactions between plant CLE peptides and the extracellular leucine-rich repeat (LRR) domains of corresponding receptors have been demonstrated (Guo, Han, Hymes, Denver, & Clark, 2010; Hirakawa et al., 2008; Ogawa, Shinohara, Sakagami, & Matsubayashi, 2008). To better understand the function of cyst nematode CLEs in parasitism, a concerted effort has been made to identify the host plant receptors of these peptides. Several LRR receptor-like kinases (RLKs) including Arabidopsis CLAVATA1 (CLV1), the CLAVATA2 (CLV2)/CORNYE (CRN) heterodimer receptor complex and receptor-like protein kinase 2 (RPK2) have been shown to be required for BCN-CLE signaling (Replogle et al., 2011, 2013). Plants harbouring knock-out mutations in any of these receptors leads to a 25% decrease in nematode development coupled with reduced syncytium size linking nematode CLE signaling with the formation and/or maintenance of syncytia (Replogle et al., 2011, 2013). When combined in a *clv1clv2rpk2* triple mutant, nematode development is reduced by up to 60% (Replogle et al., 2013) indicating that parallel signaling pathways involving CLV1, CLV2/CRN and RPK2 are important for BCN parasitism of *Arabidopsis*. Direct evidence that nematode CLEs can form receptor–ligand complexes with several of these receptors has been demonstrated (Guo et al., 2010; Guo, Ni, Denver, Wang, & Clark, 2011). Consistent with this, the plant CLE receptor proteins are constitutively expressed in the initial syncytia cell, and/or induced in feeding cells or adjacent cells to perceive CLEs secreted into root cells by cyst nematodes (Replogle et al., 2011, 2013).

The identification of multiple *CLE* genes encoding different peptides from individual species of cyst nematodes, as well as single genes encoding CLE proteins with tandem, polymorphic CLE motifs (Lu et al., 2009) supports the notion that multiple host receptors are targeted simultaneously by the nematode. Therefore, the spatial localization of the plant receptors of nematode CLEs is not only likely to be highly complex, but also crucial for proper feeding cell development. Efforts to develop improved confocal imaging methods should allow for more precise monitoring of the spatial and temporal dynamics of receptor protein expression in real time during the process of feeding cell formation (Dinh, Knoblauch, & Elling, 2014). Additionally, mapping studies that pinpoint regions within the LRR domain of receptors that are necessary for perception and binding to nematode CLE peptides should provide additional insight on function and specificity, and the transfer of this knowledge to plant species of agronomic importance is opening avenues for the design of targeted novel resistance to cyst nematodes (Chen et al., 2015; Guo et al., 2015).

Despite extensive progress in cyst nematode CLE effector biology, many crucial questions with regard to the function of CLE effectors in feeding cell formation still remain. Little is known about the molecular components that act downstream of plant and nematode CLE perception, thus future studies directed at elucidating the host signaling pathways modulated by nematode CLE peptides are critical to advancing our understanding of CLE peptides in cyst nematode parasitism and may simultaneously shed new light into plant CLE signaling pathways.

2.2 Phytohormone Balance and Signaling

Local changes in phytohormone levels and/or signaling at the site of cyst nematode feeding within roots has been well established (Absmanner, Stadler, & Hammes, 2013; Goverse et al., 2000; Grunewald et al., 2008; Grunewald, Cannoot, Friml, & Gheysen, 2009; Hewezi et al., 2014; Karczmarek, Overmars, Helder, & Goverse, 2004; Wubben, Su, Rodermel, & Baum, 2001). Not surprisingly, the deployment of stylet-secreted effectors (SSEs) by cyst nematodes to co-opt developmental programs that modulate phytohormone signaling pathways for the establishment of feeding cells is an emerging theme in parasitism. A growing body of evidence also suggests crosstalk between plant CLE–receptor–WOX and phytohormone signaling pathways for meristem regulation (Leibfried et al., 2005; Kondo, Hirakawa, Kieber, & Fukuda, 2011). Therefore, the intersection of phytohormone and nematode CLE signaling may in fact be an early requirement for the initiation of syncytia. However, further work is needed to establish these potential linkages.

The BCN *H. schachtii* dorsal gland effector protein 19C07 is the first effector implicated in the direct modulation of host transporter proteins to alter the phytohormone balance in favour of syncytium development. Protein–protein interaction studies identified the Arabidopsis LAX3 protein, a member of the AUX/LAX family of auxin influx transporters, as a host target of 19C07 (Lee et al., 2011). LAX3 mutants fail to form lateral roots due to defects in lateral root emergence. In the absence of LAX3, auxin is not pumped into cells overlaying lateral root primordia, which is required for the upregulation of cell wall loosening enzymes that allow for lateral roots to emerge. In response to nematode infection, LAX3 is expressed in initial syncytial cells and then shifts to adjacent cells prior to their incorporation into the developing syncytium. Consistently, a LAX3-regulated polygalacturonase gene shows a similar pattern of expression. Thus, LAX3 likely facilitates the incorporation of adjacent cells into the developing syncytium by increasing auxin levels, which then leads to the activation of cell wall modifying proteins necessary for wall loosening and dissolution. No nematode infection phenotype was observed on the *lax3* mutant most likely due to functional redundancy within this family of transporters and the upregulation of AUX1 and LAX1 in syncytia (Lee et al., 2011; Mazarei, Lennon, Puthoff, Rodermel, & Baum, 2003); however, significant reductions in infection were observed on the double *aux1lax3* and quadruple *aux1lax1lax2lax3* mutants confirming the requirement for auxin in feeding cell formation. Interestingly, these mutants exhibited hyperinduction of lateral roots at feeding sites suggesting that other classes of auxin influx transporters are being upregulated in the mutants or that the nematode is secreting other effectors or small molecules that bypass the auxin requirement by targeting auxin signaling to upregulate auxin-inducible genes (Lee et al., 2011). Regardless, this study highlights not only the requirement of auxin influx transporters, but also their proper regulation for feeding cell formation. When 19C07 is constitutively expressed in Arabidopsis lateral root emergence was not inhibited, instead lateral roots were observed to emerge more quickly. These data suggest that this effector is a positive regulator of LAX3 activity. Further work is needed to explore the mechanistic basis of this modulation. As more novel effectors continue to be characterized, it will be interesting to uncover those that directly target host components of auxin or other phytohormone signaling pathways.

2.3 Cell Wall Architecture

Cyst nematodes secrete a myriad of cell wall modifying enzymes to facilitate penetration and migration through root tissues and we refer readers to a recent thorough review of this topic (Bohlmann & Sobczak, 2014). However, we highlight here the putative function of a cellulose-binding protein (CBP) effector originally identified from the SCN *H. glycines* that peaks in expression at the J3 stage (Gao et al., 2003). The orthologous protein in BCN *H. schachtii* was further characterized and shown to play an important role in the early stages of syncytium formation (Hewezi et al., 2008). The CBP effector is interesting because it consists of only a SP and a cellulose-binding domain (CBD). The protein lacks hydrolytic activity, but is capable of binding cellulose in vitro. When expressed without a SP in Arabidopsis, the plants exhibited a long-root phenotype and were more susceptible to *H. schachtii*, the latter of which could not be explained by increased root area available for infective juveniles. Interestingly, the nematode CBP was found to interact with a plant pectin methyltransferase 3 (PME3) that is upregulated during nematode infection. Consistent with this, nematode infection was impaired on *pme3* mutant plants. Although the site of action of these two proteins is presumably the cell wall, the site of interaction of these two proteins requires further investigation. Direct delivery of CBP to the cytoplasm or to the apoplast remains equally plausible scenarios. Subcellular localization of PME3 was found to be in the cytoplasm and cell wall, so the possibility exists that these effectors interact in the host cytoplasm and this aids in CBP trafficking to the cell wall. Expression and activity of PME3 was increased in plants overexpressing CBP suggesting that this nematode effector may function to increase the activity of its host target for the demethylesterification of pectin in the cell wall to facilitate early stages of syncytium formation by increasing wall polymer access to other cell wall modifying enzymes (Hewezi et al., 2008).

3. MODULATION OF HOST STRESS AND DEFENCE RESPONSES

3.1 Regulators of ROS

The use of reactive oxygen species (ROS) has long been considered a central component of plant defence responses to many pathogens, nematodes included (Lamb & Dixon, 1997). In addition to the direct toxic effect of an oxidative burst on a pathogen, plants can also use these molecules as signals to regulate multiple processes including hormone signaling and development (Mittler, Vanderauwera, Gollery, & Van Breusegem, 2004). To combat this, plant-parasitic nematodes utilize a suite of specialized effectors that impact multiple steps in this pathway and subsequently protect the nematode while it proceeds through the stages of its life cycle.

One of the more recent effectors that appears to be implicated in suppressing ROS production as a defence response is the cyst nematode effector *4F01* (Patel et al., 2010). Originally isolated from a *H. glycines* oesophageal gland cell library (Gao et al., 2003), this effector shows high sequence similarity to the annexin protein family and is expressed in the dorsal gland of the parasitic juvenile stage onwards. Proteins from this family have been identified in both plants and animals and are characterized by a Ca^{2+}-/lipid-binding domain that allows them to function in a variety of different activities (Gerke & Moss, 2002). In addition, members contain an amino acid repeat of approximately 70 bases that contributes to a conserved structural fold in the protein. Plants utilize annexins for multiple functions that can be associated with a broad variety of abiotic functions such as osmotic, drought and oxidative stresses (Lee et al., 2004; Konopka-Postupolska, Clark, & Hofmann, 2011). Through the use of a yeast two-hybrid screen the effector homologue in *H. schachtii* (*Hs4F01*) was found to interact with an oxidoreductase from Arabidopsis (Patel et al., 2010). This oxidoreductase had been previously implicated in plant stress responses and resistance to the downy mildew pathogen *Hyaloperonospora parasitica* (Van Damme, Huibers, Elberse, & Van den Ackerceken, 2008). Although the consequence of this interaction has not been fully investigated as of yet, there are multiple possibilities for the function of the *Hs4F01* effector in manipulating this enzyme to either directly or indirectly benefit the nematode's survival. The first of these possibilities is triggering the oxidoreductase to catalyse the breakdown of free oxygen radicals, preventing damage to the nematode as a result of these defence compounds. Another possible option is that the targeted oxidoreductase functions in regulation of defence pathways and the nematode suppresses this signaling by targeting it, resulting in inhibition of the defence pathways involved.

Cyst nematodes have also been shown to secrete a variety of enzymes that may function in detoxifying active oxygen species within the plant. Most of the work identifying these enzymes to date has been done in the PCN *G. rostochiensis*, although there is preliminary evidence of the existence of these enzymes in other cyst nematode species. One of the first detoxifying enzymes identified in plant-parasitic nematodes was a member of the peroxiredoxin family and labelled Gr-TpX (Robertson et al., 2000). These enzymes are involved in the removal of H_2O_2 produced by host defence responses, preventing the formation of hydroxyl radicals that may harm the nematode. The Gr-TpX initially isolated from an infective J2 cDNA library shows high similarity to peroxiredoxin sequences from animal parasitic nematodes with two conserved valine–cysteine–proline motifs associated with an active enzyme site. Interestingly enough, despite lacking a predicted N-terminal SP sequence, Gr-TpX clearly exists at the surface of the nematode during infection as detected by immunolocalisation and transmission electron microscopy (Robertson et al., 2000). This observation could be the result of two possible situations, due to either an uncharacterized signal sequence or perhaps a nontraditional secretion mechanism not requiring a secretion signal. Distinguishing between these possibilities will require much further investigation before too much validity is placed in either one. Further confirmation of Gr-TpX as a peroxiredoxin was shown as it catalyses the breakdown of hydrogen peroxide, although it is unable to catalyse the breakdown of *t*-butyl or cumene hydroperoxides. This suggests a high specificity for the substrate of this enzyme, implicating the need for other nematode secretions to ensure full protection.

In addition to peroxiredoxin, glutathione peroxidases (GpX) have also been identified from the cyst nematode *G. rostochiensis* (Jones, Reavy, Smant, & Prior, 2004). Here two different GpX were identified and characterized, one with a SP and one without. The secreted glutathione peroxidase (*gr-gpx-1*) shows expression in the hypodermis of the nematode during the infective stages of the *G. rostochiensis* life cycle. While the nonsecreted GpX (*gr-gpx-2*) has a temporally similar expression pattern, it is expressed throughout the nematode, in particular in highly metabolically active tissues. The spatial expression differences between these two are an indication that the nonsecreted GpX functions in the interior of the nematode for metabolic processes while the secreted GpX acts in a manner similar to that of Gr-TpX, coating the exterior of the nematode in a protective manner. Secreted GpX on the nematode surface may interact with host plant hydrogen peroxide, removing it from the available pool to form radical oxygen species. In addition, it may also act on host lipid peroxides, preventing the lipid peroxidation that may function as a cellular signal triggering further defence responses.

The final effector implicated in cyst nematode ROS defence is superoxide dismutase (SOD). Identified in nematode secretions without a detectable secretion signal, it represents another case similar to Gr-TpX where nontraditional secretion may be involved (Robertson, Robertson, & Jones, 1999). As for its function, SOD converts the superoxide anion into hydrogen peroxide and oxygen. In this aspect, it acts upstream of the other enzymes mentioned here.

When considered together, these four secreted proteins represent multiple approaches by the nematode to undermine the defensive properties granted to plants by the production of ROS. Either through manipulating host machinery or secreting protective enzymes the nematode reduces the levels of harmful oxygen species to a more hospitable level. However, the additional functions of some of these effectors suggest they may play other roles in the nematode's life cycle, both metabolic and developmental.

Interestingly, new research has shed light on how cyst nematodes may also manipulate signaling through ROS in a more tightly controlled manner to establish a successful parasitic association. Remarkably, Arabidopsis mutants in two genes associated with cell death and ROS production in response to fungal and bacterial pathogens, *RbohD* and *RbohF*, were more resistant to infection by the cyst nematode *H. schachtii* (Siddique et al., 2014). By manipulating these genes the nematode is able to curb the plant cell death response during migration and later interfere with defences mediated by the hormone salicylic acid (SA). This mechanism introduces the concept of the nematode pitting various plant defences against each other to prevent a successful defence against the pathogen. How this interaction is mediated by the nematode and whether the other ROS-regulating effectors contribute to this will need to be examined further.

3.2 Nuclear-Targeted Effectors

Modulation of plant immune responses by pathogen effectors also involves targeting host nuclear functions. For example, AVR2 from *Fusarium oxysporum*, a xylem colonizing bacteria, interacts with the tomato I-2 resistance protein in the nucleus to elicit cell death (Ma, Cornelissen, & Takken, 2013). CRN or Crinklers are a diverse class of effectors from plant pathogenic oomycetes that exclusively target the host nucleus upon delivery to trigger cell death (Stam et al., 2013) and nuclear-targeted TAL effectors from *Xanthomonas* function as potent transcriptional activators and bind to plant resistance gene promoters (Canonne & Rivas, 2012; reviewed by Rivas & Deslandes, 2013). Bioinformatic analysis of several stylet-secreted effector proteins from cyst nematodes predicted the presence of nuclear localization signals (Elling, Davis, Hussey, & Baum, 2007; Gao et al., 2003; Jones et al., 2009) and the function of these signals in targeting effectors to the plant nucleus has been confirmed through in planta transient expression assays. This suggests that effectors may directly target essential host nuclear components and manipulate host transcription regulation to promote successful infection. Several such effectors have been studied and roles in modulating host immunity are emerging.

One such class of cyst nematode effectors with a probable role in the activation and suppression of effector-triggered immunity (ETI) is the secreted SP1a and Ryanodine receptor (SPRY) domain (SPRYSEC) protein family of PCNs *G. rostochiensis* and *G. pallida* (Carpentier et al., 2012; Cotton et al., 2014; Postma et al., 2012; Rehman et al., 2009; Sacco et al., 2009). In *G. pallida*, 299 proteins containing a SPRY/B30.2 domain were mined from the genome assembly which includes the SPRYSEC family of effectors (Cotton et al., 2014). Structurally, the SPRY/B30.2 domain is comprised of two β sheets that form a bent β-sandwich fold with structural similarity to immunoglobin fold that facilitates intermolecular protein–protein interactions and is implicated in innate and adaptive immunity in murines and humans (D'Cruz, Babon, Norton, Nicola, & Nicholson, 2013; Rhodes, De Bono, & Trowsdale, 2005). Sequence diversity and variability among SPRYSECs is evident in the protruding loops on the surface of the protein with homogeneity restricted to the 120 amino acid SPRY domain. Although some SPRYSEC family members are predicted to be cytoplasmic, many harbour predicted nuclear localization signals. Localization of several of these effectors to the nucleus, and in some cases the nucleolus, has been confirmed by transient in planta subcellular localization studies (Jones et al., 2009).

SPRYSECs are highly expressed in the dorsal gland cell during the early stages of parasitism. Several lines of evidence indicate that SPRYSECs target members of the coiled-coil nucleotide-binding leucine-rich repeat (CC-NB-LRR) class of immune receptors either for the activation or suppression of plant defence. One such example is SPRYSEC-19 of *G. rostochiensis*. When used as a bait in a yeast two-hybrid screen to identify host protein interactors from a tomato root library, SPRYSEC-19 interacted with the LRR domain of the CC-NB-LRR protein SW5F, a member of the *SW5* gene cluster which confers resistance to a broad range of tospoviruses in tomato. However, tomato plants carrying the *SW5-F* gene are fully susceptible to *G. rostochiensis*. Co-expression of SW5-F and SPRYSEC-19 in *Nicotiana benthamiana* leaves by agroinfiltration did not result in activation of ETI. Paradoxically, it was later shown to suppress it (Postma et al., 2012). Two models have been put forward by the authors regarding the role of

SPRYSEC-19 and SW5F in nematode–plant interactions. The first model suggests that SWF-5 may be an evolutionary intermediate since interaction occurs, but it does not trigger a hypersensitive response or it may possibly be suppressed by another effector. In the second model, SPRYSEC-19 targeting of SWF-5 may serve a virulence function by modulating host defence responses (Rehman et al., 2009). Additional work is needed to distinguish between these possibilities.

Another SPRYSEC protein RBP-1 of *G. pallida* (GpRBP-1) is recognized by the CC-NB-LRR resistance protein GPA-2 which confers resistance to specific strains of *G. pallida* in potatoes (Sacco et al., 2009). This finding lends further support that the SPRYSEC family is likely to be under strong selective pressure to evolve to evade the defence mechanisms of the host plant, which may explain the expansion of this nematode effector gene family in *Globodera* (Cotton et al., 2014). In contrast to GrSPRYSEC-19 which has been shown to suppress ETI mediated by GPA-2 and several NB-LRR resistance proteins (Postma et al., 2012), GpRBP-1 elicits an hypersensitive response (HR) in transient assays with GPA-2 suggesting that GpRBP1 is the effector that triggers GPA-2-mediated nematode resistance, although this remains to be shown since GpRBP1 is also present in virulent populations of *G. pallida*. Whether or not direct binding between GpRBP-1 and GPA-2 is required for recognition has not been confirmed experimentally. GpRBP-1 has a high degree of sequence polymorphism within and between populations. In fact, a single amino acid polymorphism in the SPRY domain of GpRBP-1 governs recognition by GPA-2. A proline to serine substitution at position 187 eliminates recognition by GPA-2 (Sacco et al., 2009). This lends support to the model that SWF-5 may be an evolutionary intermediate which interacts with SPRYSEC-19, but does not elicit HR an unlike GpRBP-1. Future work directed at understanding the virulence functions of SPRYSECs will help better understand the various roles this class of effectors may play in host immune modulation.

The cyst nematode ubiquitin carboxyl extension proteins (UBCEPs) represent a unique class of nuclear-targeted effectors produced in the dorsal oesophageal gland cell of parasitic life stages of *Heterodera* and *Globodera* species (Chronis et al., 2013; Gao et al., 2003; Tytgat et al., 2004) that may have a dual function in co-opting host nuclear functions and influencing the host 26S proteosome to suppress plant defence and promote successful infection (Chronis et al., 2013). Ubiquitin is a small 76 amino acid protein that is highly conserved in eukaryotes. In eukaryotic genomes, two classes of ubiquitin encoding genes are present. One class encodes for polyubiquitin and the other for the UBCEPs which are comprised of a single ubiquitin domain and a carboxyl extension protein (CEP) domain of variable length (52–80 amino acids) that is proteolytically cleaved following translation. While the ubiquitin domain plays a role in protein degradation, CEPs are associated with the ribosomes. Cyst nematode UBCEPs are unique in that they also contain predicted N-terminal SPs that target these effectors for secretion from the dorsal gland through the nematode stylet into plant tissues (Chronis et al., 2013). The ubiquitin domain of *Heterodera* and *Globodera* UBCEPs shares significant similarity to plant ubiquitin; however, the nematode CEP domains which range from 12 to 19 amino acids in length differ significantly between genera and lack similarity to CEPs from other eukaryotes. The CEP domains of *Heterodera* UBCEPs contain a predicted nuclear localization signal and the Hs-UBI1 protein localized to the nucleolus in transient assays of tobacco BY-2 cells (Tytgat et al., 2004).

The potential dual function of UBCEPs in plant parasitism was demonstrated through studies of GrUBCEP12 from the PCN *G. rostochiensis* (Chronis et al., 2013). GrUBCEP was shown to be cleaved by endogenous plant enzymes to release functional ubiquitin and a 12 amino acid carboxyl extension peptide (CEP12). Unlike *Heterodera* CEPs, GrCEP12 does not contain a predicted nuclear localization signal; however, interestingly, subcellular localization studies demonstrated that both GrUBCEP12 and its derived GrCEP12 peptide have characteristics of a dual nucleocytoplasmic-localized effector. In support of an important role in parasitism, transgenic potato lines expressing GrUBCEP12 increased plant susceptibility to *G. rostochiensis*, whereas plant host-derived RNAi of GrUBCEP12 led to a reduction in infection.

The growing body of evidence indicating a role of the ubiquitin/proteasome system (UPS) in plant defence along with the observed differential regulation of components of the UPS in various plant–pathogen interactions (Dielen, Badaoui, Candresse, & German-Retana, 2010), including during syncytium formation (Ithal et al., 2007; Kandoth et al., 2011), led the authors to evaluate a suite of UPS-related genes for altered expression in response to *G. rostochiensis*. One of these, the *RPN2a* gene encoding a subunit of the 26S proteasome was downregulated in both potato roots infected with *G. rostoschiensis* as well as in GrUBCEP12 expressing potato lines, but not GrCEP12 expressing lines. The *RPN2a* gene was found to be upregulated in syncytia induced by *H. glycines* on resistant soybean plants (Kandoth et al., 2011), which suggests that it may play a role in plant defence against nematodes. Therefore, free ubiquitin released from nematode-secreted

UBCEPs may be important for suppressing plant defence by affecting the host 26S proteasome (Chronis et al., 2013) although more in depth studies are needed to test this possibility.

Using agroinfiltration assays, it was further demonstrated that GrCEP12 plays a role in modulating host immunity. In transient assays, GrCEP12 could suppress ETI by Gpa2/RBP-1 and Rx2/CP in *N. benthamiana*. In addition, transgenic potato plants expressing GrCEP12 were found to be more susceptible to both *G. rostochiensis* and *Streptomyces scabies*, which suggested that this peptide may play a role in suppression of basal plant defence against pathogens. When GrCEP12 was transiently expressed in *N. benthamiana* by agro-infiltration it dramatically suppressed flg22-induced ROS production. Additionally, the expression of PAMP-triggered immunity (PTI) marker genes was significantly reduced by GrCEP12 compared to controls indicating PTI suppression by GrCEP12 (Chen, Chronis, & Wang, 2013). The identification of the host target(s) of GrCEP12 should provide additional insight into the role of this peptide in modulating host immunity.

3.3 Apoplastic and Cytoplasmic Effectors

Several SSEs localized in the cytoplasm and apoplast have also been implicated in modulation of plant immune responses. The first nematode effector with a potential role in immune modulation and feeding cell formation was chorismate mutase. Originally identified from the root-knot nematode *Meloidogyne javanica*, chorismate mutase has since been identified as a critical effector from most plant–parasitic nematodes (PPNs) examined to date (Gao et al., 2003; Haegeman, Joseph, & Gheysen, 2011; Jones et al., 2003; Lambert, Allen, & Sussex, 1999; Lu, Tian, Borchardt-Wier, & Wang, 2008; Nyaku et al., 2013; Vanholme et al., 2009; Yu, Chronis, Lu, & Wang, 2011) representing the spectrum of feeding lifestyles. Chorismate mutase is an enzyme of the shikimate pathway that converts chorismate to prephenate (Romero et al., 1995). Chorismate is a precursor in the biosynthesis of aromatic amino acids such as tryptophan, which serves as a precursor of indole-3-acetic acid, and phenylalanine, a precursor of SA and phytoalexins, which are molecules known to mediate defences against pathogens. Most of what we know about the function of chorismate mutase in pathogenicity is based on studies of the root-knot nematode *M. javanica* and the smut fungus *Ustilago maydis* (Djamei et al., 2011; Doyle & Lambert, 2003). Research on *M. javanica* led to the suggestion of the chorismate competition model, where subcellular chorismate levels are altered by the pathogen, resulting in a change in the production of phenolic compounds in the cell. This may include altering the auxin balance or reducing the availability of chorismate for conversion to SA (Djamei et al., 2011; Lambert et al., 1999) thereby suppressing plant defence.

In cyst nematodes, chorismate mutase has been cloned from several species of *Heterodera* (Bekal, Niblack, & Lambert, 2003; Vanholme et al., 2009) and *Globodera* (Jones et al., 2003; Lu et al. 2008; Yu et al., 2011). In *Heterodera*, chorismate mutase is encoded by a multigene family, however, in *Globodera* no related sequences were detected by Southern blotting (Jones et al., 2003). Nematode chorismate mutases exhibit strong sequence similarity to bacterial proteins as opposed to plant proteins suggesting that this gene was likely acquired from microbes via horizontal gene transfer similar to nematode endoglucanases (Yan et al., 1998).

In *Globodera*, the spatial expression pattern of the *CM* genes localized to the subventral oesophageal gland cells of the nematode (Jones et al., 2003; Lu et al., 2008). In contrast, the hybridization signal for expression of *CM* genes in *Heterodera* species was detected in the subventral and dorsal gland cells during the preparasitic stages and early parasitic stages (Bekal et al., 2003; Lambert et al., 2005; Vanholme et al., 2009). The additional expression of *CM* genes in the dorsal gland of *Heterodera* species may indicate a further function in parasitism besides whatever role CM plays in phytonematode infection. The fact that *CM* genes have been identified in such a broad range of phytonematodes that vary greatly in their lifestyles is a possible indication of a broad requirement for this effector in order to parasitize effectively, perhaps suppressing some core nematode-triggered defence response or helping the nematodes acquire key nutrients needed to moult and complete their life cycle.

In multiple species of *Globodera*, alternatively spliced versions of chorismate mutase have been identified, each showing unique expression profiles during development of the nematode (Lu et al., 2008). This may be a way for the nematode to modulate the interaction with the host plant in order to compensate for changing requirements during development. Investigating the different role these isoforms have within the plant cell could lead to insight into the details of parasitism and changing modulation of plant defences. Furthermore, in *H. glycines* a link has been shown between polymorphisms in chorismate mutase and virulence on some resistant cultivars (Bekal et al., 2003). This strengthens the implication of chorismate mutase as a modulator of defence responses. The presence of large-scale polymorphisms implies a lack of a fitness requirement on this effector and fits the profile of a suppressor of defence, changing sequence to avoid recognition by the plant.

Together with the broad distribution of the CM effector family across diverse phytoparasitic nematodes, this evidence could also indicate that chorismate mutase was one of the early evolutionary changes required for effective parasitism of plants.

Venom allergen-like proteins (VAPs) represent one of the most intriguing classes of effector proteins found in cyst nematodes. Although their role in nematode parasitism is largely unknown, these proteins most likely play an important role in establishment of the parasitic relationship between plants and nematodes as modulators of host immunity. VAPs are a class of cysteine-rich secretory proteins (CRISPs) belonging to the sperm-coating protein (SCP)-like extracellular protein family, also called SCP/Tpx-1/Ag5/PR-1/Sc7 (SCP/TAPS) and are assigned to the CRISP 'superfamily' (Chalmers & Hoffmann, 2012, Cantacessi et al., 2009). SCP/TAPS proteins play key roles in a number of fundamental biological processes, such as host–pathogen interactions and modulation of host defence mechanisms (Cantacessi & Gasser, 2012). Gene sequences of VAPs have been identified in numerous animal and plant-parasitic nematodes as well as the free-living nematode *Caenorhabditis elegans*. In animal parasitic nematodes, VAPs have been identified from *Ancylostoma caninum* (ASP-2) (Hawdon, Jones, Hoffman, & Hotez, 1996), filarial nematodes, *Onchocerca volvulus* (Tawe, Pearlman, Unnasch, & Lustigman, 2000) and *Brugia malayi* (Murray, Gregory, Gomez-Escobar, Atmadja, & Maizels, 2001), and the trematodes *Schistosoma mansoni* (Chalmers et al., 2008) among others. VAPs secreted by animal parasitic nematodes elicit IgE-mediated immune responses in their hosts (Fitzsimmons, Falcone, & Dunne, 2014). In plant-parasitic cyst nematodes, VAP genes have been isolated from pine wilt nematode *Bursaphelenchus xylophilus* (Lin, Jian, Zhao, Yang, & Liu, 2011), several species of root-knot nematode *Meloidogyne* (Ding, Shields, Allen, & Hussey, 2000; Wang, Li, Hu, Fu, & Xu, 2007), peanut pod nematode *Ditylenchus africanus* (Haegeman et al., 2009) and the lesion nematode *Pratylenchus coffeae* (Haegeman et al., 2011). The identification of VAPs from very different orders of the phylum Nematoda is strikingly different from many other effectors. They offer an interesting example of divergent evolution of nematode effectors to facilitate effective parasitism in a variety of feeding styles. Also of note is the inclusion of these proteins in a free-living nematode, *C. elegans*. What purpose these proteins serve in a nonpathogenic organism and how they became adapted to manipulate host defences is completely unknown.

With regard to their function in plant-parasitic nematodes, VAPs have been implicated in the early stages of infection as high transcript levels were observed in the subventral glands of preparasitic and early parasitic juveniles, while low or no transcripts were observed in the older juveniles and adults (Gao et al., 2001; Lozano-Torres et al., 2012). In *H. glycines*, two venom allergen-like genes *VAP-1* and *VAP-2* have been cloned, but beyond the initial characterization and developmental expression in *H. glycines* little else is currently known about the molecular function and impact of these effectors on parasitism. By contrast, much more is known about the VAPs in *G. rostochiensis*, particularly GrVAP1. In *G. rostochiensis*, VAP1 appears to function during the early stages of parasitism and feeding site formation and is presumably secreted along with plant cell wall modifying proteins into the apoplast, although direct detection in planta has not been demonstrated. In yeast two-hybrid analyses with a tomato root cDNA library, GrVAP1 interacted with an apoplastic papain-like cysteine protease RCR3. RCR3 is guarded by the resistance protein Cf-2 in tomato. Cf-2 confers resistance to the fungal pathogen *Cladosporium fulvum* by sensing perturbations to RCR3 by the fungal protease inhibitor AVR2 (reviewed by Wulff, Chakrabarti, & Jones, 2009). GrVAP1 specifically interacts with $RCR3^{pim}$ of *Solanum pimpinellifolium* and this interaction perturbs the active site of $RCR3^{pim}$. The changes in RCR3 trigger Cf-2-mediated defence responses leading to programmed cell death and nematode resistance (Lozano-Torres et al., 2012). Plants harboring $RCR3^{pim}$ are more susceptible to *G. rostochiensis* in the absence of Cf-2, underscoring the role of RCR3 as a common virulence target of two different types of pathogens that secrete structurally different effectors. Thus, investigations of this nematode effector highlight some intriguing interactions of nematode effectors with host plant proteins. Moreover, the guarding of RCR3 by Cf-2 represents one of the first resistance models of this kind implicated in resistance to a plant-parasitic nematode.

Another effector that increases host susceptibility to nematode infection by targeting specific plant pathogenesis-related (PR) proteins with a role in host defence is 30C02. This novel effector protein is expressed in the dorsal esophageal gland cell of *H. schachtii* and *H. glycines* (Hamamouch et al., 2012). Expression of *30C02* in Arabidopsis plants increased plant susceptibility to infection by the cyst nematode *H. schachtii* but not by the root-knot nematode *Meloidogyne incognita* suggesting a specific role of this effector in cyst nematode infection. Yeast two-hybrid screening identified β-1,3-endoglucanase as a host target of 30C02. β-1,3-endoglucanases play an active role in defence against fungal pathogens and are grouped in the PR-2 protein family (Van Loon, Rep, & Pieterse, 2006). In cyst nematodes, the level of expression of 30C02 is highest during the early stages of infection, which correlates with high expression of β-1,3-endoglucanases in the plant. This implies a counter mechanism to combat the increased level of plant defence genes and to promote successful infection by cyst nematode. Furthermore, a β-1,3-endoglucanase mutant exhibited increased susceptibility to nematodes and overexpression lines exhibited reduced infection. Taken together, this evidence presents a case for 30C02 as a unique cyst nematode effector whose function in modulating plant immunity may be to block secretion or inhibit the activity of PR proteins involved in plant defence. This finding certainly warrants further exploration to shed additional mechanistic insight into this interaction.

A novel stylet-secreted effector 10A06 expressed in the dorsal gland of *H. glycines* and *H. schachtii*, is most likely secreted into the developing syncytium during early plant parasitism to modulate plant immunity through its interaction with spermidine synthase (SPDS), a host target enzyme involved in polyamine biosynthesis (Hewezi et al., 2010). Polyamines (PAs) are ubiquitous biogenic amines that play a key role in defence responses against abiotic and biotic stress. Similar to several other cyst nematode effectors discussed here, constitutive expression of this nematode effector in Arabidopsis led to an increased susceptibility not only to *H. schachtii* but also to *Pseudomonas syringae* and yellow strain of *Cucumber mosaic virus*. The presumed modulation of SPDS activity by 10A06 leads to elevated SPDS mRNA, spermidine content and enhanced polyamine oxidase activity, the latter of which is important for the induction of cellular antioxidant machinery (Hewezi et al., 2010). Consistent with this, a series of antioxidant genes encoding catalase, glutathione peroxidase and ascorbate peroxidase were upregulated following nematode infection of plants expressing either 10A06 or SPDS2 suggesting that this effector may function as a key modulator of ROS levels to promote successful infection. In addition, 10A06 overexpression downregulated SA responsive genes (*PR-1*, *PR-2* and *PR-5*) suggesting that increased susceptibility of the transgenic lines may be due to suppression of SA signaling. Taken together, these data suggest that cyst nematodes deploy the 10A06 effector to modulate host defences by regulating ROS levels and suppressing SA signaling.

4. GENOME-ENABLED EFFECTOR DISCOVERY

In the 16 years, since the first phytonematode SSEs were identified from the cyst nematodes *G. rostochiensis* and *H. glycines* (Smant et al.,

1998) methods of identifying novel effectors have progressed rapidly. Beginning with a focus on isolating secretions that trigger an effect in host cells, study soon shifted to a more focused examination of expressed sequences associated with parasitism, in particular those sequences expressed in the oesophageal gland cells (Popeijus et al., 2000). Following this, the first microaspirated gland cell libraries were generated from *H. glycines* and the root-knot nematode *M. incognita*, and when coupled with SP prediction software and in situ hybridization techniques allowed for the identification of putative nematode effectors involved in parasitism (Gao et al. 2003; Huang et al., 2003). These new techniques allowed for a higher throughput examination of nematode secretions with an emphasis on the gland cells. More global examinations were performed using expressed sequence tag (EST) data from the accumulated cDNA libraries, resulting in the identification of more effector families, including the SPRYSEC family in *G. pallida* (Jones et al., 2009). In addition, recent work has generated a transcriptome of the cereal cyst nematode *Heterodera avenae* and provided a thorough examination of expressed secretory proteins (Kumar, Gantasala, Roychowdhury, Thakur, & Banakar, 2014). Homologous sequences were identified for several known effectors from other species of phytoparasitic nematodes, including SPRYSECs and VAPs. To improve on these methods even further would require the inclusion of secretions originating from other sources in the nematode, such as the cuticle and amphids.

With the sequencing of the root-knot nematode *Meloidogyne hapla* and *M. incognita* genomes representing the first phytoparasitic nematode genomes (Abad et al., 2008; Opperman et al., 2008) it was only a matter of time before the first cyst nematode genome was published, that of the PCN *G. pallida* (Cotton et al., 2014). The genome is similar in size with that of the free-living nematode *C. elegans* while nearly 25 Mb larger than that of any other sequenced phytoparasitic nematodes. Despite this much larger genome size, the predicted genes for *G. pallida* fall very close to those of the two root-knot nematode genomes, an indication of a relatively non-compact genome in the cyst nematode that is uncharacteristic of obligate parasites (Keeling, 2004). Utilizing the available genome and transcriptome data, a bioinformatics approach was taken to identify novel effectors in *G. pallida*. Transcriptome data provided information on secreted proteins upregulated in the parasitic stages of infection, followed by an examination of the proteins for known functional roles based on sequence homology. Using this approach, a total of 117 novel effectors previously unidentified in *G. pallida* were characterized (Cotton et al., 2014). This is an impressive number that may hold many key insights into the mechanisms of phytonematode parasitism once they are investigated and fully characterized. Given the large amount of novel effectors identified and variation from other types of phytonematodes, it will be interesting to see how other cyst nematode genomes such as those of *Heterodera* spp. stack up against the *G. pallida* genome once they are completed.

Besides interspecies comparisons, another area where advances in sequencing can contribute is examining effector repertoires of different populations within the same species that trigger very different host responses. It is well documented that resistance to the SCN *H. glycines* is incomplete, with some individuals within a population being able to propagate and over time generate an adapted population (Young, 1984). In the past, research approaches have attempted to cross individual nematodes of *H. glycines* and determine inheritance of nematode virulence, resulting in the identification of the reproduction on resistance or *ror* genes inherited in both a dominant and recessive manner (Dong & Opperman, 1997). However, since the initial discovery of these genes there has been no further information published concerning their sequence identity or mechanism in granting nematode virulence. Other studies as mentioned above have looked at individual effectors as sources of virulence such as chorismate mutase in *H. glycines*, where different variants have been identified that may correlate with virulence (Bekal et al., 2003; Lambert et al., 2005). Unfortunately, these studies are limited by having to focus on individual effectors instead of taking a more comprehensive global view of what is going on within the nematode. Using a comprehensive examination of the nematode may lead to a broader understanding of what is occurring that allows one individual to survive where another cannot. As sequencing improves and more cyst nematode genomes are published, whole genome resequencing becomes a more practical option for examining populations. With this approach individuals can be compared to the sequenced population and genetic differences can be narrowed in on for further investigation. Comparative analyses of effectors across species may also elucidate the basis of cyst nematode's narrow host range. Previous research using the effector orthologues of 10A06 in *H. glycines* and *H. schachtii* showed differences in the strength of their interaction with Arabidopsis SPDS2 suggesting host-specific differences in effector recognition (Hewezi & Baum, 2010). Furthermore, effector sequence polymorphisms among species of *Globodera* and subspecies of the tobacco cyst nematode *Globodera tabacum* are diagnostic, providing a tool for further investigation into what determines host range (Alenda, Gallot-Legrand, Fouville, & Grenier, 2013;

Chronis, Chen, Skantar, Zasada, & Wang, 2014; Yu et al., 2011). Combining this observation with the ability to sequence effectors across cyst nematode species may lead to an understanding of how effectors modulate interactions between various host plants, and as a result how this impacts infection by the nematode. In addition to providing insight on the evolution of these nematodes under selection pressure and the basis for host range, this information also serves a very practical application in leading to a better understanding of nematode reproduction on resistant hosts and may identify key targets to control these pathogens. Already, a better understanding of effectors has been applied to aid in the development and screening of new resistance genes against other types of pathogens (Vleeshouwers & Oliver, 2014). Advances in bioinformatics and progress towards more sequence data on cyst nematodes should speed up the pace of research in the field of effector biology considerably. Either through the identification of new effectors with genome and transcriptome mining or by applying these tools to tackle the challenge of population differentiation, underlying questions of mechanism and function can begin to be addressed.

ACKNOWLEDGEMENTS

The authors would like to thank lab members and collaborators through the years for their stimulating discussions and contributions to this work. The authors gratefully acknowledge funding for their work on cyst nematode effectors from the United States Department of Agriculture (USDA)-National Research Initiative Competitive Grants Program, United Soybean Board, Pioneer Hi-Bred, a USDA SCN Special Grant to the University of Missouri (MU), the Experiment Station of MU and an MU Life Sciences Graduate Research Fellowship to Michael Gardner.

REFERENCES

Abad, P., Gouzy, J., Aury, J. M., Castagnone-Sereno, P., Danchin, E. G., Deleury, E., et al. (2008). Genome sequence of the metazoan plant-parasitic nematode *Meloidogyne incognita*. *Nature Biotechnology, 26*, 909–915.

Absmanner, B., Stadler, R., & Hammes, U. Z. (2013). Phloem development in nematode-induced feeding sites: the implications of auxin and cytokinin. *Frontiers in Plant Science, 4*.

Alenda, C., Gallot-Legrand, A., Fouville, D., & Grenier, E. (2013). Sequence polymorphism of nematode effectors highlights molecular differences among subspecies of the tobacco cyst nematode complex. *Physiological and Molecular Plant Pathology, 84*, 107–114.

Bakhetia, M. B., Urwin, P. E., & Atkinson, H. J. (2007). qPCR analysis and RNAi define pharyngeal gland cell–expressed genes of *Heterodera glycines* required for initial interactions with the host. *Molecular Plant-Microbe Interactions, 20*, 306–312.

Bekal, S., Niblack, T. L., & Lambert, K. N. (2003). A chorismate mutase from the soybean cyst nematode *Heterodera glycines* shows polymorphisms that correlate with virulence. *Molecular Plant-Microbe Interactions, 16*, 439–446.

Bohlmann, H., & Sobczak, M. (2014). The plant cell wall in the feeding sites of cyst nematodes. *Frontiers in Plant Science, 5*.

Canonne, J., & Rivas, S. (2012). Bacterial effectors target the plant cell nucleus to subvert host transcription. *Plant Signaling Behavior, 7*, 217–221.
Cantacessi, C., Campbell, B. E., Visser, A., Geldhof, P., Nolan, M. J., Nisbet, A. J., et al. (2009). A portrait of the "SCP/TAPS" proteins of eukaryotes—developing a framework for fundamental research and biotechnological outcomes. *Biotechnology Advances, 27*, 376–388.
Cantacessi, C., & Gasser, R. B. (2012). SCP/TAPS proteins in helminths–Where to from now? *Molecular and Cellular Probes, 26*, 54–59.
Carpentier, J., Esquibet, M., Fouville, D., Manzanares-Dauleux, M. J., Kerlan, M. C., & Grenier, E. (2012). The evolution of the *Gp-Rbp-1* gene in *Globodera pallida* includes multiple selective replacements. *Molecular Plant Pathology, 13*, 546–555.
Chalmers, I. W., & Hoffmann, K. F. (2012). Platyhelminth Venom Allergen-Like (VAL) proteins: revealing structural diversity, class-specific features and biological associations across the phylum. *Parasitology, 139*, 1231–1245.
Chalmers, I. W., McArdle, A. J., Coulson, R. M., Wagner, M. A., Schmid, R., Hirai, H., et al. (2008). Developmentally regulated expression, alternative splicing and distinct sub-groupings in members of the *Schistosoma mansoni* venom allergen-like (SmVAL) gene family. *BMC Genomics, 9*, 89.
Chen, S., Chronis, D., & Wang, X. (2013). The novel GrCEP12 peptide from the plant-parasitic nematode *Globodera rostochiensis* suppresses flg22-mediated PTI. *Plant Signaling & Behavior, 8*, e25359.
Chen, S., Lang, P., Chronis, D., Zhang, S., De Jong, W. S., Mitchum, M. G., et al. (2015). In planta processing and glycosylation of a nematode CLAVATA3/ENDOSPERM SURROUNDING REGION-like effector and its interaction with a host CLAVATA2-like receptor to promote parasitism. *Plant physiology, 167*, 262–272.
Chronis, D., Chen, S., Lu, S., Hewezi, T., Carpenter, S. C. D., Loria, R., et al. (2013). A ubiquitin extension protein secreted from a plant-parasitic nematode *Globodera rostochiensis* is cleaved in planta to promote plant parasitism. *Plant Journal, 74*, 185–196.
Chronis, D., Chen, S., Skantar, A. M., Zasada, I. A., & Wang, X. (2014). A new chorismate mutase gene identified from *Globodera ellingtonae* and its utility as a molecular diagnostic marker. *European Journal of Plant Pathology, 139*, 239–246.
Cotton, J. A., Lilley, C. J., Jones, L. M., Kikuchi, T., Reid, A. J., Thorpe, P., et al. (2014). The genome and life-stage specific transcriptomes of *Globodera pallida* elucidate key aspects of plant parasitism by a cyst nematode. *Genome Biology, 15*, R43. http://dx.doi.org/10.1186/gb-2014-15-3-r43.
D'Cruz, A. A., Babon, J. J., Norton, R. S., Nicola, N. A., & Nicholson, S. E. (2013). Structure and function of the SPRY/B30. 2 domain proteins involved in innate immunity. *Protein Science, 22*, 1–10.
Davis, E. L., Hussey, R. S., Mitchum, M. G., & Baum, T. J. (2008). Parasitism proteins in nematode–plant interactions. *Current Opinion in Plant Biology, 11*, 360–366.
Dielen, A.-S., Badaoui, S., Candresse, T., & German-Retana, S. (2010). The ubiquitin/26S proteasome system in plant-pathogen interactions: a never-ending hide-and-seek game. *Molecular Plant Pathology, 11*, 293–308.
Ding, X., Shields, J., Allen, R., & Hussey, R. S. (2000). Molecular cloning and characterisation of a venom allergen AG5-like cDNA from *Meloidogyne incognita*. *International Journal for Parasitology, 30*, 77–81.
Dinh, P. T. Y., Knoblauch, M., & Elling, A. A. (2014). Nondestructive imaging of plant-parasitic nematode development and host response to nematode pathogenesis. *Phytopathology, 104*, 497–506.
Djamei, A., Schipper, K., Rabe, F., Ghosh, A., Vincon, V., Kahnt, J., et al. (2011). Metabolic priming by a secreted fungal effector. *Nature, 478*, 395–398.
Dong, K., & Opperman, C. (1997). Genetic analysis of parasitism in the soybean cyst nematode *Heterodera glycines*. *Genetics, 146*, 1311–1318.

Doyle, E. A., & Lambert, K. N. (2003). *Meloidogyne javanica* chorismate mutase 1 alters plant cell development. *Molecular Plant-Microbe Interactions, 16*, 123–131.

Elling, A. A., Davis, E. L., Hussey, R. S., & Baum, T. J. (2007). Active uptake of cyst nematode parasitism proteins into the plant cell nucleus. *International Journal for Parasitology, 37*, 1269–1279.

Fitzsimmons, C. M., Falcone, F. H., & Dunne, D. W. (2014). Helminth allergens, parasite-specific IgE, and its protective role in human immunity. *Frontiers in Immunology, 5*.

Gao, B., Allen, R., Maier, T., Davis, E. L., Baum, T. J., & Hussey, R. S. (2001). Identification of putative parasitism genes expressed in the esophageal gland cells of the soybean cyst nematode *Heterodera glycines*. *Molecular Plant-Microbe Interactions, 14*, 1247–1254.

Gao, B., Allen, R., Maier, T., Davis, E., Baum, T., & Hussey, R. (2003). The parasitome of the phytonematode *Heterodera glycines*. *Molecular Plant-Microbe Interactions, 16*, 720–726.

Gerke, V., & Moss, S. (2002). Annexins: from structure to function. *Physiology Review, 82*, 331–371.

Gheysen, G., & Mitchum, M. G. (2011). How nematodes manipulate plant development pathways for infection. *Current Opinion in Plant Biology, 14*, 415–421.

Goverse, A., Overmars, H., Engelbertink, J., Schots, A., Bakker, J., & Helder, J. (2000). Both induction and morphogenesis of cyst nematode feeding cells are mediated by auxin. *Molecular Plant-Microbe Interactions, 13*, 1121–1129.

Grunewald, W., Karimi, M., Wieczorek, K., Van de Cappelle, E., Wischnitzki, E., Grundler, F., et al. (2008). A role for AtWRKY23 in feeding site establishment of plant-parasitic nematodes. *Plant physiology, 148*, 358–368.

Grunewald, W., Cannoot, B., Friml, J., & Gheysen, G. (2009). Parasitic nematodes modulate PIN-mediated auxin transport to facilitate infection. *PLoS Pathogens, 55*, e1000266.

Guo, X., Chronis, D., de la Torre, C., Smeda, J., Wang, X., & Mitchum, M. G. (2015). Enhanced resistance to soybean cyst nematode *Heterodera glycines* in transgenic soybean by silencing putative CLE receptors. *Plant Biotechnology Journal*. http://dx.doi.org/10.1111/pbi.12313.

Guo, Y., Han, L., Hymes, M., Denver, R., & Clark, S. E. (2010). CLAVATA2 forms a distinct CLE-binding receptor complex regulating *Arabidopsis* stem cell specification. *Plant Journal, 63*, 889–900.

Guo, Y., Ni, J., Denver, R., Wang, X., & Clark, S. E. (2011). Mechanisms of molecular mimicry of plant CLE peptide ligands by the parasitic nematode *Globodera rostochiensis*. *Plant Physiology, 157*, 476–484.

Haegeman, A., Jacob, J., Vanholme, B., Kyndt, T., Mitreva, M., & Gheysen, G. (2009). Expressed sequence tags of the peanut pod nematode *Ditylenchus africanus*: the first transcriptome analysis of an Anguinid nematode. *Molecular & Biochemical parasitology, 167*, 32–40.

Haegeman, A., Joseph, S., & Gheysen, G. (2011). Analysis of the transcriptome of the root lesion nematode *Pratylenchus coffeae* by 454 sequencing technology. *Molecular and Biochemical Parasitology, 178*, 7–14.

Hamamouch, N., Li, C., Hewezi, T., Baum, T. J., Mitchum, M. G., Hussey, R. S., et al. (2012). The interaction of the novel 30C02 cyst nematode effector protein with a plant β-1, 3-endoglucanase may suppress host defence to promote parasitism. *Journal of Experimental Botany, 63*, 3683–3695.

Hawdon, J. M., Jones, B. F., Hoffman, D. R., & Hotez, P. J. (1996). Cloning and characterization of *Ancylostoma*-secreted protein a novel protein associated with the transition to parasitism by infective hookworm larvae. *Journal of Biological Chemistry, 271*, 6672–6678.

Hewezi, T., & Baum, T. (2010). Sequence divergences between cyst nematode effector protein orthologs may contribute to host specificity. *Plant Signaling & Behavior, 5*, 187–189.

Hewezi, T., Howe, P., Maier, T. R., Hussey, R. S., Mitchum, M. G., Davis, E. L., et al. (2008). Cellulose binding protein from the parasitic nematode *Heterodera schachtii* interacts with *Arabidopsis* pectin methylesterase: cooperative cell wall modification during parasitism. *The Plant Cell, 20*, 3080–3093.

Hewezi, T., Howe, P. J., Maier, T. R., Hussey, R. S., Mitchum, M. G., Davis, E. L., et al. (2010). Arabidopsis spermidine synthase is targeted by an effector protein of the cyst nematode *Heterodera Schachtii*. *Plant physiology, 152*, 968–984.

Hewezi, T., Priya, S., Richard, G., & Rice, J. H. (2014). Spatial and temporal expression patterns of auxin response transcription factors in the syncytium induced by the beet cyst nematode *Heterodera Schachtii* in Arabidopsis. *Molecular Plant Pathology, 15*, 730–736.

Hirakawa, Y., Shinohara, H., Kondo, Y., Inoue, A., Nakanomyo, I., Ogawa, M., et al. (2008). Non-cell-autonomous control of vascular stem cell fate by a CLE peptide/receptor system. *Proceedings of the National Academy of Sciences of the United States of America, 105*, 15208–15213.

Huang, G., Gao, B., Maier, T., Allen, R., Davis, E., Baum, T., et al. (2003). A profile of putative parasitism expressed in the esophageal gland cells of the root-knot nematode *Meloidogyne incognita*. *Molecular Plant-Microbe Interactions, 16*, 376–381.

Hussey, R. S., Davis, E. L., & Baum, T. J. (2002). Secrets in secretions: genes that control nematode parasitism of plants. *Brazilian Journal of Plant Physiology, 14*, 183–194.

Ithal, N., Recknor, J., Nettleton, D., Maier, T., Baum, T. J., & Mitchum, M. G. (2007). Developmental transcript profiling of cyst nematode feeding cells in soybean. *Molecular Plant-Microbe Interactions, 20*, 510–525.

Johnson, N., Powis, K., & High, S. (2013). Post-translational translocation into the endoplasmic reticulum. *Biochimica et Biophysica Acta, 1833*, 2403–2409.

Jones, J. T., Furlanetto, C., Bakker, E., Banks, B., Blok, V., Chen, Q., et al. (2003). Characterization of a chorismate mutase from the potato cyst nematode *Globodera pallida*. *Molecular Plant Pathology, 4*, 43–50.

Jones, J. T., Kumar, A., Pylypenko, L. A., Thirugnanasambandam, A., Castelli, L., Chapman, S., et al. (2009). Identification and functional characterization of effectors in expressed sequence tags from various life cycle stages of the potato cyst nematode *Globodera pallida*. *Molecular Plant Pathology, 10*, 815–828.

Jones, J., Reavy, B., Smant, G., & Prior, A. (2004). Glutathione peroxidases of the potato cyst nematode *Globodera rostochiensis*. *Gene, 324*, 47–54.

Kandoth, P. K., Ithal, N., Recknor, J., Maier, T., Nettleton, D., Baum, T. J., et al. (2011). The soybean *Rhg1* locus for resistance to the soybean cyst nematode *Heterodera glycines* regulates the expression of a large number of stress- and defense-related genes in degenerating feeding cells. *Plant Physiology, 155*, 1960–1975.

Karczmarek, A., Overmars, H., Helder, J., & Goverse, A. (2004). Feeding cell development by cyst and root-knot nematodes involves a similar early, local and transient activation of a specific auxin-inducible promoter element. *Molecular Plant Pathology, 5*, 343–346.

Keeling, P. (2004). Reduction and compaction in the genome of the apicomplexan parasite *Cryptosporidium parvum*. *Developmental Cell, 6*, 614–616.

Kondo, Y., Hirakawa, Y., Kieber, J. J., & Fukuda, H. (2011). CLE peptides can negatively regulate protoxylem vessel formation via cytokinin signaling. *Plant Cell Physiology, 52*, 37–48.

Konopka-Postupolska, D., Clark, G., & Hofmann, A. (2011). Structure, function and membrane interactions of plant annexins: an update. *Plant Science, 181*, 230–241.

Kumar, M., Gantasala, N. P., Roychowdhury, T., Thakur, P. K., & Banakar, P. (2014). *De novo* transcriptome sequencing and analysis of the cereal cyst nematode, *Heterodera avenae*. *PLoS One, 9*, e96311. http://dx.doi.org/10.1371/journal.pone.0096311.

Lamb, C., & Dixon, R. (1997). The oxidative burst in plant disease resistance. *Annual Review of Plant Physiology and Molecular Biology, 48*, 251--275.
Lambert, K. N., Allen, K. D., & Sussex, I. M. (1999). Cloning and characterization of an esophageal-gland-specific chorismate mutase from the phytoparasitic nematode *Meloidogyne javanica. Molecular Plant-Microbe Interactions, 12*, 328–336.
Lambert, K., Bekal, S., Domier, L., Niblack, T., Noel, G., & Smyth, C. (2005). Selection of *Heterodera glycines chorismate mutase-1* alleles on nematode-resistant soybean. *Molecular Plant-Microbe Interactions, 18*, 593–601.
Lee, C., Chronis, D., Kenning, C., Peret, B., Hewezi, T., Davis, E. L., et al. (2011). The novel cyst nematode effector protein 19C07 interacts with the *Arabidopsis* auxin influx transporter LAX3 to control feeding site development. *Plant Physiology, 155*, 866–880.
Lee, S., Lee, E. J., Yang, E. J., Lee, J. E., Park, A. R., Song, W. H., et al. (2004). Proteomic identification of annexins, calcium-dependent membrane binding proteins that mediate osmotic stress and abscisic acid signal transduction in *Arabidopsis. Plant Cell Online, 16*, 1378–1391.
Leibfried, A., To, J. P., Busch, W., Stehling, S., Kehle, A., Demar, M., et al. (2005). WUSCHEL controls meristem function by direct regulation of cytokinin-inducible response regulators. *Nature, 438*, 1172–1175.
Lin, S., Jian, H., Zhao, H., Yang, D., & Liu, Q. (2011). Cloning and characterization of a venom allergen-like protein gene cluster from the pinewood nematode *Bursaphelenchus xylophilus. Experimental Parasitology, 127*, 440–447.
Lozano-Torres, J. L., Wilbers, R. H., Gawronski, P., Boshoven, J. C., Finkers-Tomczak, A., Cordewener, J. H., et al. (2012). Dual disease resistance mediated by the immune receptor Cf-2 in tomato requires a common virulence target of a fungus and a nematode. *Proceedings of the National Academy of Sciences of the United States of America, 109*, 10119–10124.
Lu, S. W., Chen, S., Wang, J., Yu, H., Chronis, D., Mitchum, M. G., et al. (2009). Structural and functional diversity of *CLAVATA3/ESR (CLE)*-like genes from the potato cyst nematode *Globodera rostochiensis. Molecular Plant-Microbe Interactions, 22*, 1128–1142.
Lu, S.-W., Tian, D., Borchardt-Wier, H. B., & Wang, X. (2008). Alternative splicing: a novel mechanism of regulation identified in the chorismate mutase gene of the potato cyst nematode *Globodera rostochiensis. Molecular and Biochemical Parasitology, 162*, 1–15.
Ma, L., Cornelissen, B. J. C., & Takken, F. L. W. (2013). A nuclear localization for Avr2 from *Fusarium oxysporum* is required to activate the tomato resistance protein I-2. *Frontiers in Plant Science, 4*, 94.
Mazarei, M., Lennon, K. A., Puthoff, D. P., Rodermel, S. R., & Baum, T. J. (2003). Expression of an *Arabidopsis* phosphoglycerate mutase homologue is localized to apical meristems, regulated by hormones, and induced by sedentary plant-parasitic nematodes. *Plant Molecular Biology, 53*, 513–530.
Mitchum, M. G., Hussey, R. S., Baum, T. J., Wang, X., Elling, A. A., Wubben, M., et al. (2013). Nematode effector proteins: an emerging paradigm of parasitism. *New Phytologist, 199*, 879–894.
Mitchum, M. G., Wang, X., Wang, J., & Davis, E. L. (2012). Role of nematode peptides and other small molecules in plant parasitism. *Annual Review of Phytopathology, 50*, 175–195.
Mittler, R., Vanderauwera, S., Gollery, M., & Van Breusegem, F. (2004). Reactive oxygen gene network of plants. *Trends in Plant Science, 9*, 490–498.
Murray, J., Gregory, W. F., Gomez-Escobar, N., Atmadja, A. K., & Maizels, R. M. (2001). Expression and immune recognition of *Brugia malayi* VAL-1, a homologue of vespid venom allergens and *Ancylostoma* secreted proteins. *Molecular and Biochemical Parasitology, 118*, 89–96.
Nyaku, S. T., Sripathi, V. R., Wiley, G., Najar, F. Z., Cseke, L. J., Sharma, G. C., et al. (2013). The expressed parasitism genes in the reniform nematode (*Rotylenchulus reniformis*). *American Journal of Plant Sciences, 4*.

Ogawa, M., Shinohara, H., Sakagami, Y., & Matsubayashi, Y. (2008). *Arabidopsis* CLV3 peptide directly binds CLV1 ectodomain. *Science, 319*, 294.

Ohyama, K., Shinohara, H., Ogawa-Ohnishi, M., & Matsubayashi, Y. (2009). A glycopeptide regulating stem cell fate in *Arabidopsis thaliana*. *Nature Chemical Biology, 5*, 578–580.

Olsen, A. N., & Shriver, K. (2003). Ligand mimicry? Plant-parasitic nematode polypeptide with similarity to CLAVATA3. *Trends in Plant Science, 8*, 55–57.

Opperman, C., Bird, D., Williamson, V., Rokhsar, D., Burke, M., Cohn, J., et al. (2008). Sequence and genetic map of *Meloidogyne hapla*: a compact nematode genome for plant parasitism. *Proceedings of the National Academy of Sciences of the United States of America, 105*, 14802–14807.

Patel, N., Hamamouch, N., Li, C., Hewezi, T., Hussey, R., Baum, T., et al. (2010). A nematode effector protein similar to annexins in host plants. *Journal of Experimental Botany, 61*, 235–248.

Patel, N., Hamamouch, N., Li, C., Hussey, R., Mitchum, M., Baum, T., et al. (2008). Similarity and functional analyses of expressed parasitism genes in *Heterodera schachtii* and *Heterodera glycines*. *Journal of Nematology, 40*, 299–310.

Popeijus, H., Overmars, H., Jones, J., Blok, V., Goverse, A., Helder, J., et al. (2000). Degradation of plant cell walls by a nematode. *Nature, 406*, 36–37.

Postma, W. J., Slootweg, E. J., Rehman, S., Finkers-Tomczak, A., Tytgat, T., van Gelderen, K., et al. (2012). The effector SPRYSEC-19 of *Globodera rostochiensis* suppresses CC-NB-LRR-mediated disease resistance in plants. *Plant Physiology, 160*, 944–954.

Rehman, S., Postma, W., Tytgat, T., Prins, P., Qin, L., Overmars, H., et al. (2009). A secreted SPRY domain-containing protein (SPRYSEC) from the plant-parasitic nematode *Globodera rostochiensis* interacts with a CC-NB-LRR protein from a susceptible tomato. *Molecular Plant-Microbe Interactions, 22*, 330–340.

Replogle, A., Wang, J., Bleckmann, A., Hussey, R. S., Baum, T. J., Sawa, S., et al. (2011). Nematode CLE signaling in Arabidopsis requires CLAVATA2 and CORYNE. *Plant Journal, 65*, 430–440.

Replogle, A., Wang, J., Paolillo, V., Smeda, J., Kinoshita, A., Durbak, A., et al. (2013). Synergistic interaction of CLAVATA1, CLAVATA2, and RECEPTOR-LIKE PROTEIN KINASE 2 in cyst nematode parasitism of *Arabidopsis*. *Molecular Plant-Microbe Interactions, 26*, 87–96.

Rhodes, D. A., De Bono, B., & Trowsdale, J. (2005). Relationship between SPRY and B30. 2 protein domains. Evolution of a component of immune defence? *Immunology, 116*, 411–417.

Rivas, S., & Deslandes, L. (2013). Nuclear components and dynamics during plant innate immunity. *Frontiers in Plant Science, 4*, 481.

Robertson, L., Robertson, W., & Jones, J. (1999). Direct analysis of the secretions of the potato cyst nematode *Globodera rostochiensis*. *Parasitology, 119*, 167–176.

Robertson, L., Robertson, W., Sobczak, M., Helder, J., Tetaud, E., Ariyanayagam, M., et al. (2000). Cloning, expression and functional characterization of a peroxiredoxin from the potato cyst nematode *Globodera rostochiensis*. *Molecular and Biochemical Parasitology, 111*, 41–49.

Romero, R. M., Roberts, M. F., & Phillipson, J. D. (1995). Chorismate mutase in microorganisms and plants. *Phytochemisry, 40*, 1015–1025.

Sacco, M. A., Koropacka, K., Grenier, E., Jaubert, M. J., Blanchard, A., Goverse, A., et al. (2009). The cyst nematode SPRYSEC protein RBP-1 elicits Gpa2- and RanGAP2--dependent plant cell death. *PLoS Pathogens, 5*, e1000564. http://dx.doi.org/10.1371/journal.ppat.1000654.

Siddique, S., Matera, C., Radakovic, Z. S., Shamin Hasan, M., Gutbrod, P., Rozanska, E., et al. (2014). Parasitic worms stimulate host NADPH oxidases to produce reactive oxygen species that limit plant cell death and promote infection. *Science Signaling, 7*, ra33.

Smant, G., Stokkermans, J. P. W. G., Yan, Y., de Boer, J. M., Baum, T. J., Wang, X., et al. (1998). Endogenous cellulases in animals: isolation of ß-1,4-endoglucanase genes from two species of plant-parasitic cyst nematodes. *Proceedings of the National Academy of Sciences of the United States of America, 95*, 4906–4911.

Stam, R., Howden, A. J. M., Delgado-Cerezo, M., M M Amaro, T. M., Motion, G. B., & Pham, J. (2013). Characterization of cell death inducing *Phytophthora capsici* CRN effectors suggests diverse activities in the host nucleus. *Frontiers in Plant Science, 4*, 387.

Szakasits, D., Heinen, P., Wieczorek, K., Hofmann, J., Wagner, F., Kreil, D. P., et al. (2009). The transcriptome of syncytia induced by the cyst nematode *Heterodera schachtii* in Arabidopsis roots. *Plant Journal, 57*, 771–784.

Tawe, W., Pearlman, E., Unnasch, T. R., & Lustigman, S. (2000). Angiogenic activity of *Onchocerca volvulus* recombinant proteins similar to vespid venom antigen 5. *Molecular and Biochemical Parasitology, 109*, 91–99.

Tytgat, T., Vanholme, B., De Meutter, J., Claeys, M., Couvreur, M., Vanhoutte, I., et al. (2004). A new class of ubiquitin extension proteins secreted by the dorsal pharyngeal gland in plant parasitic cyst nematodes. *Molecular Plant-Microbe Interactions, 17*, 846–852.

Van Damme, M., Huibers, R., Elberse, J., & Van den Ackerceken, G. (2008). Arabidopsis *DMR6* encodes a putative 2OG-Fe(II) oxygenase that is defense-associated but required for susceptibility to downy mildew. *Plant Journal, 54*, 785–793.

Van Loon, L. C., Rep, M., & Pieterse, C. M. J. (2006). Significance of inducible defense-related proteins in infected plants. *Annual Review of Phytopathology, 44*, 135–162.

Vanholme, B., Kast, P., Haegeman, A., Jacob, J., Grunewald, W. I. M., & Gheysen, G. (2009). Structural and functional investigation of a secreted chorismate mutase from the plant-parasitic nematode *Heterodera schachtii* in the context of related enzymes from diverse origins. *Molecular Plant Pathology, 10*, 189–200.

Vleeshouwers, V., & Oliver, R. (2014). Effectors as tools in disease resistance breeding against biotrophic, hemibiotrophic, and necrotrophic pathogens. *Molecular Plant-Microbe Interactions, 27*, 196–206.

Wang, X., Allen, R., Ding, X., Goellner, M., Maier, T., De Boer, J. M., et al. (2001). Signal peptide-selection of cDNA cloned directly from the esophageal gland cells of the soybean cyst nematode *Heterodera glycines*. *Molecular Plant-Microbe Interactions, 14*, 536–544.

Wang, J., Joshi, S., Korkin, D., & Mitchum, M. G. (2010). Variable domain I of nematode CLEs directs post-translational targeting of CLE peptides to the extracellular space. *Plant Signaling & Behavior, 5*, 1–3.

Wang, J., Lee, C., Replogle, A., Joshi, S., Korkin, D., Hussey, R., et al. (2010). Dual roles for the variable domain in protein trafficking and host-specific recognition of *Heterodera glycines* CLE effector proteins. *New Phytologist, 187*, 1003–1017.

Wang, X., Li, H., Hu, Y., Fu, P., & Xu, J. (2007). Molecular cloning and analysis of a new venom allergen-like protein gene from the root-knot nematode *Meloidogyne incognita*. *Experimental Parasitology, 117*, 133–140.

Wang, X., Mitchum, M. G., Gao, B., Li, C., Diab, H., Baum, T. J., et al. (2005). A parasitism gene from a plant-parasitic nematode with function similar to CLAVATA3/ESR (CLE) of *Arabidopsis thaliana*. *Molecular Plant Pathology, 6*, 187–191.

Wang, J., Replogle, A., Hussey, R., Baum, T., Wang, X., Davis, E. L., et al. (2011). Identification of potential host plant mimics of CLV3/ESR (CLE)-like peptides from the plant-parasitic nematode *Heterodera schachtii*. *Molecular Plant Pathology, 12*, 177–186.

Wubben, M. J., Su, H., Rodermel, S. R., & Baum, T. J. (2001). Susceptibility to the sugar beet cyst nematode is modulated by ethylene signal transduction in *Arabidopsis thaliana*. *Molecular Plant-Microbe Interactions, 14*, 1206–1212.

Wulff, B. B. H., Chakrabarti, A., & Jones, D. A. (2009). Recognitional specificity and evolution in the tomato-*Cladosporium fulvum* pathosystem. *Molecular Plant-Microbe Interactions, 22*, 1191–1202.

Yan, Y., Smant, G., Stokkermans, J., Qin, L., Helder, J., Baum, T., et al. (1998). Genomic organization of four ß-1,4-endoglucanase genes in plant-parasitic cyst nematodes and its evolutionary implications. *Gene, 220*, 61–70.

Young, L. (1984). Changes in the reproduction of *Heterodera glycines* on different lines of Glycine max. *Journal of Nematology, 16*, 304–309.

Yu, H., Chronis, D., Lu, S., & Wang, X. (2011). Chorismate mutase: an alternatively spliced parasitism gene and a diagnostic marker for three important *Globodera* nematode species. *European Journal of Plant Pathology, 129*, 89–102.

CHAPTER TWELVE

Function of Root-Knot Nematode Effectors and Their Targets in Plant Parasitism

Nhat My Truong[a], Chinh-Nghia Nguyen[a], Pierre Abad, Michaël Quentin, Bruno Favery[1]

INRA-CNRS-Université de Nice Sophia-Antipolis, Institut Sophia Agrobiotech, UMR 1355-7254, Sophia-Antipolis, France

[1]Corresponding author: E-mail: favery@sophia.inra.fr

Contents

Abstract

Root-knot nematodes (RKNs), genus *Meloidogyne*, are widespread, polyphagous endoparasites which present a serious and growing threat to agriculture. They have evolved sophisticated strategies for exploiting plants, and research on nematode parasitism therefore tackles fundamental questions in plant development and host defence. These obligate biotrophic pathogens establish and maintain an intimate relationship with their host plants. They induce major changes in plant cells, leading to the development of the specialized hypertrophied and multinucleate feeding cells essential for nematode growth and development. RKNs have a wide host range, suggesting that they can

[a]These authors contributed equally to this work.

Advances in Botanical Research, Volume 73
ISSN 0065-2296
http://dx.doi.org/10.1016/bs.abr.2014.12.010

probably hijack fundamental host functions. Nematode-secreted proteins (effectors) are thought to be instrumental in the manipulation of developmental and defence signalling pathways in host cells, particularly during the initiation and maintenance of nematode feeding cells. Three oesophageal secretory gland cells that undergone considerable adaptation to increase their secretory activity are the main source of these effector proteins. Several molecular approaches have been used to identify and functionally characterize the effectors produced by nematode secretory organs. About 100 RKN parasitism genes have been cloned and effectors have been shown to be delivered to the apoplasm, cytoplasm and nucleus of host cells. The first host targets and functions of secreted effector proteins have recently been elucidated. In this chapter, we describe the complex interactions between RKNs and their infected hosts and highlight the progress made in the molecular and functional characterization of RKN effector proteins and our understanding of host plant cell manipulation by these parasites.

1. INTRODUCTION

Root-knot nematodes (RKNs; *Meloidogyne* spp.) are amongst the most devastating plant pathogens. Economic losses to agriculture due to the damage caused by these microscopic worms have been estimated at about 10 billion Euros per year (Trudgill & Blok, 2001). RKNs are widespread in temperate and tropical regions worldwide and are able to infect thousands of plant species, including almost all crops (Abad & Williamson, 2010; Blok, Jones, Phillips, & Trudgill, 2008). More than 80 RKN species have been described, and those reproducing asexually by mitotic parthenogenesis (*Meloidogyne incognita, Meloidogyne javanica, Meloidogyne arenaria* and *Meloidogyne enterolobii*) are the most polyphagous and damaging pests. In contrast those reproducing sexually are specialists with a more reduced host range (Blok et al., 2008; Castagnone-Sereno, 2006). They induce typical root deformations, known as galls, and greatly decrease plant growth.

Nematode control is becoming more difficult for growers in many regions of the world. In Europe, most of the chemical nematicides previously available have been banned in recent years, due to their intrinsic toxicity, and human health and environmental concerns (European Community directive 2007/619/EC). Natural plant resistance is a safe alternative, but it is limited by the availability of resistance genes. Some plant families have no known RKN resistance genes (e.g. Cucurbitaceae). In addition, the occurrence of RKN populations able to overcome and reproduce on monogenic resistance (i.e. virulent *M. incognita* populations or *M. enterolobii* species), such as the tomato *Mi* gene, has significantly decreased the value of previously resistant elite crop lines (Castagnone-Sereno, 2006; Williamson

& Kumar, 2006). Biological control provides another alternative, but its inability to produce consistent, robust, long-term results on a large scale has restricted its use to highly managed situations. Furthermore, recent studies have demonstrated that climate change may increase nematode infection capacity. With increasing temperatures the risk of more severe infestation increases because higher temperatures enable the nematodes to reproduce more rapidly (Ghini, Hamada, José, Júnior, & Marengo, 2008). New nematode control options are therefore urgently required, to provide producers with alternative control strategies enabling them to help secure global food production in the future. Molecular mechanisms of plant resistance to pathogens have been studied in detail and are now being used in pest management, but our knowledge of disease development remains limited. It is now widely agreed that pathogens can modify basic plant functions and influence defence responses. The identification of RKN pathogenicity factors and plant targets essential to the pathogen for host exploitation is opening up new possibilities for the development of new approaches to the control of these plant-parasitic nematodes.

2. COMPATIBLE INTERACTION AND LIFE CYCLE

RKN vermiform second-stage juveniles (J2) hatch from eggs in the soil to infect roots. These mobile J2s penetrate the root, usually in the root elongation zone, subsequently migrating between cells to reach the root apex, where they enter the vascular cylinder of the plant. Once in the stele, J2s select five to seven root parenchyma cells and induce their transformation into specialized nematode feeding cells known as giant cells (Figure 1). These giant cells are hypertrophied and multinucleate. They are generated by repeated nuclear divisions and cell growth in the absence of cell division (Caillaud, Lecomte, et al., 2008; Jones & Payne, 1978). In vivo confocal microscopy studies of giant cells revealed that the failure of cytokinesis after nuclear division is due to restricted outgrowth of the phragmoplast microtubule (MT) array (Caillaud, Abad, et al., 2008; Caillaud, Lecomte, et al., 2008). The cells surrounding the giant cells and the nematode undergo classical mitosis to form a typical root gall. The giant cells constitute the sole source of nutrients for the nematode, which punctures the cells to gain access to their cytoplasmic content (Abad & Williamson, 2010). Like all other plant-parasitic nematodes, RKNs use a hollow, protrusible syringe-like stylet to penetrate the wall of plant cells, to secrete proteins from their oesophageal

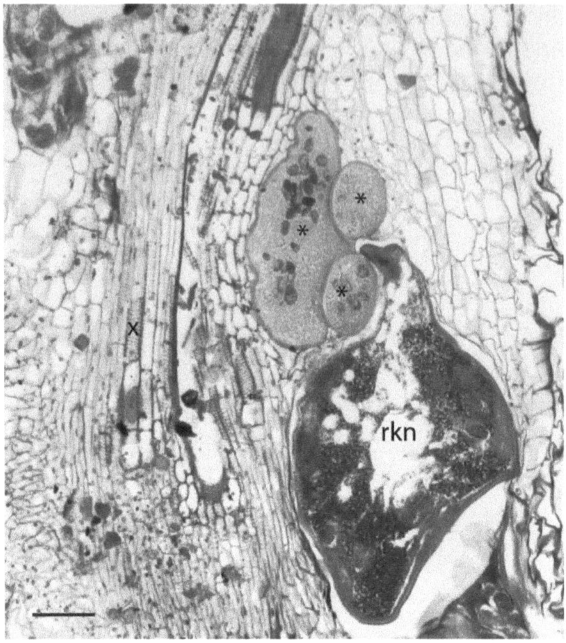

Figure 1 ***Multinucleate giant cells induced by Meloidogyne incognita in tomato.*** Section through a gall, 15 days postinfection. Asterisks, giant cells; rkn, young female root-knot nematode; x, xylem. Scale bar = 40 μm. (See colour plate)

glands into the cell, and to withdraw nutrients from the cytoplasm. The nematodes remain at the feeding site and develop through three further moults into adult males or females. After 3–10 weeks, depending on the host and temperature, the pear-shaped female produces and exudes egg masses into the rhizosphere in which several hundred eggs are embedded.

Most plant molecular biology studies have focused on giant cell formation, which is essential for nematode development and reproduction. Mature giant cells reach 100 times the size of a normal root vascular cell and can contain more than 100 enlarged nuclei. Mature giant cell nuclei are highly amoeboid and have dispersed chromatin, reflecting intensive gene transcription (Jones & Payne, 1978). Giant cells contain small vacuoles, and display proliferation of the endoplasmic reticulum, ribosomes, mitochondria and plastids. Giant cells have several features typical of highly metabolic transfer cells, such as wall ingrowths developing in contact with the xylem elements and increasing the contact area for exchange at the associated membrane. The complex changes in cell structure and physiology leading to feeding cell establishment (reviewed in chapter "Overview of root-knot nematodes and giant cells") result from profound changes in gene expression in the infected root cells (Barcala et al., 2010; Caillaud, Dubreuil, et al., 2008; Damiani et al., 2012; Gheysen & Fenoll, 2002). Genome-wide expression profiling of isolated giant cells and galls has led to the identification of thousands of genes involved in diverse processes, including cell cycle activation, cell wall modification, and hormone and defence responses, as differentially expressed during feeding cell formation. Functional analyses of these differentially expressed genes have been carried out, to identify nematode susceptibility genes essential for the development of RKNs and their feeding sites. Only a few mutations that impair nematode infection have been characterized to date (Caillaud, Dubreuil, et al., 2008). Unique defects in giant cell ontogenesis have been described in the absence of regulators of MT or microfilament dynamics, highlighting the importance of changes to cytoskeleton architecture for correct giant cell development (Caillaud, Lecomte, et al., 2008; Clement et al. 2009). However, the molecular events underlying the formation and development of nematode feeding cells remain to be identified.

3. IDENTIFICATION OF NEMATODE-SECRETED EFFECTORS

Our knowledge of the dialogue between plants and RKNs remains fragmentary, but nematode-secreted proteins, named effectors, are thought to be instrumental in manipulating developmental and defence signalling pathways in host cells. In particular, effectors produced by three oesophageal gland cells, from which they are secreted into the host through the stylet, may play an important role in the induction and maintenance of the giant cells (Davis, Hussey, & Baum, 2004; Hewezi & Baum, 2013; Mitchum et al. 2013). Secreted proteins are also known to play an important role in other aspects of the host parasite interaction, including invasion, migration and protection against host defence responses (Abad & Williamson, 2010; Mitchum et al. 2013). The three oesophageal gland cells are large and complex secretory cells that have undergone adaptation to increase their secretory activity. Two are in a subventral location and the third is in a dorsal location. Interestingly in the early stages of parasitism, the subventral glands are highly active. Following the onset of parasitism and throughout the rest of the parasitic cycle, the dorsal gland cell becomes the leading source of effector proteins. Most studies have focused on secretions originating from these glands, but putative effector proteins may also be produced by other secretory organs, such as the cuticle, the chemosensory amphids, the excretory/secretory system and the rectal glands. Like the oesophageal glands, these organs probably undergo changes in function on adoption of the parasitic stage of the life cycle (Jones, Perry, & Johnston, 1993).

The biotrophic life cycle and lack of sexual reproduction of the main *Meloidogyne* species (*M. incognita, M. javanica* and *M. arenaria*) preclude the development of forward genetic screens for identifying nematode parasitism genes. Research has thus focused on the cloning and characterization of nematode-secreted proteins with functions likely to promote the parasitism of plants by nematodes (candidate approach), the presence of these proteins in secretions (proteomics) and gene expression patterns in secretory organs or for genes containing predicted specific secretory signals in their sequences (transcriptomics/genomics). The sequencing of the full genomes of two RKN species has opened up new opportunities for studying plant–nematode interactions (Bird et al. 2009).

3.1 From Secretions...

Most effectors are secreted into host cells and tissues through the stylet. Direct qualitative analysis of the proteins secreted via the stylet by J2s was therefore considered a logical way to identify effectors. However, this approach has been of limited success, because of the small size of the nematodes, the minute amounts of secretory material recovered from J2s and their obligate biotrophy. Early attempts to analyse proteins secreted via the stylet were limited initially to one-dimensional electrophoresis (Veech, Starr, & Nordgren, 1987). Studies of proteins secreted via the stylet advanced with the advent of a strategy based on monoclonal antibodies (MAbs). MAbs were developed against secretory granules formed in the oesophageal glands of *M. incognita,* as a first step towards the identification of biologically important secretions (Hussey, 1989). MAbs have been used to isolate two high-molecular weight secretory glycoproteins from the oesophageal glands of RKNs (Hussey, Paguio, & Seabury, 1990). The amino acid sequence of the *M. incognita* 6D4 protein has yet to be determined (Vieira et al., 2011). MAbs binding to specific structures in RKNs have also been developed by an immunization procedure involving the injection of homogenates of the anterior regions or stylet secretions from *M. incognita* adult females (Davis, Allen, Pratt, & Hussey, 1992). Nine MAbs have been shown to bind to secretory granules formed in the dorsal oesophageal gland and two have been shown to bind to such granules in the subventral glands. One MAb, MGR48, developed from J2s of the potato cyst nematode, *Globodera rostochiensis,* binds specifically to the subventral oesophageal glands and was used for the immunopurification of the first parasitism protein from a plant-parasitic nematode ever identified, a β-1,4-endoglucanase (Smant et al., 1998). Cyst nematode cellulases were the first endogenous cellulases to be identified in animals. All previously identified cellulases from the digestive systems of animals originate from symbiotic microorganisms. A first cellulase gene, *Mi-ENG-1*, was cloned by PCR from *M. incognita* J2s and shown, by mRNA in situ hybridization (ISH), to be expressed in the subventral glands (Rosso et al., 1999). The nematode cellulases identified were thought to facilitate migration through plant roots by mediating the partial degradation of the plant cell wall. These enzymes were the first of a long list of cell wall-degrading or cell wall-modifying effectors to be identified in RKNs and other plant-parasitic nematodes (Table 1).

Jaubert, Ledger, et al. (2002) established a new procedure for the direct qualitative analysis of stylet-secreted proteins from infective juveniles of *M. incognita*. A large-scale procedure was established for the production of stylet secretions in semi-sterile conditions by the incubation of J2s in a solution with resorcinol. Resorcinol is a neurostimulant that stimulates stylet thrusting and the accumulation of secretions in the lip region of J2s without impairing the ability of the nematode to infect plants (Jaubert, Ledger et al., 2002; Rosso et al., 1999). The purified proteins were separated by two-dimensional electrophoresis and the seven most abundant proteins were identified by microsequencing. Genes encoding a calreticulin (CRT) and 14-3-3-like proteins were identified and shown to be expressed in the oesophageal glands of infective juveniles (Jaubert et al., 2004; Jaubert, Ledger, et al., 2002) (Table 1). The *M. incognita* secretome was then explored in greater detail, by nano-electrospray ionization and tandem mass spectrometry (nanoLC ESI MS/MS; Bellafiore et al., 2008). These sensitive methods for high-throughput proteomics-based liquid chromatography led to the identification of 486 proteins secreted by *M. incognita*. These secreted proteins were then annotated to indicate their functions and classified according to their potential roles in disease development. ISH showed that most of the secreted proteins analysed were produced by the subventral glands (Table 1). A new bacterial contamination-resistant method for collecting soluble proteins directly from the oesophageal gland cells of female *M. incognita* nematodes has recently been developed (Wang et al., 2012). In this method, the oesophageal gland cells are pierced to facilitate the adsorption of the soluble proteins before mass spectrometry. This approach has proved successful for female *M. incognita* and opens up new possibilities for identifying RKN effectors at different stages of the life cycle. Indeed, the combination of proteomics

Table 1 Root-knot Nematode Proteins Produced in Secretory Organs and Predicted to be Involved in Parasitism

Effector	Predicted Function	Organs	ISH	IL	References
Meloidogyne incognita					
Mi-PEL-1	Pectate lyase	SvG	✓		Huang et al. (2003), Huang et al. (2005b)
Mi-PEL-2	Pectate lyase	SvG	✓		Huang et al. (2004), Huang et al. (2005b)
Mi-PEL-3	Pectate lyase	SvG		✓	Vieira et al. (2011), Vieira et al. (2012)
Mi-ENG-1	Beta-1,4-endoglucanase	SvG	✓		Huang et al. (2004), Yang et al. (2013), Bellafiore et al. (2008)
5A12B	Beta-1,4-endoglucanase	SvG	✓		Huang et al. (2004)
8E08B	Beta-1,4-endoglucanase	SvG	✓		Huang et al. (2004)
Mi-PG-1	Polygalacturonase	SvG	✓		Jaubert, Laffaire, Abad, and Rosso (2002)
Mi-XYL-1	Beta-1,4-endoxylanase	SvG	✓		Dautova et al. (2001)
Mi-CBP-1	Cellulose-binding protein	SvG	✓		Huang et al. (2003),trun -1 Ding et al. (1998)
Mi-CM-1	Chorismate mutase	SvG	✓		Huang et al. (2004), Huang et al. (2005a)
Mi-CM-2	Chorismate mutase	SvG	✓		Huang et al. (2004), Huang et al. (2005a)
Mi-ASP2	Aspartyl protease-like	SvG		✓	Neveu et al. (2003)
Mi-GST-1	Glutathione-S-transferase	SvG	✓	✓	Dubreuil et al. (2007)
16D10	CLE-like peptide	SvG	✓	✓	Huang et al. (2003)
Mi-SXP-1	SXP/Ral-2 protein	SvG	✓		Tytgat et al. (2005)
Mi-VAP-2	Venom allergen-like protein	SvG	✓		Wang, Li, Hu, Fu, and Xu (2007)
Mi-MSP-1	Venom allergen-like protein	SvG	✓		Ding, Shields, Allen, and Hussey (2000)
5G05	Zinc metallopeptidase	SvG	✓		Huang et al. (2003)
30G11	Acid phosphatase	SvG	✓		Huang et al. (2003)
10A07	Sodium/calcium/potassium exchanger	SvG	✓		Huang et al. (2003)

Table 1 Root-knot Nematode Proteins Produced in Secretory Organs and Predicted to be Involved in Parasitism—cont'd

Effector	Predicted Function	Organs	ISH	IL	References
CL5 Contig2_1	Sec-2 protein	SvG	✓		Bellafiore et al. (2008)
CL2552 Contig1_1	Transthyretin-like protein	SvG	✓		Bellafiore et al. (2008)
CL321 Contig1_1	Translationally controlled tumour protein	SvG	✓		Bellafiore et al. (2008)
CL480 Contig2_1	Triosephosphate isomerase	SvG	✓		Bellafiore et al. (2008)
Minc01696	Protein kinase	SvG	✓		Rutter et al. (2014)
Minc03866	C-type lectin	SvG	✓		Danchin et al. (2013)
CL312 Contig1_1	Unknown	SvG	✓		Bellafiore et al. (2008)
Minc00344	Unknown	SvG	✓		Rutter et al. (2014)
Minc04584	Unknown	SvG	✓		Rutter et al. (2014)
Minc18033	Unknown	SvG	✓		Rutter et al. (2014)
Minc13292	Unknown	SvG	✓		Rutter et al. (2014)
Minc08073	Unknown	SvG	✓		Rutter et al. (2014)
Minc00469	Unknown	SvG	✓		Rutter et al. (2014)
Minc15401	Unknown	SvG	✓		Rutter et al. (2014)
Minc10418	Unknown	SvG	✓		Rutter et al. (2014)
Minc03328	Unknown	SvG	✓		Rutter et al. (2014)
Minc03325	Unknown	SvG	✓		Rutter et al. (2014)
Minc18636	Unknown	SvG	✓		Rutter et al. (2014)
Minc08146	Unknown	SvG	✓		Jaouannet et al. (2012)
2G02	Unknown	SvG	✓		Huang et al. (2003)
4D01	Unknown	SvG	✓		Huang et al. (2003)
8D05	Unknown	SvC	✓	✓	Huang ct al. (2003)
8H11	Unknown	SvG	✓		Huang et al. (2003)
8E10B	Unknown	SvG	✓		Huang et al. (2003)
30H07	Unknown	SvG	✓		Huang et al. (2003)
31H06	Unknown	SvG	✓		Huang et al. (2003)
35A02	Unknown	SvG	✓		Huang et al. (2003)
HM1	Unknown	SvG	✓		Neveu et al. (2003)
6D4	Unknown	SvG and DG		✓	Vieira et al. (2012), Davis et al. (1992)
Mi-CRT	Calreticulin	SvG and DG		✓	Jaubert et al. (2005)
Mi-14-3-3-b	14-3-3	DG	✓	✓	Vieira et al. (2012), Jaubert et al. (2004)

(Continued)

Table 1 Root-knot Nematode Proteins Produced in Secretory Organs and Predicted to be Involved in Parasitism—cont'd

Effector	Predicted Function	Organs	ISH	IL	References
10G02	Thioredoxin	DG	✓		Huang et al. (2003)
Minc00108	Metallopeptidase	DG	✓		Jaouannet et al. (2012)
Minc02097	Unknown	DG	✓		Rutter et al. (2014)
Mi-EFF1/ Minc17998	Unknown	DG	✓	✓	Jaouannet et al. (2012)
Minc18861	Unknown	DG	✓		Rutter et al. (2014)
Minc12639	Unknown	DG	✓		Rutter et al. (2014)
Minc11817	Unknown	DG	✓		Rutter et al. (2014)
Minc01595	Unknown	DG	✓		Rutter et al. (2014)
1C05B	Unknown	DG	✓		Huang et al. (2004)
1D08B	Unknown	DG	✓		Huang et al. (2004)
2E07	Unknown	DG	✓		Huang et al. (2003)
2G10	Unknown	DG	✓		Huang et al. (2003)
4D03	Unknown	DG	✓		Huang et al. (2003)
4F05B	Unknown	DG	✓		Huang et al. (2004)
5C03B	Unknown	DG	✓		Huang et al. (2004)
6F07	Unknown	DG	✓		Huang et al. (2003)
6G07	Unknown	DG	✓		Huang et al. (2003)
7A01	Unknown	DG	✓		Huang et al. (2003)
7E12	Unknown	DG	✓		Huang et al. (2003), Souza et al. (2011)
7H08	Unknown	DG	✓		Huang et al. (2003)
9H10	Unknown	DG	✓		Huang et al. (2003)
11A01	Unknown	DG	✓		Huang et al. (2003)
12H03	Unknown	DG	✓		Huang et al. (2003)
13A12	Unknown	DG	✓		Huang et al. (2003)
17H02	Unknown	DG	✓		Huang et al. (2003)
25B10	Unknown	DG	✓		Huang et al. (2003)
14E06	Unknown	DG	✓		Huang et al. (2003)
16E05	Unknown	DG	✓		Huang et al. (2003)
21E02	Unknown	DG	✓		Huang et al. (2003)
34D01	Unknown	DG	✓		Huang et al. (2003)
34F06	Unknown	DG	✓		Huang et al. (2003)
35F03	Unknown	DG	✓		Huang et al. (2003)
35E04	Unknown	DG	✓		Huang et al. (2003)
28B04	Unknown	DG	✓		Huang et al. (2003)
HM7	Unknown	DG	✓		Neveu et al. (2003)
HM12	Unknown	DG	✓		Neveu et al. (2003)
Mi-MAP1	Unknown	Amphids	✓	✓	Vieira et al. (2012), Semblat et al. (2001)

Table 1 Root-knot Nematode Proteins Produced in Secretory Organs and Predicted to be Involved in Parasitism—cont'd

Effector	Predicted Function	Organs	ISH	IL	References
CL1191 Contig1_1	CDC48-like	Phasmids	✔		Bellafiore et al. (2008)
Minc00801	Unknown	RG	✔		Rutter et al. (2014)
Meloidogyne javanica					
Mj-PEL-1	Pectate lyase	SvG	✔	✔	Doyle and Lambert, 2002
Mj-eng-3	Beta-1,4-endoglucanase	SvG	✔		Hu et al. (2013)
Mj-CM-1	Chorismate mutase	SvG		✔	Doyle and Lambert, (2003)
Mj-NULG1a	Unknown	DG	✔		Lin et al. (2013)
Mj-FAR-1	Fatty acid- and retinol-binding protein	Cuticle		✔	Iberkleid et al. (2013)
Meloidogyne graminicola					
Contig06513	Pectate lyase	SvG	✔		Haegeman et al. (2013)
Contig01965	C-type lectin	SvG	✔		Haegeman et al. (2013)
Contig03718	C-type lectin	SvG	✔		Haegeman et al. (2013)
Contig01318	C-type lectin	SvG	✔		Haegeman et al. (2013)
Contig06883	Ancylostoma-secreted protein	SvG	✔		Haegeman et al. (2013)
Contig01796	Unknown	SvG	✔		Haegeman et al. (2013)
Contig06410	19F07-like	DG	✔		Haegeman et al. (2013)
Contig05581	Unknown	DG	✔		Haegeman et al. (2013)
Contig03015	Unknown	DG	✔		Haegeman et al. (2013)
Contig09259	Unknown	Amphids	✔		Haegeman et al. (2013)
Meloidogyne chitwoodi					
MC01180	Pectate lyase	SvG	✔		Roze et al. (2008)
MC01203	Beta-1,4-endoglucanase	SvG	✔		Roze et al. (2008)
Mc-16D10	16D10-like	SvG	✔		Dinh et al. (in press)
MC01222	1D08B-like	SvG	✔		Roze et al. (2008)
MC01209	C-type lectin	SvG	✔		Roze et al. (2008)
MC01214	Unknown	SvG	✔		Roze et al. (2008)
MC01206	Unknown	SvG	✔		Roze et al. (2008)
MC01194	Unknown	SvG	✔		Roze et al. (2008)
MC01229	Unknown	DG	✔		Roze et al. (2008)

ISH, in situ hybridization; IL, immunolocalization; SvG, subventral glands; DG, dorsal gland; RG, rectal gland. Effectors in grey were identified by proteomics.

and bioinformatics approaches could provide an experimentally verified essential tool for biologically meaningful discovery-based proteomic analysis of nematode parasitism (Mbeunkui, Scholl, Opperman, Goshe, & Bird, 2010), which would expand our knowledge of RKN effector repertoires.

3.2 ...To Secretory Organs

Many studies have focused on the oesophageal cells as a target for the identification of expressed genes encoding secretory proteins (Davis et al., 2004). Micro-aspiration of the cytoplasm of oesophageal gland cells from parasitic stage nematodes has been used to generate cDNA libraries for gland cell-expressed genes. This approach has been useful for identifying candidate parasitism genes (Davis et al., 2004; Davis, Hussey, Mitchum, & Baum, 2008; Huang et al., 2003). Extensive expressed sequence tag (EST) analyses of gland cell libraries from *M. incognita*, together with secretion signal peptide prediction and high-throughput ISH on 185 cDNAs, led to the identification of 37 unique clones encoding parasitism effectors which specifically hybridized to transcripts accumulating in the subventral (13 clones) or dorsal (24 clones) oesophageal gland cells of *M. incognita* (Huang et al., 2003). Interestingly, 73% of the predicted proteins were novel. Those with similarities to known proteins included a pectate lyase, acid phosphatase and hypothetical proteins similar to proteins from other organisms (Table 1). In addition, a cDNA library constructed from the oesophageal gland region of *M. javanica* was used to identify genes differentially expressed in these glands, to characterize a potentially secreted chorismate mutase (CM) and pectate lyase (Doyle & Lambert, 2002; Lambert, Allen, & Sussex, 1999) (Table 1). More recently mRNA extracted from micro-aspirated gland cells has been analysed by a next-generation sequencing (NGS) approach using Roche 454 technology (Rutter et al. 2014). By combining in silico analysis and ISH on 91 cDNAs, Rutter et al. (2014) identified 17 *M. incognita* genes encoding putative effector proteins that are expressed specifically in the oesophageal gland cells. Most of these candidate effectors are pioneers with no significant sequence similarity to any proteins in the databases (Table 1). Finally, a new technique has been developed to separate and isolate individual oesophageal gland cells from different species of plant-parasitic nematodes (Maier, Hewezi, Peng, & Baum, 2013). The isolated gland cells can then be used for transcriptomic analyses by NGS. This technique has been used successfully on several nematode species and will facilitate the identification of effectors not only from RKNs, but also from nematodes with different modes of parasitism.

3.3 Differential Gene Expression

Genes differentially expressed between specific stages in the nematode life cycle, and particularly those differentially expressed between the preparasitic exophytic stage and the endoparasitic stages, have been characterized by various transcriptomic strategies, including the global analysis of gene expression based on the ESTs generated from nematodes, RNA fingerprinting, cDNA-AFLP (amplification fragment length polymorphism), suppression subtractive hybridization (SSH)-based strategies or real-time Q-PCR analysis (Dautova et al., 2001; Ding, Shields, Allen, & Hussey, 1998; Jaouannet et al., 2012; Neveu, Abad, & Castagnone-Sereno, 2002; Neveu, Jaubert, Abad, & Castagnone-Sereno, 2003; Rutter et al., 2014; Tytgat et al., 2005). A cDNA encoding a secretory cellulose-binding protein (CBP-1) was cloned from *M. incognita* by RNA fingerprinting (Ding et al., 1998). The cDNA-AFLP method was used to identify genes differentially expressed between two pairs of near-isogenic *M. incognita* lines (NIL) or between two *M. javanica* strains, one avirulent and the other virulent against the tomato *Mi-1* resistance gene (Neveu et al., 2002; Gleason & Williamson, 1999; Gleason, Liu, & Williamson, 2008). Gene expression profiles were compared for the infective *M. incognita* J2s, and 22 of the 24,025 transcript-derived fragments (TDF) generated were found to display differential expression (i.e. present in avirulent NIL and absent from both virulent NILs or vice versa). Fourteen of the TDF sequences displayed no significant similarity to known proteins, whereas eight matched reported sequences from nematodes and other invertebrates. ISH on five of the sequences showed that two were specifically expressed in the intestinal cells (HM10), one in the subventral oesophageal glands (HM1), and two in the dorsal oesophageal gland of J2s (HM7 and HM12) (Neveu et al., 2003). The *Cg-1* cDNA fragment, which was present in the avirulent *M. javanica* strain but not in the virulent strain, was found to encode a small nematode protein required for *Mi-1*-mediated resistance (Gleason et al., 2008). Genes upregulated during the endophytic stage were isolated by SSH (Dubreuil, Magliano, Deleury, Abad, & Rosso, 2007; Huang, Dong, & Maier, 2004). Upregulation was demonstrated for genes involved in detoxification (e.g. glutathione-S-transferase, *GST-1*) and protein degradation, for a gene encoding a putative secreted protein and for genes of unknown function. Secreted GSTs may protect the parasite against reactive oxygen species or modulate the plant responses triggered by pathogen attack (Dubreuil et al., 2007). Using RT-Q-PCR, Rutter et al. (2014) showed transcripts of newly discovered putative effectors to be specifically upregulated during different stages of the nematode's life cycle, indicating that they function at specific stages during *M. incognita* parasitism.

3.4 Genome and Secretome Mining

The genomes of RKNs were first compared by the AFLP fingerprinting of three pairs of *M. incognita* NILs, to identify genetic markers displaying differential expression between nematode genotypes avirulent or virulent against the tomato *Mi-1* resistance gene. For the avirulent genotypes, a cDNA encoding a secretory protein with a sequence characterized by internal repeat motifs, named MAP-1, was cloned (Semblat, Rosso, Hussey, Abad, & Castagnone-Sereno, 2001). In 2008, the sequencing of the complete genomes of two RKN species provided new opportunities for studying plant–nematode interactions and initiating comparative genomics studies (Abad et al., 2008; Bird et al., 2009; Opperman et al., 2008). The genome of *M. hapla* contains only 54 Mbp, and was the smallest nematode genome to have been sequenced (Opperman et al., 2008). The complete assembled genome sequence of *M. incognita* revealed a more complex structure with homologous but divergent segment pairs potentially derived from former alleles in this species (Abad et al., 2008). In total, 19,212 and 16,676 protein-coding genes were identified in *M. incognita* and *M. hapla*, respectively. InterPro pro tein domains were identified in 55% of *M. incognita* proteins and 22% of these proteins were predicted to be secreted. Interestingly, 2578 secreted proteins without known domains were predicted to be specific to *M. incognita. Meloidogyne incognita* has an unprecedented set of 61 plant cell wall-degrading, carbohydrate-active enzymes (CAZymes), including 21 cellulases, six xylanases, two polygalacturonases and 30 pectate lyases (Abad et al., 2008). Striking similarities between RKN and bacterial proteins led to the proposal that these cell wall-degrading or cell wall-modifying enzymes, which are generally absent from other metazoans, were acquired by multiple independent lateral gene transfers from different bacterial sources (Danchin et al., 2010). In addition, four CMs, 20 cysteine proteases of the C48 SUMO (small ubiquitin-like modifier) deconjugating enzyme family were shown to be specifically present or more abundant in *M. incognita* genomes than in those of free-living nematodes. Twenty-seven previously described *M. incognita*-restricted pioneer genes expressed in oesophageal glands were retrieved from the genome and a further 11 copies were identified; all of these genes are specific to *Meloidogyne* spp. (Abad et al., 2008).

The increasing availability of full-genome sequences for nematodes and of NGS transcriptomes and functional information databases, such as that of the Uniprot project, has made it possible to use genome mining as an approach for the identification of effectors. This bioinformatic analysis makes use of several filter steps, in which genes that do not correspond to predefined criteria, such as the presence of orthologues in nonpathogenic species or the presence of transmembrane domains, are eliminated (or vice versa). A method for the identification of discriminant motifs in biological sequences has been used to define a set of motifs specifically present in known secreted effectors but absent from evolutionarily conserved house-keeping proteins (Vens, Rosso, & Danchin, 2011). The proteome of *M. incognita* has been shown to include 2579 proteins containing specific MERCI (motif—emerging and with classes—identification) motifs, some of which can be considered to be new putative effectors. Comparative genomic studies led to the identification of a set of genes in RKN genomes that were conserved during the evolution of plant-parasitic nematodes and are only found in organisms parasitic on plants (Danchin et al., 2013). A combination of genome mining and protein function analysis identified large panels of putative effector-like proteins, which generally present an N-terminal signal peptide and no transmembrane domain (Danchin et al., 2013; Haegeman, Bauters, Kyndt, Rahman, & Gheysen, 2013; Roze, Hanse, & Mitreva, 2008). In addition, bioinformatic analyses of predicted effectors revealed the presence of nuclear localization signals (NLS) in several RKN-secreted proteins (Huang et al., 2003; Quentin, Abad, & Favery, 2013; Roze et al., 2008), suggesting that these molecules may target the host cell nucleus.

None of these approaches, including genome/secretome mining, can identify the complete set of effectors when used alone. Effectors not satisfying the predefined criteria may be missed (Mitchum et al., 2013). For example, several effector candidates without a detectable signal peptide have been reported to be secreted but are not picked up by current genome mining strategies due to the lack of a signal peptide. Indeed, although most studies focus on protein effectors, nonprotein molecules, such as plant hormones and carbohydrate effectors, have also been reported to play an important role in parasitism (McCarter et al., 2003). Nevertheless, the characterization of genes specifically expressed in secretory organs is a first step towards understanding their function in plant–RKN interactions. If a gene is identified as 'effector-like' by various approaches, extensive analyses are required, together with a detailed study of its pattern of expression in cells, to confirm its role as a key parasitism effector and for the accurate dissection of gene function during disease development.

4. FUNCTIONAL ANALYSES OF EFFECTORS

The repertoire of putative RKN effectors appears to be large: about a hundred RKN proteins can be localized within secretory organs by ISH and/or immunolocalization (Table 1) and Bellafiore et al. (2008) have suggested that the *M. incognita* secretome contains about 486 proteins. The expression of parasitism genes appears to be regulated developmentally during the parasitic cycle. However, a precise function in parasitism has been attributed to only a very small number of secreted proteins. The contribution to parasitism of some of the effectors identified is easy to deduce. Indeed, cell wall-degrading enzymes have been shown to play a role in nematode penetration and intercellular migration and in cell wall expansion and thickening associated with giant cell formation (Davis, Haegeman, & Kikuchi, 2011). It has also been suggested that nematode CMs, which act on the plant shikimate pathway, thereby decreasing the synthesis of salicylic acid and phytoalexins through competition with chorismate, may prevent the triggering of host defences (Doyle & Lambert, 2003). However, most candidate effectors have no clearly identified function or display no sequence similarity to genes in databases. Detailed functional analyses are therefore required to elucidate their role in parasitism. We will review here the approaches that have been (and are) used to elucidate the function of RKN effectors, particularly 'pioneers' (Table 2).

4.1 Effector Localization

Identification of the plant cell compartments for which nematode-secreted proteins are destined will provide compelling evidence about the molecular functions of these proteins in parasitism. Only a few RKN effectors are known to be secreted into host tissues and, for most, it remains unclear in which compartment of the plant cell they act during parasitism.

The expression *in planta* of RKN fusion proteins tagged with fluorescent proteins e.g. GFP has made it possible to localize effectors without signal peptides within the cytoplasmic and/or nuclear compartments of plant cells, leading to speculations about the functions of RKN effectors (Jaouannet et al., 2012, 2013; Jaubert et al., 2004; Lin et al., 2013). Transient expression of effectors using infiltration of *Agrobacterium tumefaciens* into leaves provides a fast and convenient strategy, but does not allow the localization of RKN effectors during infection. However, plants stably transformed with translational fusions have not yet been used to localize effectors in giant cells during plant–RKN interactions.

Table 2 Functional Analyses of Root-knot Nematode-secreted Proteins Involved in Parasitism and their Known Plant Targets

Effector	Predicted Function	Secretory Organ	*In planta* Localization	Soaking or *in planta* RNAi	*In planta* Overexpression	Host Target	References
Meloidogyne incognita							
Mi-PEL-3	Pectate lyase	SvG	Apoplasm (IL)				Vieira et al. (2011)
Mi-ASP2	Aspartyl protease-like	SvG	Apoplasm (IL)				Vieira et al. (2011), Neveu et al. (2003)
Mi-GST-1	Glutathion-S-transferase	SvG		Soaking dsRNA decreased egg masses production			Dubreuil et al. (2007)
16D10	Cle-like peptide	SvG		In planta RNAi reduced nematode virulence	Root growth promoted	Scarecrow-like transcription factors	Huang et al. (2003), Huang, Dong, et al. (2006), Yang et al. (2013)
Minc03866	C-type lectin	SvG		Soaking siRNA reduced nematode virulence			Danchin et al. (2013)
8D05	Unknown	SvG		In planta RNAi reduced nematode virulence	Accelerated shoot growth, increased suceptibility	Aquaporin tonoplast intrinsic protein 2 (TIP2)	Huang et al. (2003), Xue et al. (2013)

(*Continued*)

Table 2 Functional Analyses of Root-knot Nematode-secreted Proteins Involved in Parasitism and their Known Plant Targets—cont'd

Effector	Predicted Function	Secretory Organ	*In planta* Localization	Soaking or *in planta* RNAi	*In planta* Overexpression	Host Target	References
Minc13292	Unknown	SvG	Cytoplasmic (TE)				Rutter et al. (2014)
Minc18033	Unknown	SvG	Cytoplasmic (TE)				Rutter et al. (2014)
Minc01696	Protein kinase	SvG	Cytoplasmic (TE)				Rutter et al. (2014)
Minc00344	Unknown	SvG	Cytoplasmic (TE)				Rutter et al. (2014)
Minc00469	Unknown	SvG	Cytoplasmic (TE)				Rutter et al. (2014)
Minc15401	Unknown	SvG	Cytoplasmic (TE)				Rutter et al. (2014)
Minc10418	Unknown	SvG	Cytoplasmic (TE)				Rutter et al. (2014)
Minc03328	Unknown	SvG	Cytoplasmic (TE)				Rutter et al. (2014)
Mi-CRT	Calreticulin	SvG&DG	Cytoplasm and nucleus (TE); apoplasm (IL)	Soaking dsRNA, siRNA and in planta RNAi Reduced nematode virulence	Increased susceptibility, PTI defense suppression		Arguel et al. (2012), Jaubert et al. (2005), Jaouannet et al. (2013)

6D4	Unknown	SvG and DG	Apoplasm (IL)			Vieira et al. (2011), Davis et al. (1992), Vieira et al. (2012)
7E12	Unknown	DG			A gall formation	Souza et al. (2011)
Mi-EFF1/ Minc17998	Unknown	DG	Nucleus (TE); giant cell nuclei (IL)			Jaouannet et al. (2012)
Mi-14-3-3-b	14-3-3	DG	Cytoplasmic and nuclear (TE)			Jaubert et al. (2004)
Minc18861	Unknown	DG	Cytoplasmic (TE)			Rutter et al. (2014)
Minc02097	Unknown	DG	Cytoplasmic (TE)			Rutter et al. (2014)
Mi-MAP1	Unknown	Amphids	Apoplasm (IL)			Vieira et al. (2011), Vieira et al. (2012), Semblat et al. (2001)
Meloidogyne javanica						
Mj-eng-3	Beta-1,4-endoglucanase	SvG		In planta RNAi reduced nematode virulence		Hu et al. (2013)

(*Continued*)

Table 2 Functional Analyses of Root-knot Nematode-secreted Proteins Involved in Parasitism and their Known Plant Targets—cont'd

Effector	Predicted Function	Secretory Organ	*In planta* Localization	Soaking or *in planta* RNAi	*In planta* Overexpression	Host Target	References
Mj-CBP-1	Cellulose-binding protein	SvG (Mi-CBP-1)		Soaking dsRNA reduced root penetration and egg masses production			Adam et al. (2008)
Mj-CM-1	Chorismate mutase	SvG			Reduced and aborted lateral roots		Doyle & Lambert (2003)
Mj-NULG1a	Unknown	DG	Nucleus (TE), giant cell nuclei (IL)	In planta RNAi reduced nematode virulence	Increased suceptibility		Lin et al. (2013)
Mj-FAR-1	Fatty acid- and retinol-binding	Cuticle	Apoplasm (IL)	In planta RNAi reduced nematode virulence	Larger feeding cells		Iberkleid et al. (2013)
Meloidogyne chitwoodi							
Mc-16D10	16D10-like	SvG		In planta RNAi reduced nematode virulence			Dinh et al. (in press)

IL, immunolocalization; SvG, subventral glands; DG, dorsal gland; TE, transient expression in plant cells of the effector without its signal peptide using *Agrobacterium tumefaciens*.

Immunolocalization techniques have provided the most reliable data concerning the distribution of effectors within host cells. Mi-CRT was the first molecule shown to be secreted into the feeding site via the nematode stylet and to accumulate in large amounts at the cell wall of giant cells (Jaubert et al., 2005). The apoplasm was subsequently confirmed as an important destination compartment for nematode-secreted proteins during migration and feeding cell formation in the host plant (Rosso, Vieira, de Almeida-Engler, & Castagnone-Sereno, 2011; Vieira et al., 2011). Three putative effectors secreted by the subventral glands—a pectate lyase (Mi-PEL3), an aspartyl protease (Mi-ASP2) and the 6D4 protein of unknown function—and the amphidial protein Mi-MAP1 were shown to be secreted into the apoplasm during the intercellular migration of infective juveniles. Mi-6D4 and Mi-MAP1 have also been detected in the apoplasm at the feeding site of sedentary parasitic juveniles, indicating a possible role for these secreted proteins in giant cell formation and/or maintenance (Vieira et al., 2011).

It has been shown recently that two RKN effectors are injected into the cytoplasm of the giant cells, and that these proteins then target the nuclei. Jaouannet et al. (2012) carried out immunolocalization experiments on infected tomato roots and confirmed the secretion *in planta* of Mi-EFF1, a 122-amino acid protein with a signal peptide for secretion and an NLS. Despite the detection of Mi-EFF1 in the dorsal gland of migratory juveniles, no signal was observed along the migratory path of the nematode, suggesting that this protein is not secreted during migration. By contrast, Mi-EFF1 was secreted via the stylet, by sedentary nematodes settled at the feeding site. The secreted Mi-EFF1 was located in the nuclei of giant cells. Mi-EFF1 seems to be specific to early steps in the plant–nematode interaction, but its function during parasitism remains to be determined. Similarly, the *M. javanica* Mj-NULG1a protein is a 274-amino acid pioneer protein with a signal peptide for secretion and two NLS motifs. This effector is produced in the dorsal gland of the nematode and is injected into the cytoplasm of the giant cells. It subsequently accumulates within the nuclei, as demonstrated by immunocytochemical studies of infected tomato roots (Lin et al. 2013). Giant cell ontogenesis and maintenance requires the manipulation of host nuclear processes, such as the cell cycle and transcriptional regulation, and undoubtedly involves the targeting of the host nucleus by secreted effectors. It is also clear that RKN effectors must target various nuclear processes to redirect plant cell differentiation and immunity in ways similar to those observed for other plant pathogens (Deslandes & Rivas, 2012; Rivas, 2012; Rivas & Génin, 2011).

4.2 RNA Interference-Mediated Gene Silencing

It is not currently possible to transform RKNs, but RNA interference (RNAi) approaches have been developed for studies of the role of candidate effectors (Rosso, Jones, & Abad, 2009). In RNAi approaches, small interfering RNAs (siRNAs) are generated from a double-stranded RNA (dsRNA) by Dicer-mediated cleavage and are processed via an RNA-induced silencing complex that drives the degradation of the targeted transcript (Fire et al., 1998; Tabara, Yigit, Siomi, & Mello, 2002). The dsRNA or siRNAs inactivating the effector genes can be delivered to nematodes by in vitro 'soaking' of preparasitic J2s in exogenous dsRNA or siRNA molecules or *in planta*, by transgenic methods involving the expression of hairpin RNA (hpRNA) or through host/virus-induced gene silencing (HIGS/VIGS) (Rosso et al. 2009). Correlations between candidate effector gene silencing and the reduced parasitic success of the nematode suggest that secreted effectors do indeed play a key role in parasitism.

The silencing of a GST-1, Mi-GST-1, and of a CBP-1, Mj-CBP-1, has been achieved by soaking parasites in a solution of long dsRNA molecules homologous to the targeted transcripts of the genes to be silenced (400 and 255 nt, respectively), and was shown to lead to lower levels of parasitic success. The targeting of Mi GST 1 with dsRNA led to a 90% decrease in *gst* transcript abundance in the treated nematodes and decreased the ability of the nematode to develop and reproduce on tomato plants infected with the treated parasitic juveniles (Dubreuil et al., 2007), whereas the silencing of Mi-CBP-1 decreased the penetration success of *M. javanica* (Adam, Phillips, Jones, & Block, 2008). Mi-CBP-1 silencing, however appeared rather heterogeneous between treated nematode lines (Adam et al., 2008), and such a result may depend on the targeted gene, and could limit the use of RNAi through soaking. However, impressive results were obtained with a synthetic siRNA only 21 nucleotides in length. With this siRNA, Arguel, Jaouannet, Magliano, Abad, and Rosso (2012) validated this strategy by silencing Mi-CRT in nematode oesophageal glands, thereby decreasing nematode virulence. In vitro RKN gene silencing by siRNA soaking has since been used successfully to demonstrate the function in parasitism of new RKN target genes including an effector expressed in *M. incognita* oesophageal glands encoding a C-type lectin (Danchin et al. 2013). C-type lectins appear to be secreted by various plant-parasitic nematodes, including cyst nematodes and RKNs, and may be involved in modulating plant immune responses (Danchin et al., 2013; Ganji, Jenkins, & Wubben, 2014; Haegeman et al., 2013; Roze et al., 2008; Urwin, Lilley, & Atkinson, 2002).

Alternatively, siRNA can be delivered to nematodes by feeding on transgenic plants expressing a dsRNA (usually an hpRNA). This technology was developed in the model plant *Arabidopsis* and used to demonstrate the function of five candidate effectors (16D10, 8D05, Mi-CRT, fatty acid- and retinol-binding Mj-FAR-1 and Mj-NULG1a) in parasitism (Table 2, Huang, Dong, et al., 2006; Iberkleid et al., 2013; Jaouannet et al., 2013; Lin et al., 2013; Xue et al., 2013). *Meloidogyne incognita* 16D10 is a small peptide displaying sequence similarity to the proteins of the plant CLE protein family. The in vivo expression of 16D10 dsRNA in *Arabidopsis* resulted in resistance effective against the four major RKN species, *M. incognita, M. javanica, M. arenaria* and *M. hapla*, suggesting that this effector is conserved among RKNs (Huang, Allen, et al., 2006). Indeed, *M. chitwoodi* has also been shown to secrete a 16D10-like protein (Dinh, Brown, & Elling, in press). The *in planta* silencing of 16D10 could also be applied to crop plants such as grapes and potato, and this approach may provide interesting opportunities to generate crops with broad resistance to RKNs (Yang et al., 2013; Dinh et al., in press). Interestingly, Souza Junior et al. (2013) showed that this technology could be used to silence three *M. incognita* proteases simultaneously, with the use of transgenic RNAi tobacco lines. This ability to silence several parasitic genes simultaneous makes this strategy promising for the study of effector functions and the development of new tools for nematode control. Indeed it will allow silencing of multigenic families or genes with redundant functions and would allow better durability of the resistance. Again, efficiency of *in planta* RNAi may vary considerably, depending on the level and pattern of target gene expression, size and sequence composition of the dsRNA segment and its position in the target gene. Silencing of a targeted RKN gene may not be achieved similarly in every generated transgenic line, and questions remain regarding the stability and inheritance of this resistance (Gheysen & Vanholme, 2007; Rosso et al., 2009). Viruses have also been used to deliver dsRNA fragments to host plants, and appear to be effective for delivery of RNAi triggers to feeding nematodes. Tobacco rattle virus-mediated silencing of Mi-CRT was correlated with the decreased ability of the nematode to induce disease (Dubreuil et al., 2009). Nevertheless, viruses do not infect all root cells, and this limits the use of VIGS for functional screens (Dubreuil et al., 2009; Rosso et al., 2009; see Chapter 9, Hewezi & Baum for a detailed explanation). Finally, tools developed to express artificial microRNAs *in planta* (Carbonnell et al. 2014; Schwab, Ossowski, Riester, Warthmann, &

Weigel, 2006) will undoubtedly prove useful in the near future for developing new types of resistance to nematodes based on effector silencing.

4.3 In Planta Effector Overexpression

In addition to RNAi analyses, *in planta* gene overexpression approaches have also been used to determine functions of effectors in plant cell physiology and pathogenicity, based on the phenotypes exhibited by transgenic plants. Feeding cell development involves impressive effects of parasitism proteins on root cell physiology, and the *in planta* overexpression of RKN effectors can result in major changes to plant phenotype. The inhibition of vascularization and lateral root initiation resulting from expression of Mj-CM-1 in soybean hairy roots provided the first clear example of an oesophageal gland-specific protein altering plant development (Doyle & Lambert, 2003) (Table 2). The phenotype of Mj-CM-1-overexpressing plants can be rescued by adding indole-2-acetic acid. Doyle and Lambert (2003) hypothesized that RKN CM is active in the cytoplasm, where it alters the shikimate pathway to prevent auxin formation. Mj-CM-1 thus has the potential to act as a multifunctional enzyme in the promotion of nematode pathogenicity. Mj-CM-1 can assist in the developmental reprogramming required to generate giant cells, by interfering with the auxin signalling pathway and/or inhibiting the production of plant defence compounds, as discussed above. The overexpression of 16D10 in both tobacco and *Arabidopsis* results in greater root growth and the extensive production of lateral roots; it also induces the formation of callus on tomato roots (Huang, Dong, et al., 2006), demonstrating the ability of this effector to alter plant development significantly. The constitutive expression of 8D05 in *Arabidopsis* plants greatly increases shoot growth and early flowering, but has no effect on root growth. Susceptibility to *M. incognita* infection was found to be greater in all lines overexpressing Mi8D05 than in wild-type *Arabidopsis*, leading to the conclusion that this protein also plays a key role in parasitism (Xue et al., 2013). Lin et al. (2013) demonstrated the key role of Mj-NULG1a in parasitism in a similar manner. They showed a clear effect of transgene expression on susceptibility to nematode attack, with more nematodes in the infected *Arabidopsis* roots and larger numbers of galls formed when Mj-NULG1a was overexpressed. By contrast, Mj-NULG1a had no apparent effect on the growth of transgenic plants (Lin et al., 2013). Tomato roots constitutively expressing Mj-FAR-1 were more susceptible to RKN infection and displayed faster gall induction and expansion, with more nematodes developing into mature females than observed in control roots.

Further histological analysis of the infected Mj-FAR-1-overexpressing plants indicated that the galls contained larger feeding cells potentially able to support faster nematode development and maturation. Nevertheless, a phenotypic analysis of Mj-FAR-1-overexpressing root lines revealed no significant change in root development and growth (Iberkleid et al., 2013). An analysis of defence-related target gene expression in the Mj-FAR-1-overexpressing plants led the authors to conclude that this effector might facilitate infection by manipulating host lipid-based defences (Iberkleid et al., 2013). The putative effector 7E12, expressed in the *M. incognita* dorsal gland, has also been shown to disrupt host root physiology to favour parasitism (Souza et al., 2011). Gall formation and egg hatching occur more rapidly in 7E12-overexpressing tobacco plants than in wild-type tobacco plants. The morphology of the giant cells was shown to be affected, with more vacuoles and wall ingrowths, and the apparent proliferation of neighbouring cells. Souza et al. (2011) suggested that 7E12 may act by regulating host cell division. Finally, Jaouannet et al. (2013) recently showed that Mi-CRT overexpression increases susceptibility to *M. incognita* in *Arabidopsis*, probably by interfering with the triggering of defence responses.

4.4 Defence Suppression Assays

In general, the first line of plant defence against the invading nematode is triggered by damage-associated molecular patterns or conserved pathogen-associated molecular patterns (PAMPS). These patterns induce various defence mechanisms referred to as 'basal resistance' or 'PAMP-triggered immunity' (PTI). Plants can also recognize specific pathogen-derived effectors and have acquired a highly specific defence response known as effector-triggered immunity (ETI). Together, PTI and ETI limit the entry of microbes, restrict pathogen propagation or kill pathogens within the host plant. These immune signalling pathways are common targets of many plant pathogen effectors (Dou & Zhou, 2012; Jones & Dangl, 2006). Several assays have been developed and used to demonstrate the functions of bacterial, fungal and oomycete effectors in defence suppression. However, such approaches have not been applied to RKNs until recently. Using *Arabidopsis* lines overexpressing Mi-CRT, Jaouannet et al. (2013) demonstrated that this effector was able to suppress the defences induced by the PAMP elf18 (N-terminal 18 amino acids of elongation factor Tu), which usually triggers callose deposition and the expression of defence marker genes of salicylic acid, jasmonate and ethylene pathways (see Chapters 11 and 13 for further details).

4.5 Search for the Host Targets of Effectors

RKN effectors must manipulate host cellular processes through specific interactions with certain host proteins, to favour parasitism. We currently know almost nothing about the host targets of RKN effectors. Only two RKN effectors have been identified to date, but yeast two-hybrid (Y2H) screens have been used successfully to identify potential host targets (Huang, Dong, et al., 2006; Xue et al., 2013; Quentin, unpublished results). Studies with a tomato root cDNA library have shown that two scarecrow-like (SCL) plant transcription factors are targeted by the *M. incognita* 16D10 effector (Huang, Dong, et al., 2006). The 16D10 effector can also interact with *Arabidopsis SCL6* and *SCL21* in yeast. These interactions have been confirmed in co-immunoprecipitation experiments. In plants, SCL transcription factors regulate root meristem identity and root development. The identification of interactions with these proteins thus provided the first evidence that plant-parasitic nematode-secreted peptides may function as signalling molecules, inducing root proliferation by specifically targeting host proteins regulating transcription. However the function of these transcriptions factors in giant cell ontogenesis has not yet been demonstrated. Recently, a Y2H screen in which *M. incognita* 8D05 was used as bait revealed the occurrence of multiple interactions with the plant aquaporin TIP2 (tonoplast intrinsic protein 2) in tomato (Xue et al., 2013). This interaction suggests a potential role for the 8D05 effector protein in regulating water and solute transport within giant cells, promoting their enlargement and nematode feeding.

5. CONCLUSIONS

The rapid characterization of oesophageal gland secretions with emerging 'omics' technologies and the increasing availability of genome sequences for RKNs and now also cyst nematodes, has expanded our knowledge of the nematode effector repertoire (Bird et al., 2009; Cotton et al., 2014). Transcriptomic methods have focused on nematode genes regulated during parasitism, particularly those expressed in the oesophageal glands. Analysis of the functions of these RKN effectors remains challenging. Little is known about their role in manipulating host processes to induce the transformation of root cells into hypertrophied and multinucleate feeding cells. RNAi has proved to be an effective approach to deciphering the function of RKN effectors in parasitism and is expected to be a valuable tool for developing new strategies for plant-parasitic nematode control. One of the key unresolved questions relating to the RKN effectors identified concerns the nature of the host targets. RKNs clearly manipulate plant immune pathways, but other host processes must also be targeted for giant cell ontogenesis and nematode settlement. The identification and disabling of essential plant target genes rendering plants resistant to disease would be a promising strategy for the achievement of durable, broad-spectrum control of these pests.

ACKNOWLEDGEMENTS

This work was funded by INRA and by the French Government (National Research Agency, ANR) through the 'Investments for the Future' LABEX SIGNALIFE: program reference # ANR-11-LABX-0028-01. N.-M. T. and C.-N. N. are supported by USDH fellowships from the Ministère de l'Education et de la Formation de la République Socialiste du Vietnam.

REFERENCES

Abad, P., Gouzy, J., Aury, J. M., Castagnone-Sereno, P., Danchin, E. G. J., Deleury, E., et al. (2008). Genome sequence of the metazoan plant-parasitic nematode *Meloidogyne incognita*. *Nature Biotechnology, 26*, 909–915.

Abad, P., & Williamson, V. M. (2010). Plant nematode interaction, a sophisticated dialogue. In *Advances in botanical research* (Vol. 53, pp. 147–192). Elsevier.

Adam, M. A. M., Phillips, M. S., Jones, J. T., & Block, V. C. (2008). Characterisation of a cellulose-binding protein Mj-cbp-1 of the root knot nematode, *Meloidogyne javanica*. *Physiological and Molecular Plant Pathology, 72*, 21–28.

Arguel, M. J., Jaouannet, M., Magliano, M., Abad, P., & Rosso, M.-N. (2012). siRNAs trigger efficient silencing of a parasitism gene in plant parasitic root-knot nematodes. *Genes, 3*, 391–408.

Barcala, M., García, A., Cabrera, J., Casson, S., Lindsey, K., Favery, B., et al. (2010). Early transcriptomic events in microdissected *Arabidopsis* nematode-induced giant cells. *Plant Journal, 61*, 698–712.

Bellafiore, S., Shen, Z., Rosso, M.-N., Abad, P., Shih, P., & Briggs, S. P. (2008). Direct identification of the *Meloidogyne incognita* secretome reveals proteins with host cell reprogramming potential. *PLoS Pathogens, 4*, e1000192.

Bird, D. M., Williamson, V. M., Abad, P., McCarter, J., Danchin, E. G., Castagnone-Sereno, P., et al. (2009). The genomes of root-knot nematodes. *Annual Review of Phytopathology, 47*, 333–351.

Blok, V. C., Jones, J. T., Phillips, M. S., & Trudgill, D. L. (2008). Parasitism genes and host range disparities in biotrophic nematodes, the conundrum of polyphagy versus specialisation. *Bioessays, 30*, 249–259.

Caillaud, M. C., Abad, P., & Favery, B. (2008). Cytoskeleton reorganization, A key process in root-knot nematode-induced giant cell ontogenesis. *Plant Signaling and Behavior, 3*, 816–818.

Caillaud, M.-C., Dubreuil, G., Quentin, M., Perfus-Barbeoch, L., Lecomte, P., de Almeida Engler, J., et al. (2008). Root-knot nematodes manipulate plant cell functions during a compatible interaction. *Journal of Plant Physiology, 165*, 104–113.

Caillaud, M. C., Lecomte, P., Jammes, F., Quentin, M., Pagnotta, S., Andrio, E., et al. (2008). MAP65-3 microtubule-associated protein is essential for nematode-induced giant cell ontogenesis in *Arabidopsis*. *Plant Cell, 20*, 423–437.

Carbonell, A., Takeda, A., Fahlgren, N., Johnson, S. C., Cuperus, J. T., & Carrington, J. C. (2014). New generation of artificial microRNA and synthetic trans-acting small interfering RNA vectors for efficient gene silencing in *Arabidopsis*. *Plant Physiology, 165*, 15–29.

Castagnone-Sereno, P. (2006). Genetic variability and adaptive evolution in parthenogenetic root-knot nematodes. *Heredity, 96*, 282–289.

Clement, M., Ketelaar, T., Rodiuc, N., Banora, M., Smertenko, A., Engler, G., et al. (2009). Actin-depolymerizing factor 2-mediated actin dynamics are essential for root-knot nematode infection of *Arabidopsis*. *Plant Cell, 21*, 2963–2979.

Cotton, J. A., Lilley, C. J., Jones, L. M., Kikuchi, T., Reid, A. J., Thorpe, P., et al. (2014). The genome and life-stage specific transcriptomes of *Globodera pallida* elucidate key aspects of plant parasitism by a cyst nematode. *Genome Biology, 15*, R43.

Damiani, I., Baldacci-Cresp, F., Hopkins, J., Andrio, E., Balzergue, S., Lecomte, P., et al. (2012). Plant genes involved in harbouring symbiotic rhizobia or pathogenic nematodes. *New Phytologist, 194*, 511–522.

Danchin, E. G. J., Arguel, M.-J., Campan-Fournier, A., Perfus-Barbeoch, L., Magliano, M., Rosso, M.-N., et al. (2013). Identification of novel target genes for safer and more specific control of root-knot nematodes from a pan-genome mining. *PLoS Pathogens, 9*, e1003745.

Danchin, E. G. J., Rosso, M. N., Vieira, P., de Almeida-Engler, J., Coutinho, P. M., Henrissat, B., et al. (2010). Multiple lateral gene transfers and duplications have promoted plant parasitism ability in nematodes. *Proceedings of the National Academy of Sciences of the United States of America, 107*, 17651–17656.

Dautova, M., Rosso, M.-N., Abad, P., Gommers, J. F., Bakker, J., & Smant, G. (2001). Single-pass cDNA sequencing—a powerful tool to analyse gene expression in preparasytic juveniles of the southern root-knot nematode *Meloidogyne incognita*. *Nematology, 3*, 129–139.

Davis, E. L., Allen, R., Pratt, L. H., & Hussey, R. S. (1992). Novel immunization procedures used to develop monoclonal antibodies that bind to specific structures in *Meloidogyne* spp. *Phytopathology, 82*, 1244–1250.

Davis, E., Haegeman, A., & Kikuchi, T. (2011). Degradation of the plant cell wall by nematodes. In J. Jones, G. Gheysen, & C. Fenoll (Eds.), *Genomics and molecular genetics of plant-nematode interactions* (pp. 225–272). Berlin: Springer.

Davis, E. L., Hussey, R. S., & Baum, T. J. (2004). Getting to the roots of parasitism by nematodes. *Trends in Parasitology, 20*, 134–141.

Davis, E. L., Hussey, R. S., Mitchum, M. G., & Baum, T. J. (2008). Parasitism proteins in nematode-plant interactions. *Currnt Opinion in Plant Biology, 11*, 360–366.

Deslandes, L., & Rivas, S. (2012). Catch me if you can, bacterial effectors and plant targets. *Trends in Plant Science, 17*, 644–655.

Ding, X., Shields, J., Allen, R., & Hussey, R. S. (1998). A secretory cellulose-binding protein cDNA cloned from the root-knot nematode (*Meloidogyne incognita*). *Molecular Plant-Microbe Interactions, 11*, 952–959.

Ding, X., Shields, J., Allen, R., & Hussey, R. S. (2000). Molecular cloning and characterisation of a venom allergen AG5-like cDNA from *Meloidogyne incognita*. *International Journal of parasitology, 30*, 77–81.

Dinh, P. T. Y., Brown, C. R., & Elling, A. A. (2014). RNA interference of effector gene Mc16D10L confers resistance against Meloidogyne chitwoodi in Arabidopsis and potato. *Phytopathology, 104*, 1098–1106.

Dou, D., & Zhou, J.-M. (2012). Phytopathogen effectors subverting host immunity, different foes, similar battleground. *Cell Host and Microbe, 12*, 484–495.

Doyle, E. A., & Lambert, K. N. (2002). Cloning and characterization of an esophageal-gland-specific pectate lyase from the root-knot nematode *Meloidogyne javanica*. *Molecular Plant-Microbe Interactions, 15*, 549–556.

Doyle, E. A., & Lambert, K. N. (2003). *Meloidogyne javanica* chorismate mutase 1 alters plant cell development. *Molecular Plant-Microbe Interactions, 16*, 123–131.

Dubreuil, G., Magliano, M., Deleury, E., Abad, P., & Rosso, M. N. (2007). Transcriptome analysis of root-knot nematode functions induced in the early stages of parasitism. *New Phytologist, 176*, 426–436.

Dubreuil, G., Magliano, M., Dubrana, M. P., Lozano, J., Lecomte, P., Favery, B., et al. (2009). Tobacco rattle virus mediates gene silencing in a plant parasitic root-knot nematode. *Journal of Experimental Botany, 60*, 4041–4050.

Fire, A., Xu, S., Montgomery, M. K., Kostas, S. A., Driver, S. E., & Mello, C. C. (1998). Potent and specific genetic interference by double-stranded RNA in *Caenorhabditis elegans*. *Nature, 391*, 806–811.

Ganji, S., Jenkins, J. N., & Wubben, M. J. (2014). Molecular characterization of the reniform nematode C-type lectin gene family reveals a likely role in mitigating environmental stresses during plant parasitism. *Gene, 537*, 269–278.

Gheysen, G., & Fenoll, C. (2002). Gene expression in nematode feeding sites. *Annual Review of Phytopathology, 40*, 191–219.

Gheysen, G., & Vanholme, B. (2007). RNAi from plants to nematodes. *Trends in Biotechnology, 25*, 89–92.

Ghini, R., Hamada, E., José, M., Júnior, P., & Marengo, J. A. (2008). Risk analysis of climate change on coffee nematodes and leaf miner in Brazil. *Pesquisa Agropecuária Brasileira, 43*, 187–194.

Gleason, C., Liu, Q. L., & Williamson, V. M. (2008). Silencing a candidate nematode effector gene corresponding to the tomato resistance gene *Mi-1* leads to acquisition of virulence. *Molecular Plant-Microbe Interactions, 21*, 576–585.

Gleason, C., & Williamson, V. M. (1999). The identification of molecular differences between virulent and avirulent root-knot nematodes. *Journal of Nematology, 31*, 536.

Haegeman, A., Bauters, L., Kyndt, T., Rahman, M. M., & Gheysen, G. (2013). Identification of candidate effector genes in the transcriptome of the rice root knot nematode *Meloidogyne graminicola*. *Molecular Plant Pathology, 14*, 379–390.

Hewezi, T., & Baum, T. J. (2013). Manipulation of plant cells by cyst and root-knot nematode effectors. *Molecular Plant-Microbe Interactions, 26*, 9–16.

Huang, G., Allen, R., Davis, E. L., Baum, T. J., & Hussey, R. S. (2006). Engineering broad root-knot resistance in transgenic plants by RNAi silencing of a conserved and essential root-knot nematode parasitism gene. *Proceedings of the National Academy of Sciences of the United States of America, 103*, 14302–14306.

Huang, G., Dong, R., Allen, R., Davis, E. L., Baum, T. J., & Hussey, R. S. (2005a). Two chorismate mutase genes from the root-knot nematode *Meloidogyne incognita*. *Molecular Plant Pathology, 6*, 23–30.

Huang, G., Dong, R., Allen, R., Davis, E. L., Baum, T. J., & Hussey, R. S. (2005b). Developmental expression and molecular analysis of two *Meloidogyne incognita* pectate lyase genes. *International Journal of Parasitology, 35*, 685–692.

Huang, G., Dong, R., Allen, R., Davis, E. L., Baum, T. J., & Hussey, R. S. (2006). A root-knot nematode secretory peptide functions as a ligand for a plant transcription factor. *Molecular Plant-Microbe Interactions, 19*, 463–470.

Huang, G., Dong, R., & Maier, T. (2004). Use of solid phase subtractive hybridization for the identification of parasitism gene candidates from the root-knot nematode *Meloidogyne incognita*. *Molecular Plant Pathology, 5*, 217–222.

Huang, G., Gao, B., Maier, T., Allen, R., Davis, E. L., Baum, T. J., et al. (2003). A profile of putative parasitism genes expressed in the esophageal gland cells of the root-knot nematode *Meloidogyne incognita*. *Molecular Plant-Microbe Interactions, 16*, 376–381.

Hu, L., Cui, R., Sun, L., Lin, B., Zhuo, K., & Liao, J. (2013). Molecular and biochemical characterization of the β-1,4-endoglucanase gene *Mj-eng-3* in the root-knot nematode *Meloidogyne javanica*. *Experimental Parasitology, 135*, 15–23.

Hussey, R. S. (1989). Monoclonal antibodies to secretory granules in esophageal glands of *Meloidogyne* species. *Journal of Nematology, 21*, 392–398.

Hussey, R. S., Paguio, O. R., & Seabury, F. (1990). Localization and purification of a secretory protein from the esophageal glands of *Meloidogyne incognita* with a monoclonal antibody. *Phytopathology, 80*, 709–714.

Iberkleid, I., Vieira, P., de Almeida-Engler, J., Firester, K., Spiegel, Y., & Horowitz, S. B. (2013). Fatty acid-and retinol-binding protein, Mj-FAR-1 induces tomato host susceptibility to root-knot nematodes. *PLoS One, 8*, e64586.

Jaouannet, M., Magliano, M., Arguel, M. J., Gourgues, M., Evangelisti, E., Abad, P., et al. (2013). The root-knot nematode calreticulin Mi-CRT is a key effector in plant defense suppression. *Molecular Plant-Microbe Interactions, 26*, 97–105.

Jaouannet, M., Perfus-barbeoch, L., Deleury, E., Magliano, M., Engler, G., Vieira, P., et al. (2012). A root-knot nematode-secreted protein is injected into giant cells and targeted to the nuclei. *New Phytologist, 194*, 924–931.

Jaubert, S., Laffaire, J. B., Abad, P., & Rosso, M. N. (2002a). A polygalacturonase of animal origin isolated from the root-knot nematode *Meloidogyne incognita*. *FEBS Letters, 522*, 109–112.

Jaubert, S., Laffaire, J. B., Ledger, T. N., Escoubas, P., Amri, E. Z., Abad, P., et al. (2004). Comparative analysis of two 14-3-3 homologues and their expression pattern in the root-knot nematode *Meloidogyne incognita*. *International Journal of Parasitology, 34*, 873–880.

Jaubert, S., Ledger, T. N., Laffaire, J. B., Piotte, C., Abad, P., & Rosso, M.-N. (2002). Direct identification of stylet secreted proteins from root-knot nematodes by a proteomic approach. *Molecular & Biochemical Parasitology, 121*, 205–211.

Jaubert, S., Milac, A. L., Petrescu, A. J., de Almeida-Engler, J., Abad, P., & Rosso, M. N. (2005). *In planta* secretion of a calreticulin by migratory and sedentary stages of root-knot nematode. *Molecular Plant-Microbe Interactions, 18*, 1277–1284.

Jones, J. D. G., & Dangl, J. L. (2006). The plant immune system. *Nature, 244*, 323–329.

Jones, M. G. K., & Payne, H. L. (1978). Early stages of nematode-induced giant-cell formation in roots of *Impatiens balsamina*. *Journal of Nematology, 10*, 70–84.

Jones, J. T., Perry, R. N., & Johnston, M. R. L. (1993). Changes in the ultrastructure of the cuticle of the potato cyst nematode, *Globodera rostochiensis*, during development and infection. *Fundamental and Applied Nematology, 16*, 433–445.

Lambert, K. N., Allen, K. D., & Sussex, I. M. (1999). Cloning and characterization of an esophageal-gland-specific chorismate mutase from the phytoparasitic nematode *Meloidogyne javanica*. *Molecular Plant-Microbe Interactions, 12*, 328–336.

Lin, B., Zhuo, K., Wu, P., Cui, R., Zhang, L.-H., & Liao, J. (2013). A novel effector protein, MJ-NULG1a, targeted to giant cell nuclei plays a role in *Meloidogyne javanica* parasitism. *Molecular Plant-Microbe Interactions, 26*, 55–66.

Maier, T., Hewezi, T., Peng, J., & Baum, T. J. (2013). Isolation of whole esophageal gland cells from plant-parasitic nematodes for transcriptome analyses and effector identification. *Molecular Plant-Microbe Interactions, 26*, 31–35.

Mbeunkui, F., Scholl, E. H., Opperman, C. H., Goshe, M. B., & Bird, D. M. (2010). Proteomic and bioinformatic analysis of the root-knot nematode *Meloidogyne hapla*, the basis for plant parasitism research articles. *Journal of Proteome Research, 9*, 5370–5381.

McCarter, J. P., Mitreva, M. D., Martin, J., Dante, M., Wylie, T., Rao, U., et al. (2003). Analysis and functional classification of transcripts from the nematode *Meloidogyne incognita*. *Genome Biology, 4*, R26.

Mitchum, M. G., Hussey, R. S., Baum, T. J., Wang, X., Elling, A. A., Wubben, M., et al. (2013). Nematode effector proteins, an emerging paradigm of parasitism. *New Phytologist, 199*, 879–894.

Neveu, C., Abad, P., & Castagnone-Sereno, P. (2002). Differential gene expression between avirulent and virulent *Meloidogyne incognita* isogenic lines by cDNA-AFLP. *Nematology, 4*, 234–235.

Neveu, C., Jaubert, S., Abad, P., & Castagnone-Sereno, P. (2003). A set of genes differentially expressed between avirulent and virulent *Meloidogyne incognita* near-isogenic lines encode secreted proteins. *Molecular Plant-Microbe Interactions, 16*, 1077–1084.

Opperman, C. H., Bird, D. M., Williamson, V. M., Rokhsar, D. S., Burke, M., Cohn, J., et al. (2008). Sequence and genetic map of *Meloidogyne hapla*, A compact nematode genome for plant parasitism. *Proceedings of the National Academy of Sciences of the United States of America, 105*, 14802–14807.

Quentin, M., Abad, P., & Favery, B. (2013). Plant parasitic nematode effectors target host defense and nuclear functions to establish feeding cells. *Frontiers of Plant Science, 4*, 53.

Rivas, S. (2012). Nuclear dynamics during plant innate immunity. *Plant Physiology, 158*, 87–94.

Rivas, S., & Génin, S. (2011). A plethora of virulence strategies hidden behind nuclear targeting of microbial effectors. *Frontiers of Plant Science, 2*, 1–20.

Rosso, M. N., Favery, B., Piotte, C., Arthaud, L., De Boer, J. M., Hussey, R. S., et al. (1999). Isolation of a cDNA encoding a beta-1,4-endoglucanase in the root-knot nematode *Meloidogyne incognita* and expression analysis during plant parasitism. *Molecular Plant-Microbe Interactions, 12*, 585–591.

Rosso, M. N., Jones, J. T., & Abad, P. (2009). RNAi and functional genomics in plant parasitic nematodes. *Annual Reviews of Phytopathology, 47*, 207–232.

Rosso, M. N., Vieira, P., de Almeida-Engler, J., & Castagnone-Sereno, P. (2011). Proteins secreted by root-knot nematodes accumulate in the extracellular compartment during root infection. *Plant Signaling & Behavior, 6*, 1232–1234.

Roze, E., Hanse, B., & Mitreva, M. (2008). Mining the secretome of the root-knot nematode *Meloidogyne chitwoodi* for candidate parasitism genes. *Molecular Plant Pathology, 9*, 1–10.

Rutter, W. B., Hewezi, T., Abubucker, S., Maier, T. R., Huang, G., Mitreva, M., et al. (2014). Mining novel effector proteins from the esophageal gland cells of *Meloidogyne incognita*. *Molecular Plant-Microbe Interactions, 27*, 965–974. http://dx.doi.org/10.1094/MPMI-03-14-0076-R.

Schwab, R., Ossowski, S., Riester, M., Warthmann, N., & Weigel, D. (2006). Highly specific silencing by artificial microRNAs in *Arabidopsis*. *Plant Cell, 18*, 1121–1133.

Semblat, J. P., Rosso, M. N., Hussey, R. S., Abad, P., & Castagnone-Sereno, P. (2001). Molecular cloning of a cDNA encoding an amphid-secreted putative avirulence protein from the root-knot nematode *Meloidogyne incognita*. *Molecular Plant-Microbe Interactions, 14*, 72–79.

Smant, G., Stokkermans, J. P., Yitang, Y., de Boer, J. M., Baum, T., Wang, X., et al. (1998). Endogenous cellulases in animals, isolation of b-1, 4-endoglucanase genes from two species of plant-parasitic cyst nematodes. *Proceedings of the National Academy of Sciences of the United States of America, 95*, 4906–4911.

Souza Junior, J. D., Ramos Coelho, R., Tristan Lourenço, I., da Rocha Fragoso, R., Barbosa Viana, A. A., Lima Pepino de Macedo, L., et al. (2013). Knocking-down *Meloidogyne incognita* proteases by plant-delivered dsRNA has negative pleiotropic effect on nematode vigor. *PLoS One, 8*, e85364.

Souza, J., Souza, J. D., Jr., Grossi-de-Sá, M., Lima Rocha, T., Fragoso, R., Barbosa, A. E., et al. (2011). Ectopic expression of a *Meloidogyne incognita* dorsal gland protein in tobacco accelerates the formation of the nematode feeding site. *Plant Science, 180*, 276–282.

Tabara, H., Yigit, E., Siomi, H., & Mello, C. C. (2002). The dsRNA binding protein RDE-4 interacts with RDE-1, DCR-1, and a DExH-box helicase to direct RNAi in *C. elegans*. *Cell, 109*, 861–871.

Trudgill, D. L., & Blok, V. C. (2001). Apomictic, polyphagous root-knot nematodes, exceptionally successful and damaging biotrophic root pathogens. *Annual Review of Phytopathology, 39*, 53–77.

Tytgat, T., Vercauteren, I., Vanholme, B., De Meutter, J., Vanhoutte, I., Gheysen, G., et al. (2005). An SXP/RAL-2 protein produced by the subventral pharyngeal glands in the plant parasitic root-knot nematode *Meloidogyne incognita*. *Parasitology Research, 95*, 50–54.

Urwin, P. E., Lilley, C. J., & Atkinson, H. J. (2002). Ingestion of double-stranded RNA by preparasitic juvenile cyst nematodes leads to RNA interference. *Molecular Plant-Microbe Interactions, 15*, 747–752.

Veech, J. A., Starr, J. N., & Nordgren, R. M. (1987). Production and partial characterization of stylet exudate from adult females of *Meloidogyne incognita*. *Journal of Nematology, 19*, 463–468.

Vens, C., Rosso, M.-N., & Danchin, E. G. J. (2011). Identifying discriminative classification-based motifs in biological sequences. *Bioinformatics, 27*, 1231–1238.

Vieira, P., Banora, M. Y., Castagnone-Sereno, P., Rosso, M.-N., Engler, G., & de Almeida Engler, J. (2012). An immunocytochemical procedure for protein localization in various nematode life stages combined with plant tissues using methylacrylate-embedded specimens. *Phytopathology, 102*, 990–996.

Vieira, P., Danchin, E. G. J., Neveu, C., Crozat, C., Jaubert, S., Hussey, R. S., et al. (2011). The plant apoplasm is an important recipient compartment for nematode secreted proteins. *Journal of Experimental Botany, 62*, 1241–1253.

Wang, X., Li, H., Hu, Y., Fu, P., & Xu, J. (2007). Molecular cloning and analysis of a new venom allergen-like protein gene from the root-knot nematode *Meloidogyne incognita*. *Experimental Parasitology, 117*, 133–140.

Wang, X.-R., Moreno, Y., Wu, H.-R., Ma, C., Li, Y.-F., Zhang, J.-A., et al. (2012). Proteomic profiles of soluble proteins from the esophageal gland in female *Meloidogyne incognita*. *International Journal of Parasitology, 42*, 1177–1183.

Williamson, V. M., & Kumar, A. (2006). Nematode resistance in plants, the battle underground. *Trends in Genetics, 22*, 396–403.

Xue, B., Hamamouch, N., Li, C., Huang, G., Hussey, R. S., Baum, T. J., et al. (2013). The 8D05 parasitism gene of *Meloidogyne incognita* is required. *Phytopathology, 103*, 175–181.

Yang, Y., Jittayasothorn, Y., Chronis, D., Wang, X., Cousins, P., & Zhong, G.-Y. (2013). Molecular characteristics and efficacy of 16D10 siRNAs in inhibiting root-knot nematode infection in transgenic grape hairy roots. *PLoS One, 8*, e69463.

CHAPTER THIRTEEN

Suppression of Plant Defences by Plant-Parasitic Nematodes

Sophie Mantelin[a], Peter Thorpe[a], John T. Jones[1]
Cell and Molecular Sciences Group, The James Hutton Institute, Invergowrie, Dundee, UK
[1]Corresponding author: E-mail: john.jones@hutton.ac.uk

Contents

Abstract

Some plant-parasitic nematodes need to spend a significant part of their life cycles within plants while others have long-lasting biotrophic interactions with their hosts. In each case, these nematodes will need to suppress host defences. Large numbers of candidate effectors – proteins that nematodes introduce into their hosts – have been identified from a variety of nematode species as a result of genome and transcriptome projects. There is now a focus on functional characterization of these proteins. This has led to the identification of nematode proteins that suppress different types of host defence responses. Here we review progress towards understanding how nematodes manipulate host defences and how hormone signalling related to defence responses is targeted by nematodes.

1. INTRODUCTION

Plant-parasitic nematodes show a variety of different interactions with their hosts (reviewed by Gheysen & Jones, 2013). While some nematodes (the migratory ectoparasites) have very short-lived interactions with plants, others have much more intimate interactions and rely on prolonged periods of biotrophy to complete their life cycles. These nematodes, like other

[a]These authors contributed equally to this work.

Advances in Botanical Research, Volume 73
ISSN 0065-2296
http://dx.doi.org/10.1016/bs.abr.2014.12.011

biotrophic plant pathogens, will need to suppress host defences. In addition, migratory endoparasitic nematodes, which do not induce a feeding structure and do not have biotrophic interactions with plants, spend a large proportion of their life cycles within the host. These nematodes will also be exposed to host defence responses. Therefore, it is likely that a wide range of plant-parasitic nematodes will need to suppress host defences either to keep their feeding structures alive or to negate the effects of the induced responses.

2. PLANT DEFENCES

Plants are exposed to attacks from a wide range of pathogens. Most of these are defeated following detection of conserved pathogen molecules (Pathogen-Associated Molecular Patterns – PAMPs) by host cell surface pattern recognition receptors (PRRs) which activate pattern-triggered immunity (PTI) (Zipfel, 2009). PAMPs are frequently parts of structural molecules that are essential for the pathogens and that cannot be readily changed to avoid their detection. Consequently PAMPs are often conserved across taxa and PTI is effective against a wide range of related pathogens. Innate immunity can also be activated by the detection of the breakdown products of plant cell walls or peptides deriving from cleaved and degraded proteins (Damage-Associated Molecular Patterns – DAMPs) that are produced as a result of the activities of invading pathogens (Albert, 2013; Boller & Felix, 2009). Although PAMPs have been characterized from bacterial, fungal and oomycete plant pathogens (reviewed by Smant & Jones, 2011) no nematode PAMPs have been identified to date. However, it is known that a variety of cell wall degrading enzymes are secreted by nematodes during their migration through plants (reviewed by Haegeman, Jones, & Danchin, 2011), suggesting that DAMPs could provoke defence signalling responses in plant–nematode interactions.

Successful pathogens, particularly biotrophic pathogens, deliver effectors that suppress PTI. In order to counter this, plants have a second layer of immune receptors encoded by resistance (*R*) genes. These proteins detect effectors or, more frequently, their activity, leading to effector-triggered immunity (ETI). The ETI response is accompanied in most cases by a hypersensitive response (HR) – a localized programmed cell death which is believed to prevent the spread of biotrophic pathogens from the infection site. In plant–nematode interactions the HR may be targeted at the developing feeding structure (as seen in the response induced by *Mi-1* to root-knot nematodes in tomato (Milligan et al., 1998)) or restricted to the cells surrounding the developing feeding site (as seen in the response induced by *Hero A* to potato cyst nematodes in tomato (Sobczak et al., 2005)). The various layers of plant defences and the manner in which effectors and resistance genes co-evolve in pathogens and plants are summarized by the zig-zag model (Jones & Dangl, 2006). In addition to these locally restricted responses, systemic responses can also be initiated by the plant, such as systemic acquired resistance (SAR) which relies on the accumulation and translocation of the defence hormone salicylic acid (SA) (Fu & Dong, 2013). During SAR, localized infection by a primary pathogen results in the induction of a broad-spectrum and long-lasting resistance to secondary pathogens in the systemic tissues of the plant.

The fact that diverse pathogens can suppress PTI in any given host has led to the suggestion that pathogens may target a limited number of host proteins that act as essential regulators of plant defence responses. Consistent with this hypothesis, effectors from bacterial and oomycete pathogens have been shown to target highly interconnected proteins from *Arabidopsis* preferentially (Mukhtar et al., 2011). A logical extension of this idea is that resistance proteins will guard these 'hubs' and, in keeping with this, several *R* genes have been identified that provide protection against diverse pathogens, including plant-parasitic nematodes, such as *Mi-1* (Rossi et al., 1998) and *Cf2* (Lozano-Torres et al., 2012). Many resistance genes encode (nucleotide binding) NB-LRR (leucine-rich repeat) proteins that contain a LRR domain involved in the recognition of either a pathogen effector or a disturbance in the guarded protein, and a nucleotide-binding (NB) domain required for the activation and downstream signalling of the resistance (Elmore, Lin, & Coaker, 2011). However, recent discoveries of *R* genes that provide resistance against cyst nematodes show that in some cases different resistance mechanisms may be deployed against plant-parasitic nematodes. For example, resistance provided by the *Rhg4* gene in soybean against *Heterodera glycines* is due to the presence of a particular variant of a serine hydroxymethyl transferase (Liu et al., 2012) while variations in the copy numbers of several genes underlie resistance to *H. glycines* at the *Rhg1* locus (Cook et al., 2012). This needs to be borne in mind when considering how widely processes characterized in other pathosystems might apply to plant–nematode interactions.

3. NEMATODE EFFECTORS

Candidate effectors have been identified from a wide range of biotrophic and migratory endoparasitic nematodes (reviewed by Haegeman,

Mantelin, Jones, & Gheysen, 2012). Genome scale analyses of several plant-parasitic nematodes, and proteomic analysis of collected secretions, have allowed a more thorough analysis of exhaustive putative effector complements, showing that, as is the case for other plant pathogens, several hundred candidate effectors may be produced by some plant-parasitic nematodes (Abad et al., 2008; Bellafiore et al., 2008; Cotton et al., 2014; Kikuchi et al., 2011; Opperman et al., 2008; Thorpe et al., 2014).

Nematode effectors can originate from a variety of body structures. The majority of characterized effectors are expressed in the subventral or dorsal gland cells and are secreted into the host through the stylet. It is likely that this is the only route by which nematode effectors can be introduced directly into host cells. However, other nematode organs may also synthesize effectors. Some plant-parasitic nematodes have been shown to coat their surfaces with antioxidant proteins that metabolize reactive oxygen species produced as a result of host defence responses (Robertson et al., 2000) or lipid binding proteins that may influence the jasmonate defence signalling pathway (Iberkleid et al., 2013; Prior et al., 2001). These proteins are produced in the hypodermis from where they are transported to the parasite surface (Jones, Reavy, Smant, & Prior, 2004; Prior et al., 2001). In addition, the amphids may produce proteins that are delivered to the apoplastic space at the anterior end of the feeding nematode; there is therefore the potential for these proteins to act as effectors (Eves van den Akker, Lilley, Jones, & Urwin, 2014; Semblat, Rosso, Hussey, Abad, & Castagnone-Sereno, 2001). Effectors produced in the oesophageal glands can also be secreted into the apoplast, as seen for the venom allergen-like effector protein Gr-VAP1 of *Globodera rostochiensis* (Lozano-Torres et al., 2012). In this case the Gr-VAP1 virulence target is apoplastic and is guarded by the extracellular plant immune receptor protein Cf-2 which confers resistance to *G. rostochiensis* in *Solanum pimpinellifolium*. The apoplast has also been shown to be a major target destination for root-knot nematode effectors (Vieira et al., 2011).

4. SUPPRESSION OF PLANT DEFENCES

Functional studies on effectors have mainly been focused on those derived from the sedentary endoparasitic cyst and root-knot nematodes. Effectors have now been identified from these nematodes that suppress both PTI and ETI, confirming the importance to nematodes of coping with host defence responses. In addition, a nematode effector, *Gp*-RBP-1 that is recognized by the *Gpa2 R* gene in potato has been identified from the cyst nematode *Globodera pallida*, providing the first report of a characterzsed nematode avirulence gene (Sacco et al., 2009).

The best characterized suppressor of PTI from a nematode is a secreted calreticulin of *Meloidogyne incognita* (Jaouannet et al., 2012). Calreticulins are highly conserved calcium binding proteins that are present in both plants and animals. Their usual role is regulation of Ca^{2+} storage and signalling in the endoplasmic reticulum. However, a secreted calreticulin was identified from *M. incognita* that is expressed in the subventral gland cells of preparasitic J2s and in the dorsal gland cell of later stages, demonstrating the importance of this protein throughout the parasitic stages of the nematode life cycle. The protein has been shown to accumulate to high levels at the interface between the nematode and the feeding site (Jaubert et al., 2005). Silencing of the calreticulin gene by RNAi delivered from a Tobacco rattle virus vector (Dubreuil et al., 2009) or through hairpin constructs in transgenic plants (Jaouannet et al., 2012) results in greatly reduced nematode infectivity demonstrating the importance of the calreticulin to the nematodes. In addition, overexpression of calreticulin in plants increased susceptibility of plants to root-knot nematodes and to a plant pathogenic fungus. These experiments also demonstrated that apoplastic localization of the calreticulin is required for function as constructs lacking a signal peptide gave no enhancement of susceptibility to nematodes. In addition, direct evidence supporting a role for the calreticulin in suppression of PTI was obtained by demonstrating that the presence of the calreticulin suppressed normal elf18-induced deposition of callose in *Arabidopsis* and reduced activation of defence-related genes in response to this PAMP (Jaouannet et al., 2012). Intriguingly, it is known that some bacterial plant pathogens produce exopolysaccharides that suppress PTI by chelating apoplastic Ca^{2+}, thus preventing the normal influx of this ion required for PTI signalling (Aslam et al., 2008). Clearly it is possible that the root-knot nematode calreticulin functions in an analogous manner.

A substantial family of SPRYSEC effectors has been identified in potato cyst nematodes (Cotton et al., 2014) and it has been shown that at least one of these effectors suppresses ETI in plants. The SPRYSEC-19 effector from *G. rostochiensis* suppresses ETI induced by Gpa2 and the related Rx in the presence of the respective avirulence factor recognized by these R proteins (Postma et al., 2012). However, cell death mediated by other more distantly related *R* genes including *Cf4, Cf9, R3a* and *Rpi-blb2* is not suppressed by

SPRYSEC-19. The biological relevance of this activity was demonstrated by the fact that replication of a virus strain that is avirulent against *Rx* was enabled in the presence of SPRYSEC-19. The presence of SPRYSEC-19 also increased susceptibility of otherwise resistant potatoes to *Verticillium dahliae* (Postma et al., 2012). Although the mechanisms by which SPRYSEC-19 suppresses ETI remain unclear, the ability to suppress host defences is not restricted to this specific protein as other SPRYSECs from *G. pallida* also suppress ETI (Mei et al., 2015). The size of the SPRYSEC gene family and the fact that at least one SPRYSEC is recognized by an R protein (Sacco et al., 2009) suggest that this gene family is under strong diversifying selection and may, as a consequence, have a variety of roles in plant–nematode interactions.

A further effector from *G. rostochiensis* has been identified that suppresses both PTI and ETI. This effector, a unique ubiquitin carboxyl extension protein (GrUBCEP12), consists of a signal peptide, a highly conserved ubiquitin domain and a C-terminal extension (Chronis et al., 2013). The first characterized effector of this type (HsUBI1) was identified from *Heterodera schachtii* (Tytgat et al., 2004) and orthologues were subsequently identified from *H. glycines* and *G. rostochiensis*. However, although the overall domain arrangements in the proteins from the different nematodes, as well as the sequence of the ubiquitin domain itself, are conserved, the sequence of the C-terminal extension is highly variable, although it is only 12 amino acids long. This may have an influence on the function of the protein since, for example, the C-terminal extension of the *H. schachtii* effector is translocated to the nucleolus of host plant cells (Tytgat et al., 2004) while this is not the case for the *G. rostochiensis* homologue despite its active accumulation in the nucleus (Chronis et al., 2013). The *G. rostochiensis* effector has been studied in some detail. This protein is cleaved *in planta* and the released GrCEP12 peptide suppresses cell death induced by both Gpa2/Gp-RBP-1 and Rx2/Potato Virus X coat protein in *Nicotiana benthamiana*. In addition, constitutive expression of the GrCEP12 peptide in potato increases susceptibility to nematode infection and to the unrelated bacterial pathogen *Streptomyces scabies*, suggesting that GrCEP12 has a role in suppressing basal plant defences or PTI (Chronis et al., 2013). Further studies showed that the peptide also suppresses flg22-induced responses including induction of marker genes and production of reactive oxygen species, demonstrating its role in PTI suppression (Chen, Chronis, & Wang, 2013). Given the range of defence responses apparently targeted by this specific effector, it is possible that the protein targets a downstream part of defence signalling pathways and is a general suppressor of cell death or the pathways that lead to generation of active oxygen species. In addition, the variety of C-terminal extensions found on these effectors suggests that nematodes may use ubiquitin as a carrier for short peptides that are destined for the plant; such ubiquitin extension proteins will be cleaved by de-ubiquitination enzymes inside the host cell releasing the biologically active peptide in situ.

If the ubiquitination system plays a central role in mediating plant defence signalling, it is also widely exploited by pathogens in order to suppress plant defences (Trujillo & Shirasu, 2010). The ubiquitin–proteasome pathway is used to regulate to abundance and activity of proteins within the plant cell (reviewed by Vierstra, 2009). The ubiquitination process requires the sequential action of E1 (ubiquitin activating enzymes), E2 (ubiquitin-conjugating enzyme) and E3 (ubiquitin ligase) proteins. The E3 enzymes recognize target substrates and confer specificity in the transfer of the ubiquitin from the UBI-E2 complex to the target protein, therefore determining the protein destined for degradation. The *Pseudomonas syringae* AvrPtoB effector mimics an E3 ubiquitin ligase that targets the Fen kinase to suppress defences in tomato (Rosebrock et al., 2007). AvrPtoB has also been shown to suppress PTI by the ubiquitination of the PAMP receptor FLS2 (Goehre et al., 2008) and the co-receptor CERK1 (Gimenez-Ibanez et al., 2009). It is fascinating to note that bacteria do not have their own endogenous ubiquitination pathways but have still evolved effectors that exploit this biochemical system. Exploitation of the ubiquitination pathway is also seen in eukaryotic plant pathogens; the *Phytophthora infestans* effector AVR3a targets the host E3 ligase CMPG1, which is essential for the downstream signal transduction pathway induced upon recognition of INF1 (Bos et al., 2010). It is also possible that plant-parasitic nematodes target the ubiquitination pathway to a greater extent than is currently appreciated as two of the 'pioneer' effectors identified by Gao et al. (2003) are similar to E3 ligases or individual proteins that can be part of multicomponent E3 ligase complexes.

Other effectors have been identified that may suppress host defences but for which less direct evidence of this role is available. The *H. schachtii* homologue of the *H. glycines* effector 10A06 is a cytoplasmic effector that interacts with spermidine synthase 2 (SPDS2) in *Arabidopsis* (Hewezi et al., 2010). Constitutive expression of either the effector or SPDS2 in plants increased susceptibility to nematodes by up to 80% and led to upregulation of antioxidant proteins. The expression of Hs10A06 in *Arabidopsis* also promoted host susceptibility to unrelated bacterial and viral pathogens.

In addition, SA signalling was disrupted in plants over expressing the effector protein suggesting a role in defence (Hewezi et al., 2010). A similar approach has led to the *H. schachtii* homologue of the *H. glycines* 30C02 effector being identified as a suppressor of host defences (Hamamouch et al., 2012). The 30C02 effectors from both species interact with a host beta-1,3-endoglucanase and may interfere with this protein's biological role as a pathogenesis-related protein. Constitutive expression of Hg30C02 in plants increases susceptibility to nematode infection. However, the effector protein specifically promotes host susceptibility to infection by cyst nematodes in *Arabidopsis*.

5. HORMONE SIGNALLING AND PLANT DEFENCES

Various hormones are involved in plant–nematode interactions (see other chapters in this volume). Hormonal signalling pathways are essential for building up defences in plants, but they are also manipulated by nematodes to promote susceptibility. However, very few examples of how hormone signalling is manipulated by nematodes to suppress defences have been described in detail. Nematodes may manipulate their host to suppress SA-mediated defence pathways, as described by Uehara, Sugiyama, Matsuura, Arie, and Masuta (2010). In plants, SA-induced responses are associated with the transcriptional upregulation of a number of genes, including those encoding pathogenesis-related (PR) proteins such as PR-1 (Boatwright & Pajerowska-Mukhtar, 2013). In tomato, PR-1(P4) is a marker for resistance conferred by *Hero A* against *G. rostochiensis* but nematode parasitism resulted in the inhibition of the SA signalling pathway in susceptible cultivars (Uehara et al., 2010). Suppression of host defences by migratory endoparasites has been less intensively studied compared to this process in cyst and root-knot nematodes. However, it has recently been shown that two migratory endoparasites, *Hirschmanniella oryzae* and *Pratylenchus coffeae* both produce a chorismate mutase (Bauters, Haegeman, Kyndt, & Gheysen, 2014; Haegeman et al., 2011). This protein was previously thought to be present only in sedentary endoparasitic nematodes, where a role in feeding site formation had been proposed (Doyle & Lambert, 2003). Its presence in both sedentary and migratory nematodes argues for a role in a process conserved between these two groups such as invasion of the host or suppression of host defences. Chorismate mutase converts chorismate to prephenate and may therefore prevent the metabolism of chorismate to SA, which is the major biosynthetic pathway involved in pathogen-induced SA production (Boatwright & Pajerowska-Mukhtar, 2013). Intriguingly, *H. oryzae* also produces a protein similar to isochorismatase (Bauters et al., 2014), an enzyme which is also involved in SA metabolism (Dempsey, Vlot, Wildermuth, & Klessig, 2011), further suggesting that targeting this process may be important for both sedentary and migratory nematodes.

The best characterized mechanism by which nematodes suppress hormone mediated signalling processes is probably the suppression of the jasmonic acid (JA)-mediated response to both root-knot (*Meloidogyne javanica*) and cyst nematodes (*G. pallida*) by the fatty acid-and retinol-binding protein FAR-1 (Iberkleid et al., 2013; Prior et al., 2001). In both nematodes *far-1* is expressed in all life cycle stages tested but is particularly strongly induced in infective juveniles. The FAR-1 protein accumulates on the nematode cuticle surface. The secretion of this effector is important for parasitism since host-mediated silencing of *Mj-far-1* attenuates nematode development while constitutive expression in tomato hairy roots increases susceptibility to infection by *M. javanica,* promoting nematode development and giant cell formation (Iberkleid et al., 2013). Recombinant Gp-FAR-1 was found to bind retinol and a broad range of fatty acids, including linolenic and linoleic acids that are precursors of plant defence compounds and components of the JA signalling pathway (Prior et al., 2001). Moreover, Gp-FAR-1 inhibits the lipoxygenase-mediated modification of these substrates (but not JA itself) in vitro, suggesting that sequestration of these compounds by the FAR-1 proteins could be the mode of action of the nematode effectors. Further work on Mj-FAR-1 demonstrated that increased susceptibility of *Mj-far-1* expressing roots is most likely associated with suppression of JA pathway, as indicated by downregulation of two major JA responsive genes (Iberkleid et al., 2013). Together, these studies indicate that FAR-1 might regulate jasmonate-dependent defence responses to nematodes to promote susceptibility in the host plant.

The ability to avoid or suppress host defences is clearly essential for a wide range of plant-parasitic nematodes. The importance of these processes to nematodes is illustrated by the devastating effects (from the nematode's point of view) of successful induction of an HR in a resistant plant. It is also worth noting that many sedentary endoparasitic nematodes have a longer period of biotrophy than almost all other plant pathogens. It is therefore likely that current studies in this area have only scratched the surface of the range of mechanisms used by nematodes to manipulate host defences.

ACKNOWLEDGEMENTS

The James Hutton Institute receives funding from the Rural and Environment Science and Analytical Services Division (RESAS) of the Scottish Government. This work was partly funded by a BBSRC CASE studentship between Leeds University and Mylnefield Research Services Ltd and by BBSRC grant BB/F00334X/1 and Scottish Government Contract Research Fund project CRF0926.

REFERENCES

Abad, P., Gouzy, J., Aury, J.-M., Castagnone-Sereno, P., Danchin, E. G. J., Deleury, E., et al. (2008). Genome sequence of the metazoan plant-parasitic nematode *Meloidogyne incognita*. *Nature Biotechnology, 26*, 909–915.

Albert, M. (2013). Peptides as triggers of plant defence. *Journal of Experimental Botany, 64*, 5269–5279.

Aslam, S., Newman, M., Erbs, G., Morrissey, K., Chinchilla, D., Boller, T., et al. (2008). Bacterial polysaccharides suppress induced innate immunity by calcium chelation. *Current Biology, 18*, 1078–1083.

Bauters, L., Haegeman, A., Kyndt, T., & Gheysen, G. (2014). Analysis of the transcriptome of *Hirschmanniella oryzae* to explore potential survival strategies and host–nematode interactions. *Molecular Plant Pathology, 15*, 352–363.

Bellafiore, S., Shen, Z. X., Rosso, M. N., Abad, P., Shih, P., & Briggs, S. P. (2008). Direct identification of the *Meloidogyne incognita* secretome reveals proteins with host cell reprogramming potential. *PLoS Pathogens, 4*, e1000192.

Boatwright, J. L., & Pajerowska-Mukhtar, K. (2013). Salicylic acid: an old hormone up to new tricks. *Molecular Plant Pathology, 14*, 623–634.

Boller, T., & Felix, G. (2009). A renaissance of elicitors: perception of microbe-associated molecular patterns and danger signals by pattern-recognition receptors. *Annual Review of Plant Biology, 60*, 379–406.

Bos, J. I., Armstrong, M. R., Gilroy, E. M., Boevink, P. C., Hein, I., Taylor, R. M., et al. (2010). *Phytophthora infestans* effector AVR3a is essential for virulence and manipulates plant immunity by stabilizing host E3 ligase CMPG1. *Proceedings of the National Academy of Sciences of the United States of America, 107*, 9909–9914.

Chen, S., Chronis, D., & Wang, X. (2013). The novel GrCEP12 peptide from the plant-parasitic nematode *Globodera rostochiensis* suppresses flg22-mediated PTI. *Plant Signaling and Behavior, 8*, e25359.

Chronis, D., Chen, S., Lu, S., Hewezi, T., Carpenter, S. C. D., Loria, R., et al. (2013). A ubiquitin carboxyl extension protein secreted from a plant-parasitic nematode *Globodera rostochiensis* is cleaved *in planta* to promote plant parasitism. *The Plant Journal, 74*, 185–196.

Cook, D. E., Lee, T. G., Guo, X. L., Melito, S., Wang, K., Bayless, A. M., et al. (2012). Copy number variation of multiple genes in *Rhg1* mediates nematode resistance in soybean. *Science, 338*, 1206–1209.

Cotton, J. A., Lilley, C. J., Jones, L. M., Kikuchi, T., Reid, A. J., Thorpe, P., et al. (2014). The genome and life-stage specific transcriptomes of *Globodera pallida* elucidate key aspects of plant parasitism by a cyst nematode. *Genome Biology, 15*, R43.

Dempsey, D. A., Vlot, A. C., Wildermuth, M. C., & Klessig, D. F. (2011). Salicylic acid biosynthesis and metabolism. Arabidopsis *Book, 9*, e0156.

Doyle, E. A., & Lambert, K. N. (2003). *Meloidogyne javanica* chorismate mutase 1 alters plant cell development. *Molecular Plant-Microbe Interactions, 16*, 123–131.

Dubreuil, G., Magliano, M., Dubrana, M., Lozano, J., Lecomte, P., Favery, B., et al. (2009). Tobacco rattle virus mediates gene silencing in a plant parasitic root-knot nematode. *Journal of Experimental Botany, 60*, 4041–4050.

Elmore, J. M., Lin, Z.-J. D., & Coaker, G. (2011). Plant NB-LRR signaling: upstreams and downstreams. *Current Opinion in Plant Biology, 14*, 365–371.

Eves-van den Akker, S., Lilley, C. J., Jones, J. T., & Urwin, P. E. (2014). Identification and characterisation of a hyper variable apoplastic effector gene family of G. pallida. *PLoS Pathogens, 10*(9), e1004391.

Fu, Z. Q., & Dong, X. (2013). Systemic acquired resistance: turning local infection into global defense. *Annual Review of Plant Biology, 64*, 839–863.

Gao, B., Allen, R., Maier, T., Davis, E. L., Baum, T. J., & Hussey, R. S. (2003). The parasitome of the phytonematode *Heterodera glycines*. *Molecular Plant-Microbe Interactions, 16*, 720–726.

Gheysen, G., & Jones, J. T. (2013). Molecular aspects of plant-nematode interactions. In R. N. Perry, & M. Moens (Eds.), *Plant nematology* (2nd ed.). Wallingford, UK: CAN International.

Gimenez-Ibanez, S., Hann, D. R., Ntoukakis, V., Petutschnig, E., Lipka, V., & Rathjen, J. P. (2009). AvrPtoB targets the LysM receptor kinase CERK1 to promote bacterial virulence on plants. *Current Biology, 19*, 423–429.

Goehre, V., Spallek, T., Haeweker, H., Mersmann, S., Mentzel, T., Boller, T., et al. (2008). Plant pattern-recognition receptor FLS2 is directed for degradation by the bacterial ubiquitin ligase AvrPtoB. *Current Biology, 18*, 1824–1832.

Haegeman, A., Jones, J. T., & Danchin, E. (2011). Horizontal gene transfer in nematodes: a catalyst for plant parasitism? *Molecular Plant-Microbe Interactions, 24*, 879–887.

Haegeman, A., Mantelin, S., Jones, J. T., & Gheysen, G. (2012). Secretions of plant parasitic nematodes. *Gene, 492*, 19–31.

Hamamouch, N., Li, C., Hewezi, T., Baum, T. J., Mitchum, M. G., Hussey, R. S., et al. (2012). The interaction of the novel 30C02 cyst nematode effector protein with a plant β-1,3-endoglucanase may suppress host defence to promote parasitism. *Journal of Experimental Botany, 63*, 3683–3695.

Hewezi, T., Howe, P. J., Maier, T. R., Hussey, R. S., Mitchum, M. G., Davis, E. L., et al. (2010). *Arabidopsis* spermidine synthase is targeted by an effector protein of the cyst nematode *Heterodera schachtii*. *Plant Physiology, 152*, 968–984.

Iberkleid, I., Vieira, P., de Almeida Engeler, J., Firester, K., Spiegel, Y., & Brown Horowitz, S. (2013). Fatty acid- and retinol- binding protein MjFAR-1 induced tomato host susceptibility to root-knot nematodes. *PLoS One, 8*(5), e64586.

Jaouannet, M., Maglianao, M., Arguel, M. J., Gourges, M., Evangelisti, E., Abad, P., et al. (2012). The root-knot nematode calreticulin Mi-CRT is a key effector in plant defense suppression. *Molecular Plant-Microbe Interactions, 26*, 97–105.

Jaubert, S., Milac, A. L., Petrescu, A. J., de Almeida-Engler, J., Abad, P., & Rosso, M. N. (2005). *In planta* secretion of a calreticulin by migratory and sedentary stages of root-knot nematode. *Molecular Plant-Microbe Interactions, 18*, 1277–1284.

Jones, J. D. G., & Dangl, J. L. (2006). The plant immune system. *Nature, 444*, 323–329.

Jones, J.. T., Reavy, B., Smant, G., & Prior, A.. E. (2004). Glutathione peroxidases of the potato cyst nematode *Globodera rostochiensis*. *Gene, 324*, 47–54.

Kikuchi, T., Cotton, J. A., Dalzell, J. J., Hasegawa, K., Kanzaki, N., McVeigh, P., et al. (2011). Genomic insights into the origin of parasitism in the emerging plant pathogen *Bursaphelenchus xylophilus*. *PLoS Pathogens, 7*, e1002219.

Liu, S., Kandoth, P. K., Warren, S. D., Yeckel, G., Heinz, R., Alden, J., et al. (2012). A soybean cyst nematode resistance gene points to a new mechanism of plant resistance to pathogens. *Nature, 492*, 256–260.

Lozano-Torres, J. L., Wilbers, R. H. P., Gawronski, P., Boshoven, J. C., Finkers-Tomczak, A., Cordewener, J. H. G., et al. (2012). Dual disease resistance mediated by the immune receptor Cf-2 in tomato requires a common virulence target of a fungus and a nematode. *Proceedings of the National Academy of Sciences of the United States of America, 109*, 10119–10124.

Mei, Y., Thorpe, P., Guzha, A., Haegeman, A., Blok, V. C., MacKenzie, K., et al. (2015). Only a small subset of the SPRY domain gene family in Globodera pallida is likely to encode effectors, two of which suppress host defences induced by the potato resistance gene Gpa2. *Nematology –in press.*

Milligan, S. B., Bodeau, J., Yaghoobi, J., Kaloshian, I., Zabel, P., & Williamson, V. M. (1998). The root knot nematode resistance gene *Mi* from tomato is a member of the leucine zipper, nucleotide binding, leucine- rich repeat family of plant genes. *Plant Cell, 10*, 1307–1319.

Mukhtar, M. S., Carvunis, A.-R., Dreze, M., Epple, P., Steinbrenner, J., Moore, J., et al. (2011). Independently evolved virulence effectors converge onto hubs in a plant immune system network. *Science, 333*, 596–601.

Opperman, C. H., Bird, D. M., Williamson, V. M., Rokhsar, D. S., Burke, M., Cohn, J., et al. (2008). Sequence and genetic map of *Meloidogyne hapla*: a compact nematode genome for plant parasitism. *Proceedings of the National Academy of Sciences of the United States of America, 105*, 14802–14807.

Postma, W. J., Slootweg, E. J., Rehman, S., Finkers-Tomczak, A., Tytgat, T. O. G., van Gelderen, K., et al. (2012). The effector SPRYSEC-19 of *Globodera rostochiensis* suppresses CC-NB-LRR mediated disease resistance in plants. *Plant Physiology, 160*, 944–954.

Prior, A. E., Jones, J. T., Blok, V. C., Beauchamp, J., McDermott, L., Cooper, A., et al. (2001). A surface-associated retinol- and fatty acid-binding protein (GpLBP20) from the potato cyst nematode *Globodera pallida* – lipid binding activities, structural analysis and expression pattern. *Biochemical Journal, 356*, 387–394.

Robertson, L., Robertson, W.. M., Sobczak, M., Bakker, J., Tetaud, E., Arinagayayam, M.. R., et al. (2000). Cloning, expression and functional characterisation of a thioredoxin peroxidase from the potato cyst nematode *Globodera rostochiensis*. *Molecular and Biochemical Parasitology, 111*, 41–49.

Rosebrock, T. R., Zeng, L., Brady, J. J., Abramovitch, R. B., Xiao, F., & Martin, G. B. (2007). A bacterial E3 ubiquitin ligase targets a host protein kinase to disrupt plant immunity. *Nature, 448*, 370–374.

Rossi, M., Goggin, F. L., Milligan, S. B., Kaloshian, I., Ullman, D. E., & Williamson, V. M. (1998). The nematode resistance gene *Mi* of tomato confers resistance against the potato aphid. *Proceedings of the National Academy of Sciences of the United States of America, 95*, 9750–9754.

Sacco, M. A., Koropacka, K., Grenier, E., Jaubert, M. J., Blanchard, A., Goverse, A., et al. (2009). The cyst nematode SPRYSEC protein RBP-1 elicits Gpa2-and RanGAP2-dependent plant cell death. *PLoS Pathogens, 5*, e1000564.

Semblat, J. P., Rosso, M. N., Hussey, R. S., Abad, P., & Castagnone-Sereno, P. (2001). Molecular cloning of a cDNA encoding an amphid-secreted putative avirulence protein from the root-knot nematode *Meloidogyne incognita*. *Molecular Plant-Microbe Interactions, 14*, 72–79.

Smant, G., & Jones, J. T. (2011). Suppression of plant defences by nematodes. In J. T. Jones, G. Gheysen, & C. Fenoll (Eds.), *Genomics and molecular genetics of plant-nematode interactions* (pp. 273–286). Springer Academic Publishers.

Sobczak, M., Avrova, A., Jupowicz, J., Phillips, M. S., Ernst, K., & Kumar, A. (2005). Characterization of susceptibility and resistance responses to potato cyst nematode

(*Globodera* spp.) infection of tomato lines in the absence and presence of the broad-spectrum nematode resistance *Hero* gene. *Molecular Plant-Microbe Interactions, 18*, 158–168.

Thorpe, P., Mantelin, S., Cock, P. J. A., Blok, V. C., Coke, M. C., Cotton, J. A., et al. (2014). Characterisation of the full effector complement of the potato cyst nematode *Globodera pallida*. *BMC Genomics, 15*, 923.

Trujillo, M., & Shirasu, K. (2010). Ubiquitination in plant immunity. *Current Opinion in Plant Biology, 13*, 402–408.

Tytgat, T., Vanholme, B., De Meutter, J., Claeys, M., Couvreur, M., Vanhoutte, I., et al. (2004). A new class of ubiquitin extension proteins secreted by the dorsal pharyngeal gland in plant parasitic cyst nematodes. *Molecular Plant-Microbe Interactions, 17*, 846–852.

Uehara, T., Sugiyama, S., Matsuura, H., Arie, T., & Masuta, C. (2010). Resistant and susceptible responses in tomato to cyst nematode are differentially regulated by salicylic acid. *Plant and Cell Physiology, 51*, 1524–1536.

Vieira, P., Danchin, E. G. J., Neveu, C., Crozat, C., Jaubert, S., Hussey, R. S., et al. (2011). The plant apoplasm is an important recipient compartment for nematode secreted proteins. *Journal of Experimental Botany, 62*, 1241–1253.

Vierstra, R. D. (2009). The ubiquitin–26S proteasome system at the nexus of plant biology. *Nature Reviews Molecular Cell Biology, 10*, 385–397.

Zipfel, C. (2009). Early molecular events in PAMP-triggered immunity. *Current Opinion in Plant Biology, 12*, 414–420.

CHAPTER FOURTEEN

Application of Biotechnology for Nematode Control in Crop Plants

John Fosu-Nyarko*, Michael G.K. Jones[†,1]
*Nemgenix Pty Ltd, WA State Agricultural Biotechnology Centre, Murdoch University, Perth, WA, Australia
[†]School of Veterinary and Life Sciences, WA State Agricultural Biotechnology Centre, Murdoch University, Perth, WA, Australia
[1]Corresponding author: E-mail: m.jones@murdoch.edu.au

Contents

Advances in Botanical Research, Volume 73
ISSN 0065-2296
http://dx.doi.org/10.1016/bs.abr.2014.12.012

Abstract

Effective control of plant parasitic nematodes in crop plants will contribute hundreds of millions of dollars to global agriculture and help underpin future food security. Natural nematode resistance genes present in gene pools of crop species and their relatives have long been exploited with the aim of transferring such traits into economically important crops where effective resistance is lacking. Biotechnology also contributes to this process via marker-assisted selection to identify and combine the best nematode resistance genes, and increasingly in providing new knowledge of target genes, and the potential to exploit this knowledge using transgenic technology. Thus recent advances now make it possible to exploit specific aspects of nematode-host plant interactions to design control strategies that include enabling plants to prevent nematode invasion, reducing effectiveness of nematode migration through tissues, preventing successful establishment or reducing feeding ability or nematode fecundity. The knowledge of what genes are vital for successful nematode parasitism can also be used to develop new chemical control agents. These new strategies may either be available for public use or be delivered commercially. For transgenic technologies, both modes of delivery face the same issues in terms of deployment, such as substantial field testing, meeting environmental and human safety regulations, adequate funding to complete statutory requirements, and public acceptance of GMOs when the product is to be marketed. However, as technology develops, new strategies for nematode control are emerging, both for transgenic approaches and in genome editing, which should be regarded by regulators as a form of mutation rather than 'genetic modification'. With such advances in biotechnology, the release of commercial varieties of major crops with new forms of nematode resistance, or new modes of delivery of control agents, is likely to become a commercial reality. To improve durability, transgenic traits could be based on resistance with different modes of action: for example, RNAi-based technology combined with expression of peptides which disrupt sensory activities. Ideally such traits would be added to existing crop genotypes with the best conventional or natural nematode resistance, to increase the effectiveness and durability of the nematode resistance trait. Biotech trait expression could also be limited to roots to minimise expression in harvested parts, and this could improve public acceptability.

1. INTRODUCTION

The current status of molecular understanding of nematode–plant interactions is described in earlier chapters in this volume, and it is clear that rapid advances are being made in unravelling the mechanisms which enable plant parasitic nematodes to be such successful plant pests. The question addressed in this chapter is how this new information can be translated to practical application, and used to reduce crop losses caused by these devastating parasites. If this can be achieved it will be a significant contribution to future crop security and increased productivity in a sustainable manner. The requirement to find new ways of controlling plant nematodes is all the more pressing because some of the older chemical nematicides have been withdrawn or are now under restricted use mandates: this has led to a renewed interest in developing new strategies to control plant parasitic nematodes based on genetic, chemical or integrated approaches to manage nematode pests.

The academic advances in knowledge are now impressive, and it is clear that research of excellent quality is being done to understand nematode–plant interactions: in particular such research is leading to identifying what effectors they secrete to be able to avoid or neutralize host plant defences, detect gradients, migrate within roots and, depending on the species, induce the formation of long-term feeding sites. However, there is still a gap between this basic research and its practical application to control these pests. As concluded by McCarter (2009), 'the future of plant nematology as a discipline is dependent on the value of commercial solutions delivered to growers'. Such advances are likely to come from both conventional and genetic approaches: McCarter also emphasized that economically and environmentally sound methods to control nematodes which contribute a commercial increase in crop yields will result in more investment in the field. A summary of the biotechnology-based strategies now available for nematode control, which include both established breeding technologies and transgenic approaches, is provided in Table 1, with brief explanations of the strategies and of their current status.

2. EARLY SELECTION FOR PLANTS WITH NEMATODE RESISTANCE; SUSCEPTIBILITY, RESISTANCE AND TOLERANCE

The earliest reports of selection for plant resistance to nematodes date back to the late nineteenth century, and were based on phenotypic selection for plants which had fewer galls on roots when infected with root knot nematodes. From these selections, varieties of cowpea, sugar beet, cotton and coffee were reported with improved resistance to root knot nematodes (Ware, 1936; Webber & Orton, 1902; Wilfarth, 1900).

With a better understanding of nematode–plant interactions, plant nematologists now describe host interactions as compatible when a plant supports reproduction of the parasite, in which case the host is either susceptible or tolerant to infestation, and incompatible when the host is resistant to nematodes, and cannot be invaded successfully or only supports very limited or no growth and reproduction by the parasite. Plants susceptible or resistant to

Table 1 Biotechnology-Based Strategies for Nematode Control

Target for Control	Considerations	Status/Example
Major or minor natural resistance genes	The introgression and combination of natural resistance genes, for example from related or wild species, has been the mainstay of resistance breeding strategies	Marker-assisted breeding for nematode resistance has become routine in many breeding programs, although effective resistance genes are not available for all crops
Nematode migration in the rhizosphere and root entry	Disruption of sensory functions	Peptide(s) that inhibit reception of gradients by amphids RNAi disruption of amphid proteins/function
Migration in the root	Wall-degrading enzymes may be required for migration, e.g. *Endoparasites* Positional gradients in roots detected for migration to the required site in the root	RNAi downregulation of nematode expression of cell wall-degrading enzymes Inhibition of sensing gradients in roots
Avoiding host defences	Effectors that enable nematodes to evade or neutralize host defences	RNAi downregulation of expression of effectors involved in avoiding host defences
Disruption of feeding site formation or function	Effectors enable sedentary endoparasites to induce giant cells and syncytia. Disrupt feeding site formation, triggered by nematode-responsive promoter(s)	RNAi downregulation of expression of key effector(s) required for feeding site formation Nematode responsive promoter(s) linked to 'cell death' gene, e.g. barnase

Disrupting vital genes	Disrupt expression of genes vital for the nematode life cycle	RNAi downregulation of expression of vital nematode genes
Overexpression of host genes with modified expression in feeding cells	Many genes in nematode feeding cells are up-or downregulated	Overexpression of some host genes with altered expression in nematode feeding sites reduce nematode parasitism
Modify genes for host plant susceptibility to nematodes	New approaches for genome editing now available	New technologies not necessarily regarded as genetic modification, more acceptable in some jurisdictions
Delivery of toxic compounds to the nematodes	Make use of basic work on nematode effectors and genes vital for their survival: these can define new targets for control	Use bioinformatics filters to identify new targets for chemical control. Design new nematicides to these targets
Develop new nematicides and modes of delivery; new biological control agents	There is a need to develop new more environmentally friendly forms of chemical control and delivery, and new forms of biological control	A series of new nematicides are now available commercially, based on biological and chemical control, separately or in combination, e.g. using delivery by drip irrigation or seed coating

nematodes can also exhibit varying degrees of tolerance to infestation, when they can support a level of nematode infestation without showing severe symptoms. When applying nematode control strategies, the aim is to reduce nematode reproduction and thereby the level of infestation, resulting in a decrease in symptoms of root damage and associated susceptibility to abiotic stresses and secondary attack by soil pathogens.

3. BIOTECHNOLOGICAL APPROACHES TO PLANT PARASITIC NEMATODE CONTROL

Research on biotechnological approaches to nematode control aims either to exploit natural resistance present in gene pools of crop species and their relatives or to employ synthetic forms of resistance, such as those based on disruption of feeding cells, expression of specific proteins or peptides, on gene silencing (RNAi) or on delivery of toxic compounds to the invading nematode (Table 1). To exploit natural variation for resistance, large-scale screening of germplasm is often employed, together with molecular markers and/or positional cloning to identify resistance (R) genes or metabolites that confer resistance to particular nematodes in a wide range of germplasm of crop plants and their wild relatives. Identified sources of resistance are then introgressed into the desired germplasm. In contrast, transgenic approaches to nematode control exploit knowledge of nematode–host interactions and can be directed to targeting the nematode, including disorientating the infective stages to prevent them from finding host roots, reducing the effectiveness of migration through host tissues, reducing successful establishment in host cells or reducing feeding ability and fecundity of nematodes on a susceptible or tolerant host (Table 1).

4. NATURAL RESISTANCE APPROACH TO PLANT PARASITIC NEMATODE CONTROL

Effective resistance against plant parasitic nematodes is not available in all economically important crops. It has been argued that deploying sources of natural resistance against pests and pathogens is the most cost-effective and environmentally sustainable method of reducing crop losses resulting from infection by diseases and pests. It is therefore not surprising that earlier efforts to control nematodes focussed on using marker-assisted breeding methods to identify sources of nematode resistance. This usually involves large-scale screening for resistance in germplasm from wild ancestors or progenitors of cultivars of particular crop plants, mapping of quantitative trait loci, positional cloning and perhaps isolation and characterization of the genes responsible for conferring resistance. Molecular methods used for mapping and fine mapping of populations have included RFLPs (Restriction Amplified Length Polymorphisms), AFLPs (Amplified Fragment Length Polymorphisms), RAPDs (Random Amplified Polymorphic DNA), SCAR (Sequenced Characterised Amplified Regions)- and STS (Sequence Tagged Site)-based methods, and more recently deep sequencing technologies.

Marker-assisted selection for nematode resistance has been a major focus to improve crops affected by nematodes. Because cyst nematodes attack and can cause major losses to most of the world's important stable crops including potato, soybean and wheat, it is not surprising that substantial breeding efforts have been undertaken to identify stable sources of resistance to different species and pathotypes of cyst nematodes. For example both polygenic and monogenic genes for resistance to potato cyst nematodes have been identified and markers closely linked to these alleles have since been developed for use in potato resistance breeding programmes (Table 2) (Niewöhner, Salamini, & Gebhardt, 1995). Similarly, different types of resistance genes have been identified, mapped and/or cloned from host plants that confer near complete and partial resistances to *Heterodera glycines*, *Heterodera avenae* and *Heterodera schachtii* including the map-based cloning of a gene encoding a serine hydroxymethyl transferase, at the *Rhg4* locus, that confers resistance to soybean cyst nematode race 4 (Table 2) (Liu et al., 2012). In Australia, where wheat and barley crops suffer losses from infestation with the cereal cyst nematode *H. avenae*, characterized nematode resistance loci *Ha1* and *Ha2* (allelic to *Ha3*) on chromosome 2, the gene *Ha4* (chromosome 5) in barley and the *Cre1* locus on chromosome 2B, the *Cre3* (*Ccn D1*) from *Triticum tauschii* in wheat, and other *Cre* genes have been deployed widely in cereal breeding programmes (Eastwood, Lagudah, & Appels, 1994; Kretschmer, et al., 1997; Lagudah, Moullet, & Appels, 1997; Williams, Fisher, & Langridge, 1996).

For root knot nematodes, five resistant genes have been identified of which the well-characterized *Mi* gene, isolated from the wild relative of tomato, *Solanum peruvianum*, induces a hypersensitive response on infection with *Meloidogyne* spp. (*Meloidogyne incognita*, *Meloidogyne javanica* and *Meloidogyne arenaria*) which results in the death of infective juveniles, and has been incorporated successfully into many cultivars of tomato (Table 2). The *Mi* gene is unique in that it also confers resistance to the potato aphid

Table 2 Summary of Natural Resistance Genes to Cyst and Root Knot Nematodes, and Major QTLs Associated with Resistance to *Pratylenchus* spp

Nematode Species	Resistance Genes	Crop or Source of Resistance	References
Cyst Nematodes			
Globodera rostochiensis	*Gro1*	Potato	Ballvora et al. (1995), Leister et al. (1996), Kreike et al. (1993)
	H1	Potato	Niewöhner et al. (1995)
	Hero	Tomato	Ganal et al. (1995)
Globodera pallida	*Gpa2*	Potato	van der Voort et al. (1997)
Heterodera glycines	*rhg1*	Soybean	Concibido, Diers, and Arelli (2004)
	Rhg4	Soybean	Webb et al. (1995)
Heterodera avenae	*Ha2*, *Ha3*	Barley	Kretschmer et al. (1997)
	Ha4	Barley	Barr et al. (1998)
	Cre1	Wheat	Eastwood et al. (1994), Williams et al. (1996)
	Cre3	Wheat	Laguduah et al. (1997)
	Cre8	Wheat	Lewis et al. (2009), Ogbonnaya et al. (2009)
Heterodera schachtii	$Hs1^{pro-1}$	Sugar beet	Cai et al. (1997)
	Hs2	Sugar beet	Heller, Schondelmaier, Steinrücken, and Jung (1996)

Root Knot Nematodes			
Meloidogyne arenaria	*Mae, Mag, Rma*	Peanut	Garcia, Stalker, Shroeder, and Kochert (1996), Chu et al. (2011)
Meloidogyne incognita	*Mi-1*	*Solanum peruvianum*	Ganal and Tanksley (1996)
	Mi-3 on chromosome 12	*S. peruvianum*	Yaghoobi, Kaloshian, Wen, and Williamson (1995)
	Mi-9	*S. peruvianum*	Ammiraju, Veremis, Huang, Roberts, and Kaloshian (2003)
	Mi-1 and Mi-9 on chromosome 6	Tomato	Klein-Lankhorst et al. (1991), Messeguer et al. (1991), Ammiraju et al. (2003)
	Me3 on chromosome P9	Pepper	Djian-Caporalino et al. (2007)
Root Lesion Nematodes	***Major QTLs Identified on Chromosomes***		
Pratylenchus thornei	Examples of QTLs on 2BS, 6DS and 6DL, 6D, 1B, 2B, 3B, 4D, 6D, 7A		Thompson, Brennan, Clewett, Sheedy, and Seymour (1999), Totkay, McIntyre, Nicol, Ozkan, and Elekcioglu (2006), Schmidt, McIntyre, Thompson, Seymour, and Liu (2005), Zwart, Thompson, and Godwin (2005)
	QRlnt.lrc-6D.2, QRlnt.lrc-6D.1	Wheat	Zwart et al. (2005)
Pratylenchus neglectus	Examples of QTLs on chromosome 2B, 4DS, 6DS, 7AL, 3, 5, 6, 7H		
	QRlnn.lrc-4D.l, QRlnn.lrc-6D.l	Wheat	Zwart et al. (2005)
	Rlnn1 resistance locus	Wheat	Williams et al. (2002)
	Pne3H-1, Pne3H-2, Pne5H, Pne6H, Pne7H	Barley	Sharma et al. (2011)
Pratylenchus penetrans	Rlnn1 resistance locus	Wheat	Williams et al. (2002)
P. neglectus & P. penetrans	QTLs on chromosome 1B, 2B and 6D	Wheat	Toktay et al. (2006)
	Rlnnp6H resistance on Chromosome 6H	Barley	Galal et al. (2014)
P. thornei & P. neglectus	Xbarc 183 on chromosome 6DS	Wheat	Zwart et al. (2005)

Macrosiphum euphorbiae and the white fly *Bemisia tabaci* (Nombela, Williamson, & Muñiz, 2003; Rossi et al., 1998). Although *Pratylenchus* species are often regarded as being less damaging in terms of crop losses, they can be the most economically important nematode pests in areas of low rainfall such as the Australian wheatbelt. Not surprisingly, the most detailed research on breeding for tolerance and resistance to *Pratylenchus* spp. has been carried out in Australian cereal breeding programs (Table 2) (Jones & Fosu-Nyarko, 2014). Genotypes with high tolerance to infestation with *Pratylenchus thornei* and medium tolerance to *Pratylenchus penetrans* have been identified among wheat cultivars, although they are not necessarily resistant or tolerant to other *Pratylenchus* species (Smiley & Nicol, 2009). Also major Quantitive Trait Loci (QTLs) for *P. thornei*, *P. penetrans* and *Pratylenchus neglectus*, some of which have polygenic and additive resistance effects, have been used routinely to select for resistance for these nematodes in Australian and CIMMYT (International Maize and Wheat Improvement Center) wheat breeding programs (Table 2) (Williams et al., 2002).

Study of resistance to *Pratylenchus* species is also important in barley because losses caused can also be substantial. To date, five QTL loci contributing to resistance to *P. neglectus* in barley germplasm have been identified on chromosomes 3H, 5H, 6H and 7H, and these may be useful for marker-assisted selection for resistance in barley (Table 2) (Sharma et al., 2011). Although the resistance conferred by some of these genes is useful in improving resistance to *Pratylenchus* species in commercial crop varieties, there is still a need to identify new, more effective and durable sources of natural resistance to nematodes in most major crop species.

4.1 Transfer of Natural Resistance Genes to Different Species

A major aim of identifying nematode resistance genes is to introduce them into other susceptible crops of economic importance, to enhance crop yield and quality and, where relevant, to reduce costs and reliance on chemical nematicides. While there has been successful deployment of crops with a series of nematode resistance genes (e.g. tomato cultivars with the *Mi* gene, potato cultivars with the H1 gene), there have been few reports of successful transfer of characterized R genes into new species. It appears that the efficacy of these genes in heterologous systems is genotype and/or species dependent and may require several elements for effective signalling in the pathways that induce a hypersensitive response, and the required interactions with proteins may not be present in a different species. For example, transfer of the *Mi* gene to eggplant confers resistance to *M. javanica* but not to the potato aphid, *M. euphorbiae*. Similarly, the transfer of the tomato *Hero A* gene into tomato cultivars confers desirable levels of resistance to potato cyst nematode in tomato, but not in potato (Sobczak et al., 2005). Even in tomato cultivars carrying the *Mi* gene there is variation in resistance to *M. incognita* attributed to their genotypic background (Jacquet et al., 2005). A better understanding of the mechanisms of nematode resistance offered by this group of Nucleotide Binding Site - Leucine Rich Repeat (NBS-LRR) class of plant R genes should make their introduction into other commercial crops more effective.

5. TRANSGENIC APPROACHES TO PLANT PARASITIC NEMATODE CONTROL

5.1 Disruption of Feeding Site Formation or Function

Since the discovery that reproduction of sedentary endoparasitic nematodes (*Heterodera* spp., *Gobodera* spp., *Meloidogyne* spp., *Rotylenchulus* spp., *Nacobbus* spp. and *Tylenchulus* spp.) depends on successful formation and function of giant cells, syncytia or similarly modified host cells (Jones, 1981), strategies which can disrupt feeding site formation have been investigated. RNAi-based methods that target the nematodes' ability to induce feeding sites are discussed below: here we consider plant processes involved in feeding site formation and function. Success with this type of approach very much depends on identifying plant promoters which are specifically or highly upregulated in feeding cells, and which can be linked to expression of a cytotoxic gene which when expressed in feeding cells results in cell death or impairment. The first example of this approach was by Opperman, Taylor, and Conkling (1994), who reported that the truncated (Δ0.3 kb) promoter of the water channel protein TobRB7 was expressed specifically in root knot giant cells, and when linked to the cytotoxic ribonuclease *barnase* resulted in cell death. However, unintended or 'leaky' expression of such a cytotoxic gene in other cells is a serious drawback to this approach. Even when combined with constitutive expression of the gene *barstar* which can neutralize the activity of *barnase* (Sijmons, Atkinson, & Wyss, 1994), unless it is highly upregulated there is danger of unintended side effects on the plant. Although a series of genes highly upregulated or downregulated in nematode feeding cells have since been identified, such as the heat shock promoter *Hahsp17.7G4* (Escobar et al., 2003), it appears that none of these promoters alone is sufficiently tightly expressed in the feeding cells to link them to a cytotoxic gene without collateral damage elsewhere in the plant.

An alternative approach, based on using two nematode responsive promoters, both of which must be upregulated in nematode feeding cells for expression of a cytotoxic gene to occur, may overcome this issue of target cell specificity of expression (Wang, Shuie, & Jones, 2008 and unpublished data).

5.2 Overexpression of Host Genes with Modified Expression in Feeding Cells

For a substantial time it had been predicted that there would be many changes in the metabolism of giant cells, syncytia and other feeding cells induced in hosts by endoparasitic nematodes (Jones, 1981). As technology advanced it has become possible to analyze changes in patterns of expression of genes in nematode feeding cells in ever greater detail, for example by differential display, microaspiration of feeding cell contents, laser microdissection and capture, microarrays and making use of new deep sequencing technologies (e.g. Alkhaouf et al., 2006; Fosu-Nyarko, Jones, & Wang, 2009; Ibrahim et al., 2011; Ramsay, Wang, & Jones, 2004; Wang, Potter, & Jones, 2003; Szakasits et al., 2009; Barcala et al., 2010; Portillo et al., 2013). Much of the research has focused on syncytia induced in soybean by *H. glycines* because of the economic importance of this nematode. Matthews et al. (2012) selected 100 soybean genes with expression modified in syncytia, identified using microarrays, for overexpression in a composite hairy root soybean system. Of these, nine reduced the number of females by 50% or more when overexpressed; conversely some enhanced the number of females. The challenge here is that the genes overexpressed would be expected to play a role in normal plant metabolism, and so overexpression may well confer a level of nematode resistance, but there is the risk that in a field situation an abnormal phenotype or reduced yield may result. It may be possible to choose a target gene whose high expression is vital for feeding site formation or metabolic function, but select a level of modified expression which interferes with feeding cell formation without adversely affecting any other parameter of plant growth.

5.3 RNAi-Based Nematode Resistance

Since the discovery of RNAi in nematodes, the potential to develop plants which produced double-stranded RNA to nematode target genes and so to silence expression of genes vital for their development or infection processes has been proposed as a sustainable, environmentally friendly strategy to add to current methods used for nematode control (Fire et al., 1998; Tan, Jones, & Fosu-Nyarko, 2013; Urwin, Lilley, & Atkinson, 2002). The first question was how to make plant parasitic nematodes take up dsRNA from external solutions, and this was solved by the pioneering work of Urwin et al. (2002), who showed that upon addition of neurostimulants to the 'soaking' solution for *H. glycines* J2s they could be induced to take up sufficient dsRNA to induce RNAi. Since that time, dsRNA feeding/soaking has been used to assess the effects of downregulation of over 30 essential and parasitism genes of various plant nematode species, including cyst nematodes (*H. glycines, H. schachtii, Gobodera pallida, Gobodera rostochiensis*), root knot nematodes (*M. incognita, M. javanica, Meloidogyne hapla, M. arenaria* and *Meloidogyne artiellia*), root lesion nematodes (*Pratylenchus zeae, P. thornei, Pratylenchus coffeae*) and other ectoparasitic nematodes (*Radopholus similis*, and *Bursaphelenchus xylophilus*) (Joseph, Gheysen, & Subramaniam, 2012; Li, Todd, Oakley, Lee, & Trick, 2011; Lilley, Bakhetia, Charlton, & Urwin, 2007; Tan et al., 2013) (Reviewed for RKNs in chapter "Function of Root-Knot Nematode Effectors and Their Targets in Plant Parasitism"). However, it has since been demonstrated that neurostimulants and other chemicals are not necessarily needed to induce RNAi using dsRNA (e.g. Fanelli, Di, Jones, & Giorgi, 2005; Kimber et al., 2007). In *H. glycines* and for some *Pratylenchus* spp, it appears that, for some genes at least, silencing resulting from soaking in dsRNA does not always produce stable phenotypic effects, since it appears that the effects of RNAi can wear off hours or days after the initial effect, leading to nematode recovery or regaining of function. Nevertheless the soaking method has been shown to be an effective method for initial screening of gene function and for discovery of candidate target genes suitable for plant-delivered RNAi for nematode control.

In contrast to soaking plant nematodes in solutions containing dsRNA, host (*in planta*) delivery provides dsRNA continuously if expressed in host cells from a constitutive promoter. This mode of delivery of dsRNA appears to be an ideal and economical approach to control obligate parasites such as plant parasitic nematodes. *In planta* delivery of dsRNA of two target genes (an integrase and a pre-mRNA splicing factor) was first demonstrated by Yadav, Veluthambi, and Subramaniam (2006) to reduce replication of *M. incognita* on transgenic tobacco plants, and this was quickly followed by the work of Huang, Allen, Davis, Baum, and Hussey (2006) who expressed dsRNA to an *M. incognita* effector protein in transgenic plants and also showed reduced nematode reproduction. Since then a series of experimental and crop plants have been engineered to generate inverted repeats

Table 3 Host-Delivered RNAi for Parasitism and Essential Genes of Cyst and Root Knot Nematodes: Summary of Reduced Infectivity on Model and Crop Plants

Nematode	Gene Silenced	Plant/Crop	Major Phenotype	References
Root Knot Nematodes				
Meloidogyne incognita	SNF (Sucrose Non Fermentable) chromatin remodelling Complex component (*snfc-5*)	Tobacco	>90% reduction in established nematodes	Yadav et al. (2006)
	Pre-mRNA splicing factor (*prp-21*)	Tobacco	>90% reduction in established nematodes	Yadav et al. (2006)
	Secreted peptide (*16D10*)	*Arabidopsis*	69–83% reduction in the number of eggs per gram root, >63% reduction in galls and gall size	Huang et al. (2006)
	Troponin C (*tnc*)	Tomato	59% reduction in hatching rate of J2s	Dubreuil et al. (2009)
	Secreted peptide (*16D10*)	Grapes	General reduction in number of eggs per gram of hairy root	Yang et al. (2013)
	Calreticulin (*crt*)	Tomato	J2s recovered from silenced progeny induces up to 84% fewer galls	Dubreuil et al. (2009)
	L-Lactate dehydrogenase	Soybean	57% reduction in galls per plant root, 77% reduction in RNAi nematode diameter	Ibrahim et al. (2011)
	Mitochondrial stress-70 protein	Soybean	92% reduction in galls per plant root, 85% reduction in RNAi nematode diameter	Ibrahim et al. (2011)

	ATP synthase beta-chain mitochondrial precursor	Soybean	64% reduction in galls per plant root, 62% reduction in RNAi nematode diameter	Ibrahim et al. (2011)
	Tyrosine phosphatase	Soybean	95% reduction in galls per plant root, 82% reduction in RNAi nematode diameter	Ibrahim et al. (2011)
	Dual oxidase	Tomato	52% reduction in saccate nematodes, 61% reduction in total nematodes	Charlton et al. (2010)
	Signal peptidase complex 3	Tomato	63% reduction in saccate nematodes, 52% reduction in total nematodes	Charlton et al. (2010)
Meloidogyne javanica	Nematode effector protein (NULG1a)	*Arabidopsis*	Up to 88% reduction in number of nematodes in roots	Lin et al. (2013)
	Secreted peptide (*16D10*)	*Arabidopsis*	90–93% reduction in the number of eggs per gram root, >63% reduction in galls and gall size	Huang et al. (2006)
Meloidogyne arenaria	Secreted peptide (*16D10*)	*Arabidopsis*	84–92% reduction in the number of eggs per gram root, >63% reduction in galls and gall size	Huang et al. (2006)
Meloidogyne hapla	Secreted peptide (*16D10*)	*Arabidopsis*	69–73% reduction in the number of eggs per gram root, >63% reduction in galls and gall size	Huang et al. (2006)
Meloidogyne chitwoodi	Secreted peptide (*16D10*)	*Arabidopsis*	57% and 67% reduction in egg masses and eggs	Dinh et al. (2014)
	Secreted peptide (*16D10*)	Potato	71% and 63% reduction in egg masses and eggs	Dinh et al. (2014)

(*Continued*)

Table 3 Host-Delivered RNAi for Parasitism and Essential Genes of Cyst and Root Knot Nematodes: Summary of Reduced Infectivity on Model and Crop Plants—cont'd

Nematode	Gene Silenced	Plant/Crop	Major Phenotype	References
Cyst Nematodes				
Heterodera glycines	Major sperm protein	Soybean	Up to 68% reduction in female cysts	Steeves et al. (2006)
	Ribosomal protein 3a (*rps-3a*)	Soybean	87% reduction in female cysts	Klink et al. (2009)
	Ribosomal protein 4 (*rps-4*)	Soybean	81% reduction in female cysts	Klink et al. (2009)
	Spliceosomal SR protein (*spk-1*)	Soybean	88% reduction in female cysts	Klink et al. (2009)
	Synaptobrevin (*snb-1*)	Soybean	93% reduction in female cysts	Klink et al. (2009)
	Beta subunit of the COPI complex (*Y25*)	Soybean	81% reduction in female cysts	Li et al. (2010)
	Pre-mRNA splicing factor (*prp-17*)	Soybean	79% reduction in nematode	Li et al. (2010)
	Uncharacterized protein (*cpn-1*)	Soybean	95% reduction in nematode	Li et al. (2010)
Heterodera schachtii	Ubiquitin-like protein (*4G06*)	*Arabidopsis*	23–64% reduction in developing females	Sindhu et al. (2009)
	Cellulose binding protein (*3B05*)	*Arabidopsis*	12–47% reduction in developing females	Sindhu et al. (2009)
	SKP1-like protein (*8H07*)	*Arabidopsis*	>50% reduction in developing females	Sindhu et al. (2009)
	Zinc finger protein (*10A06*)	*Arabidopsis*	42% reduction in developing females	Sindhu et al. (2009)
	Nematode secreted peptide, *Hssyv46*	*Arabidopsis*	36% reduction in female cysts	Patel et al. (2008)
	Nematode secreted peptide, *Hs5d08*	*Arabidopsis*	Up to 20% reduction in female cysts	Patel et al. (2008)
	Nematode secreted peptide, *Hs4e02*	*Arabidopsis*	Up to 20% reduction in female cysts	Patel et al. (2008)
	Nematode secreted peptide, *Hs4F01*	*Arabidopsis*	Up to 55% reduction in female cysts	Patel et al. (2008)
	30C02 effector protein	*Arabidopsis*	Up to 92% reduction in female cysts	Hamamouch et al. (2012)

of dsRNA targeting genes expressed in pharyngeal gland cells or those essential for development and reproduction in cyst and root knot nematodes (Table 3). To date, reports of significant reductions in the number of females of soybean cyst nematode (81–93%) and eggs (68–95%) produced by female cysts developing on hairy roots or composite transgenic soybean expressing dsRNA of genes involved in RNA and protein synthesis provide a level of confidence that RNAi can be an important tool for nematode control (Klink et al., .2009; Li, Todd, Oakley, Lee, & Trick, 2010; Li et al., 2011; Steeves, Todd, Essig, & Trick, 2006).

5.4 Differences in Responses to RNAi in Different Nematode Species

Depending on the target gene and the experimental procedures used (e.g. model or crop plant species and genotype, target gene silenced, dsRNA sequences used, number of events generated and studied, the methods and nematode genotypes used for screening, quantifying and analysis of results), there is a wide range of reports of the efficacy of RNAi when used to reduce nematode reproduction. As a general observation it would appear that *Meloidogyne* spp. are more susceptible to RNAi than *Heterodera/Gobodera* species, but with more limited data from *Pratylenchus* species it would appear that these are highly amenable to control by RNAi. For example, high levels of resistance were reported in tobacco and *Arabidopsis* producing dsRNA to genes of root knot nematodes, including the parasitism gene *16D10*, a gene expressed in the subventral gland cells of *M. incognita*, a pre-mRNA splicing factor and an integrase gene: their expression resulted in an inability of >90% for J2s to establish feeding sites (Yadav et al., 2006). (RNAi of the *M. incognita 16D10* gene also confers resistance to transgenic *Arabidopsis* infected with four other *Meloidogyne* species: *M. hapla*, *M. javanica*, *M. chitwoodi* and *M. arenaria*, and RNAi of this gene has since been demonstrated to provide a level of resistance to several important crops such as grapes and potato (Dinh, Brown, & Elling, 2014; Huang et al., 2006; Yang et al., 2013)).

There may also be differences in the effectiveness of this approach between species of the same genus: in some publications it seems that in planta RNAi of *H. schachtii* may not be as effective as for *H. glycines*, because only a 12–64% reduction in female nematodes (except for the 92% reported for 30C02 effector protein) was observed when genes encoding putative secreted effector proteins, a ubiquitin-like gene, and those of a cellulose binding protein, SKP1, and a zinc finger protein were silenced (Patel et al., 2008; Sindhu et al., 2009; Patel et al., 2010; Hamamouch et al., 2012). However, silencing seven genes (elongation factor 1α, two vacuolar H ATPases, integrase, pre-mRNA splicing factor, troponin C and tropomyosin) of *H. schachtii* via transgenic *Arabidopsis* resulted in up to 98% reduction in adult females (Fosu-Nyarko & Jones, 2013, 2014).

The differences in effectiveness of RNAi may relate to differences in biology of host–parasite interactions such as presence or absence of feeding tubes: *Meloidogyne* feeding tubes appear to be larger and more regular in structure than those formed by feeding cyst nematodes, and this may influence the ability of uptake of dsRNAs, whereas *Pratylenchus* species do not form feeding tubes. Alternatively there may be fundamental differences in RNAi pathways in different nematodes, or in systemic movement of siRNAs in the nematodes. The reasons for such differences, or whether current reports reflect more differences in experimental procedures used, remain to be demonstrated experimentally.

5.5 Factors that Affect the Efficacy of RNAi Traits

Since there are now many examples in which RNAi has been used to confer varying degrees of resistance to root knot and cyst nematodes (Table 3), it is important to consider what factors influence the effectiveness of this strategy. The first factor is choice of target gene – is it an effector vital for successful parasitism, or a gene whose expression is vital for completing some aspect of the nematode's life cycle? Other considerations are the length of dsRNA used, the specific sequence chosen, whether the target gene is a member of a multigene family, and whether there are compensating pathways for loss of a particular function. In terms of acceptability, the target sequence chosen should preferably be unique to the nematode specie(s) of interest, or at least not be present in mammals, beneficial organisms and all non-target species for which sequence data are available. The shorter the sequence chosen, the less chance there is of off-target effects, and so the use of an artificial miRNA vector, in which only 20–24 bases of target sequence may be used, should reduce possible off-target effects (but the most effective sequence from the target gene should then be used). Even when these selection criteria are applied, it seems that most transgenic RNAi experiments give varying levels of effectiveness, with none 100% effective. This observation may reflect variability in the populations of target nematode species rather than efficacy of RNAi per se. In any case, it is well known that, depending on the site of transgene insertion, copy number, promoter strength and construct design, any set of transgenic plant events will exhibit a range of the desired property, and only the most effective events will be chosen for progression through a commercialization pipeline.

5.6 Differences in Results between Model and Crop Plants

Another factor that is often overlooked is the difference between using model experimental plants such as *Arabidopsis* for nematode challenge experiments compared to crop plants. Model plants such as *Arabidopsis* have not been selected for their resistance to plant nematodes, and so are likely to be more susceptible, whereas in many cases crop varieties have been selected for resistance or tolerance to nematodes, even if only partial. As a result, promising results from model species do not always map over to crop species, since the percentage improvement in nematode resistance is that conferred over and above the selected resistance, rather than against a highly susceptible host.

5.7 Broad Resistance to Different Plant Nematodes

One of the attractions of developing transgenic resistance to plant nematodes using RNAi technology is the potential to confer broader resistance to several species in one construct, in contrast to the more specific resistance conferred by natural resistance genes. The principle is that hairpin dsRNAs to a number of different target genes can be made either from the same or different species, or to target different populations of the same nematode species. When *P. thornei* and *P. zeae* were soaked in dsRNA sequences of two target genes from each species, there was a reduction in subsequent reproduction on carrot discs irrespective of the target gene source (Tan et al., 2013). However, so far there are no convincing reports from transgenic in planta experiments that two different nematode species can be controlled with one hybrid dsRNA construct (Charlton et al., 2010). Perhaps the RNAi mechanism can be overwhelmed if too many siRNAs are generated, and with more subtle expression or choice of target sequence the potential for broad resistance to different nematode species may be achieved.

6. 'TRANSGENIC' TECHNOLOGY ADVANCES

There continue to be advances in the technology of genetic modification (GM) which challenge the current regulatory definitions of a genetically modified organism, because it is not always clear whether the products obtained using these techniques are subject to the prevailing GM legislation or not (Breyer et al., 2009). Examples of new technologies or concepts include:

- Cisgenesis – this involves introduction of DNA from the same or a compatible species
- RNAi – downregulation of expression of existing genes
- Reverse breeding
- Genome editing – directed mutation or 'precision gene editing' – introduction of targeted changes to nucleotides in the genome, such as oligonucleotide-mediated mutagenesis (e.g. Cibus), 'zinc finger'/ designer nucleases (ZFNs) – gene disruption or precise insertion of DNA sequences (e.g. EXACT™), CRISPR-Cas systems (clustered regularly interspaced short palindromic repeats)
- Virus-delivered ZFN genome editing
- Epigenetics – induced differentially methylated regions
- Agroinfection
- Virus-induced gene silencing
- Genomics-enabled technologies, e.g. ectopic delivery of dsRNA (e.g. Biodirect™ Technology)
- Grafting nontransgenic scions onto GM rootstocks (e.g. for vines or fruit trees)

With advances in biotechnology, new techniques of genome editing have emerged, that is, the ability to make tailored changes to a genome sequence. These techniques can enable modification of expression of existing genes or introduction of targeted changes to nucleotides in the genome (e.g. oligonucleotide-mediated mutagenesis). Such techniques began with methods based on ZFN to define their binding site on a DNA sequence, linked to Fok1 endonuclease to generate double-stranded breaks in the DNA at the defined sequence. DNA repair mechanisms are then recruited either by nonhomologous end joining pathways or homologous repair pathways, to generate mutations or insert exogenously supplied sequences with flanking sequences homologous to the insert (Lozano-Juste & Cutler, 2014). Tailored ZFNs are expensive to make, and other developments such as transcription activator-like effectors (TALEs; DNA-binding proteins produced and secreted by plant pathogens into plant cells, which bind specific DNA sequences and alter transcription on endogenous genes) are easier to modify. TALEs have many copies of a 33–35 amino acid repeat, with DNA recognition dependent on two variable amino acids in the repeats, and can be modified with addition of nuclease to make TALE nucleases (TALENs), that can cut DNA at specific sites. The CRISPR/Cas9 system has been developed which is technically simpler to use for genome editing (Lozano-Juste & Cutler, 2014). Its advantage over ZFNs and TALENs is that it uses synthetic guide-RNAs (gRNAs) rather than synthetic DNA-binding domains, to define the cleavage site. The CRISPR/Cas9 system is based on a bacterial antiviral and transcriptional regulation system, modified such that two RNA components have been combined into a single gRNA which is transcribed from a construct containing a user-defined target sequence of 20 nucleotides complementary to the desired target sequence. Guided by the gRNA, the Cas9 nuclease protein binds and nicks the defined sequence, which as above, can be modified by nonhomologous end joining or homologous repair pathways, to generate mutations or insert exogenously supplied sequences. This approach has been used to generate transgenic *Arabidopsis thaliana* plants with mutations in the PDS (phytoene desaturase) locus (Nekrasov, Staskawicz, Weigel, Jones, & Kamoun, 2013), although using standard *Agrobacterium* transformation and selection. These developing techniques of gene editing could be used to modify host plant genes to confer nematode resistance, for example by disrupting or modifying expression of genes vital for feeding site formation for sedentary endoparasites. However, there is still some uncertainty about the extent of off-target effects for this technology.

7. FROM THE LABORATORY TO THE MARKET – COMMERCIALIZATION OF PLANT PARASITIC NEMATODE-RESISTANCE TRAITS

7.1 Patenting

Before a nematode resistance or other trait can be commercialized, unless the information is for public good and free use, the trait needs to be protected by patenting. In an academic situation the approach is usually to submit a provisional patent via the University's Commercialization Office. There is 1 year to provide additional supporting data if needed before full patent application at which stage the costs increase substantially. In this period the Commercialization Office usually looks for an industry partner who is interested in taking on the costs of full patenting and may provide additional funds, in return for first use in licensing and exploiting the trait. An alternative strategy is to establish a company to raise funds for further development of the trait, and be responsible for patenting and licensing one or more traits. In the latter case, company investors will seek a way to recoup their investment (an 'exit strategy'), either by trade sale to a larger company, by public listing or by raising further investment to exploit the trait directly. The further the product is progressed along the developmental pathway, the higher the expected returns will be.

7.2 Commercialization Pathway

An overview of the pathway to commercialization of a biotechnology trait conferring nematode resistance is provided in Figure 1. From an initial idea the basic discovery research is undertaken, and if that is promising it moves from the discovery phase to proof-of-concept, then early and advanced stages of product development, to a prelaunch phase, and finally to commercial release to growers. The activities to be undertaken in each phase are indicated in Figure 1, as well as indicative timescales and the probability of success in progressing to commercial release. Basic discovery research is more the realm of public research organizations such as universities and government-funded research institutes, but these are often viewed as poor at commercialization activities. As a result the discovery or trait moves along the pipeline often via a 'start-up' or 'expansion stage' company, and at some stage for biotech traits, the trait is either licensed to a large corporation or multinational company or the company is bought by such companies for advanced development, prelaunch and commercial release to growers.

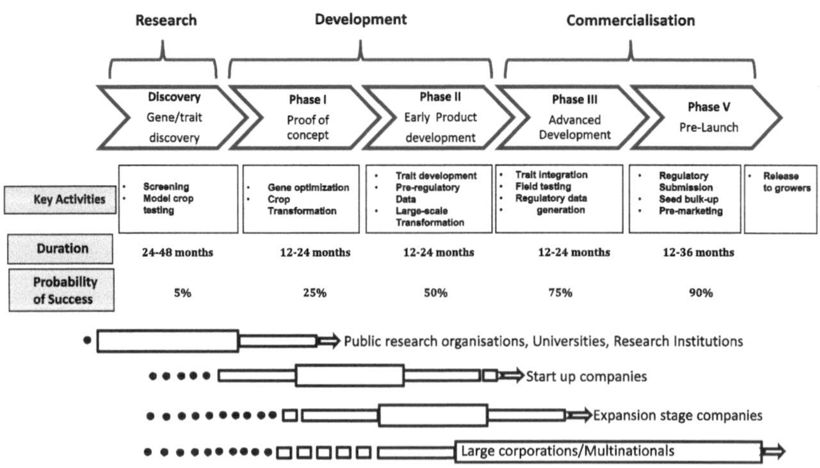

Figure 1 Pathway to commercialization of a biotechnology trait conferring resistance to plant parasitic nematodes.

As a product moves along the pipeline the costs of development and commercialization increase, and it is for this reason that most biotech traits for large-scale commodity world crops (e.g. soybean, corn, cotton, canola; possibly wheat and rice in the future), such as nematode resistance, are necessarily deployed by multinational corporations. Increasingly new traits will be stacked with other biotech traits, requiring coordination of their development and introduction into the best available germplasm for a particular crop.

To estimate the costs of trait deployment, a recent consultancy study by Phillips McDougall for Crop Life International (September 2011) was undertaken on 'The cost and time involved in the discovery, development and authorization of a new plant biotechnology trait'. It was based on the responses of the following multinational companies: BASF, Bayer CropScience, Dow AgroSciences, DuPont/Pioneer Hi-Bred, Monsanto Company and Syngenta AG on costs involved in introducing a new GM crop trait over the period 2008–2012. The costs reported are shown in Table 4.

The study revealed that the mean cost associated with the discovery, development and authorization of a new biotechnology-derived crop trait introduced in the 2008–2012 timeframe, including associated international market approvals required for a grain crop to enter the global grain trade, was US$136.0 million. However, with deregulation of traits and possible relaxation of safety and environmental testing based on history of safe usage, the cost of trait deployment is likely to decrease in the future. Among other findings was that the mean time taken for all crops from initial research and development until commercial sales was 13.1 years.

Table 4 The Cost Involved in Various Stages of Development of a Biotechnology Trait for Grain Crops

Category		Cost ($ million)	No of Responses
Discovery	Early discovery	17.6	5
	Late discovery	13.4	5
	Total cost	31.0	5
Construct optimization		28.3	5
Commercial Event production & selection		13.6	6
Introgression Breeding & Wide Area Testing		28.0	6
Regulatory Science		17.9	6
Regulatory & Regulatory Affairs		17.2	6
Total		136.0	

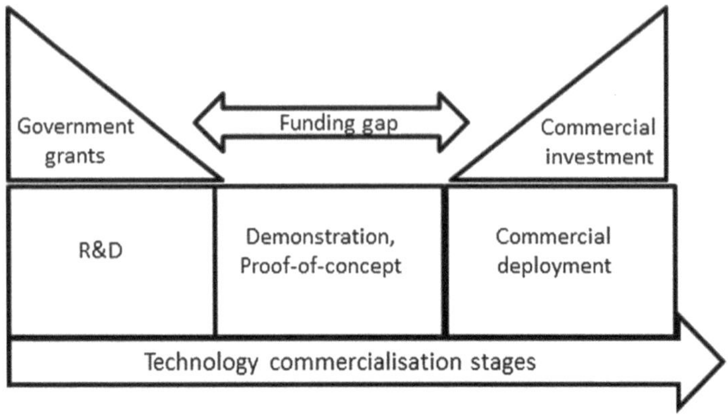

Figure 2 Stages and sources of funding for technology commercialization.

7.3 The Funding Gap for Early Stages of Commercialization

A major problem for early stage commercialization of discoveries is known as the 'funding gap' (Figure 2), and this is particularly relevant to translation of research from universities. The basic research may be funded by a variety of government or competitive grants, but without good proof-of-concept data there is a funding gap in which commercialization of university research is lacking. However, once that gap is bridged, with suitable evidence of efficacy, commercial investment is more readily available.

7.4 The Commercial Value of Nematode Resistance Traits

The commercial value of a biotech trait conferring nematode resistance depends on many factors, and on a case-by-case basis the following aspects need to be considered: the value of the crop, where it is grown (this may be limited by the robustness of intellectual property (IP) and patenting regimes in a particular jurisdiction), the area grown, the percentage of that area that is affected by nematodes, the extent of losses caused by specific nematodes, the degree of protection provided by the biotech trait, the mode of delivery, the cost of alternative methods of control, the addressable market, the expected time course of uptake and percentage of the market that can be accessed, the cost of meeting regulatory requirements and the added value provided by the trait. Since a nematode-resistance trait will be delivered via appropriate germplasm, the main value of the germplasm containing the trait will go to the breeders and seed marketers as is standard, with the technology developers of the trait receiving a small percentage of the overall value of seed sales based on the value added by the trait. This may be in the order of 1–5%, depending on the factors to be considered indicated above, but in the future may be less, if a nematode-resistance trait is delivered as one of a set of stacked biotech traits. Conversely, a biotech nematode resistance trait could differentiate one variety of seeds for a crop from those of another supplier that lacks them, and so have a greater value as a result of improved seed marketing.

7.5 Specialist/Small-Scale Commercialization of Nematode Resistance Traits

The costs of deploying a trait in a major grain crop may appear daunting, but there are small companies, such as the Canadian biotech company, Okanagan Specialty Fruits (www.okspecialtyfruits.com), which is developing transgenic fruit tree products themselves, at a fraction of the cost indicated above for major grain crops. Okanagan Specialty Fruits are using RNAi technology to downregulate expression of a polyphenol oxidase (PPO) gene in apples to reduce browning. The company has applied for regulatory approval in the United States and Canada for commercial growth of two GM apple varieties ('Arctic Granny' and 'Arctic Golden'). This company is also using 'transient gene silencing' to modify the expression of genes in existing apple cultivars: they argue that 'Transient gene silencing is not genetic modification in the traditional sense'. What this means is that they are developing RNAi-based transgenic apple rootstocks, modified to suppress the expression of PPO, on which commercial cultivars of apple are grafted. They state that 'the trait can be transferred from the donor to the recipient though small interfering RNA (siRNA) that migrates through the plant'. The result is the production of nonbrowning fruit on the nontransgenic recipient. They suggest that 'transient gene silencing' of PPO in apples produced on the scion can be triggered in the right conditions.

The strategy of using transgenic rootstocks grafted to nontransgenic scions is clearly relevant to nematode control, in which the transgenic rootstock is nematode resistant, whereas the produce harvested from the scion is not.

8. TRANSGENIC NEMATODE RESISTANCE FOR PUBLIC GOOD

The alternative to commercial deployment of a nematode resistance trait is to bypass all the issues of IP and costs of patenting a trait, and to provide it for public good, often funded by overseas philanthropic organizations. Nevertheless, significant funds are still needed to meet safety, human health, environmental and other regulatory requirements before deployment. One such example of this approach is the work of Atkinson and colleagues (Atkinson, Lilley, & Urwin, 2012; Atkinson 2014 personal communication). They employed two biotech approaches to develop nematode resistance in plants – overexpression of cysteine proteinase inhibitors (cystatins) which interfere with intestinal digestion of dietary protein ingested from the plant, and synthetic peptides expressed or secreted from roots which interfere with nematode chemoreception by binding to either acetylcholinesterase or nicotinic acetylcholine receptors, which are both targets in the nematode cholinergic nervous system. The peptide inhibits nematode chemoreception after uptake by chemosensory sensilla in the amphid pouches and transports along chemoreceptive neurons to their cell bodies.

Plant parasitic nematodes cause an average of 12.75% losses to five staple crops in Africa (maize, sugarcane, banana and plantain, yam and cassava) (FAOSTAT, 2012), with particular losses in banana and plantain (up to 70%) in some regions. In the latter case several nematode species are responsible: control is inadequate, nematicides are not appropriate due to cost and hazards of application. In field experiments with transgenic plantain events expressing the 7-mer repellent peptide and/or the cystatin protease inhibitor, promising resistance has been obtained (89–98% reduction in nematodes recovered), especially for plants expressing the repellent peptide (Tripathi, Roderick, Babirye, & Atkinson, 2014). Biosafety assessments accompanying this work indicate safety based on the fact that cystatin is a normal part of the human diet, is not allergenic and is rapidly degraded by gastric juices: the peptide is too small to be allergenic and is degraded in the human small intestine. Regarding environmental safety, no adverse effects of cystatin expression were evident on the range of non-target species tested, the peptide is rapidly degraded in the soil and does not affect the soil microflora or other non-target species tested. There is therefore no evidence for safety concerns: in addition the expression of the cystatin and peptide genes could be driven by a root-specific promoter, limiting their presence to the roots.

In discussing de-regulated release of such nematode-resistant crops, Atkinson et al. (2012), note that countries with future food security concerns are most likely to adopt transgenic resistance, particularly for crops like cooking bananas, plantains or yams which cannot be improved readily using other approaches. For wider acceptance, effective policies must be developed to engage consumers and the food industry as well as growers (farmers).

9. REGULATION AND PUBLIC ACCEPTANCE OF GM TRAITS

The issues involving regulation and acceptance of GM crops are well known, and some of the developing technologies and concepts have the potential to improve public acceptance. For example, there have been a number of publications arguing for the exemption of cisgenic plants from the scope of GM regulations (e.g. Jacobsen & Schouten, 2008). In the current debate on how regulators may deal with newer GM technologies, directed mutation may be treated like mutagenesis, treatment of cisgenic plants depends on the method of transformation, treatment of non-GM grafts on GM rootstocks depends on the safety assessment of the GM rootstock; reverse breeding and agro-inoculation may be non-GM or exempt (J Dunlop, University of Reading, personal communication). The regulatory and acceptance aspects of applying new technologies to nematode control is very important when considering trait commercialization, since the complexities of regulations and public opinion affect the cost of deployment. If it is too expensive in relation to the added value of the trait, then commercial deployment may not be undertaken.

Of particular relevance to nematode control, using RNAi technology, inserted DNA does not encode message for a functional protein, and so should be in a lower risk category. Extending this to transgenic rootstocks with a nematode resistance trait (e.g. RNAi), with harvested produce from a nontransgenic scion, any risk factors are again reduced. However, in France vines with transgenic rootstocks were destroyed, in contrast, in Canada as discussed above, 'Artic Apples', grown on transgenic rootstocks, are close to commercialization, and the potential movement of siRNAs from rootstock to scion needs to be considered.

10. SAFETY OF RNAi-BASED TRAITS

Since RNAi technology has been used widely in functional genomics studies of nematode effectors or vital genes, and is a potential route for commercialization of research findings, the safety of food and feed with RNAi-based traits needs particular attention. A review on this subject relating to human and animal health (Petrick, Brower-Toland, Jackson, & Kier, 2013) considered these aspects in relation to molecular mediators of RNAi – long dsRNAs, small interfering RNAs (siRNAs) and micro RNAs (miRNAs). They reviewed available data including that on comparative safety assessments, mice fed on wheat with RNAi-mediated traits, the fundamental differences between biotech crops expressing heterologous proteins and those with RNAi-mediated gene suppression cassettes, the potential for unintended effects, a long safe history of ingestion of naturally occurring dsRNAs in plants and foods, reports of plant-derived miRNA in mice after oral ingestion (Zhang et al., 2012), mammalian and human studies on siRNA uptake, RNA molecule specificity, half-life, secretion and the barriers to oral ingestion. They concluded that 'available data strongly support the conclusion that biotechnology-derived crops employing RNA-mediated gene regulation are safe for human and animal consumption'.

11. GENOME-ENABLED DEVELOPMENT OF NOVEL CHEMICAL NEMATICIDES

As has been discussed by Jones and Fosu-Nyarko (2014) in the short to medium term at least, it is unlikely that a transgenic or equivalent biotechnology-based approach can be deployed to protect all commonly grown crops from nematode attack. This view is based both on the costs of developing and implementing such approaches, and public acceptance considerations. Nevertheless, information on genes whose products are vital for different processes of nematode root location, invasion, host defence evasion, general metabolic and developmental processes, and feeding or feeding site formation, can be used to inform the development of new, environmentally friendly nematicides (Danchin et al., 2013). The approach was to use a bioinformatics pipeline to filter potential gene targets based on genomic information from *M. incognita*. It involved the application of a stepwise set of rigorous criteria in which all the genes present in the genome, of known or unknown function, were assessed, for example to reduce the possibility of off-target effects and sequences potentially common to non-target organisms, candidates from multigene families and known effectors with deleterious RNAi phenotypes. This strategy led to a shortlist of high-quality target genes, which had the potential to serve as leads for development of new chemical nematicides. Functional analysis was in the form of feeding experiments in vitro, in which siRNAs designed to target each candidate gene was assessed for its effect on phenotype or ability of the nematode to infect host roots. Once appropriate vital nematode target genes have been identified, then targeted development or screening for chemicals which can inhibit such functions can be undertaken to develop novel nematicides.

12. ECTOPIC DELIVERY OF dsRNA - NONTRANSGENIC RNAi

Using a similar approach to identify target genes for nematode control, there is also the exciting possibility of using ectopic delivery of dsRNA (i.e. by spraying) onto plants. This strategy is being followed in the development of BioDirect™ Technology as a nontransgenic alternative mode of delivery of RNAi for crop protection against herbicides, insects and viruses (http://www.monsanto.com/investors/documents/whistle%20stop%20tour%20vi%20-%20aug%202012/wst-biodirect_posters.pdf). The challenge here for nematode control is to develop stable forms of dsRNA and their mode of delivery, such that they are taken up systemically when sprayed onto crop plant leaves, and move via the plant vasculature to the roots where they can be ingested by nematodes when feeding, resulting in inhibition of a vital process and control of the nematodes.

13. OTHER NEW NEMATODE CONTROL AGENTS

With the loss or reduced use mandates of more than two-thirds of the 'classical' nematicides, because of their toxicity, persistence in the environment or capacity to deplete the ozone layer, industry has been active in developing the next generation of nematode control agents. These are mainly targeted at horticultural crops, since a genetic/biotech strategy is more likely to be developed for major broadacre crops such as soybean, corn and cotton. Although there is substantial literature on the search for new phytochemical control agents (Chitwood, 2002; Taylor, Belton, Beta, & Duodu, 2014), very few have shown promise – the reasons for this include low activity, phytotoxicity, poor stability, leaching to ground water, too expensive to produce or mammalian toxicity.

In some cases industry has brought in new technologies, with the intention of applying integrated pest management programs. New chemical nematicide discovery usually involves synthesis of new compounds, modifying existing compounds or testing compounds identified for other activities, running them through a high-throughput screen to identify hits, then lead compounds, then glasshouse and field assessments for efficacy, environmental aspects, cost of production, meeting market demand, etc. Alternatively a target-based approach can be employed, using data such as that gained from gene silencing experiments (Danchin et al., 2013) to design molecules that will fit the active site of established target gene products and inhibit their vital function.

One focus is on an integrated pest management strategy to control nematodes, which may combine both biological and chemical agents, agents such as Bayer CropSciences's *Bacillus firmus* (Votivo), *Paecilomyces lilacinus* (Bioact) combined with chemical control, e.g. fluopyram (Velum/Verango), or using *Pasteuria* technology (e.g. CALARIVA) with AVICTA (Syngenta) and other new products on the market. These can be combined with more restricted methods of delivery, such as via drip irrigation or seed coating. These delivery methods are more suited to nematode control in horticultural crops, and have the benefit of providing control agents only where they are needed, thus enabling more expensive and less stable chemicals to be applied, so reducing costs and any detrimental environmental effects.

14. CONCLUSIONS

The exploitation of natural resistance genes using genotype screening and marker-assisted selection will continue to be the standard approach to improving resistance and/or tolerance to nematodes for many horticultural

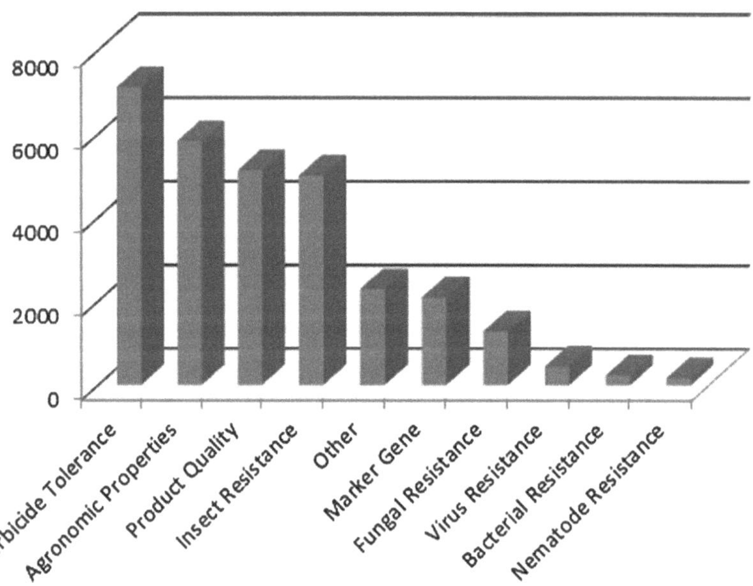

Figure 3 Number of approved releases agronomic traits by phenotype (permits and notifications). *http://www.isb.vt.edu/release-summary-data.aspx.*

and staple food crops. However, for many crops this approach is slow, and in most cases adequate broad-spectrum resistance is not available. This in part accounts for the lag in approved release of nematode resistance traits compared to other important agronomic traits as compiled by Information Systems for Biotechnology (Figure 3, http://www.isb.vt.edu/release-summary-data.aspx). With advances in transgenic technologies the release of commercial varieties of major crops with transgenic nematode resistance traits can be predicted to arrive soon. Already maize with eight stacked genes (SmartStax 'Geniuty'corn, with two herbicide tolerances and six *Bacillus thuringiensis*-based insect resistance traits) is commercially available, and varieties with similar stacked traits will soon be available for cotton, soybean and other crops. The addition of a nematode resistance trait would differentiate the first producer of such a trait, and would provide a competitive advantage. Such a biotech trait could be based on RNAi or on expression of other synthetic resistance genes, and be added to genotypes with the best conventional nematode resistance. In addition, synthetic resistance genes with different modes of action could be stacked to increase the efficacy and durability of the nematode resistance trait, and their expression limited to roots to improve acceptability and unnecessary expression in harvested parts of the plant.

As emphasized by Atkinson et al. (2012), crop losses to nematode infestation is often greatest in developing countries, where the need for food security is greatest, and nematodes attack both commodity and staple crops. In these countries it may not be a commercial proposition to develop nematode-resistant transgenic crops: rather a humanitarian approach may be adopted in which such crops are developed for the public good. However, following this approach will still require detailed studies to be undertaken to ensure the food is safe for humans and the environment, and the political and public support for their implementation, and in country support for their further development and deployment.

Since two-thirds of the world's population already lives in countries where GM crops are grown, and concern for world food security is now increasing, it is likely that the pace for adoption of transgenic crops will increase as one of the mix of measures that will be needed to feed over 9 billion people in a sustainable manner. Transgenic nematode resistance traits will undoubtedly be used to help ensure global food security in the future, although the path to reach this stage will have some issues to address, particularly in relation to public perception rather than science.

That is not to say that there have not been practical successes in nematode control, as can be seen when considering advances in control for different crops and nematode pests on a case-by-case basis. For example, real benefits have been derived through conventional plant breeding practices by introduction of the H1 major resistance gene against the potato cyst nematode *Globodera rostochiensis* (Finkers-Tomczak et al., 2011) and the *Mi* gene conferring host resistance to root knot (*Meloidogyne*) species in tomato (Williamson & Kumar, 2006), which has been transferred to many different tomato varieties throughout the world. A notable success has arisen from the study of natural genes for resistance to the cereal cyst nematode (*H. avenae*), in which effective control has been developed for wheat initially using large-scale screening of genotypes with resistance to *H. avenae*, followed by identification of NBS-LRR type resistance genes from wheat and its wild relatives, and their implementation in breeding programs. A series of 11 *Cre* genes have been identified and introgressed into wheat cultivars (in particular the genes *Cre 1*, *Cre3* and *Cre 8*) to confer resistance, particularly in Australia (Lewis, Matic, & McKay, 2009; Ogbonnaya, Eastwood, & Lagudah, 2009). Marker-assisted selection is now used to pyramid some of these resistance genes, and this approach is used routinely in many wheat breeding programs worldwide (Ogbonnaya et al., 2009) to provide effective control of *H. avenae*.

Marker-assisted selection can also be used to pyramid and combine different desired traits into improved germplasm, and this has been an ongoing activity undertaken by plant breeders over many years. In addition to major gene resistances, minor (multigene-based) resistance has been used incrementally to improve host resistance to nematode infestation in many crops, such as the rht genes in soybean improving resistance to the soybean cyst nematode (*H. glycines*), and a range of minor genes contributing to increased resistance to root lesion nematodes (*Pratylenchus* spp.) in cereals (Jones & Fosu-Nyarko, 2014). The more conventional application of biotechnology to nematode control by marker-assisted selection is now being complemented by the developing biotechnological approaches of transgenic plants, genome editing, developing new nematode control agents and new modes of delivery of control agents. The commercial implementation and delivery of these biotechnology-based technologies will undoubtedly contribute to future global food security.

REFERENCES

Alkharouf, N. W., Klink, V. P., Chouikha, I. B., Beard, H. S., MacDonald, M. H., Meyer, S., et al. (2006). Time course microarray analyses reveals global changes in gene expression of susceptible *Glycine max* (soybean) roots during infection by *Heterodera glycines* (soybean cyst nematode). *Planta, 224*, 838–852.

Ammiraju, J., Veremis, J., Huang, X., Roberts, P., & Kaloshian, I. (2003). The heat-stable root-knot nematode resistance gene *Mi-9* from *Lycopersicon peruvianum* is localized on the short arm of chromosome 6. *Theoretical and Applied Genetics, 106*, 478–484.

Atkinson, H. J., Lilley, C. J., & Urwin, P. E. (2012). Strategies for transgenic nematode control in developed and developing world crops. *Current Opinion in Biotechnology, 23*, 251–256.

Atkinson, H. (2014). Transgenic nematode resistance for African food security: nematode resistant bananas as a case study. Proceeding of the 6th International Congress of Nematology, Cape Town, May 2014.

Ballvora, A., Hesselbach, J., Niewoehner, J., Leister, D., Salamini, F., & Gebhardt, C. (1995). Marker enrichment and high resolution map of the segment of potato chromosome VII harbouring the nematode resistance gene Gro1. *Molecular & General Genetics, 249*, 82–90.

Barcala, M., Garcia, A., Cabrera, J., Casson, S., Lindsey, K., Favery, B., Garcia-Casado, G., Solano, R., Fenoll, C., Escobar, C., et al. (2010). Early transcriptomic events in microdissected Arabidopsis nematode-induced giant cells. *Plant J, 61*, 698–712.

Barr, A. R., Chalmers, K. J., Karakousis, A., Kretschmer, J. M., Manning, S., Lance, R. C. M., et al. (1998). RFLP mapping of a new cereal cyst nematode resistance locus in barley. *Plant Breeding, 117*, 185–187.

Breyer, D., Herman, P., Brandenburger, A., Gheysen, G., Remaut, E., Soumillion, P., et al. (2009). Genetic modification through oligonucleotide-mediated mutagenesis. A GMO regulatory challenge? *Environmental Biosafety Research*, 1–8.

Cai, D., Kleine, M., Kifle, S., Harloff, H. J., Sandal, N. N., Marcker, K. A., et al. (1997). Positional cloning of a gene for nematode resistance in sugar beet. *Science, 275*, 832–834.

Charlton, W. C., Harel, H. Y. M., Bakhetia, M., Hibbard, J. K., Atkinson, H. J., & McPherson, M. J. (2010). Additive effects of plant expressed double-stranded RNAs on root-knot nematode development. *International Journal for Parasitology, 40*, 855–864.

Chitwood, D. J. (2002). Phytochemical based strategies for nematode control. *Annual Review of Phytopathology, 40*, 221–249.

Chu, Y., Wu, C. L., Holbrook, C. C., Tillman, B. L., Person, G., & Ozias-Akins, P. (2011). Marker-assisted selection to pyramid nematode resistance and the high oleic trait in peanut. *The Plant Genome, 4*, 110–117.

Concibido, V. C., Diers, B. W., & Arelli, P. R. (2004). A decade of QTL mapping for cyst nematode resistance in soybean. *Crop Science, 44*, 1121–1131.

Danchin, E. G. J., Arguel, M.-J., Campan-Fouriner, A., Perfus-Barbeoch, L., Magliano, M., Rosso, M. N., et al. (2013). Identification of novel target genes for safer and more specific control of root-knot nematodes from a pan-genome mining. *PLoS Pathogens, 9*(10), e1003745.

Dinh, P. T. Y., Brown, C. R., & Elling, A. A. (2014). RNA interference of effector gene Mc16D10L confers resistance against *Meloidogyne chitwoodi* in *Arabidopsis* and potato. *Phytopathology, 104*, 1098–1106.

Djian-Caporalino, C., Fazari, A., Arguel, M. J., Vernie, T., VandeCasteele, C., Faure, I., et al. (2007). Root-knot nematode (*Meloidogyne* spp.) *Me* resistance genes in pepper (*Capsicum annuum* L.) are clustered on the P9 chromosome. *Theoretical and Applied Genetics, 114*, 473–486.

Dubreuil, G., Magliano, M., Dubrana, M. P., Lozano, K., Lecomte, P., Favery, B., Abad, P., & Rosso, M. N. (2009). Tobacco rattle virus mediates gene silencing in a plant parasitic root-knot nematode. *Journal of Experimental Botany, 60*, 4041–4050.

Eastwood, R. F., Lagudah, E. S., & Appels, R. (1994). A directed search for DNA sequences tightly linked to cereal cyst nematode resistance genes in *Triticum tauschii*. *Genome, 37*, 311–319.

Escobar, C., Barcala, M., Portillo, M., Almoguera, C., Jordano, J., & Fenoll, C. (2003). Induction of the *Hahsp17.7G4* promoter by root-knot nematodes: involvement of heat-shock elements in promoter activity in giant cells. *Molecular Plant-Microbe Interaction, 16*, 1062–1068.

Fanelli, E., Di, V. M., Jones, J. T., & Giorgi, C. (2005). Analysis of chitin synthase function in a plant parasitic nematode, *Meloidogyne artiellia*, using RNAi. *Gene, 349*, 87–95.

FAOSTAT. (2012). *FAO Statistical Year Book, 2012*. Rome: Food and Agriculture Organization of the United Nations, ISBN 978-92-5-106913-4.

Finkers-Tomczak, A., Bakker, E., Boer, J. D., Van der Vossen, E. A., Achenbach, U., Golas, T., et al. (2011). Comparative sequence analysis of the potato cyst nematode resistance locus H1 reveals a major lack of co-linearity between three haplotypes in potato (*Solanum tuberosum* ssp). *Theoretical Applied Genetics, 122*, 595–608.

Fire, A., Xu, S., Montgomery, M. K., Kostas, S. A., Driver, S. E., & Mello, C. C. (1998). Potent and specific genetic interference by double-stranded RNA in *Caenorhabditis elegans*. *Nature, 391*, 806–811.

Fosu-Nyarko, J., & Jones, M. G. K. (2013). *Target genes for control of plant parasitic nematodes and use of same*. Patent Publication Number US20130269057.A1, PCT number PCT/IB2011/002130.

Fosu-Nyarko, J., & Jones, M. G. K. (2014). *Target genes for control of plant parasitic nematodes and use of same*. Patent Publication Number WO2014001893.A3, PCT number PCT/IB2013/001385.

Fosu-Nyarko, J., Jones, M. G. K., & Wang, Z. (2009). Functional characterisation of transcripts in early stage *Meloidogyne javanica*-induced giant cells isolated by laser microdissection. *Molecular Plant Pathology, 10*, 237–248.

Galal, A., Sharma, S., Abou-Elwafa, S. F., Sharma, S., Kopisch-Obuch, F., Laubach, E., et al. (2014). Comparative QTL analysis of root lesion nematode resistance in barley. *Theoretical and Applied Genetics, 127*, 1399–1407.

Ganal, M. W., Simon, R., Brommonschenkel, S., Arndt, M., Phillips, M. S., Tanksley, S. D., et al. (1995). Genetic mapping of a wide spectrum nematode resistance gene (*Hero*) against *Globodera rostochiensis* in tomato. *Molecular Plant Microbe Interactions, 8*, 886–891.

Ganal, M. W., & Tanksley, S. D. (1996). Recombination around the *Tm2a* and *Mi* resistance genes in different crosses of *Lycopersicon peruvianum*. *Theoretical and Applied Genetics, 92*, 101–108.

Garcia, G. M., Stalker, H. T., Shroeder, E., & Kochert, G. (1996). Identification of RAPD, SCAR, and RFLP markers tightly linked to nematode resistance genes introgressed from *Arachis cardenasii* into *Arachis hypogaea*. *Genome, 39*, 836–845.

Hamamouch, N., Li, C., Hewezi, T., Baum, T. J., Mitchum, M. G., Hussey, R. S., Vodkin, L. O., & Davis, E. L. (2012). The interaction of the novel 30C02 cyst nematode effector protein with a plant beta-1,3 endoglucanase may suppress host defence to promote parasitism. *Journal of Experimental Botany, 63*, 3683–3695.

Heller, R., Schondelmaier, J., Steinrücken, G., & Jung, C. (1996). Genetic localization of four genes for nematode (*Heterodera schachtii* Schm.) resistance in sugar beet (*Beta vulgaris* L.). *Theoretical and Applied Genetics, 92*, 991–997.

Huang, G., Allen, R., Davis, E. L., Baum, J. T., & Hussey, S. R. (2006). Engineering broad root-knot resistance in transgenic plants by RNAi silencing of a conserved and essential root-knot nematode parasitism gene. *Proceedings of National Academy of Science of the USA, 103*, 14302–14306.

Ibrahim, H. M. M., Hosseini, P., Alkharouf, N. W., Hussein, E. H. A., El Kader, A., El-Din, G., et al. (2011). Analysis of gene expression in soybean (*Glycine max*) roots in

response to the root knot nematode *Meloidogyne incognita* using microarrays and KEGG pathways. *BMC Genomics, 12*, 220.
Jacobsen, E., & Schouten, H. J. (2008). Cisgenesis, a new tool for traditional plant breeding, should be exempted from the regulation on genetically modified organisms in a step by step approach. *Potato Research, 51*, 75–88.
Jacquet, M., Bongiovanni, M., Martinez, M., Verschave, P., Wajnberg, E., & Castagnone-Sereno, P. (2005). Variation in resistance to the root-knot nematode *Meloidogyne incognita* in tomato genotypes bearing the *Mi* gene. *Plant Pathology, 54*, 93–99.
Jones, M. G. K. (1981). Host cell responses to endoparasitic nematodes. *Annals of Applied Biology, 97*, 353–372.
Jones, M. G. K., & Fosu-Nyarko, J. (2014). Molecular biology of root lesion nematodes (*Pratylenchus* spp.) and their interaction with host plants. *Annals of Applied Biology, 164*, 163–181.
Joseph, S., Gheysen, G., & Subramaniam, K. (2012). RNA interference in *Pratylenchus coffeae*: knock down of Pc-pat-10 and Pc-unc-87 impedes migration. *Molecular and Biochemical Parasitology, 186*, 51–59.
Kimber, M. J., McKinney, S., McMaster, S., Day, T. A., Fleming, C. C., & Maule, A. G. (2007). *flp* Gene distribution in a parasitic nematode reveals motor dysfunction and unusual neuronal sensitivity to RNA interference. *The FASEB Journal, 21*, 1233–1243.
Klein-Lankhorst, R., Rietveld, P., Machiels, B., Verkerk, R., Weide, R., Gebhardt, C., et al. (1991). RFLP markers linked to the root knot nematode resistance gene *Mi* in tomato. *Theoretical and Applied Genetics, 81*, 661–667.
Klink, V. P., Kim, K. H., Martins, V., Macdonald, M. H., Beard, H. S., Alkharouf, N. W., et al. (2009). A correlation between host-mediated expression of parasite genes as tandem inverted repeats and abrogation of development of female *Heterodera glycines* cyst formation during infection of *Glycine max*. *Planta, 230*, 53–71.
Kreike, C. M., deKoning, J. R. A., Vinke, J. H., van Ooijen, J. W., Gebhardt, C., & Stiekema, W. J. (1993). Mapping of loci involved in quantitatively inherited resistance to the potato cyst-nematode *Globodera rostochiensis* pathotype Ro1. *Theoretical Applied Genetics, 87*, 464–470.
Kretschmer, J. M., Chalmers, K. J., Manning, S., Karakousis, A., Barr, A. R., Islam, A. K. M. R., et al. (1997). RFLP mapping of the Ha 2 cereal cyst nematode resistance gene in barley. *Theoretical and Applied Genetics, 94*, 1060–1064.
Lagudah, E. S., Moullet, O., & Appels, R. (1997). Map-based cloning of a gene sequence encoding a nucleotide-binding domain and a leucine-rich region at the Cre3 nematode resistance locus of wheat. *Genome, 40*, 659–665.
Leister, D., Ballvora, A., Salamini, F., & Gebhardt, C. (1996). A PCR—based approach for isolating pathogen resistance genes from potato with potential for wide application in plants. *Nature Genetics, 14*, 421–429.
Lewis, J. G., Matic, M., & McKay, A. C. (2009). Success of cereal cyst nematode resistance in Australia: history and status of resistance screening systems. In I. T. Riley, J. M. Nicol, & A. A. Dababat (Eds.), *Proceedings of the first workshop of the international cereal cyst nematode initiative: Cereal cyst nematode: Status, research and outlook* (pp. 137–142) (Antalya, Turkey).
Lilley, C. J., Bakhetia, M., Charlton, W. L., & Urwin, P. E. (2007). Recent progress in the development of RNA interference for plant parasitic nematodes. *Molecular Plant Pathology, 2007*, 701–711.
Li, J., Todd, T. C., Oakley, T. R., Lee, J., & Trick, H. N. (2010). Host-derived suppression of nematode reproductive and fitness genes decreases fecundity of *Heterodera glycines* Ichinohe. *Planta, 232*, 775–785.
Li, J., Todd, T. C., Oakley, T. R., Lee, J., & Trick, H. N. (2011). Biotechnological application of functional genomics towards plant-parasitic nematode control. *Plant Biotechnology Journal, 9*, 936–944.

Lin, B., Zhuo, K., Wu, P., Cui, R., Zhang, L. H., & Liao, J. (2013). A novel effector protein, MJ-NULG1a, targeted to giant cell nuclei plays a role in *Meloidogyne javanica* parasitism. *Molecular Plant-Microbe Interactions, 26*, 55–66.

Liu, S., Kandoth, P. K., Warren, S. D., Yeckel, G., Heinz, R., Alden, J., et al. (2012). A soybean cyst nematode resistance gene points to a new mechanism of plant resistance to pathogens. *Nature, 492*, 256–260.

Lozano-Juste, J., & Cutler, S. R. (2014). Plant genome engineering in full bloom. *Trends in Plant Science, 19*, 284–287.

Matthews, B. F., Beard, H., MacDonald, M. H., Kabir, S., Youssef, R. M., Hosseini, P., et al. (2012). Engineered resistance and hypersusceptibility through functional metabolic studies of 100 genes in soybean to its major pathogen, the soybean cyst nematode. *Planta, 237*, 1337–1357.

McCarter, J. P. (2009). Molecular approaches toward resistance to plant-parasitic nematodes. In R. H. Berg, & C. G. Taylor (Eds.), *Cell biology of plant nematode Paraistism* (pp. 239–268). Springer-Verlag Berlin, Heidelberg.

Messeguer, R., Ganal, M., De Vicente, M. C., Young, N. D., Bolkan, H., & Tanksley, S. D. (1991). High resolution RFLP map around the root knot nematode resistance gene (*Mi*) in tomato. *Theoretical and Applied Genetics, 82*, 529–536.

Nekrasov, V., Staskawicz, B., Weigel, D., Jones, J. D. G., & Kamoun, S. (2013). Targeted mutagenesis in the model plant *Nicotiana benthamiana* using Cas9 RNA-guided endonuclease. *Nature Biotechnology, 31*, 691–693.

Niewöhner, J., Salamini, F., & Gebhardt, C. (1995). Development of PCR assays diagnostic for RFLP marker alleles closely linked to alleles *Gro1* and *H1*, conferring resistance to the root cyst nematode *Globodera rostochiensis* in potato. *Molecular Breeding, 1*, 65–67.

Nombela, G., Williamson, V. M., & Muñiz, M. (2003). The root-knot nematode resistance gene *Mi*-1.2 of tomato is responsible for resistance against the whitefly *Bemisia tabaci*. *Molecular Plant-Microbe Interactions, 16*, 645–649.

Ogbonnaya, F. C., Eastwood, R. F., & Lagudah, E. (2009). Identification and utilisation of genes for cereal cyst nematode resistance (*Heterodera avenae*) resistance in wheat: the Australian experience. In I. T. Riley, J. M. Nicol, & A. A. Dababat (Eds.), *Proceedings of the first workshop of the international cereal cyst nematode initiative: Cereal cyst nematode: Status, research and outlook* (pp. 166–171) (Antalya, Turkey).

Opperman, C. H., Taylor, C. G., & Conkling, M. A. (1994). Root-knot nematode-directed expression of a plant root-specific gene. *Science, 263*, 221–223.

Patel, N., Hamamouch, N., Li, C., Hussey, R., Mitchum, M., Baum, T., Wang, X., & Davis, E. L. (2008). Similarity and functional analyses of expressed parasitism genes in Heterodera schachtii and Heterodera glycines. *Journal of Nematology, 40*, 299–310.

Patel, N., Hamamouch, N., Li, C., Hewezi, T., Hussey, R., Baum, T., Mitchum, M., & Davis, E. L. (2010). A nematode parasitism protein similar to annexins in host plants. *Journal of Experimental Botany, 61*, 235–248.

Petrick, J. S., Brower-Toland, B., Jackson, A. L., & Kier, L. D. (2013). Safety assessment of food and feed from biotechnology-derived crops employing RNA-mediated gene regulation to achieve desired traits: a scientific review. *Regulatory Toxicology and Pharmacology, 66*, 167–176.

Portillo, M., Cabrera, J., Lindsey, K., Topping, J., Andrés, M. F., Emiliozzi, M., Oliveros, J. C., García-Casado, G., Solano, R., Koltai, H., Resnick, N., Fenoll, C., & Escobar, C. (2013). Distinct and conserved transcriptomic changes during nematode-induced giant cell development in tomato compared with Arabidopsis: a functional role for gene repression. *New Phytol., 197*, 1276–1290. http://dx.doi.org/10.1111/nph.12121.

Ramsay, K., Wang, Z. H., & Jones, M. G. K. (2004). Using laser capture microdissection to study gene expression in early stages of giant cells induced by root-knot nematodes. *Molecular Plant Pathology, 5*, 587–592.

Rossi, M., Goggin, F. L., Milligan, S. B., Kaloshian, I., Ullman, D. E., & Williamson, V. M. (1998). The nematode resistance gene *Mi* of tomato confers resistance against the potato aphid. *Proceedings of the National Academy of Sciences of the USA, 95*, 9750–9754.

Schmidt, A. L., McIntyre, C. L., Thompson, J., Seymour, N. P., & Liu, C. J. (2005). Quantitative trait loci for root lesion nematode (*Pratylenchus thornei*) resistance in middle-eastern landraces and their potential for introgression into Australian bread wheat. *Crop and Pasture Science, 56*, 1059–1068.

Sharma, S., Sharma, S., Kopisch-Obuch, F. J., Keil, T., Laubach, E., Stein, N., et al. (2011). QTL analysis of root-lesion nematode resistance in barley: 1. *Pratylenchus neglectus*. *Theoretical and Applied Genetics, 122*, 1321–1330.

Sijmons, P. C., Atkinson, H. J., & Wyss, U. (1994). Parasitic strategies of root-knot nematodes and associated host cell responses. *Annual Review of Phytopathology, 32*, 235–259.

Sindhu, A. S., Maier, T. R., Mitchum, M. G., Hussey, R. S., Davis, E. L., & Baum, T. J. (2009). Effective and specific in planta RNAi in cyst nematodes: expression interference of four parasitism genes reduces parasitic success. *Journal of Experimental Botany, 60*, 315–324.

Smiley, R. W., & Nicol, J. M. (2009). Nematodes which challenge global wheat production. In B. F. Carver (Ed.), *Wheat science and trade* (pp. 171–187). USA: Wiley-Blackwell.

Sobczak, M., Avrova, A., Jupowicz, J., Phillips, M. S., Ernst, K., & Kumar, A. (2005). Characterization of susceptibility and resistance responses to potato cyst nematode (*Globodera* spp.) infection of tomato lines in the absence and presence of the broad-spectrum nematode resistance *Hero* gene. *Molecular Plant-Microbe Interactions, 18*, 158–168.

Steeves, R. M., Todd, T. C., Essig, J. S., & Trick, H. N. (2006). Transgenic soybeans expressing siRNAs specific to a major sperm protein gene suppress *Heterodera glycines* reproduction. *Functional Plant Biology, 33*, 991–999.

Szakasits, D., Heinen, P., Wieczorek, K., Hofmann, J., Wagner, F., Kreil, D. P., et al. (2009). The transcriptome of syncytia induced by the cyst nematode Heterodera schachtii in Arabidopsis roots. *The Plant Journal, 57*(5), 771–784. http://dx.doi.org/10.1111/j.1365-313X.2008.03727.x. [pii]: TPJ3727.

Tan, J., Jones, M. G. K., & Fosu-Nyarko, J. (2013). Gene silencing in root lesion nematodes (*Pratylenchus* spp) significantly reduces reproduction in a plant host. *Experimental Parasitology, 133*, 166–178.

Taylor, J. R. N., Belton, P. S., Beta, T., & Duodu, K. G. (2014). Increasing the utilisation of sorghum, millets and pseudocereals: developments in the science of their phenolic phytochemicals, biofortification and protein functionality. *Journal of Cereal Science, 59*, 257–275.

Thompson, J. P., Brennan, P. S., Clewett, T. G., Sheedy, J. G., & Seymour, N. P. (1999). Progress in breeding wheat for tolerance and resistance to root-lesion nematode (*Pratylenchus thornei*). *Australasian Plant Pathology, 28*, 45–52.

Toktay, H., McIntyre, C. L., Nicol, J. M., Ozkan, H., & Elekcioglu, H. I. (2006). Identification of common root-lesion nematode (*Pratylenchus thornei* Sher et Allen) loci in bread wheat. *Genome, 49*, 1319–1323.

Tripathi, L., Roderick, H., Babirye, A., & Atkinson, H.J. (2014). Transgenic nematode resistance for African food security: nematode resistance bananas as a case study. In *Proceeding of the 6th international congress on nematology, 2014* (pp. 121–122) (Cape Town, South Africa).

Urwin, P. E., Lilley, C. J., & Atkinson, H. J. (2002). Ingestion of double-stranded RNA by preparasitic juvenile cyst nematodes leads to RNA interference. *Molecular Plant-Microbe Interactions, 15*, 747–752.

van der Voort, J. R., Wolters, P., Folkertsma, R., Hutten, R., Van Zandvoort, P., Vinke, H., et al. (1997). Mapping of the cyst nematode resistance locus Gpa2 in potato using a strategy based on comigrating AFLP markers. *Theoretical and Applied Genetics, 95*, 874–880.

Wang, Z. H., Potter, R. H., & Jones, M. G. K. (2003). Differential display analysis of gene expression in the cytoplasm of giant cells induced in tomato roots by *Meloidogyne javanica. Molecular Plant Pathology, 4*, 361–371.

Wang, Z., Shuie, L., & Jones, M. G. K. (2008). Use of a double promoter system to target nematode feeding cells. In *Proceedings of the EU COST ACTION 872 workshop, exploiting genomics to understand plant-nematode interactions, Postonja, Slovenia, May 2008.*

Ware, J. O. (1936). Plant breeding and the cotton industry. In *Yearbook of agriculture* (p. 657). Washington D.C: US Department of Agriculture.

Webb, D. M., Baltazar, B. M., Rao-Arelli, A. P., Schupp, J., Clayton, K., Keim, P., et al. (1995). Genetic mapping of soybean cyst nematode race-3 resistance loci in the soybean PI 437.654. *Theoretical and Applied Genetics, 91*, 574–581.

Webber, H. J., & Orton, W. A. (1902). Some diseases of cowpea. II. A cowpea resistant to root knot nematode (*Heterodera radicola*). In *USDA bureau of plant industry Bulletin no. 17.*

Wilfarth, H. (1900). Ein neuer gesichtpunkt zur bekmpfung der nematoden. Zeitchr. d. ver d Deut. Zucker Industrie. *Lieferung, 529*, 195–204.

Williams, K. J., Fisher, J. M., & Langridge, P. (1996). Development of a PCR-based allele-specific assay from an RFLP probe linked to resistance to cereal cyst nematode in wheat. *Genome, 39*, 798–801.

Williamson, V. M., & Kumar, A. (2006). Nematode resistance in plants: the battle underground. *Trends in Genetics, 22*, 396–403.

Williams, K., Taylor, S., Bogacki, P., Pallotta, M., Bariana, H., & Wallwork, H. (2002). Mapping of the root lesion nematode (*Pratylenchus neglectus*) resistance gene *Rlnn1* in wheat. *Theoretical and Applied Genetics, 104*, 874–879.

Yadav, B. C., Veluthambi, K., & Subramaniam, K. (2006). Host-generated double stranded RNA induces RNAi in plant-parasitic nematodes and protects the host from infection. *Molecular and Biochemical Parasitology, 148*, 219–222.

Yaghoobi, J., Kaloshian, I., Wen, Y., & Williamson, V. M. (1995). Mapping a new nematode resistance locus in *Lycopersicon peruvianum. Theoretical and Applied Genetics, 91*, 457–464.

Yang, Y., Jittayasothorn, Y., Chronis, D., Wang, X., Cousins, P., & Zhong, G.-Y. (2013). Molecular characteristics and efficacy of *16D10* siRNAs in inhibiting root-knot nematode infection in transgenic grape hairy roots. *PLoS One, 8*, 69463. http://dx.doi.org/10.1371/journal.pone.0069463.

Zhang, L., Hou, D., Chen, X., Li, D., Zhu, L., Zhang, Y., et al. (2012). Exogenous plant MIR168a specifically targets mammalian LDLRAP1: evidence of cross-kingdom regulation by microRNA. *Cell Research, 22*, 107–126.

Zwart, R. S., Thompson, J. P., & Godwin, I. D. (2005). Identification of quantitative trait loci for resistance to two species of root-lesion nematode (*Pratylenchus thornei* and *P. neglectus*) in wheat. *Crop and Pasture Science, 56*, 345–352.

SUBJECT INDEX

Note: Page numbers followed by "f" and "t" indicate figures and tables respectively.

D

E

N

Q

R

V

W

X

Y

Z

AUTHOR INDEX

Note: Page numbers with "f" denote figures; "t" tables.

A

B

C

D

E

F

G

H

I

J

K

M

S

U

V

W

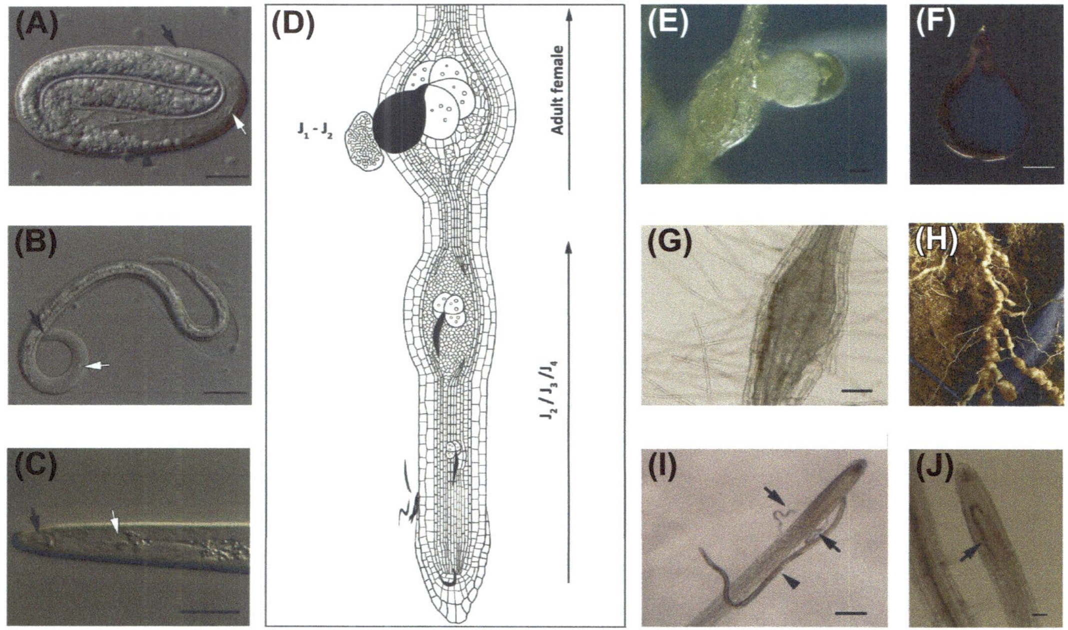

Carolina Escobar *et al.*, Figure 1 ***Root-knot nematode life cycle.*** A schematic diagram with pictures illustrating some of the key stages during the interaction. (A) Photograph of a developing *Meloidogyne javanica* J2 inside the egg. (B) Recently hatched *M. javanica* J2. (C) Close-up of an *M. javanica* J2 anterior body. For A, B and C stylet is indicated by a black arrow, median bulb by a white arrow and lipid globules by a black arrowhead. (D) Schematic diagram of an Root-knot nematode (RKN) life cycle as a time course of the progression of the infection represented in the same root. RKNs are black-coloured for easy location. Starting at the bottom of the diagram, a J2 penetrates the root at the elongation zone, migrates towards the tip and it turns 180° to enter the vascular cylinder, where it induces several giant cells (GCs). By 3 days post infection (dpi), an incipient gall has formed around the nematode and its GCs. The nematode gradually grows and develops into a female while GCs and galls enlarge, and eventually the pear-shaped mature female lays an egg mass that protrudes from the root surface. (E) Mature gall of *Arabidopsis thaliana* with adult female posterior region exposed outside root and laying eggs within the gelatinous matrix. (F) Enlarged adult female of *M. javanica* showing the typical pear-like shape. (G) Incipient gall in *A. thaliana* plant at 3 dpi. (H) Overview of tomato roots infected with *M. javanica* showing profuse galling. (I) Two infective juveniles (J2, black arrow) an adult male (black arrowhead) show *M. javanica* motile stages nearby an *A. thaliana* root tip. (J) Initial stages of an *M. javanica* J2 migration in an Arabidopsis root, turning at the root tip to enter the vascular cylinder. Scale bars in A, B, C represent 20 μm, 0.2 mm in E, F and I, and 0.1 mm G and J.

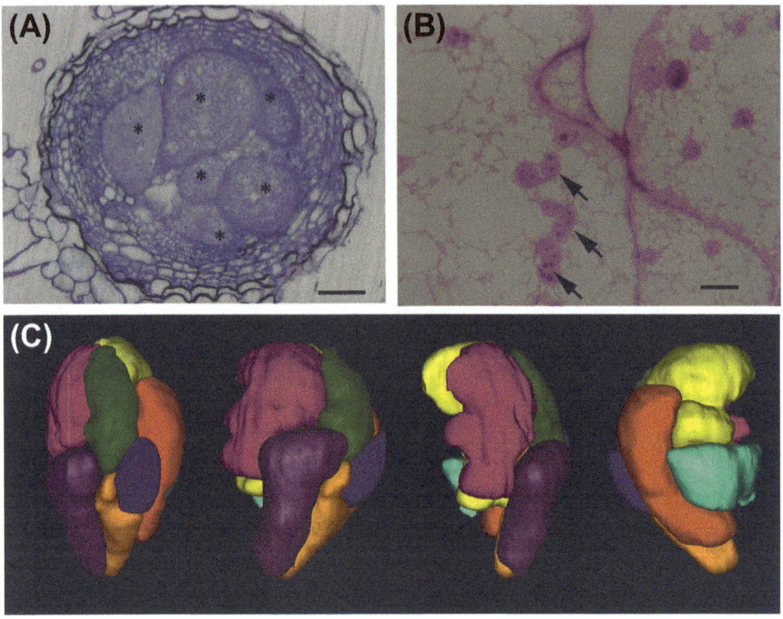

Carolina Escobar *et al.*, Figure 2 ***Morphology of giant cells induced by root-knot nematodes.*** Semi-thin sections of *Meloidogyne javanica* induced galls stained with toluidine. (A) Cross-section of a 3 days post infection (dpi) *Arabidopsis thaliana* gall showing giant cells (GCs) (*). Scale bar represents 50 μm. (B) *Nicotiana tabacum* longitudinal cross-section of a 7 dpi gall showing a partial view of three multinucleate GCs with ameboid nuclei (purple stain) and prominent nucleoli (black arrows) scale bar represents 20 μm. (C) 360° rotation views of a 3D reconstruction image of a fully developed GC system from an Arabidopsis gall 7 dpi with *M. javanica*. Note the irregular shape and ragged surface of the eight GCs that comprise this individual feeding site.

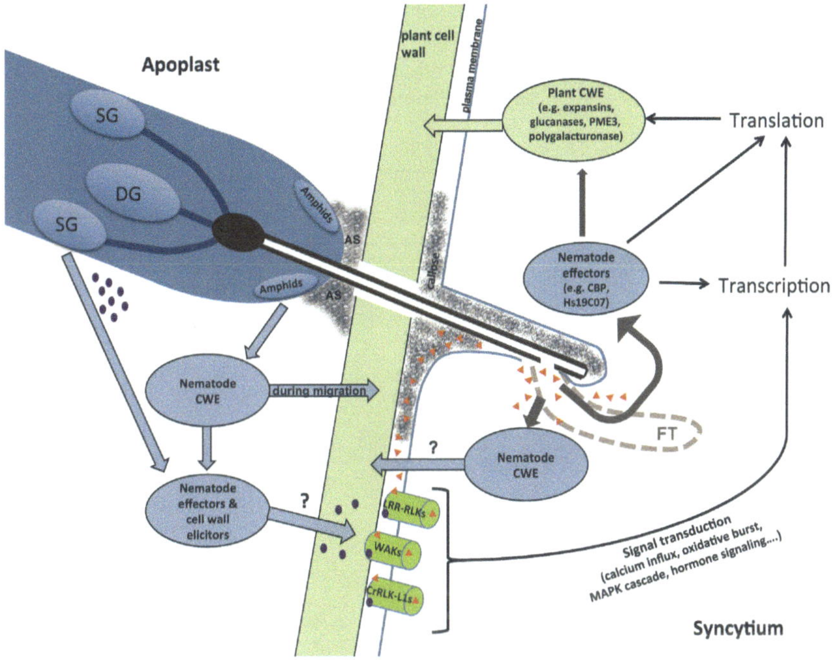

Krzysztof Wieczorek, Figure 1 Cell wall modifying and degrading enzymes, of both nematode and plant origin, as well as nematode effector proteins and their functions in plant tissue and within the syncytia induced by cyst nematodes. During migration a cocktail of cell wall-related enzymes from nematode's subventral gland cells is released through the stylet facilitating the nematode movement through the root tissue. As consequence, damage-derived cell wall elicitors might be recognized by a set of different plant receptors mediating plant defence responses (e.g. LRR-RLKs, WAKs, CrRLK-L1s). In turn, the nematode might secrete effectors involved in the suppression of this induced host plant defence. During the sedentary stage, effectors from the dorsal gland are injected into the feeding site triggering substantial cellular changes including – amongst others – alterations in the expression of cell wall-related plant genes such as expansins or glucanases. These effectors may be deposited outside the plasma membrane and interact with membrane-bound receptors or further manipulate different pathways in the feeding site promoting its development. Various effector proteins (e.g. CBP, Hs19C07) may also directly affect plant cell wall enzymes (grey arrows). Blue dots – nematode's effectors and cell wall elicitors, red triangles – nematode's effectors injected into the feeding site, CWE – cell wall enzymes, SG – subventral gland, DG – dorsal gland, AS – amphidial secretions, FT – feeding tube.

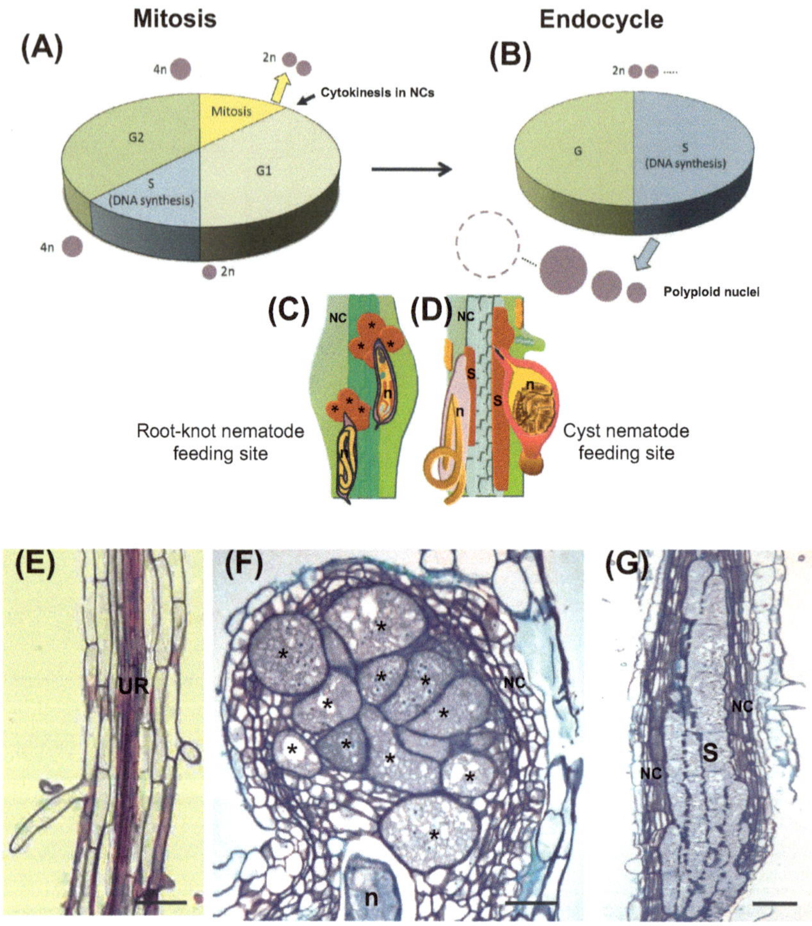

Janice de Almeida Engler *et al.*, Figure 1 From mitosis to the endocycle (A–B) and root and nematode feeding sites morphology in the model host *Arabidopsis thaliana* (C–G). (A) The mitotic cycle in giant cells is acytokinetic and cytokinetic in neighbouring cells. (B) The endocycle occurring in galls and syncytia. (C–D) Schematic representations of a root-knot nematode-induced gall and a cyst nematode-induced syncytium. (E) Toluidine blue stained longitudinal section of an uninfected root, (F) of a gall and (G) of a syncytium. *, giant cell; n, nematode; NC, neighbouring cells; S, syncytium; UR, uninfected root. *Schemes (A) and (B) are modified from de Almeida Engler and Gheysen (2013) and (C) drawing by Mansour Karimi. Bars = 50 μm.*

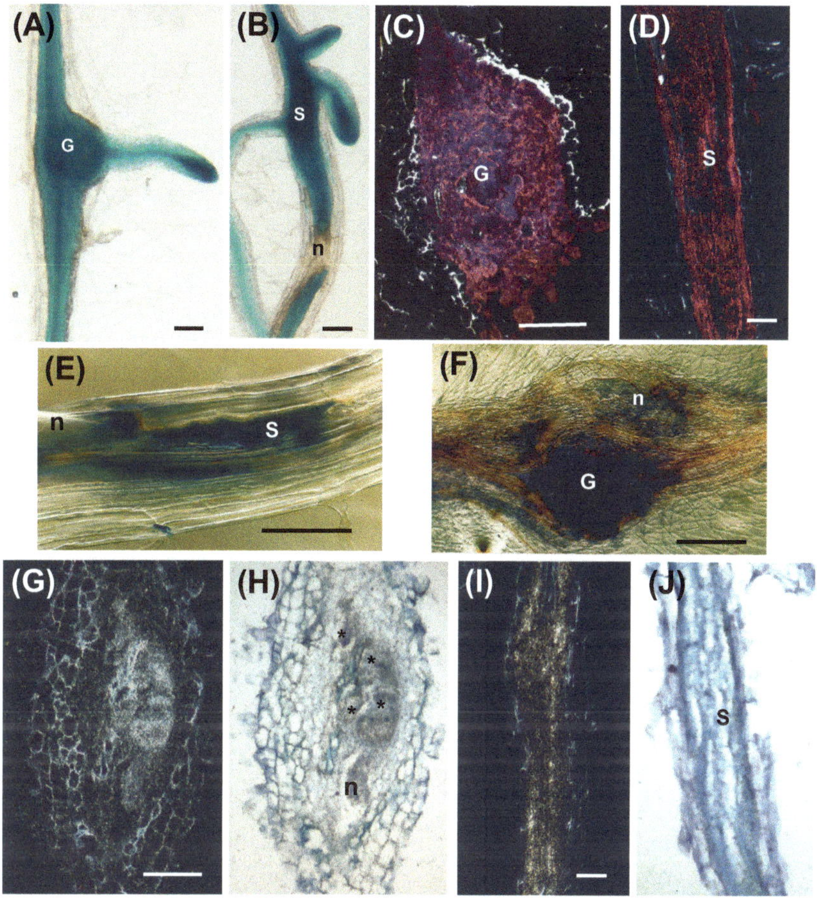

Janice de Almeida Engler *et al.*, Figure 3 Expression analysis of cell cycle genes by promoter activity and mRNA in situ hybridization. *CKSpro:GUS* lines of whole (A and B) and sectioned (C and D) galls and syncytia induced by *Meloidogyne incognita* in *Arabidopsis* roots. Bright- (A and B) and dark-field (C and D) images illustrate GUS staining in blue and red respectively. (E and F) Whole mount mRNA in situ hybridization using a *CDKA;1* non-radioactive antisense probe (purple blue colour) on a whole syncytium 7 DAI and in a thick slice of a gall respectively. (G–J) In situ localization of *CCS52B* transcripts in a gall and a syncytium 7 DAI. White dots represent the hybridization signal from an antisense *CCS52B* radioactive probe. *, giant cell; G, gall; n, nematode; S, syncytium. Bars = 50 μm.

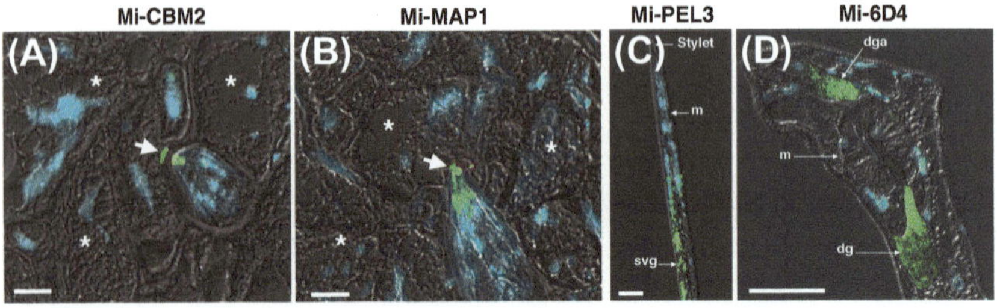

Janice de Almeida Engler *et al.*, Figure 4 Immunolocalization of *Meloidogyne incognita* secreted proteins in methacrylate sections. (A) Gall sections showing the localization of Mi-CBM2 accumulated at the tip of the stylet and along the giant cell wall (arrow), and (B) Mi-MAP1 accumulated in the nematode amphids and secreted within the apoplasm (arrow). (C) Localization of Mi-PEL3 protein in the subventral glands of pre-parasitic J2, and (D) of Mi-6D4 protein in the dorsal gland (dg) and dorsal gland ampulla (dga) of an adult female. Each figure corresponds to the overlay of images showing the antibody localization (green), DAPI-satined nuclei (blue) and differential interference contrast (grey). *, giant cells; m, metacorpus. Bars = 20 μm.

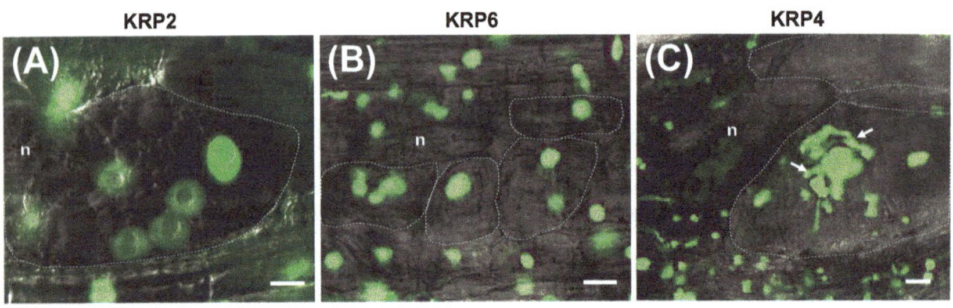

Janice de Almeida Engler *et al.*, Figure 7 In vivo nuclear localization of different members of the *KRP* family in *Arabidopsis* galls induced by *Meloidogyne incognita.* (A) A gall (14 DAI) showing nuclear localization of GFP-KRP2 within giant cell nuclei. (B) A gall (4 DAI) showing nuclear localization of GFP-KRP6 within giant cell nuclei. (C) A gall (14 DAI) showing abnormal nuclear division in (A) and interconnected nuclei within giant cells (white arrows) in an *KRP4* (*35S:GFP-KRP4*) overexpressing line. Giant cells are marked with dashed lines. n, nematode. Bars = 10 μm.

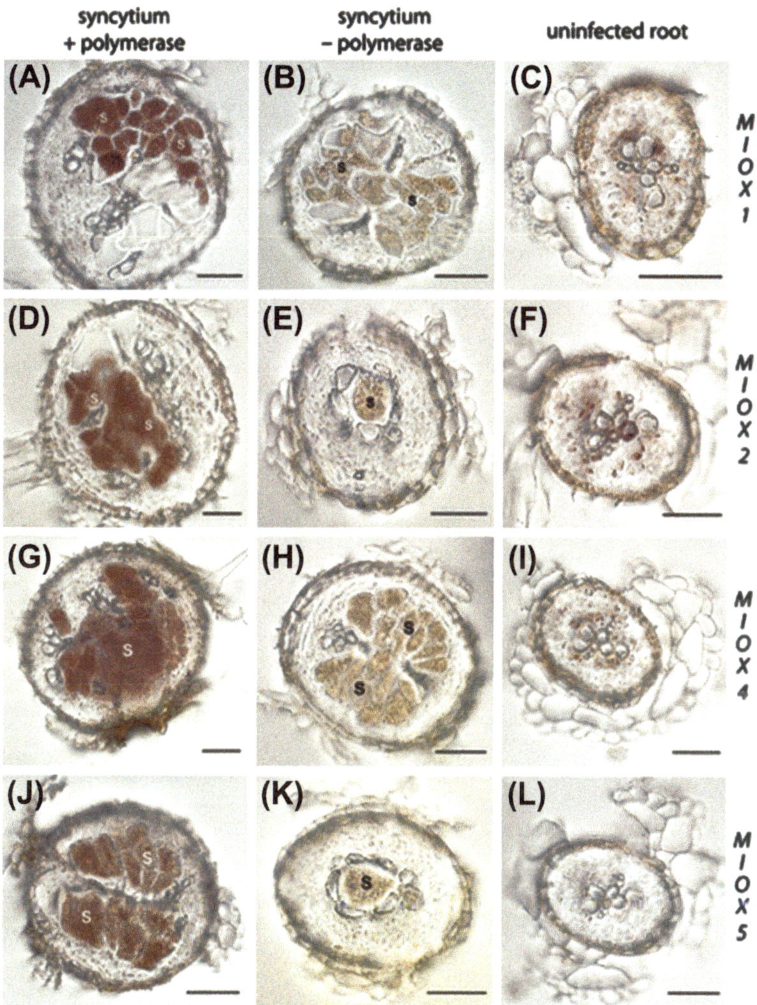

Shahid Siddique and Florian M.W. Grundler, Figure 2 In situ reverse transcriptase chain reaction (RT-PCR) of *MIOX* gene expression in syncytia (s). (A–C) *MIOX1*; (D–F) *MIOX2*; (G–I) *MIOX4*; (J–L) *MIOX5*. (A,D,G,J) specific reaction; (B,E,H,K) control without polymerase; (C,F,I,L) uninfected roots. Bar, 50 μm. *Siddique et al. (2009)*.

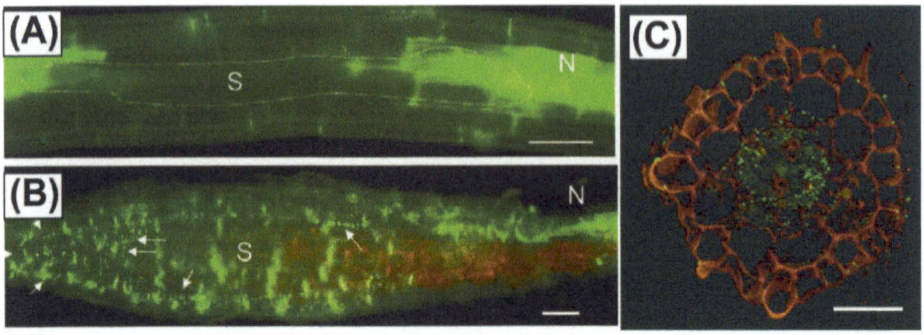

Shahid Siddique and Florian M.W. Grundler, Figure 4 (A) and (B), Localization of plasmodesmatas (PDs) in the Arabidopsis cell walls of syncytium expressing p35S:MP^{17}-GFP. Occasional PDs are present at 4 dpi (A) and numerous PDs (white arrows) are detected at 10 dpi (B). (C) p35S:MP^{17}-GFP fluorescence in a cross-section of a syncytium at 5 dpi. Many PDs are found in the area of the developing syncytium. S = syncytium, N = nematode. White arrows in (C) indicate GFP fluorescence. Bars, (A) and (B) = 20 μm, (C) = 40 μm. *(A) and (B), Hofmann, Youssef-Banora, et al. (2010); (C), Hoth et al. (2008).*

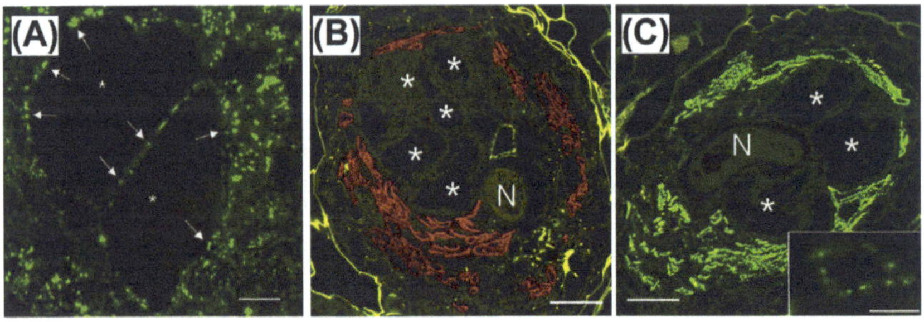

Shahid Siddique and Florian M.W. Grundler, Figure 5 Localization of plasmodesmatas (PDs) in the Arabidopsis cell walls of giant cells (GCs) expressing MP^{17}-GFP. (A) Abundant PDs (white arrows) are present at 13 dpi in walls between GCs and surrounding cells. (B) Cross-section of galls with red fluorescence result from Cy3-labelled second antibody used to detect the sieve element (SE)-specific RS6 antibodies at 18 dpi. (C) Localization of PDs in the Arabidopsis cell walls of GCs expressing Mp^{17}-GFP. Same cross-section as in (B). Green fluorescent protein (GFP) was detected by using an antiserum against it in C. Green colour results from Cy2-labelled second antibody. PDs are primarily present in SEs. N = nematode, Asterisks = GCs. Bars, (A) = 20 μm, (B) and (C) = 150 μm. *(A), Hofmann, Youssef-Banora, et al. (2010); (B) and (C), Hoth et al. (2008).*

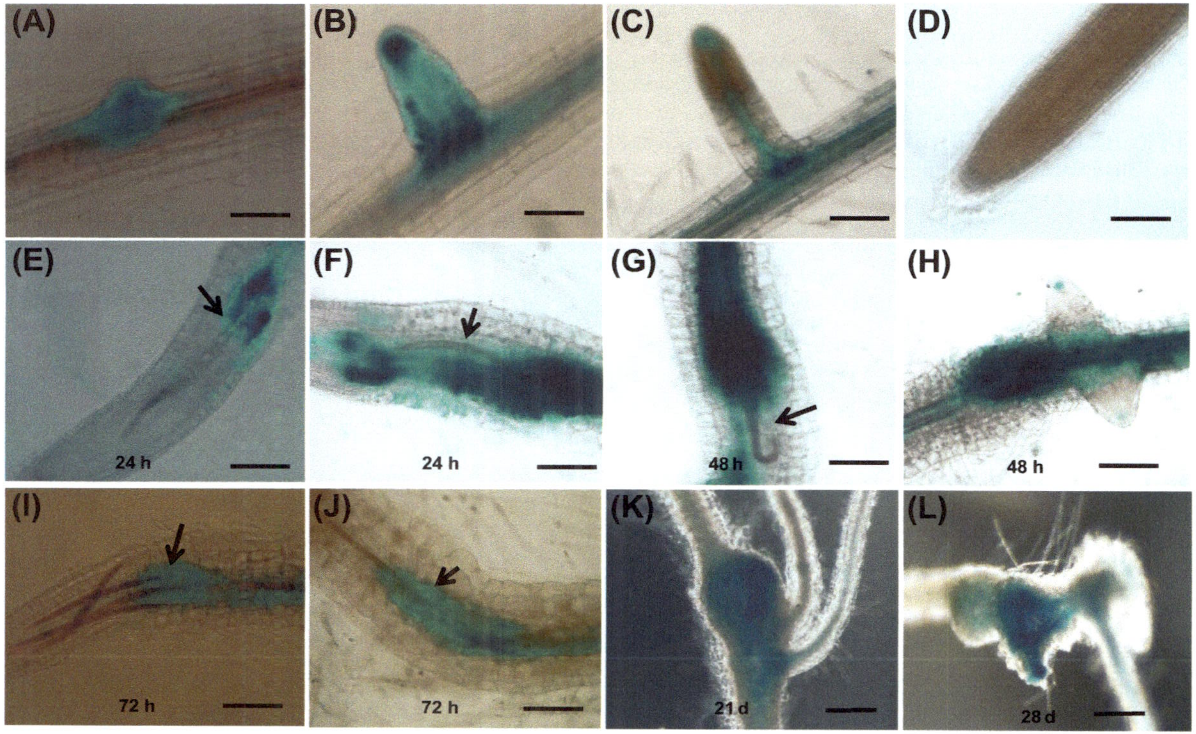

Sigal Brown Miyara *et al.*, Figure 3 ***Microscopic analysis of* GUS*-expression patterns without inoculation and upon RKN infection in* Arabidopsis *plants harbouring the* LOX3 *promoter-GUS fusion construct.*** Noninfested control root harbouring the *LOX3*-GUS fusion construct demonstrated strong GUS staining of lateral root primordia (A) and emerging lateral root (B). GUS staining decreased with time for mature lateral root and was no longer detected in the mature root tip (C, D). In *Arabidopis* plants infested with *M. javanica*, strong GUS signal was observed during J2 migration along the vascular bundle at 24 hpi (E, F) and during entrance into the sedentary phase at 48 and 72 hpi (G–J). Strong GUS staining was observed in galls 28 dpi by *M. javanica*.(K, L). (A–J) Light micrographs as viewed under light microscope. (K, L), Bright-field image of galls photographed using a stereomicroscope. Bars: (A) = 50 μm, (B–J) = 100 μm, (K, L) = 500 μm (Ozalvo et al., 2014).

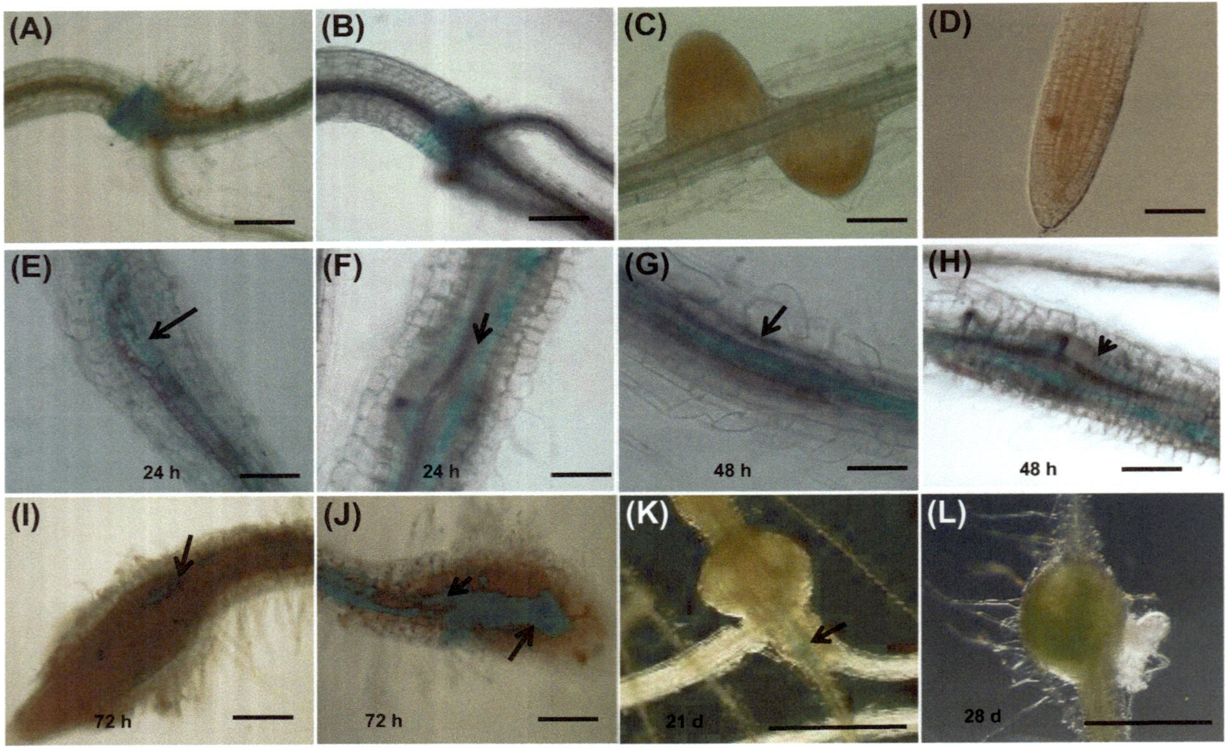

Sigal Brown Miyara *et al.*, Figure 4 ***Microscopic analysis of GUS-expression patterns upon RKN infection in* Arabidopsis *plants harbouring the* LOX4 *promoter-GUS fusion.*** Noninfested control root harbouring the *LOX4*-GUS fusion construct demonstrated consistent GUS staining at the stem–root juncture (A, B), while no GUS staining was detected in roots primordia, developing lateral roots or mature root tip (C, D). Upon nematode inoculation, GUS expression was correlated with nematode parasitic stages: a weak GUS signal was observed during the migratory phase of *M. javanica* J2 at 24 hpi (E, F), 48 hpi (G, H) and at 72 hpi (I, J). GUS signal remained faint in galls 21 and 28 dpi (K, L) by *M. javanica*. Bars: (A–J) = 100 μm, (K, L) = 500 μm (Ozalvo et al., 2014).

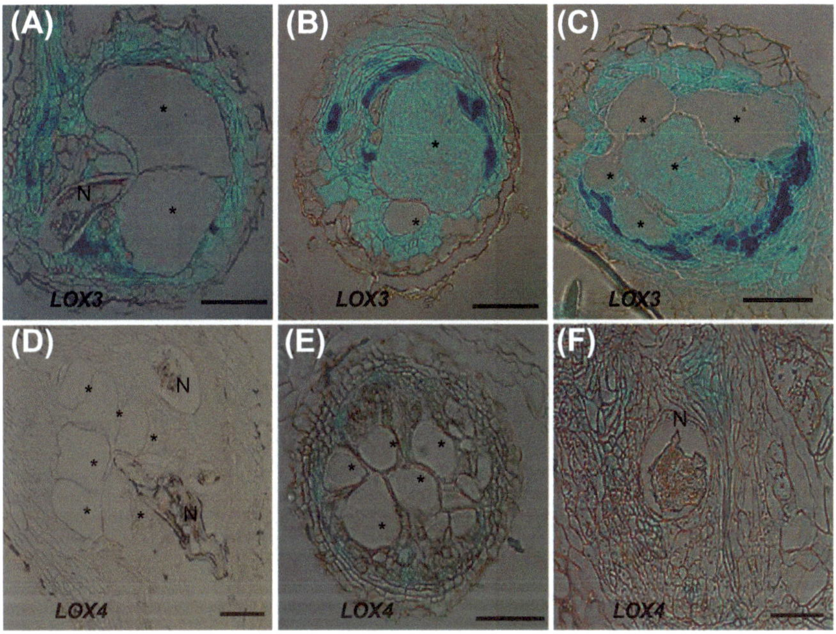

Sigal Brown Miyara *et al.*, Figure 5 ***Microscopic analysis of GUS activity in cross-section of* Arabidopsis *root gall expressing* LOX3 *and* LOX4 *promoter-GUS construct.*** At 28 dpi, all giant cells (GCs) were mature and nematodes developed into the fourth juvenile stage. Histological analysis of roots expressing the *LOX3*-GUS construct clearly showed strong GUS expression inside the cells surrounding the female body and the hyperplastic cells surrounding the GCs inside the vascular cylinder (A), and within the developed GC systems (B, C). *Arabidopsis* roots expressing *LOX4*-GUS construct still showed a very weak signal in a few cells associated with the developed GC system (D–F). The female body of the nematode (N) can be seen at the edge of the GCs (*). Bars = 200 μm (Ozalvo et al., 2014).

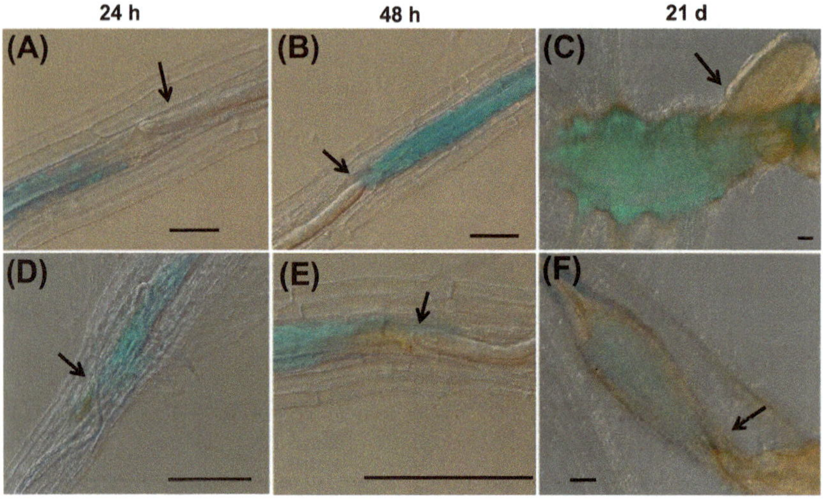

Sigal Brown Miyara *et al.*, Figure 6 ***Microscopic analysis of GUS-expression patterns upon cyst nematode infection in* Arabidopsis *plants harbouring the* LOX3 *and LOX4 promoter-GUS fusion.*** Consistent GUS staining was observed as early as 24 hpi and the signal was centred inside the vascular cylinder close to the nematode head (A). The signal increased as the infection progressed to 48 hpi (B), and GUS signal remained at late infection stages, 21 dpi (C). In *Arabidopsis* plants harbouring the *LOX4* promoter-GUS fusion, LOX4 was activated at 24 hpi around the nematode body inside the vascular cylinder (D), and a clear signal was also observed at 48 hpi (E). The activity was maintained till the late infection stages but intensity markedly decreased at 21 dpi (F). Bars = 100 μm (Ozalvo et al., 2014).

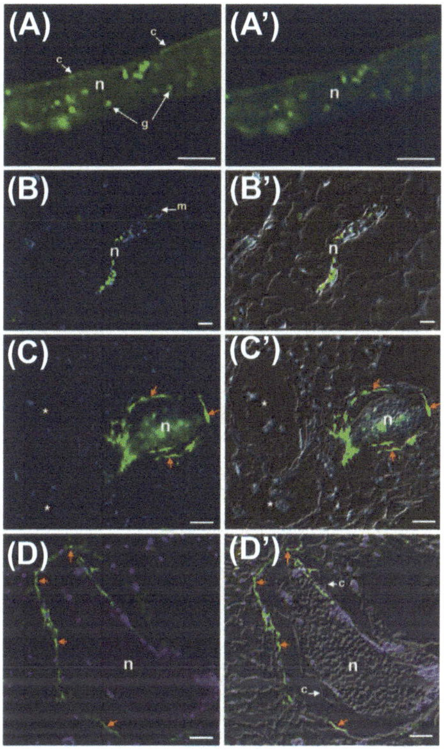

Sigal Brown Miyara *et al.*, Figure 7 ***Immunodetection of FAR proteins in* Meloidogyne incognita *pre-parasitic J2 and during parasitism of* Arabidopsis thaliana *roots.*** (A) Cross nematode sections of pre-parasitic J2 displaying the protein localization at the nematode cuticle surface and circular granules structures within the posterior nematode body. (B) Localization of FAR proteins during nematode migration (B–B′), nematode sedentary stages at 10 DAI (C–C′) and 21 DAI (D–D′) within the roots of *A. thaliana*. Arrows point out the accumulation of FAR along the nematode cuticle and adjacent cells surrounding the nematode body at 10 and 21 DAI. Micrographs on the left are overlays of Alexa-488 fluorescence (green) and DAPI-stained nuclei (blue). Micrographs on the right are overlays of an Alexa-488 fluorescence (green), DAPI-stained nuclei (blue) and differential interference contrast (grey). **c**, cuticle, **g**, granules, **n**, nematode, **m**, metacorpus, * giant cell. Bars = 10 μm (Iberkleid et al., 2013).

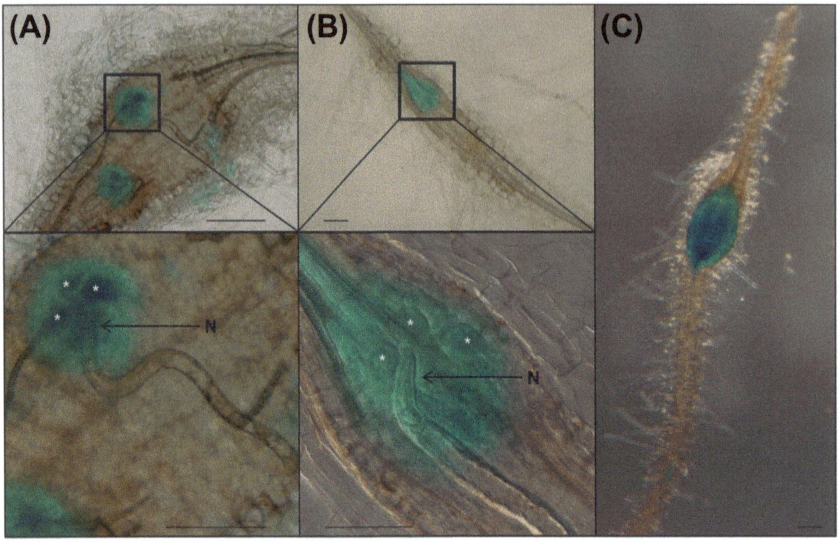

Javier Cabrera *et al.*, Figure 1 Transgenic *Arabidopsis* line *DR5::GUS* showing GUS activity in 4 dpi galls induced by *Meloidogyne javanica* at different incubation times in the GUS staining solution. (A) After 2 h incubation, GUS activity is centred in giant cells. (B) After 4 h incubation, staining spreads to the adjacent vascular tissue. (C) With overnight incubation, the signal covers most of the gall vascular tissues. N, nematode; *, giant cells. Zoom in images of (A) and (B) is shown in the panels below. Bars: 100 μm.

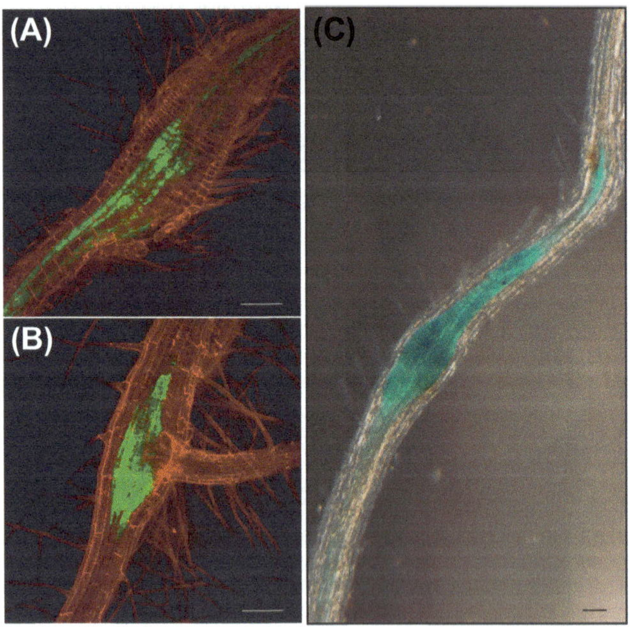

Javier Cabrera *et al.*, Figure 2 Transgenic lines of lateral root and XPP pericycle markers and *LBD16* expressing either GFP or GUS in galls formed by *Meloidogyne javanica*. (A) J0121 >> GFP at 4 dpi. (B) J0192 >> GFP at 4 dpi. (C) *pLBD16:GUS* at 7 dpi (Cabrera, Díaz-Manzano, et al., 2014). Bars: 100 μm.

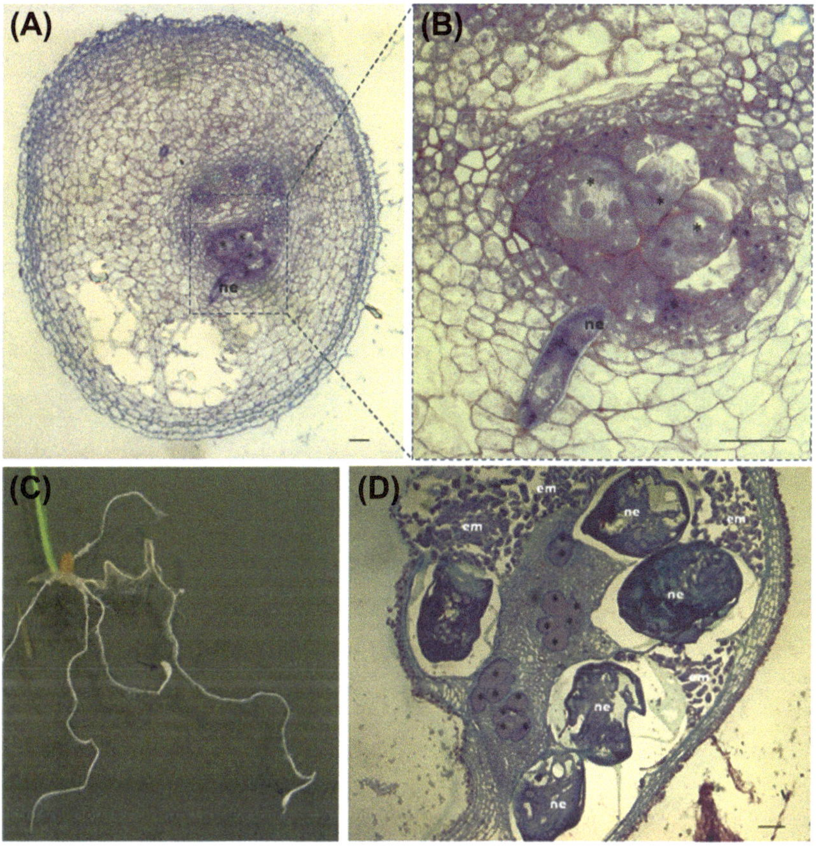

Diana Fernandez *et al.*, Figure 1 Galls induced by *Meloidogyne graminicola* in rice (*Oryza sativa*) roots. A, B and D: Histological analysis where root cross sections (10 μm) obtained at 4 (A, B) and 31 (D) days after nematode infection were stained with toluidine blue. A: 4-days-old gall showing young giant cells containing two or more nuclei, with feeding nematode. B: close-up of A. C: photo of the root system of a 14-days-old rice plantlet 4 days after inoculation. Note the characteristic hook-type galls (arrows). D: 31-days-old galls containing giant cells, with several adult females with eggs embedded in the cortex tissue of a susceptible rice root. Asterisk: giant cell; ne: nematode; em: egg mass. Bars represent 100 μm.

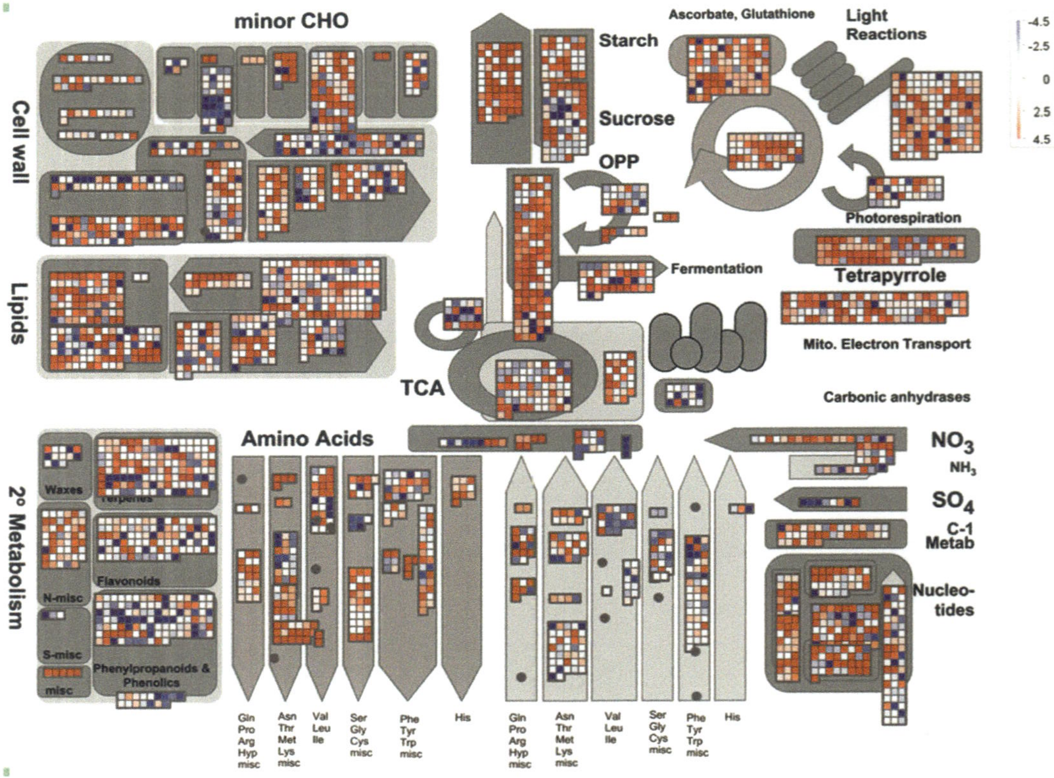

Diana Fernandez *et al.*, Figure 2 Visualization of transcriptome data of giant cells induced by *Meloidogyne graminicola* in rice (*Oryza sativa*) at 7 days after infection. The graph was obtained using MapMan (Thimm et al., 2004) and shows the expression profiles of genes involved in the general metabolism of the rice plant. The observed differential expression patterns are shown as the Log_2 fold changes of mRNA levels in giant cells versus control vascular root cells of uninfected plants. Red dots indicate that the gene is upregulated in giant cells versus the corresponding healthy control cells, while blue indicates downregulation. *Figure based on data provided by Ji et al. (2013).*

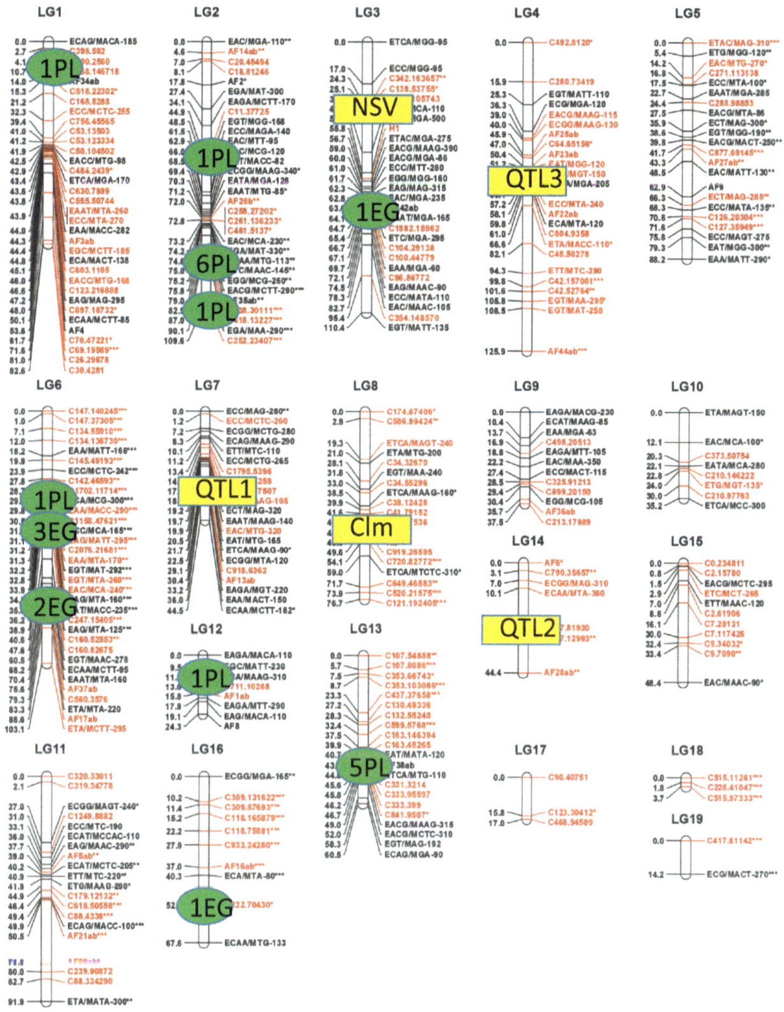

David M. Bird *et al.*, Figure 1 ***Genetic map of* Meloidogyne hapla.** This map is based on segregation of 414 polymorphic DNA markers in 183 F2 lines (Thomas et al., 2012). Green ovals indicate the approximate positions of genes encoding pectate lyase (PL) or endoglucanase (EG) with the numeral in front of the gene abbreviation indicating the number of closely linked copies at each locus. Yellow boxes indicate positions of phenotypic markers. *Clm* is a locus that modulates clumping, NSV determines ability to reproduce on common bean cultivar NemaSnap. QTL1, QTL2, QTL3, are three quantitative trait loci that modulate ability to reproduce on *Solanum bulbocastanum* SB22.

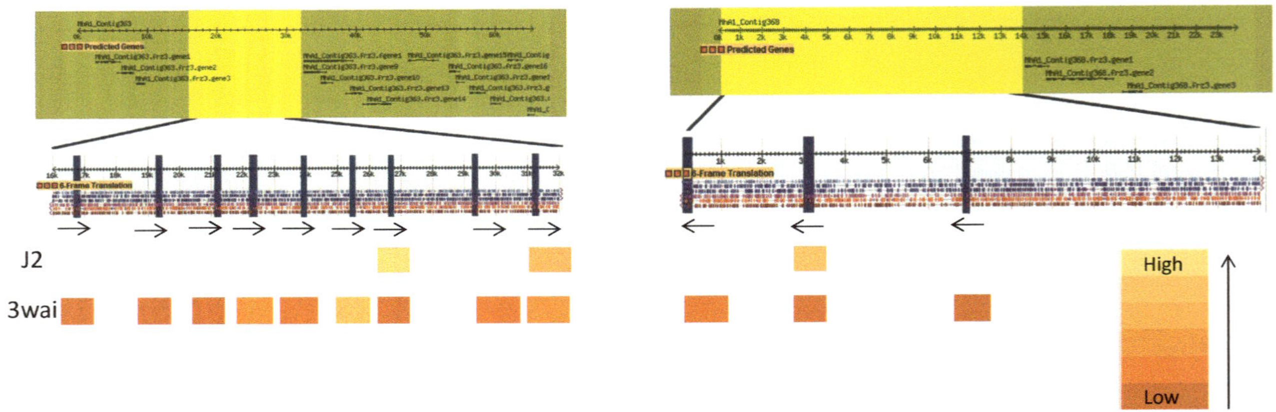

David M. Bird *et al.*, Figure 2 ***G-Browse view of the CEP super-locus in* Meloidogyne hapla.** Two contigs (olive boxes) exhibit typical gene density, except in the yellow boxes, which initially were annotated as being absent of genes, but which spans 12 CEP genes (arrows, which also indicate the direction of transcription) contigs over 122 kb. The heat map indicates transcriptional status in the pre-penetration J2, and *in planta* 3 weeks after infection.

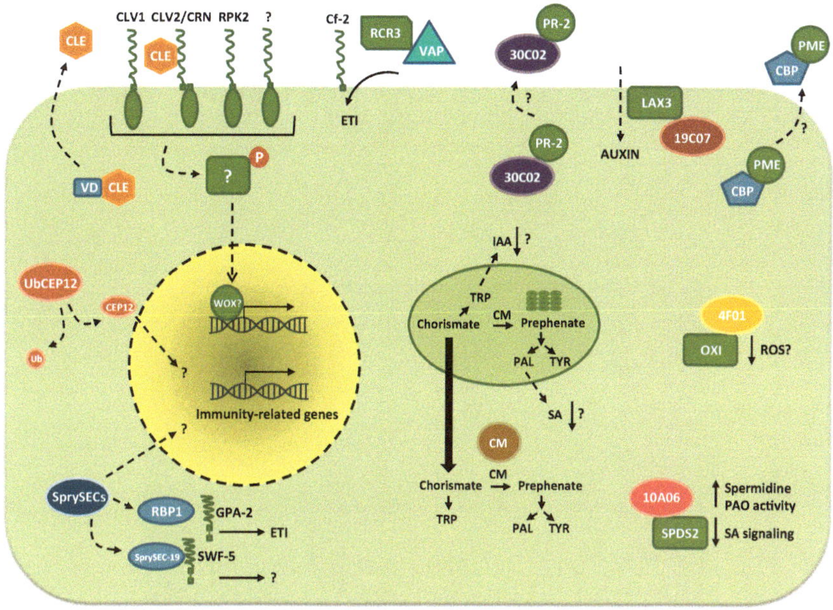

Michael Gardner *et al.*, Figure 1 A diagram depicting the various plant targets and activities of stylet-secreted effectors (SSEs) from cyst nematodes. Cyst nematodes secrete a suite of SSEs into plant cells via a hollow, protrusible mouth spear or stylet. 19C07 interacts with LAX3 to possibly modulate auxin transport. CLAVATA3/endosperm surrounding region-related (CLE) peptides are peptide hormone mimics that interact with the extracellular leucine-rich repeat domain of plant receptor-like kinases with roles in stem cell maintenance. Cellulose-binding protein (CBP) interacts with pectin methylesterase (PME3) to likely modify cell wall properties. Chorismate mutase (CM) affects the plant shikimate pathway to modulate aspects of cellular metabolism. 10A06, 4F01 and 30C02 are novel effectors that interact with spermidine synthase (SPDS2), oxidoreductase (OXI) and β-1,3-endoglucanase (PR-2), respectively and may modulate plant defense responses. GrUBCEP12 is processed in planta to release free ubiquitin and the 12 amino acid GrCEP12 peptide to possibly affect the 26S proteasome and modulate plant defense. SPRYSECs are a large class of secreted SPRY-domain containing proteins that localize to different subcellular compartments of host cells. The R protein GPA-2 mediates effector-triggered immunity (ETI) by directly or indirectly recognizing the SPRYSEC effector RBP-1. SPRYSEC-19 interacts with SWF-5; however, the biological significance of this interaction is unknown. The R protein Cf-2 causes ETI by recognizing perturbations to the active site of the apoplastic protease Rcr3 by VAP-1. Host protein targets are green; cyst effectors are coloured. VD, variable domain; VAP, venom allergen-like protein; PR, pathogenesis related; PAO, polyamine oxidase; IAA, indole-3-acetic acid; UBCEP12, ubiquitin carboxyl extension protein 12; SA, salicylic acid; ROS, reactive oxygen species; WOX, WUSCHEL-like homeobox.

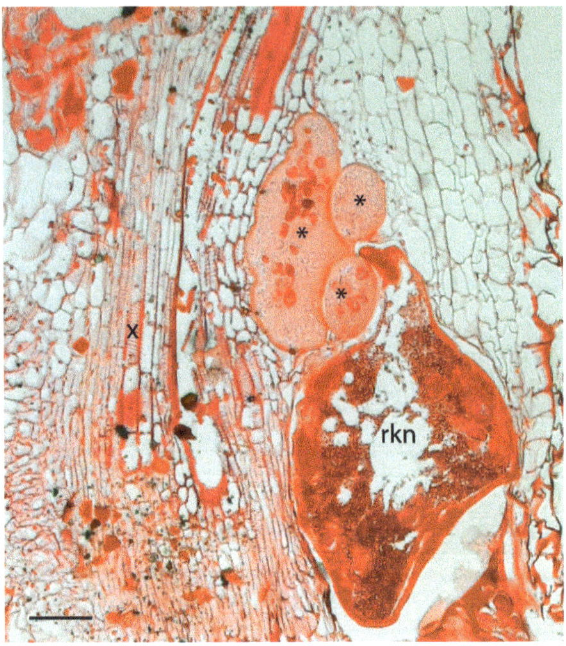

Nhat My Truong *et al.*, Figure 1 ***Multinucleate giant cells induced by Meloidogyne incognita in tomato.*** Section through a gall, 15 days postinfection. Asterisks, giant cells; rkn, young female root-knot nematode; x, xylem. Scale bar = 40 μm.

Kolibri 360 GmbH
Ferdinand-Jühlke-Straße 7
99095 Erfurt, Deutschland
produktsicherheit@kolibri360.de